AF779534

FOR REFERENCE

NOT TO BE TAKEN FROM THIS ROOM

Other Books in Chemistry

COLLEGE CHEMISTRY IN THE LABORATORY, No. 1
Lloyd E. Malm and Harper W. Frantz

GENERAL CHEMISTRY (Second Edition)
Linus Pauling

COLLEGE CHEMISTRY IN THE LABORATORY, No. 2
Lloyd E. Malm and Harper W. Frantz

A SYSTEM OF CHEMICAL ANALYSIS
Ernest H. Swift

INTRODUCTORY QUANTITATIVE CHEMISTRY
Axel R. Olson, Charles W. Koch, George C. Pimentel

A LABORATORY STUDY OF CHEMICAL PRINCIPLES (Second Edition)
Harper W. Frantz

CHEMISTRY OF THE COVALENT BOND
Leallyn B. Clapp

GENERAL CHEMISTRY WORKBOOK: How to Solve Chemistry Problems (Second Edition)
Conway Pierce and R. Nelson Smith

CHEMICAL THERMODYNAMICS: A Course of Study
Frederick T. Wall

THE HYDROGEN BOND
George C. Pimentel and Aubrey L. McClellan

ESSENTIALS OF CHEMISTRY IN THE LABORATORY
Harper W. Frantz and Lloyd E. Malm

QUALITATIVE ELEMENTAL ANALYSIS
Ernest H. Swift and William P. Schaefer

PRINCIPLES OF ORGANIC CHEMISTRY (Second Edition)
T. A. Geissman

FUNDAMENTAL EXPERIMENTS FOR COLLEGE CHEMISTRY: Twenty Selected Experiments for a One-semester or Two-quarter Course
Harper W. Frantz and Lloyd E. Malm

COLLEGE CHEMISTRY (Third Edition)
Linus Pauling

SELECTED EXPERIMENTS IN ORGANIC CHEMISTRY
George K. Helmkamp and Harry W. Johnson

TABLES OF EXPERIMENTAL DIPOLE MOMENTS

TABLES OF EXPERIMENTAL DIPOLE MOMENTS

A. L. McClellan

CALIFORNIA RESEARCH CORPORATION

W. H. FREEMAN AND COMPANY

San Francisco and London

Ref
QD
571
.M27

© *Copyright 1963 by W. H. Freeman and Company*

The publisher reserves all rights to reproduce this book in whole or in part, with the exception of the right to use short quotations for review of the book.

Library of Congress Catalog Card Number 63-14844
Printed in the United States of America

To my Mother

WHO POINTED TO THE STARS

140213

PREFACE

THIS COLLECTION is the outgrowth of an attempt to make simple comparison of some dipole moments and the search that followed. There are, of course, other tabulations. The most important are Wesson's book, *Tables of Electric Dipole Moments* (The Technology Press, Massachusetts Institute of Technology, 1948) and the yearly *Digest of Literature on Dielectrics* (compiled under the auspices of the National Academy of Sciences and the National Research Council). Neither of these is now both comprehensive and up to date—comments which too quickly apply to any tabulation. My intent here is to include all experimentally determined values reported in 1961 or earlier. Calculated values, mainly available from molecular orbital treatments, are *not* listed here.

I greatly appreciate the support of my co-workers at California Research Corporation, in particular D. H. Etzler and J. F. Senger for encouragement at the correct moments, G. G. Bejarano for his patient and talented direction of the computer calculations of selected dipole moments, Geri Esola, Nancy Haritatos, and Terry Addison for locating elusive reports, and Virginia Halstenrud and Sharon Andrews for superior typing of an exacting kind. Every author knows the debt of gratitude due my wife for suffering in not-quite-perfect silence.

May 1963 A. L. MCCLELLAN

CONTENTS

INTRODUCTION

SOME GENERAL REMARKS

The measurement of dipole moments began in the 1920's, although dielectric constant had been determined earlier. There was an immediate surge of interest and, as the graph shows, within only five years about 0.2% of the papers reported in *Chemical Abstracts* were on this subject. The next ten or fifteen years saw a decline in relative attention to dipole moment measurements. The upsurge culminating in 1949 reflects two influences: the completion of work delayed by World War II and the impetus of a new method, microwave spectroscopy. Currently, the quantity of research is still large, now making up nearly 0.1% of the enormous chemical literature reported each year.

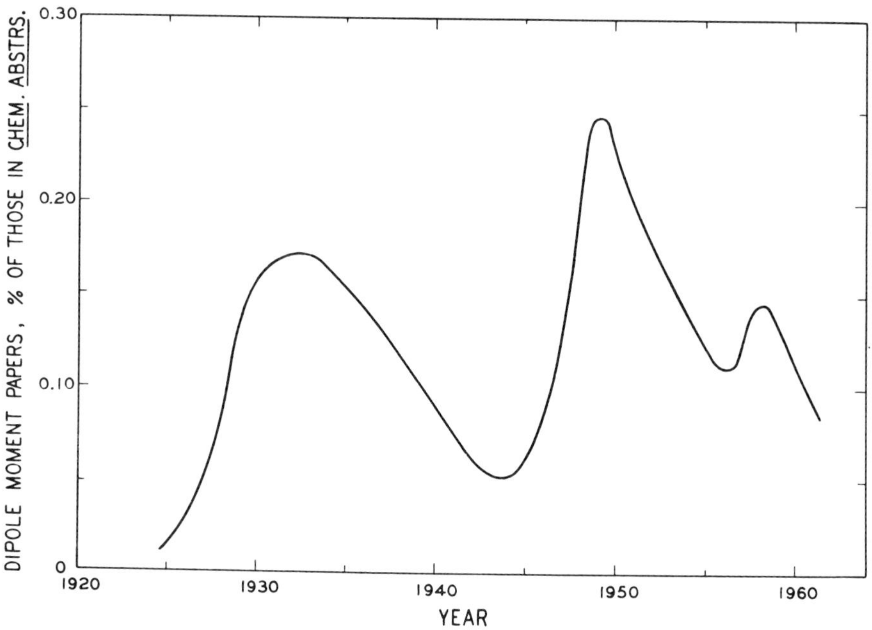

The methods for which work is reported in the tables are described by J. W. Smith[1] and the references quoted therein. There are perhaps 20 equations for converting dielectric constant data to dipole moment,[2] but most workers use the Debye or Onsager equations. The Debye equation is usually recast according to the Hedestrand or Halverstadt and Kumler modifications. Both of these achieve ease of calculation by more direct use of the measured density data. Smith gives details in his Section 2D (page 52–64). Guggenheim proposed a computational procedure which does not require accurate density data. (See Smith, page 58.)

All these methods require some decision about atomic polarization (P_a). This quantity is not readily determined. There are two major alternatives, neither completely correct. Atomic polarization may be omitted on the grounds that it is small or is partially compensated by the usual use of refractive index measurements to estimate electronic polarization. Or, P_a may be taken as an arbitrary percentage of the electronic polarization (usually 5%, 10%, or 15%). There is ample evidence that P_a is, in some cases, not small nor is it any fixed part of P_e. However, to complete the calculation, the experimenter must make some decision.

Since moment values depend on conditions and some assumptions, I voice a plea for authors and editors to provide an adequate description of the experimental and computational details. It is surprising how many values are reported without some of these important facts. A suitable statement should specify state or solvent, temperature, treatment of P_a, and values of fundamental constants. One convenient form takes only one sentence. For example, "Dipole moment values were found for hexane solutions at 25°C using the equation

$$\mu = 0.0128 \sqrt{(P_{2\infty} - 1.05\,P_e)\,T}."$$

It is well known that dipole moment values are subject to a solvent effect which depends on interaction of the test substance with the solvent. The standard for observing such an effect is the moment found for the gaseous phase. With the development of the microwave method for measuring gaseous samples, it is pertinent to ask how values determined in this way agree with dielectric tests on gases. There are about 55 substances tested by both techniques with the results given in the table below. We find generally good agreement with, however, an appreciable number of examples for which the difference is too large to be ascribed to experimental error.

MOMENT DIFFERENCE (*Dielectric-Microwave*)	EXAMPLES
0–0.05	36
0.05–0.1	8
0.1–0.2	2
>0.2	11

[1] J. W. Smith, *Electric Dipole Moments*, Chapter 3, Butterworth's Scientific Publications, London, 1955.

[2] C. P. Smyth, *Dielectric Behavior and Structure*, page 16–50, McGraw-Hill Book Co., Inc., New York, 1955.

USE OF THE TABLES

The experimental values are contained in three tables:

Table I. Compounds Without Carbon

Table II. Compounds Containing Carbon

Table III. Compounds of Unspecified Formula

The second is by far the largest, containing about 6000 entries, while the other two have roughly 500 entries each. In each table there are seven columns whose headings are largely self-explanatory; but some notes may be helpful.

Column 1, Formula

The arrangement follows that of the formula index of *Chemical Abstracts*. The first ordering is alphabetical by chemical symbol, except that in Table II (and Table III wherever a partial formula is known), *C* for carbon is placed first, then *H* for hydrogen, followed by the other elements according to the alphabetical position. The second order of classification is by number of each kind of atom—all compounds with one carbon atom preceding those with two, and so on. Each succeeding element is arranged by number in the same fashion.

Column 2, Compound Name

The compounds of a single empirical formula are in alphabetical order. Some initial symbols are ignored in the primary arrangement, but serve for secondary ordering. The principal ones used in this way are *cis*, *trans*, and *o*, *m*, and *p*. Numerical prefixes also serve the same purpose, the lowest numbers appearing first.

The problem of nomenclature forces itself into any tabulation of chemical compounds. For simple compounds that do not have well-known trivial names, the *Chemical Abstracts* system is generally used (without inversion). For more complex substances, the names used by the original authors are usually retained. There has been no attempt to be strict in naming compounds.

For about 50 of the most frequently studied compounds, a recommended value of dipole moment was derived by considering all the reported data for each compound. The data surviving this scrutiny were treated as one set for computtation by the Halverstadt-Kumler method. Appendix C (p. 581) describes the details of this treatment. The selected value, along with P_e and P_a, are given in a box under the compound name.

Column 3, Dipole Moment
Column 4, State or Solvent

The values of dipole moment in Column 3 are arranged according to the state or solvent used, the latter being indicated by code symbol in Column 4. The following list gives the symbols (and their meaning) in the *arbitrary* order of preference:

gas	gaseous
liq	liquid
solid	solid
B	benzene
D	dioxane
CCl_4	carbon tetrachloride
Hx	hexane
cHx	cyclohexane
Hp	heptane
Tol	toluene
pXy	*para*xylene
Dec	decalin
footnote	other hydrocarbons
CS_2	carbon disulfide
$CHCl_3$	chloroform
ether	ether
ClB	chlorobenzene
NB	nitrobenzene
EtOH	ethanol
footnote	others
n.s.	not specified

For a given state or solvent, the dipole moment values are in numerically increasing order.

The values of dipole moment are as given by the original authors, *with one important exception.* Some of these values have been corrected to allow for changes in Avogadro's number and Boltzmann's constant. The new values, published in 1948,[3] are different enough to make slightly less than 1% difference in the moment. All values before 1941 (and later ones for which the older constants were used) have been increased by the amount shown in the following table:

μ CALCULATED WITH OLD CONSTANTS	AMOUNT ADDED
0 – 0.799 D	Nothing
0.8– 1.699	0.01 D
1.7– 2.799	0.02
2.8– 4.299	0.03
4.3– 5.899	0.04
5.9– 7.799	0.05
7.8– 9.799	0.06
9.8–10.999	0.07
11.0–13.999	0.10
14.0–16.999	0.12

The experimental error is shown either by a lowered final digit or a plus or minus range. In either case, the error is as given in the original work.

[3] J. W. DuMond and E. R. Cohen, *Rev. Mod. Phys.* **20**, 82 (1948).

Column 5, Temperature

The values in parentheses were calculated from reported moment and polarization values.

Column 6, Method or Atomic Polarization

This column serves a double purpose. Most of the dipole moment values are from measurement of the dielectric constant. In these cases, this column has a number showing the percentage of electronic polarization used for atomic polarization in the calculation of dipole moment. A dash means this percentage was not stated.

When some other method of determination was used, a symbol from the following list indicates what it was:

CP	contact potential
DVC	dielectric virial coefficient
EB	electroöptic birefrigence
ER	electron resonance
ES	electrostriction
FL	fluorescence
IRI	infrared intensity
Kerr	Kerr effect
M	microwave, Stark effect
MA	microwave absorption
MB	molecular beam
MD	microwave dispersion
MR	microwave resonance
NRA	nonresonant absorption
PE	photoemission
REF	rotating electric field
RT	relaxation times
SPC	special calculation—see original
VS	velocity of sound

Column 7, References

Numbers refer to the Bibliography starting on page 587.

APPENDIXES

Appendix A (p. 579) contains average dipole values for 75 common substances. It can be used for approximate values or to locate a solvent of some required moment.

The table in Appendix B (p. 580) displays the available moments for twelve homologous alkane series. The values show that moment is essentially fixed by the functional group. Values for the corresponding benzene and cyclohexyl compounds illustrate the alterations found if other hydrocarbon groups are attached to the functional groups.

Appendix C (p. 581) describes the criteria and treatment of data applied during the computation of the recommended moment values. The table at the end of this Appendix also serves as a list of the compounds for which recommended values are given.

TABLE I

COMPOUNDS WITHOUT CARBON

FORMULA	COMPOUND NAME	μ, D	State, or Solv.	t, °C	Method or P_a, % of P_e	Reference
$AgClO_4$	Silver perchlorate	4.7	B	25	0?	2117
		10.7	B	25	-	759
		12.04	B	25	-	963
$AlBr_3$	Aluminum bromide	4.95	B	(20)	15	2006
		5.07	B	(20)	0	1523-4
		5.2	B	20	15	1415
		5.22	B	25-50	10	1955
		0	CS_2	(20)	0	1523-4
		0.55	CS_2	20	15	1415
		0.66	CS_2	(20)	15	2006
		0	Br_2 (liq)	(20)	15	1523-4
$AlBr_3H_2S$	Aluminum bromide - hydrogen sulfide complex	5.28	B	(20)	15	2006-7
		5.42	B	20	15	1415
$AlCl_3$	Aluminum chloride	1.97	gas	-	M	1962
		2.03±0.05	D	25	10	1038
AlI_3	Aluminum iodide	2.28	B	25-50	10	1955
		2.5	B	20	15	1415
		4.98	D	20	-	1445
		0	CS_2	20	15	1415
Al_2Br_6ClCs	Aluminum bromide - cesium chloride complex	10.9	B	20	-	1742
Al_2Br_6ClK	Aluminum bromide - potassium chloride complex	10.1	B	20	-	1742
Al_2Br_6ClLi	Aluminum bromide - lithium chloride complex	8.2	B	20	-	1742
Al_2Br_6ClNa	Aluminum bromide - sodium chloride complex	9.0	B	20	-	1742
Al_2Br_6ClRb	Aluminum bromide - rubidium chloride complex	10.5	B	20	-	1742

FORMULA	COMPOUND NAME	μ, D	State, or Solv.	t, °C	Method or P_a, % of P_e	Reference
Al_2Br_7Cs	Aluminum bromide - cesium bromide complex	10.6	B	20	-	1742
Al_2Br_7I	Aluminum bromide - iodine bromide complex	1.30	CS_2	20	-	1742
Al_2Br_7K	Aluminum bromide - potassium bromide complex	9.7	B	20	-	1742
Al_2Br_7Li	Aluminum bromide - lithium bromide complex	8.5	B	20	-	1742
Al_2Br_7Na	Aluminum bromide - sodium bromide complex	9.3	B	20	-	1742
Al_2Br_7Rb	Aluminum bromide - rubidium bromide complex	10.4	B	20	-	1742
Ar	Argon	0	gas	23-110	-	2050
		0	gas	0-110	-	2172
		<0.03	gas	-190 to 25	-	2070
$AsBr_3$	Arsenic tribromide	2.92	D	25	0	1289
		3.30	D	20	-	1445
		1.67	CCl_4	18	0	133-5
		1.61±0.05	CS_2	25	-	1252
$AsCl_3$	Arsenic trichloride	1.6	gas	130-195	-	612
		2.16	B	25	-	1791
		3.12	D	25	0	1289
		1.99	CCl_4	18	0	133-5
$AsDH_2$	Arsine-*d*	0.22±0.02	gas	-	M	1167
AsF_3	Arsenic trifluoride	2.57±0.02	gas	80-160	DVC	240
		2.815±0.025	gas	-	M	1768
		2.834±0.025	gas	-	M	1963
		2.67±0.05	B	25	-	1252

FORMULA	COMPOUND NAME	μ, D	State, or Solv.	t, °C	Method or P_a, % of P_e	Reference
AsH_3	Arsine	0.13	gas	-47 to 16	-	2067
		0.18	gas	16 to 100	-	2067
		0.22±0.02	gas	-	M	1166
		0.15	solid	-253 to -122	0	1823
AsI_3	Arsenic triiodide	1.84	D	25	0	1289
		0.97±0.1	CS_2	25	-	1252
As_2O_3	Arsenic trioxide	0.13	gas	250	MB	315
BBr_3	Boron tribromide	~0	solid, liq	-70 to 80	-	752
		0	B, cHx	25	0	493
		0.194	B	20	-	1664
BCl_3	Boron trichloride	0	gas	20-90	20	1091
		0.61	gas	35-170	-	612
		0	B	20	13	1415
		0	B	(20)	15	2006
		4.92±0.07	D	25	10?	1038
		0.21	CCl_4	18	0	133-5
BF_3	Boron trifluoride	0	gas	-80 to 25	-	2068-9
		0	gas	20-200	-	1157
BH_6N	Ammonia - borane complex	4.92±0.1	D	25	0?	2073
B_2H_6	Diborane	0	gas	23-195	-	1563
B_2O_3	Boron oxide	3.5	liq	200-600	-	981
$B_3Cl_3H_3N_3$	*B*-trichloroborazole	0.56	B	25	5	2065
$B_3H_6N_3$	Borazole	~0.67	gas	25-95	-	1563
		0.50	B	25	5	2065

FORMULA	COMPOUND NAME	μ, D	State, or Solv.	t, °C	Method or P_a, % of P_r	Reference
B_4H_{10}	Tetraborane	0.56±0.1	B	25	-	2072
B_5D_9	Pentaborane-*d*$_9$	2.16±0.04	gas	-	M	764
B_5H_9	Pentaborane	2.13±0.04	gas	-	M	765
		4.54-3.37	liq	-47 to 25	-	2131
$B_{10}H_{14}$	Decaborane	3.52±0.02	B	25	5	1043
$BeBr_3$	Beryllium bromide	~0	B	(20)	15	2006
$BeCl_3$	Beryllium chloride	~0	B	(20)	15	2006
BrCl	Bromine chloride	0.57±0.02	gas	-	M	1788
BrCs	Cesium bromide	10.7[aa]	B	20	-	1736
BrD	Hydrogen-*d* bromide	0.83±0.02	gas	-	M	244
BrF	Bromine monofluoride	1.29	gas	-	M	1787
BrF_3	Bromine trifluoride	1.0	gas	-	M	1229
		1.19±0.12	gas	140-175	43	1648
BrF_5	Bromine pentafluoride	1.51±0.10	gas	72-160	0	1649
$BrFeN_3O_3$	Trinitrosyliron bromide	3.7	B	n.s.	-	710
BrH	Hydrogen bromide	0.788	gas	-67 to 400	-	326,2159
		1.02	B	20	15	490
		1.09	B	30	-	2089
		2.88	D	30	-	2089
		0.93	CCl_4	30	-	2089
		0.97	CCl_4	30	-	490
		1.03	Hp	30	-	2089

aa) In the presence of $AlBr_3$.

FORMULA	COMPOUND NAME	μ, D	State, or Solv.	t, °C	Method or P_a, % of P_e	Reference
BrH_3Si	Bromosilane	1.32±0.03	gas	-	M	1283
BrH_4N	Ammonium bromide	6.75[a]	B	20	-	1736
BrI	Iodine bromide	1.21	Br_2	25	0?	1735
		1.31[a]	CS_2	20	-	1736
BrK	Potassium bromide	9.93±0.05[b]	gas	-	MB	485
		10.41±0.05[c]	gas	-	MB	485
		9.22[d]	gas	650	MB	1627
		9.7[a]	B	20	-	1736
BrLi	Lithium bromide	6.25±0.15	gas	680	M	758
		8.6[a]	B	20	-	1736
BrNO	Nitrosyl bromide	1.80[f]	gas	-	M	426
		1.88	CCl_4	12	-	905
BrNa	Sodium bromide	9.4[a]	B	20	-	1736
BrRb	Rubidium bromide	~0	gas	928	MB	2148a
		10.5[a]	B	20	-	1736
Br_2	Bromine	0	gas	20-140	-	1177
		0.49	gas	0-54	-	410
		0.4±0.04	liq	0-20	-	35
$Br_2Co_2N_4O_4$	Dinitrosylcobalt bromide dimer	0	B	n.s.	-	710
		1.02±0.1	B	n.s.	-	711
Br_2Hg	Mercuric bromide	0	gas	340-420	-	200
		0.96	B	25	10	1956
		1.06	D	30-50	-	1957
		1.54	D	25	-	365

a) In the presence of $AlBr_3$. b) Excited state $\nu = 2$. c) Ground state $\nu = 0$.
d) As corrected J. Chem. Phys. 4, 732 (1936). f) Along a axis only; μ_b not measured.

FORMULA	COMPOUND NAME	μ, D	State, or Solv.	t, °C	Method or P_a, % of P_e	Reference
Br_2OS	Thionyl bromide	9.11	liq	20	-	175
		1.48	B	20	-	175
Br_3HSi	Tribromosilane	0.79	Hp	25	0	1119
Br_3P	Phosphorus tribromide	0.4_9	B	40	-	1897
		0.5_2	B	25	-	1897
		0.5_8	B	10	-	1897
		1.66	D	25	0	1289
		0.58	CCl_4	-25 to 25	-	1897
		0.61	CCl_4	18	0	133-5
Br_3Sb	Antimony tribromide	3.30±0.1	B	25	-	1252
		5.04	D	25	0	1289
		2.48±0.1	CS_2	25	-	1252
		2.4	n.s.	n.s.	-	1251
Br_4Si	Silicon tetrabromide	0.0±0.1	CCl_4	25	-	899
Br_4Sn	Tin tetrabromide	0	gas	145-245	-	331
		4.13	D	20	-	1445
Br_4Ti	Titanium tetrabromide	5.05	D	20	-	1445
Br_4Zr	Zirconium tetrabromide	4.68	D	20	15	1443
ClCs	Cesium chloride	~0	gas	655	MB	2148a
		10.42±0.02[g]	gas	-	MB	1971
		10.46±0.14	gas	-	MB	1175
		10.5±0.25	gas	-	MB	1174
		11.0	B[h]	20	15	1737
ClD	Hydrogen-*d* chloride	1.089	gas	20-245	-	121
		1.12±0.04	gas	-	M	244
ClF	Chlorine fluoride	0.881±0.02	gas	-	M	580

g) For $\nu = 0$ for $\nu = 1$-4; $\mu_\nu = (10.42\pm0.02) + (0.056\pm0.002)\nu$. h) In the presence of $AlBr_3$.

FORMULA	COMPOUND NAME	μ, D	State, or Solv.	t, °C	Method or P_a, % of P_e	Reference
$ClFO_3$	Perchloryl fluoride	0.023±0.003	gas	-	M	1277
		<0.09	gas	-	M	1140
ClF_3	Chlorine trifluoride (See also Cl_2F_6)	0.554±0.002	gas	0-83	-	1227
		0.557	gas	0-83	NRA	1228
		0.65±0.12	gas	45-145	53	1648
		1.00	liq	0	0?	1651
		1.03	liq	25	0?	1651
$ClGeH_3$	Chlorogermane	2.04	gas	0-25	0	1811
		2.124±0.02	gas	-	M	1283
		2.13	gas	-	M	377
ClH	Hydrogen chloride	1.034	gas	-70 to 315	-	326,2159
	1.07	1.04	gas	20	Kerr	1556
		1.07	gas	14-100	-	199
		1.08_5	gas	20-245	-	121
		1.10	gas	30-90	-	189
		1.19	gas	-	i	1662
		1.190±0.05	gas	15-115	-	2050
		1.49	gas	20	-	552
		1.95±0.01	gas	290	MB	466
		1.22	B	10-25	-	1341
		1.26	B	30	-	2089
		1.27	B	20	22	487
		1.27	B	20	15	490
		1.29	B	25	20	486
		2.13	D	30	-	2089
		1.07	CCl_4	30	-	2089
		1.32	CCl_4	25	22	486
		1.05	Hx	10-25	-	1341
		1.33	cHx	25	22	486
		1.20	Hp	30	-	2089
		1.25	Tol	10-25	-	1341
		1.04	$CHCl_3$	10-25	-	1341

i) Infrared dispersion.

FORMULA	COMPOUND NAME	μ, D	State, or Solv.	t, °C	Method or P_a, % of P_c	Reference
ClH (contd.)	Hydrogen chloride (contd.)	2.23	Ether	10-25	-	1341
		1.03	j	20	22	487
		0.98	k	20	22	487
ClH_3Si	Chlorosilane	1.292±0.005	gas	15-130	-	220
		1.303±0.01	gas	-	M	1283
		1.31	gas	-	M	377
		1.311	gas	-	M	378
ClI	Iodine monochloride	~0.5	gas	60-160	-	1177
		0.65±0.07	gas	-	M	1960-1
		0.9-1.1	CCl_4	25	-	1252
		1.50	CCl_4	n.s.	-	492
		1.48	cHx	n.s.	-	492
ClK	Potassium chloride	6.3±0.3	gas	1024	MB	1704
		8.00[mm]	gas	675	MB	1627
		10.1±0.2	gas	-	M	1933
		10.48±0.05	gas	-	MB	1058
		10.61	gas	-	MB	484
		10.69±0.05[m]	gas	-	MB	1058
		10.82[m]	gas	-	MB	484
		10.2±0.1	B[n]	20	15	1737
ClLi	Lithium chloride	8.2-8.3±0.1	B[n]	20	15	1737
$ClLiO_4$	Lithium perchlorate	7.89±0.05	D	25	-	1252
ClNO	Nitrosyl chloride	1.28±0.04[p]	gas	-	M	1629
		1.84±0.01	CCl_4	12	-	905
$ClNO_2$	Nitryl chloride	0.42±0.01	gas	-	M	1315
		0.53	gas	-	M	316

j) Bromoethane. k) Chloroethane. mm) As corrected J. Chem. Phys., 4, 730 (1936).
m) $\nu = 2$. n) In presence of $AlBr_3$. p) μ_a.

FORMULA	COMPOUND NAME	μ, D	State, or Solv.	t, °C	Method or P_a, % of P_e	Reference
ClNa	Sodium chloride	8.5±0.2	gas	-	M	1933
		8.5±0.4	gas	800	M	758
		9.0±0.1	B[pp]	20	15	1737
ClO	Chlorine monoxide	1.70±0.09	gas	-18 to 24	-	1901
ClO_2	Chlorine dioxide	0.78±0.08	gas	-10 to 24	-	1901
ClRb	Rubidium chloride	10.6	B[pp]	20	15	1737
ClTl	Thallium chloride	4.444±0.014	gas	-	MB	270
		5.1±0.4	gas	-	M	550
Cl_2	Chlorine	0.23	liq	-65 to 8	-	2172
$Cl_2Co_2N_4O_4$	Di-μ-chloro-*bis*(dinitrosylcobalt)	1.04±0.1	B	n.s.	-	711
Cl_2CrO_2	Chromyl chloride	0.47	CCl_4	25	0	1811
Cl_2F_6	Chlorine trifluoride dimer	~0	gas	0-83	-	1227
$Cl_2F_6P_2Pt$	*Bis*(trifluorophosphine)dichloroplatinum	4.4±0.5	B	25	10	297
Cl_2GeH_2	Dichlorogermane	2.22	CCl_4	25	0	1119,1811
Cl_2H_2Si	Dichlorosilane	1.181±0.005	gas	20-125	-	220
Cl_2Hg	Mercuric chloride	0	gas	325-430	-	200
		1.24	B	25	10	1956
		1.32±0.03	D	30-50	-	1957
		1.44	D	25	-	365
Cl_2OS	Thionyl chloride	1.452±0.005	gas	15-135	15	333
		9.31	liq	20	-	175
		1.39	B	25-45	62	1792
		1.61	B	20	-	175

pp) In presence of $AlBr_3$.

FORMULA	COMPOUND NAME	μ, D	State, or Solv.	t, °C	Method or P_a, % of P_e	Reference
Cl_2OSe	Selenium oxychloride	2.64	B	25	0	1815
Cl_2O_2Cr	Chromyl chloride	0.47	CCl_4	25	0	1811
Cl_2O_2S	Sulfuryl chloride	1.810±0.005	gas	20-145	15	333
		1.65	B	25-45	87	1792
Cl_2O_7	Chlorine heptoxide	0.72	CCl_4	n.s.	-	532
Cl_2S_2	Sulfur monochloride dimer	1.07±0.02	B	20	10	1705
		1.61	B	25	0	1815
		0.92±0.02	CCl_4	20	10	1705
Cl_2Se_2	Selenium monochloride dimer	2.1	B	25	0	1815
Cl_3Fe	Ferric chloride	1.28±0.05	D	25	10	1038
Cl_3Ga	Gallium chloride	0.85	CS_2, CCl_4	13-21	0	2005
Cl_3HSi	Trichlorosilane	0.855±0.005	gas	10-140	-	220
		0.98±0.06	B	25	5	1853
Cl_3N	Nitrogen trichloride	0.6±0.2	B	20	0	6
Cl_3OP	Phosphorus oxychloride	2.37-2.44	B	10-60	0	1815
		2.41	B	20	0	1263
		2.42_5	B	25	5	644
Cl_3OV	Vanadium oxychloride	0.3±0.2	CCl_4	25	6	273
Cl_3P	Phosphorus trichloride	0.79	gas	30-90	-	613
		0.80	gas	-	M	1962
		0.70	solid	17-60	-	1170
		0.89_0	B	25	5	644
		0.91	B	25	-	1791

FORMULA	COMPOUND NAME	μ, D	State, or Solv.	t, °C	Method or P_a, % of P_e	Reference
		1.00	B	10-40	-	1897
		1.10	B	n.s.	-	57
		1.90	D	25	0	1289
		0.78	CCl_4	-25 to 25	-	1897
		0.80	CCl_4	18	0	133-5
Cl_3PS	Thiophosphoryl chloride	1.42	B, Hp	25	0	1815
Cl_3Sb	Antimony trichloride	3.93	gas	-	M	1962
		3.67	B	(20)	0	2098
		3.78	B	25	-	1791
		4.12	B	16	0	134-5
		5.19	D	25	0	1289
		3.14±0.05	CS_2	25	-	1252
		3.95	Ether	(20)	0	2098
		3.1	n.s.	n.s.	-	1251
Cl_4Ge	Germanium tetrachloride	0	gas	96-228	-	331
		0.67±0.08	D	25	10?	1038
Cl_4Si	Silicon tetrachloride	0	gas	80-140	18	1091
		0	D	25	10	1038
		0	CCl_4	19	0	133-5
Cl_4Sn	Tin tetrachloride	0	gas	90-200	-	331
		0	liq	-20 to 20	-	2004
		0	solid	-35	-	2004
		0	B	(20)	15	2006
		0.80	B	15-20	-	2004
		0.87	B	n.s.	-	592
		0.96	B	25	10	1954
		3.84±0.15	D	25	10?	1038
		5.10	D	20	-	1445
		0	CCl_4	15-19	-	2004
		0	CCl_4, Hx	25	-	1852

FORMULA	COMPOUND NAME	μ, D	State, or Solv.	t, °C	Method or P_a, % of P_e	Reference
Cl_4Sn (contd.)	Tin tetrachloride (contd.)	0.80	CCl_4	18	0	133-5
		0.86	CCl_4	25	10	1954
Cl_4Te	Tellurium tetrachloride	2.56	B	25	0	1811
		2.58	B	25	15	849
Cl_4Ti	Titanium tetrachloride	0	gas	100-208	-	331
		0	liq	-20 to 20	-	2004
		0	solid	-30	-	2004
		0	B	25	10	1954,2004
		3.48	D	20	15	1443
		0	CCl_4	20,25	0	133-5
						1954,2004
Cl_4V	Vanadium tetrachloride	0	CCl_4	25	6	273
Cl_4Zr	Zirconium tetrachloride	5.94	D	20	15	1443
Cl_5P	Phosphorus pentachloride	~0	CCl_4	9-32	-	1776
		0.8_8	CCl_4, CS_2	25	0	1976
Cl_5PPt	Phosphorus trichloride - platinum dichloride complex	3.08	B	n.s.	0?	57
Cl_5Sb	Antimony pentachloride	~0	CCl_4	7-47	-	1776
		1.15	CCl_4	16	0	133-5
Cl_5Ta	Tantalum pentachloride	1.2	CS_2	25	0?	1360
$Cl_6N_3P_3$	Phosphonitrile chloride trimer	0.51	Dec	-20 to 95	-	965
$Cl_8N_4P_4$	Phosphonitrile chloride tetramer	0.39	Dec	-20 to 95	-	965
$Cl_9N_6P_6$	Phosphonitrilohexamer chloride	0.72	Dec	-20 to 95	-	965

FORMULA	COMPOUND NAME	μ, D	State, or Solv.	t, °C	Method or P_a, % of P_e	Reference
$Cl_{10}N_5P_5$	Phosphonitrile chloride pentamer	0.47	Dec	-20 to 95	-	965
$Cl_{12}N_6P_6$	Phosphonitrile chloride hexamer	0.48	Dec	-20 to 95	-	965
$Cl_{14}N_7P_7$	Phosphonitrile chloride heptamer	0.54	Dec	-20 to 95	-	965
$Co_2I_2N_4O_4$	Dinitrosyl cobalt iodide dimer	0	B	n.s.	-	710
CsF	Cesium fluoride	7.3±0.5	gas	~575	MB	775
		7.42±0.47	gas	-	MB	776
		7.85±0.25	gas	700	M	758
		7.875±0.006[q]	gas	-	MB	1971
		7.89±0.17	gas	-	MB	1970
		7.98±0.18[r]	gas	-	MB	1970
CsI	Cesium iodide	10.2[s]	gas	600	MB	1627
DHO	Water-d_1	1.78±0.06	gas	-	M	1884
		1.84±0.01	gas	-	M	1880
DHS	Hydrogen-d_1 sulfide	1.02±0.02	gas	-	M	738
DHSe	Hydrogen-d_1 selenide	0.62	gas	-	M	2042-3
DH_2P	Phosphine-d_1	0.55	gas	-	M	1167
		0.579±0.012	gas	-	M	1780
DH_2Sb	Stibine-d_1	0.116±0.003	gas	-	M	1167
DI	Hydrogen-d_1 iodide	0.445±0.02	gas	-	M	244
D_2HP	Phosphine-d_2	0.565±0.008	gas	-	M	1780
D_2O	Water-d_2	1.85	gas	90-200	5	622
		1.87±0.02	gas	-	M	112
		2.04±0.04	gas	-	M	1879

q) For ν = 0, for ν = 1-4; $\mu_\nu = (7.875\pm0.006) + (0.0722\pm0.0003)\nu$. r) ν = 1.
s) As corrected J. Chem. Phys. 4, 730 (1936).

FORMULA	COMPOUND NAME	μ, D	State, or Solv.	t, °C	Method or P_a, % of P_e	Reference
D_2O (contd.)	Water-d_2 (contd.)	1.79±0.02	B	20	0	1364
		1.88	D	20	0	1
D_2Se	Hydrogen-d_2 selenide	0.24±0.02	gas	-	M	813
D_2Te	Hydrogen-d_2 telluride	<0.2	gas	-	M	813
D_3N	Ammonia-d_3	1.50_6	gas	0-150	-	396
D_3P	Phosphine-d_3	0.578±0.010	gas	-	M	243
FH	Hydrogen fluoride	1.736	gas	25	t	1785
		1.91	gas	30-100	-	659,1435
		1.98	B	30	-	2089
		2.35	D	30	-	2089
		2.0	CCl_4	30	-	2089
		2.2	Hp	30	-	2089
	hexamer	0	gas	11-60	t	1785
FH_3Si	Fluorosilane	1.268±0.013	gas	-	M	1728
FK	Potassium fluoride	7.33±0.24	gas	-	MB	605
		8.60±0.09[tt]	gas	875-900	MB	614
FLi	Lithium fluoride	6.6±0.3	gas	-	MB	201
		6.4±0.3[u]	gas	-	MB	201
$FMnO_3$	Permanganyl fluoride	1.5±0.2	gas	-60	M	834
FNO	Nitrosyl fluoride	1.81[v]	gas	-	M	1225
FNO_2	Nitryl fluoride	0.47	gas	-	M	1786
FO_3Re	Perrhenyl fluoride	0.85±0.05	gas	-	M	1168-9
FRb	Rubidium fluoride	8.80±0.10	gas	-	ER	1117

t) Value to fit vibration-rotation spectrum. tt) FK^{34} $\nu = 0$. u) $\nu = 1$.
v) $\mu_a = 1.70$, $\mu_b = 0.62$.

FORMULA	COMPOUND NAME	μ, D	State, or Solv.	t, °C	Method or P_a, % of P_c	Reference
FTl	Thallium fluoride	3.99±0.06	gas	-	M	550
		4.2282	gas	-	MB	606
		4.2971[w]	gas	-	MB	606
		4.2275[x]	gas	-	MB	606
		7.6±0.8	gas	-	M	102
F_2O	Oxygen difluoride	0.1759±0.001	gas	-	M	198
		0.297±0.005	gas	-	M	1507[y]
		0.4±0.010	gas	-75 to 80	-	411
F_2OS	Thionyl fluoride	1.618±0.010	gas	-	M	501
F_2O_2S	Sulfuryl fluoride	1.110±0.015	gas	-	M	1141
F_3HSi	Trifluorosilane	1.26±0.01	gas	-	M	575-6
F_3N	Nitrogen trifluoride	0.21	gas	-80 to 95	-	1563,2068
		0.234±0.004	gas	-	M	576
		0.235±0.007	gas	-	M	917
		0.25	gas	-80 to 25	-	2068-9
F_3OP	Phosphoryl fluoride	1.69±0.05	gas	-	M	687
		1.735	gas	-	M	1724
		1.77±0.02	gas	-	M	575-6
F_3P	Phosphorus trifluoride	0	gas	10-115	-	1157
		1.025±0.009	gas	-	M	1768
		1.03±0.01	gas	-	M	575-6
F_3PS	Thiophosphorus fluoride	0.633±0.02	gas	-	M	687
F_4N_2	Tetrafluorohydrazine	0.26	gas	-	M	1146
$F_4N_4S_4$	Fluorosulfur nitrile tetramer	0	n.s.	n.s.	-	584
F_4S	Sulfur tetrafluoride	1.0±0.1	gas	-30 to 80	-	411

w) $\nu = 1$.   x) $\nu = 2$.   y) Says the 0.1759 value is incorrect.

FORMULA	COMPOUND NAME	μ, D	State, or Solv.	t, °C	Method or P_a, % of P_c	Reference
F_4Si	Silicon tetrafluoride	0	gas	25	-	2071
F_5I	Iodine pentafluoride	2.18±0.12	gas	115-170	18	1648
		2.24±0.1	gas	n.s.	0?	1651
		3.62-3.69	liq	12-40	0?	1651
F_5P	Phosphorus pentafluoride	0	gas	280-390	-	1157
F_6S	Sulfur hexafluoride	0	gas	25	-	2071
		0	gas	20	-	1156
		0	gas	86-132	DVC	240
F_6Se	Selenium hexafluoride	0	gas	20	-	1156
F_6Te	Tellurium hexafluoride	0	gas	20	-	1156
F_6U	Uranium hexafluoride	0	gas	30-90	-	29, 1226, 1812
		0	gas	96	NRA	1228
$F_{10}S_2$	Disulfur decafluoride	0	gas	25	-	756
		0	liq	-10 to 20	-	756
$FeIN_3O_3$	Trinitrosyliron iodide	3.5	B	n.s.	-	710
H	Hydrogen atom (adsorbed) (See also H_2)	0.15	-	-	PE	446
		0.18	-	-	PE	444
HI	Hydrogen iodide	0.382	gas	-28 to 340	-	2159
		0.58	B	20	-	490
		0.50	CCl_4	20	-	490
HLi	Lithium hydride	5.882±0.003[z]	gas	-	MB	2103
		5.990±0.003[a]	gas	-	MB	2103
		5.884±0.003[b]	gas	-	MB	2103

z) ^{7}LiF, ν = 0. a) ^{7}LiH, ν = 1. b) ^{6}LiH, ν = 0.

FORMULA	COMPOUND NAME	μ, D	State, or Solv.	t, °C	Method or P_a, % of P_c	Reference
HNO_3	Nitric acid	2.16±0.02	gas	-	M	1313-4
HN_3	Hydrazoic acid	0.847±0.005	gas	-	M	26
HO	Hydroxyl	1.54	gas	n.s.	-	1224
		1.65±0.25	gas	-	M	1306
H_2	Hydrogen (See also H)	0	gas	0-571	-	2002,2172
		<0.015	gas	-210 to 298	-	2070
		0.038±0.016	gas	20-300	-	534
		0.043	gas	-190	-	1602
H_2F_2Si	Difluorosilane	1.54±0.02	gas	-	M	1047
$H_2N_2O_2$	Nitramine	3.78	D	20	-	793
H_2O	Water	1.84	gas	-	M	591
	1.82	1.844±0.006	gas	21-198	-	1876
		1.84_6±0.005	gas	25-100	M	180
		1.85±0.01	gas	100-200	5	622
		1.85	gas	20-70	-	352
		1.85_5±0.008	gas	120-210	-	1695
		1.856	gas	110-250	-	795
		1.86_0±0.02	gas	120-210	-	1692
		1.86_2±0.008	gas	120-7	-	1690
		1.87±0.05	gas	45-115	MR	450
		1.88	gas	115-170	-	871
		0.74	liq	18	REF	1115
		1.92	liq	15,30	-	2060
		3.12	liq	20	-	657
		1.71±0.06	B	25	-	2110,2112
		1.7_2	B	25	0	2116
		1.76	B	30	-	44
		1.77±0.02	B	20	0	1364
		1.85±0.04	B	20	3	1718

FORMULA	COMPOUND NAME	μ, D	State, or Solv.	t, °C	Method or P_a, % of P_e	Reference
H_2O (contd.)	Water (contd.)	1.87	D	20	0	1
		1.90	D	30	-	2089
		1.9	D	25	0	2116
		1.91	D	25	0	1159
		1.91	D	25	-	386
		1.93±0.03	D	30-50	0	1958
		1.91±0.12	CCl_4	20	3	1718
		1.72	Ether	10	0	1159
		1.9	Ether	20	-	717,723
		1.95	Ether	n.s.	0	684
		1.98	Ether	n.s.	-	671
		1.99	Ether	17	-	683
H_2O_2	Hydrogen peroxide	2.07-2.26	gas	-	M	1279
		2.15	D	25	0	1159
		2.08	Ether	10	0	1159
H_2S	Hydrogen sulfide	0.89_7	gas	62-160	-	44
		0.937	gas	-75 to 270	-	2171
		0.98^c	gas	30-90	-	189
		1.101±0.05	gas	15-115	-	2050
		0.78	liq	-60	-	686
		0.88	liq	-80	-	686
		0.95	solid	-150 to -90	-	686
		1.07	solid	-170 to -150	-	686
		0.85-0.97	B	30	-	44
H_2S_2	Hydrogen disulfide	1.18	B	25	0	1815
H_2Se	Hydrogen selenide	0.4	solid	-186	0	1823
H_2Te	Hydrogen telluride	<0.2	gas	-	M	813

c) Microwave frequencies.

FORMULA	COMPOUND NAME	μ, D	State, or Solv.	t, °C	Method or P_a, % of P_c	Reference
H_3N	Ammonia	1.3	gas	n.s.	IRI	531
	1.47	1.44	gas	25-95	-	2001
		1.441	gas	5-130	-	2021
		1.45	gas	100-175	-	910
		1.45	gas	-30 to 185	-	2160
		1.45	gas	-	M[d]	1959
		1.46	gas	20-145	-	1099
		1.46_8±0.009	gas	-	M	324
		1.47_4	gas	0-150	-	396
		1.49	gas	50-200	-	2002
		1.51-1.46	gas	15-100	-	2067
		1.54±0.01	gas	20-175	-	871
		0.94	liq	5-35	-	1099
		1.39	B	25	0	993a
		1.41	B	25	-	1099
		1.42	B	25	0	68
		1.44	Hp	25	0	993a
H_3P	Phosphine	0.55	gas	-50 to 100	-	2067
		0.55±0.01	gas	-	M	1166
		0.578±0.010	gas	-	M	243
H_3Sb	Stibine	0.116±0.003	gas	-	M	1166
H_4N_2	Hydrazine	3.00	liq	20	-	657
		1.84-1.86	B	18	15	76
		1.9	B	25	-	1555
		1.92±0.05	B	15	0	2008
$H_4N_2O_2S$	Sulfamide	3.9	D	20	-	403
H_4Si	Silane	0	gas	25	-	2071
H_6Si_2	Disilane	0	gas	25	-	2071

d) μ to fit microwave intensity data.

FORMULA	COMPOUND NAME	μ, D	State, or Solv.	t, °C	Method or P_a, % of P_r	Reference
He	Helium	0	gas	0	-	2172
		<0.015	gas	-215 to 25	-	2070
		0	liq	-270	-	2172
Hg	Mercury	0	gas	127-197	-	2150
HgI_2	Mercuric iodide	0	gas	295-430	-	200
		1.68	D	25	-	365
IK	Potassium iodide	6.8±0.3	gas	940	MB	1704
		9.24[e]	gas	625	MB	1627
		~10	gas	655	MB	2148a
ILi	Lithium iodide	7.07±0.20	gas	600	M	758
INa	Sodium iodide	0	gas	655	MB	2148a
		4.9±0.2	gas	925-975	MB	1704
ITl	Thallium iodide	~0	gas	655	MB	2148a
I_2	Iodine	0±0.1	B	15-70	-	2027
		0.6	B	25	0	494
		1.4	B	25	0	2117
		1.3	D	25	0	494
		1.41±0.02	D	25	0	947
		5.3	D	25	-	1072
		0	cHx	25	0	494,949
		1.1	f	25	0	494
		0.9	pXy	25	0	494
		1.5	g	25	0	494
		0±0.1	CS_2	15-35	-	2027
		0.7	Ether	20	-	717,723
I_3P	Phosphorus triiodide	0	CS_2	25	-	1252

e) As corrected J. Chem. Phys. 4, 730 (1936). f) Cyclohexene. g) Diisobutylene.

FORMULA	COMPOUND NAME	μ, D	State, or Solv.	t, °C	Method or P_a, % of P_c	Reference
I_3Sb	Antimony triiodide	0.4	B	25	-	2112
		1.59±0.1	CS_2	25	-	1252
I_4Sn	Tin tetraiodide	0	gas	253	-	331
		0	B	25	10	1229,1954
		1.55	D	20	-	1445
I_4Zr	Zirconium tetraiodide	5.36	D	20	-	1445
K	Potassium	~0	gas	n.s.	-	315
		22.0	h	210-287	CP	192
Kr	Krypton	<0.05	gas	-80 to 25	-	2070
N	Atomic nitrogen adsorbed on W	0.16[hh]	n.s.	n.s.	SPC	445
		0.73[i]	n.s.	n.s.	SPC	445
NO	Nitric oxide	0.0_7	gas	-38 to 200	-	1816
		0.148±0.002	gas	25	NRA	1278
		0.158±0.006	gas	-	M	245
		0.16	gas	-80 to 25	-	2071
NO_2	Nitrogen dioxide	0.29	gas	50-90	6	1717
		0.58-0.30	gas	25-125	0	2124
		0.3_9	gas	25-125	-	2169
N_2	Nitrogen (See N also)	0	gas	-190 to 310	-	534,925, 2071,2160, 2172
		0	liq	-210 to -195	-	2172
N_2O	Nitrous oxide	0	gas	30-125	-	573
		<0.02	gas	30-150	-	1722
		0.14±0.02	gas	20-180	0	376

h) Adsorbed on W. hh) Up to about 2/3 monolayer. i) Above 2/3 monolayer.

FORMULA	COMPOUND NAME	μ, D	State, or Solv.	t, °C	Method or P_a, % of P_e	Reference
N_2O (contd.)	Nitrous oxide (contd.)	0.16	gas	-	M	323
		0.166±0.002	gas	-	M	1768
	0.18	0.167±0.002	gas	-	M	1963
		0.17	gas	-80 to 25	-	2071
		0.249±0.05	gas	15-95	-	2050
N_2O_4	Nitrogen tetroxide	0	gas	25-125	0	2124
		0.37	gas	50-90	6	1717
		0.5_5	gas	25-125	-	2169
N_2O_5	Nitrogen pentoxide	1.40	CCl_4	25	0	1120
N_4S_4	Sulfur nitride	0.52±0.10	B	25	0?	1644
		0.56±0.10	B	45	0?	1644
		0.72	B	30	-	1500
		0.72	CS_2	25	0?	1644
Na	Sodium	~0	gas	n.s.	-	315
		22.6-3.6	ii	-	CP	192
Ne	Neon	0	gas	0	-	2172
		<0.015	gas	-210 to 20	-	2070
O_2	Oxygen	0	gas	0-25	-	925,2071, 2172
		0	liq	-218 to -183	-	2172
O_2S	Sulfur dioxide	1.47±0.02	gas	-	M	1302
		1.59±0.01	gas	-	M	351
		1.61±0.02	gas	17-80	-	1804
	1.61	1.623	gas	-5 to 170	-	2160
		1.632	gas	15 to 180	8-20	1096
		1.63	gas	30-90	-	189
		1.65±0.03	gas	20-75	-	1173

ii) Adsorbed on W.

FORMULA	COMPOUND NAME	μ, D	State, or Solv.	t, °C	Method or P_a, % of P_r	Reference
		1.7	gas	15	-	925
		1.77	gas	20-175	-	871
		1.84±0.10	gas	20	ES	551
		1.10	liq	0	0	1095
		1.16	liq	25	0	1095
		1.93	liq	-35 to 20	-	1417
		1.61	B	25	0	1095
		1.62	CCl_4	25	0	1095
		~1.4	$CHCl_3$	25	0	1095
		1.6	j	25	5	1344
O_2Si	Silica	0.48	vitreous	~25	-	828
		0.58	quartz	~25	-	828
O_3	Ozone	0.52±0.03	gas	-80 to 185	MR	450
		0.53±0.02	gas	-	M	1965
		0.58±0.05	gas	-	M	778-9
		0.65±0.05	gas	-	M	777
		0.56	liq	-183	0	702
		0.49	k	-192	0	1119
O_3S	Sulfur trioxide	0	gas	80-160	-	1804
O_4Os	Osmium tetroxide	0	gas	155,290	-	1156
		0	B	n.s.	-	1383
O_5V_2	Vanadium pentoxide	419	kk	25	0	458-9
P_4S_3	Phosphorus sesquisulfide	0.81±0.10	B	25	0?	1644
S	Sulfur	0	liq	118-350	-	374
		4.5-8.0	liq, solid	18-125	m	1935
		0	B	25-65	-	1644
		0	CS_2	8-32	-	409
Xe	Xenon	<0.05	gas	-80 to 25	-	2070

j) *trans*-Dichloroethylene. k) Liquid oxygen. kk) Colloidal.
m) Calculated from several physical properties.

TABLE II

COMPOUNDS CONTAINING CARBON

FORMULA	COMPOUND NAME	μ, D	State, or Solv.	t, °C	Method or P_a, % of P_e	Reference
$CBrCl_3$	Trichlorobromomethane	0.21	liq	0-60	MA,10	1317
$CBrF_3$	Trifluorobromomethane	0.65	gas	25-225	~20	405
$CBrN$	Bromocyanide	2.94	gas	-	M	1962
		2.97±0.02	B	20	37	2177
$CBrN_3O_6$	Bromotrinitromethane	2.58	B	25	0?	1753
CBr_2Cl_2	Dibromodichloromethane	0.25	liq	25-60	MA,10	1317
CBr_2F_2	Dibromodifluoromethane	0.66	gas	25-225	~20	405
		0.68	liq	0-20	MA,10	1317
CBr_3Cl	Tribromochloromethane	0.2	liq	60	MA,10	1317
CBr_3F	Tribromofluoromethane	0.58	liq	0-60	MA,10	1317
CBr_4	Carbon tetrabromide	0	B	25	10	1222
$CClF_3$	Chlorotrifluoromethane	0.39±0.02	gas	0-150	MR	450
		0.46	gas	25-225	~20	405
		0.47	gas	29	5	559
		0.50	gas	~25	MA	179
		0.50±0.01[n]	gas	-	M	868
$CClN$	Chlorine cyanide	2.80	gas	-	M	1962
		2.802±0.020	gas	-	M	1768
$CClN_3O_6$	Chlorotrinitromethane	2.19	B	25	0?	1753
CCl_2F_2	Dichlorodifluoromethane	0.51±0.01	gas	30-200	25	1817
		0.55±0.03	gas	20-140	MR	450
		0.55	gas	29	5	559

n) $C^{35}ClF_3$

FORMULA	COMPOUND NAME	μ, D	State, or Solv.	t, °C	Method or P_a, % of P_e	Reference
CCl_2O	Phosgene	1.19±0.01	gas	30-150	7	1820
		1.19	gas	n.s.	-	42
		1.1_3	B	25	-	42
		1.109	CCl_4	0	-	1065
CCl_2S	Thiophosgene	0.28±0.02	gas	30-140	15	333
CCl_3F	Trichlorofluoromethane	0.45±0.01	gas	25-100	15	1817
		0.53	gas	26	5	559
		0.49	liq	0-20	MA,10	1317
CCl_3NO_2	Chloropicrin	1.89±0.07	gas	70	-	1820
		1.81	B	25-50	-	1838
		1.8	B,CCl_4	25	-	2015
		1.92	Hx	n.s.	0	1347
		1.80	Hp	25-50	-	1838
CCl_4	Carbon tetrachloride	0	gas	23-95	-	1564,2069
		0	gas	70-85	8	1091
		0	gas	90-210	-	1686
		0	liq	-186 to 25	-	809
		0	B	25	0	2121
		<0.1	B	~25	-	431
CCl_4S	Trichloromethanesulfenyl chloride	0.65±0.01	B	25	0	1644
		0.56±0.01	CCl_4	25	0	1644
CDNO	Isocyanic acid-*d*	1.619±0.015	gas	-	M	1760
CFN	Fluorine cyanide	1.68±0.05	gas	-	M	1743
CF_3I	Trifluoroiodomethane	0.92	gas	25-225	~20	405
		1.0±0.1	gas	-	M	1873
CF_4	Carbon tetrafluoride	0	gas	-80 to 100	-	1563,2068
		0	gas	24	5	559

FORMULA	COMPOUND NAME	μ, D	State, or Solv.	t, °C	Method or P_a, % of P_e	Reference
$CHBrF_2$	Bromodifluoromethane	1.31	gas	27	5	559
$CHBr_2F$	Dibromofluoromethane	0.72	gas	29	5	559
$CHBr_3$	Bromoform	1.00±0.02	gas	120-235	-	237
		0.92	liq	25	-	1097
		1.21	liq	20	-	657
		1.00	B	10-70	-	1828
		1.00	B, CCl_4	25	-	2015
		1.0_0	B	25	0	1366
		1.03	B	25	-	237
		1.07	B	25	0	1097
		0.86	$CHCl_3$	25	0	1097
		0.92	ClB	25	0	1097
		1.45	NB	25	0	1097
$CHClF_2$	Chlorodifluoromethane	1.29	gas	-	M	1962
		1.40_5	gas	30-200	35	1817
		1.42	gas	25.4	5	559
$CHCl_2F$	Dichlorofluoromethane	1.30±0.01	gas	30-150	6	1817
		1.35	gas	30	5	559
$CHCl_3$	Chloroform	0.9_6	gas	70-170	-	1686
		1.00±0.01	gas	80	DVC	240
		1.03	gas	25-95	-	1564
		1.06	gas	25-140	-	1778
		1.07	gas	25	-	1097
		1.2	gas	-	M	1962
		1.86	gas	30-140	-	1275
		0	liq	-90 to 60	2	1825
		1.10	liq	25	-	1254
		1.11	liq	25	-	1097
		1.20	liq	-60 to -20	-	1402

FORMULA	COMPOUND NAME	μ, D	State, or Solv.	t, °C	Method or P_a, % of P_e	Reference
$CHCl_3$ (contd.)	Chloroform (contd.)	1.265	liq	-186 to 55	-	809
		1.55	liq	20	-	657
		1.11	B	25	0	809
		1.11	B	25	-	2121
		1.11	B	20	-	977
		1.18	B	20	RT	921
		1.19	B	25	0	1366
		1.21	B	25	0	1082
		1.22±0.01	B	25	-	2020
		1.23	B	25	0	1097
		1.23	p	20	5	1513
		1.28	B	18	0	671,683-4
		1.14	CCl_4	20	-q	882
		1.16	CCl_4	25	0	966
		1.21	CCl_4	-18 to 46	-	1418
		1.06	Hx	-90 to 60	28	1825
		1.19	Hx	20	RT	921
		1.20	Hx	-60 to -20	-	1402
		1.25	Tol	25	0	1097
		1.58	Ether	19	0	671,683-4
		0.86	CS_2	20	-q	882
		1.19	ClB	25	0	1097
		1.49	NB	25	0	1097
		1.17	r	15	-	1125
CHFO	Formyl fluoride	2.02±0.02	gas	-	M	1057
CHF_3	Fluoroform	1.59	gas	-	M	583
		1.60	gas	-80 to 95	-	1563
		1.62	gas	25-225	~20	405
		1.64±0.02	gas	-	M	575
		1.645±0.009	gas	-	M	1758
		1.65±0.01	gas	80,160	DVC	240

p) B plus H_2O. q) At 10 cm. r) Gasoline.

FORMULA	COMPOUND NAME	μ, D	State, or Solv.	t, °C	Method or P_a, % of P_e	Reference
CHI_3	Iodoform	0.8	B	10-70	-	1828
		0.803	B	n.s.	-	1663
		0.9_6	B	25	0	1366
		1.00	B, CCl_4	25	-	2015
		1.16	B	20	-	977
		1.0_1	Hx	25	0	1366
CHN	Hydrogen cyanide	2.07	gas	30-100	0	537
	2.95	2.8	gas	-	M	1784
		2.91	gas	20-150	-	199
		2.960	gas	30-200	0	1820
		2.94	gas	20-130	-	202
		2.957±0.025[s]	gas	-	M	1769, 1771
		2.986±0.004	gas	-	M	176
		3.00±0.02	gas	-	M	576
		3.06	gas	25-95	-	2069
		4.7 (dimer)	gas	20-130	-	202
		2.56	B	16-48	-	1206
		2.61	B	20	0	2096
		7.8 (tetramer)	D	35	0	2074
		2.62	pXy	16-48	-	1206
CHNO	Isocyanic acid	1.592±0.010	gas	-	M	1758
		1.592±0.015[t]	gas	-	M	1760
		1.620±0.015[t]	gas	-	M	1760
		1.760±0.015[u]	gas	-	M	1760
CHNS	Isothiocyanic acid	1.72	gas	-	M	114
	Thiocyanic acid	1.72±0.05[v]	gas	-	M	113
CHN_3O_6	Trinitromethane	2.63	B	n.s.	5	80
		2.73	CCl_4	25	0	1120
CH_2Br_2	Dibromomethane	1.43±0.02	gas	85-250	0	237
		1.5	gas	n.s.	M	1962

s) Excited state, $\nu_2 = 1$. t) Low frequency bending mode. u) High frequency bending mode.
v) Along axis of least moment of inertia.

FORMULA	COMPOUND NAME	μ, D	State, or Solv.	t, °C	Method or P_a, % of P_e	Reference
CH_2Br_2 (contd.)	Dibromomethane (contd.)	1.934	gas	50-150	-	1241
		2.27	gas	n.s.	-	1275
		1.85	liq	20	-	657
		1.39	B	25	0	1366
		1.40	B	10-70	-	1828
		1.41	B, CCl_4	25	-	2015
		1.42	B	25	0	2[illegible]7
		1.91	B	25	-	1238
		1.4_3	Hx	25	0	1366
CH_2ClF	Chlorofluoromethane	1.82	gas	30-90	w	189
CH_2ClNO_2	Chloronitromethane	2.91	gas	140-210	-	795
CH_2Cl_2	Dichloromethane	1.54	gas	24-95	-	1564
		1.57	gas	n.s.	-	98
		1.60	gas	60-140	-	1686
		1.62±0.02	gas	-	M	1379
		1.631	gas	30-150	-	1241
		2.39	gas	35-[illegible]35	-	1275
		1.80^x	liq	-10 to 30	-	601
		1.90	liq	20	-	657
		1.44	B	n.s.	-	976
		1.47	B	20	-	977
		1.55	B	25	-	98
		1.5_8	B	25	0	1366
		1.58	B, CCl_4	25	-	2015
		1.62	B	25	-	1238
		1.49^y	CCl_4	-89 to 42	43	1353
		1.53-1.59	CCl_4	-18 to 46	-	1418
CH_2Cl_2Hg	Chloromethylmercury chloride	2.84	D	(20)	-	1700
CH_2Cl_4Si	Trichloro(chloromethyl)silane	1.61	B	25	14	887
CH_2F_2	Difluoromethane	1.96±0.02	gas	-	M	1133

w) At microwave frequencies. x) Calculated from data in 1353.
y) Ref. 601 recalculates this value as 1.80.

FORMULA	COMPOUND NAME	μ, D	State, or Solv.	t, °C	Method or P_a, % of P_e	Reference
CH_2I_2	Diiodomethane	1.09	B	25-50	-	1828
		1.1_1	B	25	0	1366
		1.11	B, CCl_4	25	-	2015
		1.22	B	20	-	977
		2.14	B	25	-	1238
		1.1_5	Hx	25	0	1366
CH_2N_2	Cyanamide	4.3	gas	-	M	1997
		3.8	B	20	-	403
		4.03	B	35	0	1713
		4.55	D	35	0	1713
CH_2N_4	1,2,3,4-Tetrazole	5.15	D	25	15	859
CH_2N_4S	5-Amino-1,2,3,4-thiatriazole	5.77	n.s.	n.s.	-	1147
CH_2O	Formaldehyde	2.17±0.02	gas	-	M	196
		2.29	gas	150-250	-	796
		2.31±0.04	gas	-	M	1054-5
		2.339±0.013	gas	-	M	1758
CH_2O_2	Formic acid	1.35±0.02	gas	-	M	452
		1.52	gas	70-150	-	2163
		1.7	gas	-	M	1962
		1.20±0.02	B	25	15?	2137
		1.46	B	22	-	208
		1.78	B	30	-	1526
		2.00	D	n.s.	-	1530
		2.09	D	25	0	2126
	dimer	0	gas	35-75	-	330
		1.00	gas	70-150	-	2163
CH_3AsBr_2	Methylarsine dibromide	2.66±0.04	B	25	-	313

FORMULA	COMPOUND NAME	μ, D	State, or Solv.	t, °C	Method or P_a, % of P_e	Reference
CH_3AsCl_2	Methylarsine dichloride	2.80±0.04	B	25	-	313
		2.68±0.04	cHx	25	-	313
CH_3AsF_2	Methylarsine difluoride	2.88±0.04	B	25	-	313
CH_3AsI_2	Methylarsine diiodide	2.30±0.04	B	25	-	313
CH_3BF_2	Methylborine difluoride	1.62±0.05	gas	25-95	-	2001
		1.67±0.02	gas	-	M	1404
CH_3BO	Borine carbonyl	1.770±0.01[z]	gas	-	M	1882
		1.793±0.01	gas	-	M	1882
CH_3Br	Bromomethane	1.76	gas	30-145	-	1236
		1.797±0.015	gas	-	M	1768
		1.80±0.01	gas	30-130	-	1818
		1.81	gas	25-140	-	1235
	1.79	1.81	gas	20-140	0	623
		1.83	gas	24-95	-	1564
		1.83±0.005	gas	20-180	0	237
		1.61	liq	20	-	657
		1.62	B	25	0	237
		1.84	B,CCl_4	25	-	2015
		1.46	Hx	-102 to 38	-	1353
CH_3Br_3Si	Methyltribromosilane	1.86±0.04	B	25	5	1592
CH_3Cl	Chloromethane	1.66	gas	20	Kerr	1556
		1.70	gas	20-200	-	1778
		1.84	gas	35-140	-	1236
	1.94	1.869±0.010	gas	-	M	1768
		1.87±0.03	gas	-	M	884
		1.87	gas	25-145	-	1694
		1.871±0.005	gas	-	M	1963

z) ν_1 vibration.

FORMULA	COMPOUND NAME	μ, D	State, or Solv.	t, °C	Method or P_a, % of P_e	Reference
		1.88	gas	20-180	-	554
		1.88±0.005	gas	25-140	-	1689,1695
		1.94	gas	23-95	-	1564
		1.9_7	gas	20-140	-	1686
		2.02	gas	-150 to 130	-	1235
		1.85	liq	-60 to -20	-	1402
		1.86	liq	20	-	657
		1.69	B	25	-	98
		1.88	B, CCl_4	25	-	2015
		1.66	CCl_4	-46 to 60	-	1353
		1.57	Hx	-95 to 40	-	1353
CH_3ClHg	Methylmercury chloride	3.36	D	(20)	-	1700
CH_3ClO_2S	Methyl chlorosulfite	2.33	B	25,45	0	1974
CH_3Cl_2N	*N*,*N*-dichloromethylamine	1.80±0.02	B	20	0	6
CH_3Cl_3Ge	Trichloromethylgermane	2.63	B	25	15	886-7
CH_3Cl_3Si	Trichloromethylsilane	1.87	B	25	15	887
		1.93±0.02	B	20	-	1702
CH_3Cl_3Sn	Trichloromethyltin	3.62±0.04	B	25	-	313
CH_3F	Fluoromethane	1.79±0.02[zz]	gas	-	M	576
		1.822±0.01	gas	-50 to 225	-	1818
		1.85	gas	-80 to 95	-	1564
		1.71	CCl_4	20	-	1073
CH_3I	Iodomethane	1.32	gas	35-125	-	1240
		1.60±0.01	gas	30-220	-	1818
		1.61-4	gas	30-140	-	1235-6

zz) Ref. 167 by the same authors gives 2.07, but there are no experimental details.

FORMULA	COMPOUND NAME	μ, D	State, or Solv.	t, °C	Method or P_a, % of P_c	Reference
CH_3I (contd.)	Iodomethane (contd.)	1.64	gas	28-95	-	1564
		1.65±0.01	gas	20-65	0	623
	1.54	1.64±0.03	gas	15-180	-	1092
		1.647±0.014	gas	-	M	1768
		1.658±0.014	gas	-	M	1963
		1.48	liq	20	-	657
		1.35-1.37	B	5-28	-	1418
		1.42	B	20	0	348
		1.43	B	20	-	977
		1.6	B	25	0	2113
		1.67	B,CCl_4	25	-	2015
		1.57	CCl_4	20	5-25	77-8
		1.36	Hx	-90 to 34	89	1353
CH_3NO	Formamide	3.2_5	gas	150-175	10	2166
		3.71±0.03	gas	-	M	1023-4
		3.37	B	30	0	108,774
		3.0	D	n.s.	-	241
		3.86	D	30	0	108
CH_3NO_2	Nitromethane	3.1	gas	-	M	1962
		3.44	gas	65-220	50	1820
		3.46±0.04	gas	70-115	-	1091
		3.46±0.02	gas	-	M	1929
		3.57±0.01	gas	60-190	0	623
		4.39	liq	20	-	657
		2.85_5	B	20	-	314
		3.05	B	20	-	793
		3.08	B	n.s.	0	751
		3.13	B	25	0	1840, 2017-8
		3.16	B	25	0	2084
		3.23	B,CCl_4	25	-	2015
		3.43	B	25	5	264
		2.83	CCl_4	20	-	314
		3.13	CCl_4	25	0	1840, 2017-8

FORMULA	COMPOUND NAME	μ, D	State, or Solv.	t, °C	Method or P_a, % of P_e	Reference
		3.21	Hp	25-50	-	1838
	Methyl Nitrate	2.20-2.61	liq	-165 to -15	0?	611
		2.3	B, CCl_4	25	-	2015
CH_3NO_3	Methyl nitrate	2.88	B	20	0	338
		3.10±0.05	gas	-	M	408
CH_3N_5	5-Amino-1,2,3,4-tetrazole	5.75	D	25	-	859
CH_4	Methane	0	gas	20-140	-	1686
		0	gas	100	-	2002
		0	gas	25	-	2071
		0.048	gas	-155	-	1602
$CH_4Cl_8OSn_2$	Methanol - *bis*(tin tetrachloride) complex	1.67	B	n.s.	-	592
CH_4F_2Si	Methyldifluorosilane	2.11±0.02	gas	-	M	1918
CH_4N_2O	Formhydrazide	2.72	D	35	0	1557
	Formamideoxime	2.26	D	35	0	1557
	Urea	4.56	D	25	-	562
		4.59	D	25	0	1001
		4.20	H_2O	25	-	581
		5.68	H_2O	25	-	562
		6.25	a	25	-	581
		4.38	b	25	-	562
		5.06	c	25	-	562
		4.51	d	25	-	562
		6.30	e	25	-	562
		6.38	f	25	-	562
		4.66	g	25	-	562

a) 20 weight per cent H_2O + acetone. b) Acetone. c) Methanol. d) Ethanol.
e) Methyl cellosolve. f) Butyl cellosolve. g) Polyether.

FORMULA	COMPOUND NAME	μ, D	State, or Solv.	t, °C	Method or P_a, % of P_e	Reference
CH_4N_2S	Thiourea	4.93	D	25	0	1001
		~7.6	D	47	0	166
$CH_4N_4O_2$	Nitroguanidine	6.95	D	30	0	1012
CH_4O	Methanol	1.61	gas	80-180	-	871
		1.688±0.006	gas	25-205	-	1878
	1.71	1.69	gas	-	M	242
		1.69_2	gas	70-230	-	1311
		1.70	gas	23-95	-	1564
		1.71	gas	35-210	-	983
		2.87	liq	20	10	1295
		2.97	liq	20	-	863
		3.10	liq	30-40	-	1498
		1.60	B	n.s.	-	2039
	1.70 P_e = 8.05	1.63	B	30	0	1267
	F_a = 0.80	1.63	B	25	RT	2051
		1.66	B	10-60	-	1875
		1.67	B	n.s.	0	1665
		1.67	B	20	0	1387
		1.67	B	30	0	81
		1.676±0.02	B	22	15	2140
		1.70	B	20	15	1241a
		1.76	B	25	5	1108
		1.78	B	20	-	977
		1.86±0.01	D	12-50	-	126
		1.93	D	20	0	1387
		1.93	D	30	0	81
		1.68	CCl_4	-18 to 50	-	1875
		1.69	CCl_4	20	10	1295
		1.7_7	CCl_4	25	5	1108
		1.77	CCl_4	25	-	1078
		1.74	h	20	0	1387
CH_4S	Methylmercaptan	1.26±0.05	gas	-	M	1732

h) Petroleum benzine plus dioxare.

FORMULA	COMPOUND NAME	μ, D	State, or Solv.	t, °C	Method or P_a, % of P_e	Reference
CH_5FSi	Methylmonofluorosilane	1.71	gas	-	M	1506
CH_5N	Methylamine	1.00	gas	20-90	-	572
		1.24±0.02	gas	60-185	30?	1695,1867
	1.29	1.26	gas	23-95	-	1564
		1.27[i]	gas	-	M	1131
		1.29	gas	25	-	1099
		1.34	gas	15-145	3	624
		1.326±0.015	gas	-	M	1135
		1.09	liq	-10 to 25	-	1099
		1.47	B	25	-	1099
CH_5N_3O	Semicarbazide	3.77	D	30	0	1282
CH_5N_3S	Thiosemicarbazide	5.36	D	30	0	1282
$CH_5N_5O_2$	Aminonitroguanidine	6.13	D	30	0	1012
CH_5P	Methylphosphine	1.100±0.010	gas	-	M	939
CH_6Ge	Methylgermane	0.635±0.006	gas	-	M	1048
		0.644±0.005	gas	-	M	95
		0.67±0.01	gas	-	M	96-7
CH_6N_2	Methylhydrazine	1.69±0.14	B	15	0	2008
CH_6Si	Methylsilane	0.73±0.02	gas	-	M	912
		0.73	gas	-	M	1139
CH_6Sn	Methyltin	0.68±0.03	gas	-	M	1130
CH_8BN	Methylamine - borane complex	5.19±0.04	B	25	15	1420
CIN	Iodine cyanide	3.71	gas	-	M	1962
		3.74±0.02	B	20	44	2178

i) $\mu_a = 0.304$, $\mu_c = 1.232$ - preliminary analysis.

FORMULA	COMPOUND NAME	μ, D	State, or Solv.	t, °C	Method or P_a, % of P_e	Reference
CIN_3O_6	Iodotrinitromethane	3.82	B	25	0?	1753
CN_4O_8	Tetranitromethane	0	gas	82	-	331
		<0.2	B	25	-	2112
		0.19	CCl_4	25	0	2084
		0.58	CCl_4	25	0	1081
		0.71	CCl_4	25	0	1120
CO	Carbon monoxide	0.10	gas	-	M	962
		0.10	gas	-185 to 120	-	2171
	0.13					
		0.10	gas	25	-	2071
		0.112±0.005	gas	-	M	243
		0.1172	gas	0-100	-	2021
		0.118±0.0016	gas	n.s.	-	2075
		0.119±0.029	gas	20-300	-	534
		0.124±0.01	gas	15-110	-	2050
		0.128±0.007	gas	-190	-	1602
		0.14	gas	30-90	-	189
		~0.1	j	Rm temp	-	443
		0.72	k	Rm temp	-	443
COS	Carbonyl sulfide	0.650	gas	-70 to 90	-	2171
		0.694[mm]	gas	-	M	1881
		0.700±0.004[mm]	gas	-	M	1770
		0.7085±0.004	gas	-	M	1770
		0.712±0.004	gas	-	M	1758
		0.7124±0.0002	gas	-	M	1260
		0.720±0.005	gas	-8 to 60	-	837
		0.722±0.01[m]	gas	-	M	227, 380, 739
		0.732±0.01[n]	gas	-	M	227, 380, 739
COSe	Carbonyl selenide	0.59	gas	-54 to 10	-	586
		0.728[p]	gas	-	M	1883

j) Adsorbed - beyond first layer. k) Adsorbed - in first layer. mm) Excited state, $\nu_2 = 1$.
m) $C^{13}O^{16}S^{32}$. n) $C^{12}O^{16}S^{32}$. p) Symmetrical stretching mode.

FORMULA	COMPOUND NAME	μ, D	State, or Solv.	t, °C	Method or P_a, % of P_e	Reference
		0.732[q]	gas	-	M	1883
		0.752±0.007	gas	-	M	2093
		0.754	gas	-	M	1883
CO_2	Carbon dioxide	0	gas	25	-	795
		0	gas	-15 to 180	-	1895
	0.18	0	gas	20-140	-	199
		0	gas	0-100	-	2002
		0	gas	25	-	2071
		<0.015	gas	-190	-	2070
		0.06	gas	-70 to 200	-	2160
		0.132	gas	-75	-	1602
		0.142±0.0017	gas	n.s.	-	2075
		0.145±0.03	gas	20-110	-	2050
		0.20±0.05	gas	20	ES	925
		0.208±0.029	gas	20-200	-	534
		0.303±0.145	gas	20-150	-	871
CS	Carbon monosulfide	1.97±0.02	gas	-	M	1343
CSTe	Carbon sulfide telluride	0.172±0.002	gas	-	M	664
CS_2	Carbon disulfide	0	gas	50-145	5	1091
		0	gas	30-140	-	573
		0	gas	50-215	-	2161
		<0.020	gas	30-150	-	1722
		0.326	gas	30-215	-	2171
		0	liq	-185 to 20	-	809
		0	B, CCl_4, Hp	20	3-10	209
		0	B	25	0	1067
		0.06	B	25	-	2122
		0.4	CCl_4	n.s.	r	882

q) Double degenerate bending mode. r) Microwave frequencies.

FORMULA	COMPOUND NAME	μ, D	State, or Solv.	t, °C	Method or P_a, % of P_e	Reference
CS_2 (contd.)	Carbon disulfide (contd.)	0.08	Hx	25	-	2122
		0.49	ClB	25	0	1067
		1.21	NB	25	0	1067
$C_2Br_2Cl_4$	1,2-Dibromo-1,1,2,2-tetrachloroethane	0	B, Hx	25	0	935
C_2ClF_3	Chlorotrifluoroethylene	0.38±0.12	gas	25-100	15	1647
		0.40	gas	25-225	20-30	405
		0.61	gas	30-90	0	189
C_2ClF_5	Chloropentafluoroethane	0.14	gas	27	5	559
		0.52	gas	25-155	20-30	405
$C_2Cl_2F_4$	1,2-Dichlorotetrafluoroethane	0	gas	27	5	559
		0.46s	gas	30-90	-	189
		0.533±0.018	gas	15-132	MA	1228
		0.56	gas	n.s.	MD	190
$C_2Cl_2O_2$	Oxalyl chloride	0.93	B	20	0	1265
$C_2Cl_2O_2Pt$	Dichlorodicarbonylplatinum	4.68±0.5	B	25	10	297
		4.90±0.05	B	25	25	1222
C_2Cl_3N	Trichloroacetonitrile	1.93	liq	19	-	388
		2.0	Hx	25	-	2018
C_2Cl_4	Tetrachloroethylene	0	B	25	0	1982
		0	CCl_4	25	10?	197
C_2Cl_4O	Trichloroacetyl chloride	1.20	B	20	0	1263
		1.12	CCl_4	25	-	1392
$C_2Cl_4O_2$	Trichloromethyl chloroformate	2.18	Hx	-70	15	1347
C_2Cl_6	Hexachloroethane	0	B, Hx	15-30	0	935
		0	Hx	25	-	1334

s) Microwave frequencies.

FORMULA	COMPOUND NAME	μ, D	State, or Solv.	t, °C	Method or P_a, % of P_e	Reference
C_2DHO	Ketene-d_1	1.423±0.015	gas	-	M	867,1732
C_2D_2O	Ketene-d_2	1.442±0.013	gas	-	M	867,1732
C_2F_6	Perfluoroethane	0	gas	23	5	559
C_2F_6Hg	*Bis*(trifluoromethyl)mercury	0.97	B	20	-	1700
		0.21	CCl_4	20	-	1700
$C_2F_6N_2$	"Nitrogen-carbon-fluorine" compound	0.47	gas	25-95	-	1563
C_2F_6O	Perfluoromethyl ether	0.54	gas	25-225	20-30	405
$C_2FeN_2O_4$	Iron carbonyl nitrosyl	0.72±0.07	B	25	20	715
C_2HBr	Bromoacetylene	0	gas	15-80	-	220
C_2HBr_3O	Tribromoacetaldehyde	1.70	B	20	0	329
C_2HCl	Chloroacetylene	0.44±0.01	gas	15-90	-	220
		0.44±0.01	gas	-	M	2100
C_2HCl_2N	Dichloroacetonitrile	2.0	B,CCl_4	25	-	1840
		2.5	Hx	25	-	2018
C_2HCl_3	Trichloroethylene	0.77	liq	28-64	-	823
		0.95	B	30	0	394
		0.8	CCl_4	n.s.	-	2014
		0.85	CCl_4	25	10	197
C_2HCl_3O	Dichloroacetyl chloride	1.58	CCl_4	25	-	1392
	Trichloroacetaldehyde	1.59	B	20	0	329
		1.96±0.04	B	25	-	313
		1.96	D	25	0	391

FORMULA	COMPOUND NAME	μ, D	State, or Solv.	t, °C	Method or P_a, % of P_e	Reference
$C_2HCl_3O_2$	Trichloroacetic acid (dimer)	1.1	B	25	-	1106
C_2HCl_5	Pentachloroethane	0.92±0.01	gas	130-240	-	1942
		0.98-0.90^t	liq	25-85	-	824
		1.0	B	n.s.	-	2014
C_2HF	Fluoroacetylene	0.75±0.01	gas	-	M	1997
$C_2HF_3O_2$	Trifluoroacetic acid	2.28	gas	100-230	-	577
C_2HF_5	Pentafluoroethane	1.54	gas	25-225	20-30	405
C_2H_2	Acetylene	0	gas	-80 to 190	-	1839
		0	gas	25	-	2071
$C_2H_2AsCl_2$	*cis*-2-Chlorovinyldichloroarsine	1.79	B	20	0	1348
		2.61	B	25	5	1291
	trans-2-Chlorovinyldichloroarsine	2.21	B	25	5	1291
C_2H_2BrCl	*cis*-1,2-Chlorobromoethylene	1.56	B	n.s.	-	453,455
	trans-1,2-Chlorobromoethylene	0	B	n.s.	-	455
C_2H_2BrClO	Bromoacetyl chloride	2.14	CCl_4	25	-	1392
		2.15	Hp	25	-	1392
$C_2H_2Br_2$	*cis*-1,2-Dibromoethylene	1.35	B	n.s.	-	453,455
	trans-1,2-Dibromoethylene	0	B	n.s.	-	455
	Dibromoethylene (configuration n.s.)	1.18	B	n.s.	u	976
$C_2H_2Br_2O$	Bromoacetyl bromide	2.07	CCl_4	25.1	-	1392

t) Calculated by Jatkar's equation. u) Author states "equation doesn't apply."

FORMULA	COMPOUND NAME	μ, D	State, or Solv.	t, °C	Method or P_a, % of P_e	Reference
$C_2H_2Br_4$	1,1,2,2-Tetrabromoethane	1.30	Hx	25	0	935
		1.70	Ether	20	0	935
$C_2H_2ClF_3$	2-Chloro-1,1,1-trifluoroethane	1.65	gas	26.6	5	559
C_2H_2ClI	*cis*-1-Chloro-2-iodoethylene	0.57	B	n.s.	0	457
	trans-1-Chloro-2-iodoethylene	1.28	B	n.s.	0	457
C_2H_2ClN	Chloroacetonitrile	2.89	B,CCl_4	25	-	2015
		3.00	B,CCl_4, Hx	25	-	1840
		3.00±0.05	B	25	0	1630
		3.0	Hx	25	-	2018
$C_2H_2Cl_2$	1,1-Dichloroethylene	1.19	B	n.s.	-	455
		1.30	B	25	0	1631
		1.25	CCl_4	25	10	197
	cis-1,2-Dichloroethylene	2.95	gas	50-155	-	1275
		1.7_6	B	25	0	1366
		1.91	B	25	-	453,455
		1.80	CCl_4	25	10	446
	trans-1,2-Dichloroethylene	0	B	n.s.	-	455
		0.70	B	25	-	1366
		0	CCl_4	25	10?	197
	Dichloroethylene (configuration n.s.)	1.96	B	n.s.	-	976
$C_2H_2Cl_2F_2$	1,1-Dichloro-2,2-difluoroethane	1.34	gas	60	-	1273
		1.36	gas	84	-	1273
		1.39	gas	110	-	1273
		1.44	gas	155	-	1273
		1.45	gas	182	-	1273
		1.47	gas	200	-	1273

FORMULA	COMPOUND NAME	μ, D	State, or Solv.	t, °C	Method or P_a, % of P_e	Reference
$C_2H_2Cl_2Hg$	*cis*-1,2-Dichloro-1-mercuriethylene	2.76	n.s.	n.s.	-	1414
	trans-1,2-Dichloro-1-mercuriethylene	1.64	n.s.	n.s.	-	1414
$C_2H_2Cl_2O$	Chloroacetyl chloride	2.2_0	gas	85-255	10?	2167
		2.24	B	20	0	1263
		2.17	CCl_4	25	-	1392
		2.08	CS_2	20	0	1263
	Dichloroacetaldehyde	2.36±0.04	B	25	-	313
		2.36	D	40	0	391
$C_2H_2Cl_3F$	1,1,2-Trichloro-2-fluoroethane	1.38-1.44	gas	105-240	0	406
$C_2H_2Cl_3NO$	Trichloroacetamide	3.94±0.04	B	25	-	313
$C_2H_2Cl_4$	1,1,1,2-Tetrachloroethane	1.2	B, CCl_4	-	-	2014
	1,1,2,2-Tetrachloroethane	1.29±0.01	gas	90-205	-	1942
		1.37	gas	130-160	-	1821
		1.70-1.76	B	0-50	-	1333
		1.95	B	-	-	2014
		1.85	CCl_4	-	-	2014
		1.43-1.50	Hx	-25 to 50	-	1333
		1.45	Hx	25	-	1334
		1.93-2.00	Ether	-25 to 25	-	1333
$C_2H_2Cl_4Si$	Trichloro(chlorovinyl)silane	0.88	B	25	15	887
$C_2H_2Cl_6Si_2$	*trans*-1,2-*Bis*(trichlorosilyl)ethylene	0.5-0.6	B	25	18	885
$C_2H_2F_2$	1,1-Difluoroethylene	1.366±0.02	gas	-	M	1619
	cis-1,2-Difluoroethylene	2.42±0.03	gas	-	M	1050
$C_2H_2F_3NO$	Trifluoroacetamide	4.33±0.04	B	25	-	313

FORMULA	COMPOUND NAME	μ, D	State, or Solv.	t, °C	Method or P_a, % of P_e	Reference
$C_2H_2I_2$	*cis*-1,2-Diiodoethylene	0.71	B	n.s.	-	453
		0.76	B	n.s.	-	455
	trans-1,2-Diiodoethylene	0	B	n.s.	-	455
C_2H_2O	Ketene	1.31[v]	gas	-	M	865
		1.31[w]	gas	-	M	865
		1.35±0.05[x]	gas	-	M	865
		1.37[y]	gas	-	M	865
		1.386±0.010[z]	gas	-	M	866-7
		1.387±0.010[a]	gas	-	M	866-7
		1.414±0.01	gas	-	M	866-7
		1.430±0.012[b]	gas	-	M	866-7
		1.46	gas	125-175	0	662
		1.52	gas	60-180	0	41
		1.4_3	B	30	0	41
	dimer	3.23	B	25	0	1621
$C_2H_2O_4$	Oxalic acid	2.63±0.10	D	25	0	1635
		3.03	D	35	5	1399
$C_2H_3BCl_3N$	Boron trichloride - acetonitrile complex	7.79±0.16	B	20	15	1415, 2006-7
C_2H_3Br	Bromoethylene	1.28	gas	155-260	10	1154
		1.417±0.005	gas	20-140	-	780
$C_2H_3BrN_2O_4$	1-Bromo-1,1-dinitroethane	3.29	B	25	0?	1753
C_2H_3BrO	Acetyl bromide	2.45	B	20	0	1263
$C_2H_3Br_3$	1,1,2-Tribromoethane	1.16	Hp	-20 to 50	0	1326-7
C_2H_3Cl	Chloroethylene	1.42±0.02[c]	gas	-	M	918
		1.44	gas	-	M	1962

v) $\nu = 0$. w) $\nu_1 = 420\ cm^{-1}$. x) $\nu_2 = 460\ cm^{-1}$. y) $\nu_3 = 570\ cm^{-1}$. z) ν_7 mode.
a) ν_8 mode. b) ν_9 mode. c) μ_a ; μ_b assumed = 0.

FORMULA	COMPOUND NAME	μ, D	State, or Solv.	t, °C	Method or P_a, % of P_e	Reference
C_2H_3Cl (contd.)	Chloroethylene (contd.)	1.452±0.01	gas	15-140	-	780
		1.92	gas	130-270	10	1154
$C_2H_3ClF_2$	1-Chloro-1,1-difluoroethane	2.14	gas	27	5	559
		2.14	gas	85-235	-	1273
$C_2H_3ClN_2O_4$	1-Chloro-1,1-dinitroethane	3.09	B	25	0?	1753
C_2H_3ClO	Acetyl chloride	2.71	gas	50-110	10?	2167
		2.42	B	25	0	927
		2.47	B	20	0	1263
	Chloroacetaldehyde	1.99±0.04	B	25	-	313
		1.99	D	25	0	391
$C_2H_3ClO_2$	Chloroacetic acid	2.31	B	30	-	1526
	Methyl chloroformate	1.69	gas	207	10	1332
		2.31	gas	140	10	1332
		2.40	gas	35	10	1332
		2.42	gas	78	10	1332
		2.24	B	n.s.	0	1347
$C_2H_3Cl_2F$	1,1-Dichloro-2-fluoroethane	1.34-1.47	gas	60-200	0	406
$C_2H_3Cl_2NO$	Dichloroacetamide	3.46±0.04	B	25	-	313
$C_2H_3Cl_3$	1,1,1-Trichloroethane	1.79	gas	65-125	0	2132
		2.03	gas	35-135	-	1275
		0.88	B	25	0	1628, 1637,2034
		1.5	B	25	-	2014-5
		1.58	B	25	0	1903
		1.66	B	20	0,RT	921
		1.1	Hx	25	0	2034

FORMULA	COMPOUND NAME	μ, D	State, or Solv.	t, °C	Method or P_a, % of P_e	Reference
		1.15-1.20	Hx	n.s.	-	1628
		1.75	Hx	20	0,RT	921
	1,1,2-Trichloroethane	1.25±0.01	gas	70-240	-	1942
		1.42±0.02	gas	100-260	0	1436
		1.41-1.48	gas	90-240	0	406
		1.55	B	n.s.	-	2014
		1.15	CCl_4	n.s.	-	2014
$C_2H_3Cl_5Si$	Trichloro(2,2-dichloroethyl)silane	1.35	B	25	15	887
C_2H_3F	Fluoroethylene	1.427±0.010	gas	-	M	1325
C_2H_3FO	Acetyl fluoride	2.96±0.03	gas	-	M	1509
$C_2H_3F_2NO$	Difluoroacetamide	3.58±0.04	B	25	-	313
$C_2H_3F_3$	1,1,1-Trifluoroethane	2.27±0.04	gas	-	M	575
		2.28	gas	25	5	559
		2.321±0.034	gas	-	M	1768
		2.33	gas	-	M	1772
		2.35±0.02	gas	-	M	1963
C_2H_3I	Iodoethylene	1.27±0.015	gas	15-140	-	780
C_2H_3N	Acetonitrile	3.92±0.06	gas	-	M	575-6
		3.96±0.01	gas	80-160	DVC	240
		3.97±0.01	gas	80-190	0	623
		3.97[d]	gas	-	M	325
		3.97	gas	25	5	1802
		4.01	gas	23-95	-	1564
		3.39	liq	20	-	657
		3.14	B	20	0	2097
		3.18	B	20	0	791
		3.4	B	25	0	2113
		3.45	B,CCl_4	25	-	2015

d) ν_8 mode.

FORMULA	COMPOUND NAME	μ, D	State, or Solv.	t, °C	Method or P_a, % of P_e	Reference
C_2H_3N (contd.)	Acetonitrile (contd.)	3.47±0.02	B	20	0	342,1474
		3.47	B	20	-	969
		3.47±0.04	B	25	-	313
		3.47	B	25	5	1802
		3.48	B, CCl_4, Hx	25	-	1840
		3.51	B	18	-	439
		3.54	B	25	0	1121
		3.35	D	25	5	1802
		3.56	D	25	-	313
		3.56	D	20	-	969
		3.38	CCl_4	25	5	1073
		3.43	CCl_4	25	5	1802
		3.43	CCl_4	20	-	969
		3.43	CCl_4	25	-	313
		3.34	Hx	25	5	1802
		3.35	Hx	25	-	313
		3.4	Hx	25	-	2018
		3.48	Hp	25	0	1121
		3.21	Tol	-60	0	1121
		3.29	Tol	-30	0	1121
		3.35	Tol	0	0	1121
		3.43	Tol	30	0	1121
		3.48	Tol	60	0	1121
		3.08	e	25	5	1802
		3.08	e	25	-	313
		3.36	f	25	-	1127
		3.38	g	25	-	1127
		3.21	CS_2	25	5	1802
		3.21	CS_2	20	-	969
		3.21	CS_2	25	-	313
		3.42	h	25	5	1802
		3.51	i	25	5	1802
	Methyl isocyanide	3.83±0.06	gas	-	M	576
		3.78±0.06[j]	gas	-	M	576

e) Tetralin. f) Gasoline. g) Kerosene, petroleum ether. h) CCl_4 plus *p*-dichlorobenzene.
i) Dioxane plus *p*-dichlorobenzene. j) Bending mode excited.

FORMULA	COMPOUND NAME	μ, D	State, or Solv.	t, °C	Method or P_a, % of P_e	Reference
$C_2H_3NO_2$	Nitroethylene	3.44	B	25	-	2031-2
C_2H_3NS	Methyl isothiocyanate	3.59	B	20	0	791
	Methyl thiocyanate	3.34	B	20	0	791
$C_2H_3N_3$	1,2,3-Triazole	1.79	B	25	15	859
	1,2,4-Triazole	3.2	D	50	-	769
		3.20	D	25	15	859
		3.27	D	25	-	770
$C_2H_3N_3O$	1,2,4-Triazole-5-one	3.33	D	25	15	859
$C_2H_3N_3O_2$	Methyl azidoformate	1.73	B	25	-	1762
$C_2H_3N_3O_6$	1,1,1-Trinitroethane	3.20	B	25	0?	1753
		3.35	B	n.s.	5	80
C_2H_4	Ethylene	0	gas	-35 to 190	-	1839
		0	gas	25	-	2071
C_2H_4BrCl	1-Chloro-2-bromoethane	1.10	gas	65	-	2165
		1.15	gas	95	-	2165
		1.21	gas	130	-	2165
		1.29	gas	160	-	2165
		1.51	B	25	10	1335
		1.53	B	25	0	720
		1.54	B	50	10	1335
		0.86	Hx	-50	10	721
		0.96	Hx	-25	10	721
		1.05	Hx	0	10	721
		1.15	Hx	25	10	721
		1.18	Hx	25	0	720
		1.22	Hx	50	10	721

FORMULA	COMPOUND NAME	μ, D	State, or Solv.	t, °C	Method or P_a, % of P_e	Reference
C_2H_4BrCl (contd.)	1-Chloro-2-bromoethane (contd.)	0.93	Hp	-50	0	1809-10
		0.99	Hp	-30	0	1809-10
		1.06	Hp	-10	0	1809-10
		1.12	Hp	10	0	1809-10
		1.20	Hp	30-70	0	1809-10
$C_2H_4BrNO_2$	1-Bromo-1-nitroethane	2.97	B	25	0?	1753
$C_2H_4Br_2$	1,1-Dibromoethane	2.14	B	25	-	1238
	1,2-Dibromoethane	0.95	gas	65	-	2165
		0.98	gas	75	-	615
		1.00	gas	85	-	615
		1.00	gas	95	-	2165
		1.01	gas	35	0	184
		1.04	gas	130	-	2165
		1.05	gas	175	-	615
		1.05	gas	65	0	184
		1.09	gas	95	0	184
		1.11	gas	160	-	2165
		1.16	gas	180	-	615
		1.14	liq	25	-	707
		1.15	liq	20	0	907
		1.20	liq	45	0	907
		1.25	liq	70	0	907
		1.14	B	10	15	1586
		1.16	B	20	15	1586
		1.16	B	20	-	977
		1.19	B	20	10	1335
		1.19	B	30	15	1586
		1.21	B	50	10	1335
		1.21	B	40	15	1586
		1.23	B	20-70	0	907
		1.28	B	20	-	1410
		1.4	B	25	0	2113
		1.47	B	10	0	1813

FORMULA	COMPOUND NAME	μ, D	State, or Solv.	t, °C	Method or P_a, % of P_e	Reference
		1.53	B	30,50	0	1813
		1.56	B	70	0	1813
		1.32±0.05	D	20-70	0	907
		0.87	CCl_4	20	10	1335
		0.87	CCl_4	10	15	1586
		0.90	CCl_4	20	15	1586
		0.93	CCl_4	30	15	1586
		0.96	CCl_4	40	15	1586
		0.98±0.06	CCl_4	20-60	0	907
		0.58	Hx	-50	10	721
		0.68	Hx	-25	10	721
		0.80	Hx	0	10	721
		0.91	Hx	25	-	1334
		0.92	Hx	25	10	721
		0.93	Hx	50	10	721
		0.94±0.03	Hx	20-60	0	907
		0.93_9	cHx	20	-	1410
		0.80	Hp	-30	0	1813
		0.95	Hp	-10	0	1813
		0.99	Hp	10	0	1813
		1.05	Hp	30	0	1813
		1.06	Hp	50,70	0	1813
		0.85	cHx	10	15	1586
		0.89	cHx	20	15	1586
		0.92	cHx	30	15	1586
		0.94	cHx	40	15	1586
		0.94±0.04	cHx	20-70	0	907
		0.96	Tol	10	15	1586
		0.99	Tol	20	15	1586
		1.02	Tol	30	15	1586
		1.04	Tol	40	15	1586
		0.88	CS_2	20	10	1335
$C_2H_4Br_2O$	*Bis*(bromomethyl) ether	0.94	gas	-12	5	304
		0.97	gas	0	5	304

FORMULA	COMPOUND NAME	μ, D	State, or Solv.	t, °C	Method or P_a, % of P_e	Reference
$C_2H_4Br_2O$ (contd.)	*Bis*(bromomethyl) ether (contd.)	0.99	gas	24	5	304
		1.06	gas	50	5	304
C_2H_4ClF	1-Chloro-2-fluoroethane	1.84-1.97	gas	35-235	0	406
C_2H_4ClNO	2-Chloroacetamide	2.2-2.8	B	30	-	914
		2.97±0.04	B	25	-	313
		2.94±0.01	D	35	0	1846
$C_2H_4ClNO_2$	1-Chloro-1-nitroethane	3.33	gas	140-195	-	795
$C_2H_4Cl_2$	1,1-Dichloroethane	2.070	gas	40-140	-	574
		2.63	gas	45-150	-	1275
		1.8	B, CCl_4	n.s.	-	2014
		1.97	B	25	-	1238
		2.00	B	25	0	620
	1,2-Dichloroethane	1.13-1.55	gas	30-270	-	2164
		1.2-1.4	gas	60-180	-	1691
		1.25-1.46	gas	35-140	0	2066
		1.28-1.58	gas	25-315	-	615
		1.44	gas	35	0	184
		1.48	gas	60	0	184
		1.52	gas	100	0	184
		1.578	gas	30-120	-	574
		1.84-1.91	gas	30-65	-	832
		2.94	liq	25	-	707
		1.68	B	20	-	977
		1.70	B	50	0	720
		1.71	B	40	0	720
		1.73	B	20	0	720
		1.76_0	B	20	-	1410
		1.77	B	25	0	2113
		1.78	B	10-70	0	1866
		1.85	B	6-50	-	1304
		1.89	B	25	0	620

FORMULA	COMPOUND NAME	μ, D	State, or Solv.	t, °C	Method or P_a, % of P_e	Reference
		1.9	B	n.s.	-	2014
		1.86	D	25	0	184
		1.3	CCl_4	n.s.	-	2014
		1.34-1.46	CCl_4	-20 to 60	0	1866
		1.53-1.59	CCl_4	-18 to 46	-	1418
		1.14	Hx	-50	0	721
		1.22	Hx	-25	0	721
		1.31	Hx	0	0	721
		1.37	Hx	25	0	721
		1.37	Hx	25	-	1334
		1.43	Hx	50	0	721
		1.37_2	cHx	20	-	1410
		1.08	Hp	-70	0	1810
		1.17	Hp	-50	0	1810
		1.25	Hp	-30	0	1810
		1.32	Hp	-10	0	1810
		1.32-1.52	Hp	-10 to 90	0	1866
		1.37	Hp	10	0	1810
		1.42	Hp	30	0	1810
		1.43	Hp	50	0	1810
		1.69-1.74	Tol	-50 to 30	10	1335
		1.23	k	-50	0	720
		1.30	k	25	0	720
		1.34	k	0	0	720
		1.39	k	25	0	720
		1.23-1.41	CS_2	-70 to 30	0	1866
		1.26-1.43	$CHCl_3$	-50 to 50	0	1866
		1.24-1.52	Ether	-60 to 20	-	1331
		1.26-1.52	Ether	-60 to 20	10	1335
		1.27-1.45	Ether	-20 to 20	0	1866
		1.5	Ether	20	0	717,723
$C_2H_4Cl_2N_6$	1,1-Azo*bis*(chloroformamidine)	1.95	D	30	0	999
$C_2H_4Cl_2O$	*Bis*(chloromethyl) ether	0.98-1.09	gas	65-140	5	304
		0.70	CCl_4	30	-	1358

k) Pentane.

FORMULA	COMPOUND NAME	μ, D	State, or Solv.	t, °C	Method or P_a, % of P_e	Reference
$C_2H_4Cl_4Ge$	Trichloro(2-chloroethyl)germane	2.41	B	25	-	886
$C_2H_4Cl_4Si$	Trichloro(chloroethyl)silane[kk]	1.51	B	25	15	887
	Trichloro(1-chloroethyl)silane	2.30	B	25	15	887
C_2H_4FNO	Fluoroacetamide	2.62±0.04	B	25	-	313
$C_2H_4F_2$	1,1,-Difluoroethane	2.24 2.30±0.03	gas gas	100-230 -	25 M	577 1032
$C_2H_4I_2$	1,1-Diiodoethane	2.32	B	25	-	1238
	1,2-Diiodoethane	1.3 0.44 0.55	B Hx Hx	25 25 50	0 0 0	2113 721 721
$C_2H_4N_2O_2$	Glyoxime	1.23	D	20	0	1319
	Oxamide	9	D	30	-	108
$C_2H_4N_2O_4$	1,1-Dinitroethane	3.50	B	25	0?	1753
$C_2H_4N_2O_6$	1,2-Dinitroxyethane	3.58±0.03	B	25	0	1053
$C_2H_4N_4$	Dicyandiamide	8.22	D	35	0	1713
	1-Methyl-1,2,3,4-tetrazole	5.42	B	25	15	859
$C_2H_4N_4S$	5-Methylamino-1,2,3,4-thiatriazole	5.72	n.s.	n.s.	-	1147
$C_2H_4N_8O_2$	1,1'-Azo*bis*(nitroformamidine)	3.63	D	30	0	999
C_2H_4O	Acetaldehyde	2.68[m] 2.69 2.71	gas gas gas	- - 25-180	M M 10?	975 911 2167

kk) Position of Cl not stated. m) $\mu_a = 2.55$; $\mu_b = 0.87$.

FORMULA	COMPOUND NAME	μ, D	State, or Solv.	t, °C	Method or P_a, % of P_e	Reference
		2.74	gas	150-195	-	796
		2.51	B	20	0	1472,329
	Ethylene oxide	1.88±0.01	gas	-	M	362
		1.89	gas	15-175	-	1896
		1.89	gas	25	-	42
		1.91	gas	15-175	17	566
		1.93	gas	23-95	-	1564
		1.92	liq	-35 to 20	-	1417
		1.83	B	25	-	42
		1.90	n.s.	n.s.	-	709
$C_2H_4O_2$	Acetic acid	1.4-1.7	gas	140-200	-	2162
		1.75	gas	25-220	-	2163
		1.75	gas	20	-	1869
		1.92	liq	20	-	1495
		2.17	liq	60	-	1495
		0.74±0.02	B	25	15	2137
		0.84n (dimer)	B	20	-	882
		1.05	B	22	-	208
		1.64	B	30	-	1526
		1.76	D	25	0	2126
		1.76	D	20	-	1869
		1.77	D	n.s.	-	1530
		0.38	CS_2	10,30	-	1505
		0.73	Ether	-60	0	718,723
		0.86	Ether	-40	0	718,723
		0.94	Ether	-20	0	718,723
		1.02	Ether	0	0	718,723
		1.15	Ether	15	0	718,723
		1.18	Ether	20	0	718,723
		1.24	Ether	25	0	718,723
		1.40	Ether	0-30	-	1827
	Methyl formate	1.77±0.03	gas	-	M	364

n) 10 cm wave length.

FORMULA	COMPOUND NAME	μ, D	State, or Solv.	t, °C	Method or P_a, % of P_e	Reference
$C_2H_4O_3S$	Ethylene sulfite	3.68	B	20	5	55
C_2H_4S	Ethylene sulfide	1.84±0.02	gas	-	M	362
		1.66±0.03	liq	20	-	629
$C_2H_5AsCl_2$	Ethyldichloroarsine	2.53	B	20	0	1348
C_2H_5Br	Bromoethane	1.80	gas	35-125	-	1240
		1.80	gas	25-145	-	1235
		2.01	gas	30-150	-	1236
		2.03±0.01	gas	20-170	0	623
		2.04±0.01	gas	-35 to 260	-	1818
		1.90	liq	25	-	707
		2.04	liq	20	-	657
		1.79	B	25	11	1460
		1.85	B	20	-	977
		1.90	B	10-40	-	347
		1.90	p	20	5	1513
		1.94	B	6-28	-	1418
		1.90	CCl_4	0-40	-	347
		2.01	CCl_4	20	5-25	77
		1.88	Hx	-90 to 60	60	1825
		1.92	Hx	-23 to 40	-	347
		1.92	cHx	10-40	-	347
		1.96	Hp	0-20	-	1418
		1.75-1.78	Tol	0-20	-	1418
		1.83-1.86	Tol	-23 to 40	-	347
		1.71	CS_2	-23 to 40	-	347
C_2H_5BrO	Bromomethyl methyl ether	2.05±0.05	gas	25-135	5	304
		2.17±0.01	Hp	-5 to 60	5	304
	Ethylene bromohydrin	2.18	B	30	-	1573

p) B plus water.

FORMULA	COMPOUND NAME	μ, D	State, or Solv.	t, °C	Method or P_a, % of P_e	Reference
C_2H_5BrHg	Ethylmercury bromide	2.79±0.06	B	40	0	873
		2.96±0.08	B	30.4	0	1245
		2.79	D	n.s.	-	872
$C_2H_5Br_3Si$	Ethyltribromosilane	1.99±0.04	B	25	5	1592
C_2H_5Cl	Chloroethane	1.745±0.021[q]	gas	-	M	94
		1.76	gas	20	Kerr	1556
	2.04	1.79±0.05[q]	gas	-	M	93
		1.98±0.03	gas	20-85	-	837
		2.00	gas	25-200	-	1778
		2.01	gas	30-150	-	1235
		2.02±0.01	gas	25-145	-	1694
		2.02	gas	30-150	-	1236
		2.04_0±0.025	gas	25-145	-	1689,1695
		2.07	gas	20-180	-	554
		2.09	gas	23-95	-	1564
		1.96	liq	-35 to 20	-	1417
		2.0	B, CCl_4	-	-	2014
C_2H_5ClHg	Ethylmercury chloride	2.99±0.08	B	18.4	0	1245
		3.64	D	(20)	-	1700
C_2H_5ClO	Chloromethyl methyl ether	1.76-1.90	gas	25-195	5	304
		2.03-2.07	gas	70-230	0	406
		1.85	B	10-40	15	1587
		1.78	CCl_4	30	-	1358
		1.90	CCl_4	10-40	15	1587
	Ethylene chlorohydrin	1.7_7	gas	65-160	10	2166
		1.90	B	25-50	0	1835
		1.96	B	30	-	1573
		2.10	Hp	25-50	0	1835
$C_2H_5ClO_2S$	Ethanesulfonyl chloride	3.89	B	25	0	634

q) μ along *a* axis only.

FORMULA	COMPOUND NAME	μ, D	State, or Solv.	t, °C	Method or P_a, % of P_e	Reference
$C_2H_5ClO_2S$ (contd.)	Ethyl chlorosulfite	2.66	B	25-45	0	1974
C_2H_5ClS	Chloromethyl methyl sulfide	1.67	CCl_4	25	-	691
$C_2H_5Cl_2N$	*N*,*N*-Dichloroethylamine	2.01±0.01	B	20	0	6
$C_2H_5Cl_3Ge$	Ethylgermanium trichloride	2.28±0.09	B	25	5	1453
		2.87	B	25	-	886
$C_2H_5Cl_3Si$	Dichloro(chloromethyl)methylsilane	1.82	B	25	15	887
	Ethyltrichlorosilane	2.06	B	25	0	373
C_2H_5F	Fluoroethane	1.94±0.01	gas	-35 to 260	-	1818
		1.96	gas	-	M	961
C_2H_5HgI	Ethylmercury iodide	3.04±0.8[r]	B	18.4	0	1245
C_2H_5I	Iodoethane	1.63	gas	35-125	-	1240
		1.63	gas	30-130	-	1235
		1.77±0.1	gas	-	M	889
		1.89±0.01	gas	20-65	0	623
		1.92±0.01	gas	75-190	-	1818
		1.95	gas	30-130	-	1236
		1.69	liq	20	-	657
		1.67	B	-100 to 70	50	1830
		1.7	B	25	0	2113
		1.71	B	20	-	977
		1.80	B	20	0	348
		1.82	B	6-28	-	1418
		1.91	CCl_4	20	5-25	77-8
		1.74	Hp	0-20	-	1418

r) ±0.8 is the value given, but it seems high compared to other estimates of precision in the same work.

FORMULA	COMPOUND NAME	μ, D	State, or Solv.	t, °C	Method or P_a, % of P_e	Reference
		1.69	Tol	0-24	-	1418
		1.67	n.s.	n.s.	-	1806
C_2H_5Li	Ethyllithium	0.70±0.01	B	25	5-15	694
		0.88	B	25	0	1628, 1637,2034
		1.1	Hx	25	0	2034
		1.15-1.20	Hx	n.s.	-	1628
	hexamer	1.51	B	25	-	1637
		1.73	B	25	5-15	694
C_2H_5N	Ethylenimine	1.89±0.01	gas	-	M	869
C_2H_5NO	Acetaldoxime	0.8_3	"pure"	20	0	249
		0.9_0	B	25	0	249
		0.75	D	35	0	1557
	Acetamide	3.44	B	30	0	108
		3.60	B	30	0	744
		3.90±0.04	B	25	-	313
		3.6	D	20	-	403
		3.90	D	30	0	108
		3.92	D	30	10	998
$C_2H_5NO_2$	Ethyl nitrite	2.40	gas	25	0	376
		2.41	gas	-20 to 60	5	305
		2.22	B	20	0	338
		2.3	B, CCl_4	25	-	2015
		2.29	CCl_4	25	0	2017-8
	Glycine	20.8	H_2O	1-30	-	1153
	Nitroethane	3.61±0.01	gas	90-190	0	623
		3.70±0.04	gas	140-210	-	795
		3.22	B	20	-	793

FORMULA	COMPOUND NAME	μ, D	State, or Solv.	t, °C	Method or P_a, % of P_c	Reference
$C_2H_5NO_3$	Ethyl nitrate	2.93 2.96±0.03	B B	20 25	0 0	338 1053
C_2H_5NS	Thioacetamide	4.77±0.01	D	35	0	1846
$C_2H_5N_3$	Azidoethane	2.12 2.14	B B	25 25	5 -	1854 1762
$C_2H_5N_3O$	Azidoethanol	2.48 2.49	B B	25 25	5 -	1854 1762
$C_2H_5N_3O_2$	Biuret	3.27±0.02	D	30	0	1008
C_2H_6	Ethane	0	gas	-75 to 200	-	1839
		0	gas	25	-	2071
C_2H_6AsBr	Dimethylarsine bromide	2.71±0.04	B	25	-	313
C_2H_6AsCl	Dimethylarsine chloride	2.68±0.04	B	25	-	313
C_2H_6AsF	Dimethylarsine fluoride	2.37±0.04	B	25	-	313
C_2H_6AsI	Dimethylarsine iodide	2.54±0.04	B	25	-	313
C_2H_6BF	Fluorodimethyl borine	1.32±0.04	gas	29-95	-	2001
$C_2H_6BF_3O$	Boron trifluoride - methyl ether complex	4.38±0.12	B	25	-	1044
$C_2H_6Br_2Si$	Dimethyldibromosilane	2.45±0.04	B	25	5	1592
C_2H_6ClN	*N*-Chlorodimethylamine	1.93±0.01	B	20	0	6
$C_2H_6Cl_2Ge$	Dimethyldichlorogermane	3.11	B	25	-	886

FORMULA	COMPOUND NAME	μ, D	State, or Solv.	t, °C	Method or P_a, % of P_e	Reference
$C_2H_6Cl_2Si$	Dimethyldichlorosilane	1.89	B	25	15	887
		2.28±0.01	B	20	-	1702
	Ethyldichlorosilane	2.06±0.04	B	25	5	1853
$C_2H_6Cl_2Sn$	Dimethyltin dichloride	4.22±0.04	B	25	-	313
$C_2H_6Cl_3NSi$	Dimethylaminotrichlorosilane	2.48±0.08	B	20	-	1702
$C_2H_6Cl_8OSn_2$	Ethanol - *bis*(tin tetrachloride) complex	1.67	B	n.s.	-	592
C_2H_6Hg	Dimethylmercury	0.69	B	20	-	1700
		0.74	B	40	-	1700
		0.71	D	20	-	1700
$C_2H_6I_2Te$	Dimethyltellurium diiodide	2.28	B	25	0	2155
$C_2H_6N_2$	Diazomethane	1.4_5	gas	-	M	350
		0	Hp	25	0	2099
$C_2H_6N_2O$	Acethydrazide	3.18	D	35	0	1557
	Methyl urea	4.34±0.04	D	25	0	120
	Nitrosodimethylamine	4.01	B	20	0	339
$C_2H_6N_2O_2$	Dimethylnitramine	4.61	D	20	0	569
$C_2H_6N_2S$	Methylthiourea	4.2±0.1	D	18	0	166
C_2H_6O	Ethanol	1.1_2	gas	90-180	-	1687
		1.68	gas	35-110	-	983
		1.69_6	gas	25-180	-	926
		1.70_8	gas	80-225	-	1311
		1.7	gas	-	M	1962
		0.53	liq	18	REF	1115

FORMULA	COMPOUND NAME	μ, D	State, or Solv.	t, °C	Method or P_a, % of P_e	Reference
C_2H_6O (contd.)	Ethanol (contd.)	1.66	liq	20	VS	1701
		2.89	liq	20	10	1295
		3.43	liq	20	-	657
		1.67	B	30	0	1267
		1.68	B	20	-	977
	1.73 P_e = 12.59, P_a = 1.26	1.710±0.006	B	25	0	747
		1.71_8±0.02	B	22	15	2140
		1.75	B	25	5	1108
		1.76	B	10-70	-	1875
		1.80	B^s	20	0	1951
		1.64	CCl_4	25	0	966
		1.684±0.005	CCl_4	25	0	747
		1.73	CCl_4	20	10	1295
		1.7_9	CCl_4	25	5	1108
		1.79	CCl_4	25	-	1078
		1.52	CS_2	25	0	748
		1.5	Ether	-60	-	723
		1.70	Ether	0	-	723
		1.8	Ether	20	-	717,723
		1.88	t	25	0	2017
		1.66	u	55	-	1865
	Methyl ether	1.29	gas	23-95	-	1564
		1.29_5±0.01	gas	-55 to 105	-	1695
		1.30±0.01	gas	15-155	1	624
		1.30±0.01	gas	20-180	-	1896
		1.316±0.012	gas	n.s.	-	1689
		1.33±0.02	gas	25-145	-	1693
		1.25	B	25	-	99
C_2H_6OS	Dimethylsulfoxide	3.9±0.01	B	25	0	334a
$C_2H_6O_2$	Ethylene glycol	2.2_7	gas	145-225	10	2166
		2.20	liq	15	-	2060

s) Saturated with water. t) 2,5-Dimethylpyrazine. u) *p*-Dichlorobenzene.

FORMULA	COMPOUND NAME	μ, D	State, or Solv.	t, °C	Method or P_a, % of P_e	Reference
		3.59-3.42	liq	20-70	0	907
		4.02	liq	20	-	657
		4.87 v	liq	20	-	933
		1.5	B	25	0	2113
		2.30	B	30	0	590
		2.28-2.36	D	20-75	0	907
		2.30	D	25	0	1833
		2.32	D	50	0	1833
		2.40	D	25	10	1999
		1.2_5	Ether	-60	0	723
		1.7_0	Ether	-40	0	723
		1.9_2	Ether	-20	0	723
		2.4_0	Ether	20	0	723
$C_2H_6O_2S$	Dimethyl sulfone	4.47±0.01	gas	150-250	15	333
		4.25±0.05	B	20	0	1201
$C_2H_6O_2S_2$	Dimethyl thiosulfite	1.91±0.02	B	20	10	1705
$C_2H_6O_3S$	*sym*-Dimethyl sulfite	2.93	B	20	-	793
C_2H_6S	Dimethyl sulfide	1.50±0.01	gas	-	M	1508
		1.41	B	20	0	791
		1.45±0.02	B	20	0	1193
	Ethanethiol	1.57	gas	35-105	15	985
		1.39±0.02	liq	15	-	2060
		1.57	liq	25	2.7	1280
		1.40	B	20	0	790
		1.48	B	25	2.7	1280
$C_2H_6S_2$	Dimethyldisulfide	1.97±0.02	B	30	0	1026
$C_2H_6S_3$	Dimethyltrisulfide	1.67±0.02	B	30	0	1026
C_2H_6Se	Dimethyl selenide	1.32±0.02	B	20	0	309,310

v) $g^{1/2}\mu$ from Kirkwood's equation.

FORMULA	COMPOUND NAME	μ, D	State, or Solv.	t, °C	Method or P_a, % of P_e	Reference
C_2H_6Si	Vinylsilane	0.66±0.012	gas	-	M	1434
C_2H_6Zn	Dimethylzinc	1.00±0.02	B	25	5-15	694
$C_2H_7AlCl_3N$	Aluminum chloride - ethylamine complex	6.87-6.99	B	(20)	15	1415, 2006-7
C_2H_7N	Dimethylamine	0.91	gas	20-50	-	572
		0.97	gas	25-145	15?	1695,1869
		1.03	gas	15-155	5	624
		1.03	gas	25	-	1099
		1.14	liq	0-25	-	1099
		1.18	B	25	-	1099
	Ethylamine	1.00	gas	20-70	-	572
		1.22_5	gas	25	0	100
		1.09	liq	25	0	100
		1.28	B	15	0	100
		1.38	B	25	-	1979
		1.40	B	25	0	1977
C_2H_7NO	2-Aminoethanol	2.59±0.03	D	25	-	1485
$C_2H_7NO_2S$	Ethanesulfonamide	4.06±0.01	B	25	0	635
		4.65	D	25	0	635
$C_2H_7O_3P$	Dimethylphosphite	2.94	CCl_4	20	-	50
C_2H_7P	Ethylphosphine	1.17±0.02	gas	25-60	-	1040
C_2H_8BN	Aminodimethylborine	1.47±0.06	B	20	0	115
$C_2H_8Cl_4O_2Sn$	Tin tetrachloride - di(methanol) complex	6.18	B	20	0	1439
$C_2H_8N_2$	1,2-Diaminoethane	1.9_8	gas	80-155	10	2166
		1.92	B	25-75	0	1977,1979

FORMULA	COMPOUND NAME	μ, D	State, or Solv.	t, °C	Method or P_a, % of P_e	Reference
	N,*N*′-Dimethylhydrazine	1.36	Hp	25	0	2099
$C_2H_8O_2Si$	Dimethylsilanediol	1.94	D	25	-	307
$C_2H_{11}BN$	Dimethylamine - borane complex	4.87±0.03	B	25	15	1420
C_2I_2	Diiodoacetylene	~0.33	CCl_4	0	0	1907
C_2N_2	Cyanogen	0	gas	-80 to 100	-	2068-9
		0.3	gas	20-140	-	199
C_2N_2S	Sulfur dicyanide	2.77±0.10	B	25	0	1644
$C_2N_2Se_3$	Selenium cyanide	3.98±0.10	B	25	0	1644
$C_3Cl_3F_3$	1,1,2-Trichloro-3,3,3-trifluoropropene	1.28 1.08	gas B	100-125 20	0 8	327 1669
C_3Cl_6	Perchloropropene	0.46 0.45	B CCl_4	20 n.s.	8 -	1669 2014
C_3Cl_8	Perchloropropane	0	CCl_4	n.s.	-	2014
C_3CoNO_4	Cobalt carbonyl nitrosyl	0.43±0.1	B	25	20	715
C_3HCl_7	1,1,1,2,2,3,3,3-Heptachloropropane	1.0	CCl_4	n.s.	-	2014
C_3HF_3	3,3,3-Trifluoropropyne	2.36±0.04	gas	-	M	1759
C_3HF_7	1,1,2,2,3,3,3-Heptafluoropropane	1.62±0.12	gas	25-105	15	1647
C_3HN	Cyanoacetylene	3.6±0.2	gas	-	M	2101
$C_3H_2Cl_2O_2$	Malonyl chloride	2.81	B	20	0	1265

FORMULA	COMPOUND NAME	μ, D	State, or Solv.	t, °C	Method or P_a, % of P_e	Reference
$C_3H_2Cl_2O_3$	1,2-Dichloroethylene carbonate	3.47	B	25	5	897
$C_3H_2D_2$	Cyclopropene-1,2-d_2	0.461	gas	-	M	888
$C_3H_2N_2$	Malononitrile	3.735±0.017	gas	-	M	740
		3.59	liq,B	25,75	-	1978-9
C_3H_2O	Propiolaldehyde	2.46	gas	-	M	763
$C_3H_2O_2$	Propiolic acid	2.10	D	25	0	2128
$C_3H_2O_3$	Vinylene carbonate	4.51±0.05	gas	-	M	1783
C_3H_3Br	Bromopropadiene	1.50±0.10	B	25	0	1646
	3-Bromopropyne	1.52	gas	35-100	5	306
		1.49±0.05	B	25	0	1645
C_3H_3Cl	3-Chloropropyne	1.65	gas	20-50	5	306
		1.68±0.004	gas	-	M	741-2
$C_3H_3ClO_3$	Chloroethylene carbonate	4.02	B	25	5	897
C_3H_3D	Cyclopropene-1-d_1	0.460	gas	-	M	888
	Cyclopropene-3-d_1	0.450	gas	-	M	888
$C_3H_3F_3$	3,3,3-Trifluoropropene	2.45	gas	55-100	0	327
C_3H_3I	3-Iodopropyne	1.21±0.05	B	25	0	1645
C_3H_3N	Acrylonitrile	3.89	gas	-	M	2106
		3.91	gas	115-235	-	796
		3.54	B	25	0	1632
		3.3	CCl_4	25	0	880

FORMULA	COMPOUND NAME	μ, D	State, or Solv.	t, °C	Method or P_a, % of P_e	Reference
C_3H_3NO	Acetyl cyanide	3.45±0.06	gas	-	M	975
	Isoxazole	2.76±0.02	B	25	0	1858,1860
		2.84	B	25	-	859
		3.01±0.03	D	25	0	1860
$C_3H_3NOS_2$	4-Oxo-2-thionthiazolidene	2.22	D	25	-	859
$C_3H_3NO_2S$	2,4-Dioxothiazolidene	2.05	D	25	-	859
C_3H_3NS	Thiazole	1.65	B	25	-	859
C_3H_4	Cyclopropene	0.455±0.01	gas	-	M	888
	Propadiene	0.20	gas	25-95	-	2069
	Propyne	0.72	gas	25-95	-	2069
		0.75±0.01	gas	-	M	575-6
		0.77	gas	25-75	0	970
$C_3H_4Br_2$	*cis*-1,2-Dibromopropene	1.97	B	30	0	1790
	trans-1,2-Dibromopropene	1.16	B	30	0	1790
$C_3H_4Cl_2$	1,1-Dichlorocyclopropane	2.06	B	25	-	1650
	dl-1,2-Dichlorocyclopropane	1.19	B	25	0	1861
	1,1-Dichloropropene	1.69	B	25	0	1631
		1.74	B	30	0	394
		1.79	B	30	0	916
	cis-1,2-Dichloropropene	2.20	B	30	-	1789
	trans-1,2-Dichloropropene	0.84	B	30	-	1789

FORMULA	COMPOUND NAME	μ, D	State, or Solv.	t, °C	Method or P_a, % of P_c	Reference
$C_3H_4Cl_2$ (contd.)	*cis*-1,3-Dichloropropene	1.92	B	30	-	1789
	trans-1,3-Dichloropropene	1.73	B	30	-	1789
	1,3-Dichloropropene (mp 104°C)	1.79	gas	125-200	0	1436
	1,3-Dichloropropene (mp 112°C)	1.81	gas	125-200	0	1436
	2,3-Dichloropropene	1.75±0.02	gas	125-200	0	1436
		1.99	B	30	0	916
	3,3-Dichloropropene	1.92	B	30	0	916
$C_3H_4Cl_2O$	1,1-Dichloroacetone	1.91	gas	65	5	303
		1.94	gas	98	5	303
		1.99	gas	140	5	303
		2.05	gas	194	5	303
$C_3H_4Cl_2O_3S$	*cis*-1,2-Dichloro-2-hydroxypropanesulfonic acid sultone	5.07	B	25	-	483
	trans-1,2-Dichloro-2-hydroxypropanesulfonic acid sultone	4.46	B	25	-	483
$C_3H_4Cl_4$	1,1,1,3-Tetrachloropropane	1.22	B	25	0	1944
	1,1,2,2-Tetrachloropropane	1.53	gas	100-195	5	1329
		2.0±0.1	gas	n.s.	0	1327
	1,1,3,3-Tetrachloropropane	0.75	B	25	0	1945
$C_3H_4N_2$	Imidazole	3.87	B	25	-	859
		6.2	B	25	-	770
		6.22	B	50-70	-	769
		4.02	D	25	0	1451
		4.8	D	50	-	769

FORMULA	COMPOUND NAME	μ, D	State, or Solv.	t, °C	Method or P_a, % of P_e	Reference
	Pyrazole	4.9	D	25	-	770
		5.7	w	97	-	769
		1.46-1.50	liq	70-100	-	769
		1.47	B	20	-	769
		1.47	B	20	-	770
		1.58	B	25	-	859
		1.65	B	50	-	769
		1.78	B	70	-	769
		2.2	D	25	-	770
		2.21	D	20	-	769
		2.22-2.52	D	20-60	-	772
$C_3H_4N_2O$	2-Cyanoacetamide	4.10±0.01	D	35	0	1846
$C_3H_4N_2S$	2-Aminothiazole	1.77	B	25	-	859
C_3H_4O	*cis*-Acrolein	~2.6	gas	-	M	2056
	s-*trans*-Acrolein	3.11±0.04	gas	-	M	2056
	Acrolein	3.04	gas	105-205	0	662
		2.90±0.03	B	25	0	127
		2.91	B	20	0	329
	Ethynyl methyl ether	1.81±0.03	B	20	0	422
	2-Propyn-1-ol	1.78±0.05	B	25	0	1645
$C_3H_4O_2$	β-Propiolactone	4.18±0.04	gas	-	M	1029
		3.85	B	30	-	1318
		3.8±0.1	n.s.	n.s.	-	616
$C_3H_4O_3$	Ethylene carbonate	4.51±0.05	gas	-	M	1783
		4.60	B	25	0	1165
		4.84	B	20	5	54

w) Naphthalene.

FORMULA	COMPOUND NAME	μ, D	State, or Solv.	t, °C	Method or P_a, % of P_e	Reference
$C_3H_4O_3$ (contd.)	Ethylene carbonate (contd.)	4.87	B	25	-	1726
		4.91	B	40	5	897
$C_3H_4O_4$	Malonic acid	2.56±0.02	D	25	0	119
		2.57±0.10	D	25	0	1635
		2.66	D	35	5	1399
$C_3H_5BCl_3N$	Boron trichloride - propionitrile complex	7.69	B	(20)	-	2007
		7.80	B	(20)	-	2006
		7.83	B	20	15	1415
C_3H_5Br	Bromocyclopropane	1.69	B	25	0	1622
	cis-1-Bromopropene	1.57	B	25	0	1631
	trans-1-Bromopropene	1.69±0.01	B	25	0	1646
	2-Bromopropene	1.51	B	25	0	1631
	3-Bromopropene	1.81	B	20	0	1479
		1.82±0.10	B	25	0	1646
C_3H_5BrO	Bromoacetone	2.40	Hx	20	17	1345
$C_3H_5BrO_2$	Methyl bromoacetate	2.28	B	25	0	1176
$C_3H_5Br_3$	1,2,3-Tribromopropane	1.59±0.01	B	25-50	0	1835
		1.51±0.02	Hp	25-50	0	1835
C_3H_5Cl	Chlorocyclopropane	1.78	B	25	-	1850,1861
	cis-1-Chloropropene	1.71	gas	70-200	0	661
		1.65	B	25	0	1631
	trans-1-Chloropropene	1.97	gas	70-200	0	661

FORMULA	COMPOUND NAME	μ, D	State, or Solv.	t, °C	Method or P_a, % of P_e	Reference
	1-Chloropropene[x]	1.73	B	30	0	916
	2-Chloropropene	1.69	gas	65-195	0	661
		1.53	B	25	0	1631
	3-Chloropropene	1.90	gas	35-105	-	989
		1.98	gas	100-205	0	661
		2.03	gas	30-145	-	1235
		2.04	gas	30-145	-	1236
		1.79	B	20	0	1491
		2.01	D	40	0	1969
		2.01	n.s.	n.s.	-	1725
C_3H_5ClO	Chloroacetone	2.2_2	gas	60-180	10?	2167
		2.38	CCl_4	25	-	1392
		2.37	Hx	20	0	1345
	α-Epichlorohydrin	1.8	CCl_4	n.s.	-	2014
	Propionyl chloride	2.50	B	25	0	927
		2.63	B	20	0	1263
$C_3H_5ClO_2$	Ethyl chloroformate	1.44	gas	207	10	1332
		1.81	gas	127	10	1332
		2.49	gas	77	10	1332
		2.56	gas	35	10	1332
		2.22	B,CCl_4	25-45	-	831
$C_3H_5Cl_3$	1,2,2-Trichloropropane	1.61	gas	80-195	5	1329
$C_3H_5Cl_4NTi$	Titanium tetrachloride - propionitrile complex	6.10	B	16	0	2003
C_3H_5F	*trans*-Fluoropropene	1.85±0.02[y]	gas	-	M	1775
	2-Fluoropropene	1.60±0.02	gas	-	M	1510

w) Configuration not given. y) μ_a; μ_b estimated <0.02.

FORMULA	COMPOUND NAME	μ, D	State, or Solv.	t, °C	Method or P_a, % of P_r	Reference
C_3H_5I	3-Iodopropene	1.62±0.10	B	25	0	1646
C_3H_5N	Ethyl isocyanide	3.50	B	25	0	1416
	Propionitrile	4.02±0.04	gas	-	M	1049
		4.03	gas	120-205	-	796
		4.06±0.01	gas	80-195	0	623
		3.37	B	20	0	2096
		3.4	B	25	0	2113
		3.59	B	25,75	-	1978-9
		3.60	B	20	0	342
		3.69	B	18	-	439
		3.59-3.61	CCl_4	0-40	0	346
		3.65-3.69	Hx	-23 to 20	0	346
		3.68	cHx	20	0	346
		3.38-3.51	Tol	-79 to 20	0	346
		3.18-3.33	CS_2	-79 to 20	0	346
C_3H_5NO	Ethyl isocyanate	2.84	B	20	0	344,1475
$C_3H_5NO_2$	Oxazolidinone	5.07±0.01	D	30	0	1061
C_3H_5NS	Ethyl isothiocyanate	3.67	B	20	0	791
	Ethyl thiocyanate	3.33	B	20	0	791
$C_3H_5N_3$	3-Azidopropene	1.92	B	25	0	1762
$C_3H_5N_3O$	Azido-2-propanone	3.64	B	25	5	1854
$C_3H_5N_3O_9$	Nitroglycerin	3.38±0.03	B	25	0	1053
$C_3H_5N_5O_4$	1-Nitro-2-nitroiminoimidazolidine	7.65	D	30	0	1000
C_3H_6	Cyclopropane	0	gas	23-95	-	1564

FORMULA	COMPOUND NAME	μ, D	State, or Solv.	t, °C	Method or P_a, % of P_e	Reference
	Propene	0.34	gas	-80 to 25	-	2071
		0.35	gas	-25 to 200	-	1284
		0.364±0.003	gas	-	M	1143
$C_3H_6Br_2$	1,2-Dibromopropane	1.13-1.37	gas	80-195	5	1329
		3.10±0.12	gas	n.s.	-	1327
	1,3-Dibromopropane	2.05	liq	20-95	0	907
		2.00	B	25-50	0	1835
		2.00	B	25	0	1943
		2.03±0.03	B	20-70	0	907
		2.054	B	64	0	692
		2.066	B	25	0	692
		2.06	D	20-70	0	907
		1.96	CCl_4	20-60	0	907
		2.01	Hx	20-60	0	907
		2.04	Hp	25-50	0	1837
		2.09-2.21	Hp	-70 to 50	0	1813
$C_3H_6Br_4O_2Zr$	Ethyl formate - zirconium bromide complex	4.54	B	20	0	923
C_3H_6ClNO	*N*-Methylchloroacetamide	2.25	CCl_4	25	0	1337
		2.5-2.6	CCl_4	25	0	1391
$C_3H_6ClNO_2$	1-Chloro-1-nitropropane	3.52	gas	140-220	-	795
$C_3H_6Cl_2$	1,1-Dichloropropane	2.08	B	25	0	620
	1,2-Dichloropropane	1.46-1.68	gas	70-230	0	1436
		1.87	B	25	0	620
	1,3-Dichloropropane	2.09	gas	100-210	2	1821
		2.09	B	20	RT	921
		2.10	B	25	0	1944

FORMULA	COMPOUND NAME	μ, D	State, or Solv.	t, °C	Method or P_a, % of P_e	Reference
$C_3H_6Cl_2$ (contd.)	1,3-Dichloropropane (contd.)	2.26	B	25	0	620
		2.13	Hx	20	RT	921
	2,2-Dichloropropane	2.63	gas	30-100	-	1275
		2.0	B, CCl_4	25	-	2015
		2.16	B	20	RT	921
		2.20	B	25	0	620
		2.26	Hx	20	RT	921
$C_3H_6Cl_4Si$	Trichloro(1-chloropropyl)silane	1.97	B	25	15	887
	Trichloro(2-chloropropyl)silane	2.08	B	25	15	887
$C_3H_6I_4O_2Zr$	Ethyl formate - zirconium iodide complex	4.27	B	20	0	1444
$C_3H_6N_2O_2$	Malonamide	3.5	D	30	0	108
	N,*N*′-Methylene*bis*formamide	1.51	B	20	-	568
	Methyl glyoxime	0.892	D	20	0	1319
$C_3H_6N_2O_6$	1,2-Dinitroxypropane	3.95±0.03	B	25	0	1053
	1,3-Dinitroxypropane	3.52±0.03	B	25	0	1053
$C_3H_6N_2S$	Ethylenethiourea	$5.51_8 \pm 0.002$	D	25	5	275-6
$C_3H_6N_4$	1,5-Dimethyltetrazole	5.30	B	25	0	892
	C-Ethyltetrazole	2.65	B	25	0	892
	N-Ethyltetrazole	5.46	B	25	0	892
$C_3H_6N_4O_2$	1,4-Dimethyl-2-tetrazolin-5-one	1.14	B	25	0	892
	2-Nitroiminoimidazolidine	6.93	D	30	0	1000

FORMULA	COMPOUND NAME	μ, D	State, or Solv.	t, °C	Method or P_a, % of P_e	Reference
$C_3H_6N_4O_5$	1,3-Dinitro-1,3-diaza-5-oxacyclohexane	5.64	D	20	-	569
$C_3H_6N_4S$	5-Dimethylamino-1,2,3,4-thiatriazole	5.84	n.s.	n.s.	-	1147
$C_3H_6N_6O_6$	Hexahydro-1,3,5-trinitro-s-triazine	5.79	D	20	-	569
C_3H_6O	Acetone	2.8	gas	-	M	1962
	2.86	2.87±0.04	gas	20-180	-	1896
		2.87	gas	30-180	10?	2167
		2.88[z]	gas	30-90	-	189
		2.89±0.02	gas	85-315	-	238
		2.90	gas	-	M	1916
		2.97	gas	23-95	-	1564
		2.69	liq	20	VS	1701
		3.0[a]	liq	n.s.	-	863
		3.11	liq	20	-	2055
		2.4	B	30	RT	1375,1377
		2.41	B	28	RT	2051
		2.5	B	30	RT	2052
		2.72[z]	B	28	-	979
		2.74	B	18	0	672
		2.74	B	15	10	2141
		2.75_4±0.02	B	22	15	2140
		2.77±0.015	B	22	15	555
		2.79	B[b]	20	5	1513
		2.81	B	18	0	671,683-4
		2.81[b]	B	20	0	1951
		2.83±0.02	B	25	-	238
		2.85±0.02	B?	n.s.	-	302
		2.72	CCl_4	25	0	966
		2.85	CCl_4	15	10	2141
		2.74	Hx	15	10	2141
		2.84	Tol	16	0	671,683-4
		2.67	$CHCl_3$	18	0	671,683-4
		2.2	Ether	20	-	717,723
		2.39-2.49	c	19	0	671,684

z) Microwave frequencies. a) Onsager's equation. b) Saturated with water. c) Aniline.

FORMULA	COMPOUND NAME	μ, D	State, or Solv.	t, °C	Method or P_a, % of P_e	Reference
C_3H_6O (contd.)	Acetone (contd.)	2.78	cc	n.s.	0	683
		2.8	d	25	-	1125
	Propenol	1.64	gas	55-105	13	984
	Propionaldehyde	2.75	gas	80-235	-	796
		2.56	B	20	0	329,1472
	Propylene oxide	2.00±0.02	gas	-	M	1917
		1.90	B	25	0	5,709
		1.95	B	25	0	357
		2.00	B	25	0	1632
	Trimethylene oxide	1.93±0.01	gas	-	M	208,502
		1.94	gas	25-175	-	566
		1.92	B	25	0	357
		2.01	B	30	-	1318
		2.03	B	25	0	5,709
$C_3H_6O_2$	Dioxolane	1.47	B	25	0	357
	Ethyl formate	1.94	gas	20-160	-	2168
		1.96	liq	25	-	1679
		1.96	B	25,50	0	1832
		2.01	CCl_4	30	5	1075
	Methyl acetate	1.68	gas	55-240	-	2168
		1.706	gas	34-110	10	1332
		1.74	liq	40	-	1497
		1.45	B,CCl_4	25-45	-	831
		1.61[e]	B	28	-	979
		1.75_3±0.015	B	22	15	2140
		1.7_7	B	25	0	1366
		1.68	CCl_4	25	0?	966
	Propionic acid	1.76	gas	80-210	-	2163
		1.76	gas	n.s.	-	1869

cc) Aniline. d) Gasoline. e) Microwave frequencies.

FORMULA	COMPOUND NAME	μ, D	State, or Solv.	t, °C	Method or P_a, % of P_e	Reference
		1.23	liq	25	-	1495
		1.44	liq	60	-	1495
		0.63±0.01	B	25	15	2137
		0.89	B	22	-	208
		1.69	B	30	-	1526
		1.51	D	n.s.	-	1530
		1.77	D	23	-	1869
		1.77	D	25	0	2126
		0.63-0.65	CS_2	10-30	-	1505
$C_3H_6O_2S$	Methyl vinyl sulfone	4.82	B	30	0	88
	Trimethylene sulfide dioxide	4.49	B	25	0	357
$C_3H_6O_3$	Dimethyl carbonate	0.87	gas	55	10	990,1332
		0.90	gas	75	10	990,1332
		0.95	gas	140	10	990,1332
		1.015	gas	205	10	990,1332
		0.82	B	25	0	1165
		0.91	B	10	10	2157
		0.96	B	25	10	2157
		1.03	B	50	10	2157
		1.07	B	25	-	1946
		0.55	Hx	-25	10	2157
		0.62	Hx	0	10	2157
		0.74	Hx	25	10	2157
		0.80	Hx	50	10	2157
	s-Trioxane	2.08±0.01	gas	-	M	25
		2.16	B	25	0	248
		2.18	B	30	0	1270
$C_3H_6O_3S$	Trimethylene sulfite	3.63	B	20	5	55
C_3H_6S	Trimethylene sulfide	1.78	B	25	0	357

FORMULA	COMPOUND NAME	μ, D	State, or Solv.	t, °C	Method or P_a, % of P_e	Reference
$C_3H_7AuBr_2$	Propylgold dibromide	5-6	CCl_4	25	0?	232
		~6.0	CCl_4	25	-	578
C_3H_7Br	1-Bromopropane	2.03	gas	30-140	-	1236
		2.17±0.01	gas	75-170	0	623
		2.17	gas	n.s.	-	430
		2.02	liq	25	-	707
		2.16	liq	20	-	657
		1.95	B	20	0	348
		1.950	B	25	0	692
		1.96	B	20	0	1479
		1.960	B	64	0	692
		1.97	B	(27)	0	1469
		2.02	B	n.s.	-	379
		2.01	CCl_4	(27)	0	1469
		2.09	Hx	(27)	0	1469
		2.08	Hp	(27)	0	1469
		1.95	Tol	(27)	0	1469
		1.80	n.s.	n.s.	-	1725
	2-Bromopropane	2.21±0.01	gas	15-105	0	623
		2.10	liq	25	-	707
		2.06	B	20	0	348
		2.11	B	20	0	1479
		2.22	B	n.s.	-	379
C_3H_7BrO	3-Bromopropanol	2.21	B	25,50	0	1835
C_3H_7Cl	1-Chloropropane	1.92±0.01	gas	65-185	-	1694
		2.05	gas	30-130	-	1235
		2.06	gas	30-130	-	1236
		2.06_4±0.007	gas	65-185	-	1689, 1691,1695
		1.96	B	20	0	1479
		1.97	B	0-40	-	1418
		1.94-1.98	CCl_4	0-40	-	1418

FORMULA	COMPOUND NAME	μ, D	State, or Solv.	t, °C	Method or P_a, % of P_t	Reference
		2.04	Hp	0-40	-	1418
		1.83-1.86	CS_2	0-40	-	1418
		2.09	n.s.	n.s.	-	1725
	2-Chloropropane	2.17±0.01	gas	15-110	0	623
		2.06	B	20	0	1479
		2.12	B	20	RT	921
		2.23	Hx	20	RT	921
C_3H_7ClHg	*n*-Propylmercury chloride	3.67	B	(20)	-	1700
C_3H_7ClO	3-Chloropropanol	2.23	B	25,50	0	1835
		2.35	D	25,50	0	1835
$C_3H_7ClO_2S$	Isopropyl chlorosulfite	2.85	B	25-45	0	1975
	n-Propyl chlorosulfite	2.74	B	25-45	0	1974
$C_3H_7Cl_3Si$	Dichloro(1-chloroethyl)methylsilane	1.87	B	25	15	887
	Dichloro(2-chloroethyl)methylsilane	1.96	B	25	15	887
C_3H_7I	1-Iodopropane	1.64	gas	30-130	-	1235
		1.99	gas	30-130	-	1236
		2.03±0.01	gas	65-100	0	623
		1.86	B	20	0	348
		1.86	B	(27)	0	1469
		1.87	B	20	0	1479
		1.88	CCl_4	(27)	0	1469
		1.94	CCl_4	20	5-25	78
		1.96	Hx	(27)	0	1469
		1.94	Hp	(27)	0	1469
		1.85	Tol	(27)	0	1469
	2-Iodopropane	1.97	B	20	0	348
		2.01	B	20	0	1479
		2.10	CCl_4	20	5-25	78

FORMULA	COMPOUND NAME	μ, D	State, or Solv.	t, °C	Method or P_a, % of P_e	Reference
C_3H_7N	Allylamine	1.31±0.10	B	25	0	1646
C_3H_7NO	Acetoxime	0.89	D	35	0	1557
	N,*N*-Dimethylformamide	3.85	liq	n.s.	-	1911
		3.82	B	20	-	568
		3.86±0.01	B	25	0	798
	N-Methylacetamide	4.12	D	n.s.	-	1388
		4.40	D	20.1	-	1339
		3.5-3.6	CCl_4	25.5	0	1391
		4.43	n.s.	20.1	-	1340
	Propionamide	3.30	B	30	0	108
		3.47	B	30	0	744
		3.85	D	30	0	108
$C_3H_7NO_2$	Ethylcarbamate	2.59±0.02	D	30	0	1061
	1-Nitropropane	3.60±0.01	gas	70-190	0	623
		3.75	gas	110-175	0	2132
	2-Nitropropane	3.76	gas	120-180	0	2132
	n-Propylnitrite	2.38	liq	-165 to 35	0?	611
		2.30	B	20	0	338
		2.28	Hx	20	0?	611
	Isopropyl nitrite	2.57	gas	-20 to 60	5	305
		2.64-2.71	liq	-125 to 25	0?	611
		2.49	Hx	22	0?	611
$C_3H_7NO_3$	2-Nitro-1-propanol	3.44±0.05	B	n.s.	-	257

FORMULA	COMPOUND NAME	μ, D	State, or Solv.	t, °C	Method or P_a, % of P_e	Reference
	n-Propyl nitrate	3.01	B	20	0	338
$C_3H_7N_5$	1,4-Dimethyl-5-iminotetrazole	1.65	B	25	0	892
	1-Methyl-5-methylaminotetrazole	2.25	B	25	0	892
C_3H_8	Propane	0	gas	-45 to 215	-	1284
		0	gas	25	-	2071
		<0.05	gas	25-145	-	1040
		0-0.1	gas	25-160	-	643
		0.083±0.001	gas	-	M	1137
$C_3H_8BF_3O$	Boron trifluoride - methyl ethyl ether complex	5.11±0.08	B	25	-	1044
$C_3H_8Cl_2Si$	Dichloroethylmethylsilane	2.31	B	25	15	887
$C_3H_8N_2O$	1,1-Dimethyl urea	4.66±0.04	D	25	0	120
	1,3-Dimethyl urea	5.1	B	20	-	403
		4.60	D	25	-	562
		4.8±0.1	D	20	0	792
		6.37	a	25	-	562
		6.08	b	25	-	562
		6.44	c	25	-	562
		6.48	d	25	-	562
		7.42	e	25	-	562
		5.88	f	25	-	562
	Ethyl urea	4.40	D	25	-	562
		4.55±0.04	D	25	0	120
		5.75	a	25	-	562
		6.22	c	25	-	562
		6.81	e	25	-	562

a) Acetone. b) Ethanol. c) Methanol. d) Water. e) Methylcellosolve. f) Polyether.

FORMULA	COMPOUND NAME	μ, D	State, or Solv.	t, °C	Method or P_a, % of P_r	Reference
$C_3H_8N_5$	1,3-Dimethyl-5-aminotetrazole	4.02	B	25	0	892
C_3H_8O	Methyl ethyl ether	1.22	gas	30-100	-	1352
	1-Propanol	1.65	gas	30-110	10	984
		1.66_9	gas	100-230	-	1311
		3.09	liq	20	-	1498
		1.54	B	24-70	-	1039
		1.57	B	n.s.	0	1665
		1.63	B	20	-	977
		$1.67_0 \pm 0.02$	B	22	15	2140
		1.72	B	20	0	1241a
		1.75	B	25	5	1108
		1.76	B	10-70	-	1875
		1.73_5	CCl_4	25	-	1108
		1.54	Tol	24	-	1039
		1.68	g	55	-	1865
	2-Propanol	1.59	gas	35-210	-	983
		1.692±0.007	gas	25-190	-	1877
		1.64±0.02	B,Hx	7-30	15	697
		1.65	B	30	-	1237
		1.67	B	30	0	1267
		$1.70_9 \pm 0.03$	B	22	15	413
		1.716	B	20	-	724
		1.80	B	25	0	1234
		1.48	CS_2	20	-	724
		1.73	g	55	-	1865
$C_3H_8O_2$	Dimethoxy methane	0.74	gas	35	10	986,1334
		0.82	gas	55	10	986,1334
		0.85	gas	80	10	986,1334
		0.93	gas	110	10	986,1334
		0.98	gas	135	10	986,1334
		1.14	gas	200	10	986,1334
		0.67	Hx	25	10	2000

g) *p*-Dichlorobenzene.

FORMULA	COMPOUND NAME	μ, D	State, or Solv.	t, °C	Method or P_a, % of P_e	Reference
	2-Methoxyethanol	2.06	B	25	0	246
		2.22	B	30	-	590
	1,2-Propanediol	3.63	liq	20	-	1498
		2.2	D	n.s.	-	2105
		2.27	D	25	0	1833
		2.29	D	50	0	1833
	1,3-Propanediol	3.03	liq	70	0	907
		3.46	liq	45	0	907
		3.70	liq	20	0	907
		4.24	liq	20	-	1498
		2.37	D	n.s.	-	2105
		2.52	D	25-50	0	1833
		2.52	D	20-75	0	907
$C_3H_8O_3$	Glycerol	0.28	liq	18	-	1115
		4.21	liq	20	-	657
		2.56	D	15-45	-	383
		2.68±0.02	D	15-30	-	2060
C_3H_8S	1-Propanethiol	1.55	liq	25	2.7	1280
		1.34	B	20	0	791
		1.51	B	25	27	1280
	2-Propanethiol	1.64	liq	25	2.7	1280
		1.55	B	25	2.7	1280
C_3H_9As	Trimethylarsine	0.86±0.02	gas	-	M	1136
$C_3H_9AuClO_3P$	Dimethyl methylphosphate - gold chloride complex	7.08	B	n.s.	-	57
$C_3H_9BBr_3N$	Boron tribromide - trimethylamine complex	6.57±0.01	B	25	0	111
$C_3H_9BCl_3N$	Boron trichloride - trimethylamine complex	6.28±0.02	B	25	5	1502

FORMULA	COMPOUND NAME	μ, D	State, or Solv.	t, °C	Method or P_a, % of P_e	Reference
$C_3H_9BCl_3P$	Boron trichloride - trimethylphosphine complex	7.08±0.01	B	25	5	1502
$C_3H_9BF_3N$	Trimethylamine - boron trifluoride complex	~5 5.80±0.02	gas B	- 25	M 5	1535 1502
$C_3H_9BO_3$	Trimethyl borate	0.8₂ 0.2	CCl_4 Hp	25 20	0 15	72 1890
$C_3H_9B_3Cl_3N_3$	*B*-Trichloro-*N*-trimethylborazole	0.47	B	25	5	2065
C_3H_9BrPt	Trimethylplatinum bromide	~1	n.s.	n.s.	-	1746
C_3H_9BrSi	Trimethylbromosilane	2.36±0.04	B	25	5	1592
C_3H_9ClPb	Trimethyllead chloride	4.50±0.15	B	25	0	1118
C_3H_9ClPt	Trimethylplatinum chloride	~1	n.s.	n.s.	-	1746
C_3H_9ClSi	Trimethylchlorosilane	2.09±0.01	B	20	-	1702
C_3H_9ClSn	Trimethyltin chloride	3.50±0.04	B	25	-	313
C_3H_9IPt	Trimethylplatinum iodide	~1	n.s.	n.s.	-	1746
C_3H_9ISi	Trimethyliodosilane	2.46±0.04	B	20	-	604
C_3H_9N	*n*-Propylamine	1.17 1.18 1.17 1.26[i] 1.34 1.36 1.37 1.39	gas gas liq B B B Hx Hx	60-160 20 20 25 20 20 20 20	0 5 0 0 5 0 5 0	101 337 337 101 337 336 337 336

i) Calculated paralleling author's calculation for butylamine.

FORMULA	COMPOUND NAME	μ, D	State, or Solv.	t, °C	Method or P_a, % of P_e	Reference
	Trimethylamine	0.60±0.02	gas	65-185	15?	1695,1867
	0.63	0.60_1	gas	Rm. Temp.	MA	179
		0.612±0.003	gas	-	M	1144
		0.62	gas	15-145	8	624
		0.64	gas	25	-	1099
		0.83	gas	20-90	-	572
		0.72	liq	-10 to 25	-	1099
		0.86	B	25	0	68
		0.87	B	25	-	1099
C_3H_9NO	1-Amino-3-propanol	2.69±0.05	n.s.	n.s.	-	1720
	Trimethylamine oxide	4.87±0.15	B	25	5	1502
		5.02	B	45	0	1158
		5.03	D	25	5	644
		5.04	D	45	0	1158
$C_3H_9NO_2S$	Trimethylamine - sulfur dioxide complex	4.95	B	25	5	1344
$C_3H_9O_2P$	Methyl ethylphosphinate	3.20	B	25	5	61
$C_3H_9O_3P$	Dimethyl methylphosphonate	3.62[j]	liq	30	-	950
		2.86	CCl_4	20	-	50
	Trimethylphosphite	1.83	CCl_4	20	-	50,57
$C_3H_9O_4P$	Trimethylphosphate	3.02	CCl_4	20	-	50
C_3H_9P	Trimethylphosphine	1.192±0.005	gas	-	M	1145
$C_3H_{10}N_2$	1,3-Diaminopropane	1.96	B	25,45	0	1977,1979
$C_3H_{10}OSi$	Trimethylsilanol	2.01±0.10	liq	20	-	626
		1.53	D	25	-	307

j) Onsager's equation.

FORMULA	COMPOUND NAME	μ, D	State, or Solv.	t, °C	Method or P_a, % of P_e	Reference
$C_3H_{10}Si$	Trimethylsilane	0.525±0.005	gas	-	M	1511
$C_3H_{12}BN$	n-Propylamine - borane complex	4.68±0.045	B	25	15	1420
	Trimethylborine - ammonia complex	4.15±0.06	B	20	0	115
	Trimethylamine - borane complex	4.45±0.05 4.62±0.01	B B	25 25	15 0	1420 111
C_3N_2O	Carbonyl cyanide	1.35 1.5±0.08	B B, CCl_4	18.4 25	- -	585 1541
C_3O_2	Carbon suboxide	<0.09 0.7	gas B	0-30 25	- 0?	1040 1065
$C_3O_3S_3$	Trithioformaldehyde	2.38	B	25	0	248
C_4Br_4S	Tetrabromothiophene	0.73±0.01	B	30	0	616
$C_4Cl_2F_4$	1,1,4,4-Tetrafluoro-2,3-dichlorobutadiene	0.37	B	20	12	1669
	1,2,3,4-Tetrafluoro-1,4-dichlorobutadiene	1.02	B	20	8	1669
$C_4Cl_2F_6$	1,1,1,4,4,4-Hexafluoro-2,3-dichloro-2-butene	0.53	B	20	15	1669
$C_4Cl_3F_3$	1,1,4-Trifluoro-2,3,4-trichlorobutadiene	0	B	20	12	1669
$C_4Cl_3F_7$	2,2,3-Trichloroheptafluorobutane	0.85±0.08	gas	70-135	MA	1228
$C_4Cl_4F_2$	1,2,3,4-Tetrachloro-1,4-difluorobutadiene	1.03	B	20	7	1669
$C_4Cl_4F_6$	2,2,3,3-Tetrachloro-1,1,1,4,4,4-hexafluorobutane	0.66 0.60	B cHx	20 20	18 18	1669 1669
C_4Cl_4S	Tetrachlorothiophene	0.93	B	30	0	282

FORMULA	COMPOUND NAME	μ, D	State, or Solv.	t, °C	Method or P_a % of P_e	Reference
C_4Cl_6	Perchlorobutadiene	0.2	B	20	-	1669
$C_4Cl_6F_4$	1,1,4,4-Tetrafluoroperchlorobutane	0.99	B	20	10	1669
		0.96	cHx	20	10	1669
	1,2,3,4-Tetrafluoroperchlorobutane	0.80	B	20	12	1669
		0.77	cHx	20	12	1669
$C_4Cl_6Hg_2O_4$	Mercurous trichloroacetate	2.67±0.02	B	25	10-30	387
$C_4Cl_7F_3$	1,1,4-Trifluoroperchlorobutane	0.89	B	20	10	1669
		0.90	cHx	20	10	1669
C_4Cl_8	Perchloro-2-butene	0.38	B	20	10	1669
$C_4Cl_8F_2$	1,4-Difluoroperchlorobutane	0.79	B	20	10	1669
		0.76	cHx	20	10	1669
C_4Cl_{10}	Perchlorobutane	0.38	liq	20	6	1669
C_4F_8	Perfluorocyclobutane	0	gas	25-225	20-30	405
C_4F_8O	Perfluorotetramethylene oxide	0.56±0.12	gas	25-110	15	1647
$C_4F_{10}O$	Perfluoroethyl ether	0.42	gas	25-225	20-30	405
		0.51±0.12	gas	25-100	15	1647
$C_4FeI_2O_4$	Iron tetracarbonyldiiodide	3.60±0.05	B	25	20	715
C_4HI_4N	2,3,4,5-Tetraiodopyrrole	2.52±0.04	B	25	0	931
$C_4H_2BrNO_3$	2-Bromo-5-nitrofuran	4.20	B	25	0	1408
$C_4H_2Br_2O$	2,3-Dibromofuran	1.53	B	25	5?	1406-7
	2,5-Dibromofuran	1.63	B	25	5?	1406

FORMULA	COMPOUND NAME	μ, D	State, or Solv.	t, °C	Method or P_a, % of P_e	Reference
$C_4H_2Br_2S$	2,5-Dibromothiophene	1.09 1.13±0.01 1.08±0.02	B B CCl_4	25 25 25	0 5 5	1408 500 500
$C_4H_2Cl_2N_2$	2,3-Dichloropyrazine	2.09[k]	n.s.	n.s.	-	28
	2,5-Dichloropyrazine	1.15[k]	n.s.	n.s.	-	28
	2,5-Dichloropyrimidine	2.29	D	35	-	1712
$C_4H_2Cl_2N_2Si$	*Bis*(dimethylamino)dichlorosilane	2.64±0.03	B	20	-	1702
$C_4H_2Cl_2S$	2,5-Dichlorothiophene	1.04±0.01 1.12 1.05	B B CCl_4	25 30 25	5 0 5	500 902 500
$C_4H_2Cl_6O_2$	2,2,3,5,5,6-Hexachloro-1,4-dioxane	0	n.s.	n.s.	-	188
$C_4H_2Cl_{10}O_4Sn$	Tin chloride - di(trichloroacetic acid) complex	2.09	B	n.s.	15	1449
$C_4H_2N_2$	Fumaronitrile	0	gas,B	25	-	184
$C_4H_2N_2O_4$	Alloxan monohydrate	2.10±0.01	D	35	0	1847
$C_4H_2N_2O_4S$	2,5-Dinitrothiophene	2.32	B	25	5?	1406
$C_4H_2N_2O_5$	2,5-Dinitrofuran	3.93	B	25	5?	1406
$C_4H_2O_3$	Maleic anhydride	3.94	D	35	0	1851
C_4H_3BrO	2-Bromofuran	1.46	B	25	5?	1406
	3-Bromofuran	0.91	B	25	5?	1406
$C_4H_3BrO_2$	3-Bromocrotonolactone	4.70	B	25	0	685,1899

k) Purity low.

FORMULA	COMPOUND NAME	μ, D	State, or Solv.	t, °C	Method or P_a, % of P_e	Reference
	4-Bromocrotonolactone	3.86	B	25	0	685,1899
$C_4H_3BrO_3$	α-Bromotetronic acid	6.05	D	25	0	994
C_4H_3BrS	2-Bromothiophene	1.34	B	25	5	500
		1.37	B	25	0	1408
		1.37±0.01	B	30	0	902
		1.39±0.10	B	25	0	1641
		1.38±0.01	CCl_4	25	5	500
$C_4H_3ClN_2$	4-Chloropyrimidine	0.9	B	20	-	772
$C_4H_3ClN_2S$	2-Mercapto-5-chloropyrimidine	0	D	35	-	1712
$C_4H_3ClO_2$	3-Chlorocrotonolactone	4.83	B	25	0	685,1899
	4-Chlorocrotonolactone	3.57	B	25	0	685,1899
$C_4H_3ClO_3$	α-Chlorotetronic acid	5.73	D	25	0	994
C_4H_3ClS	2-Chlorothiophene	1.60	B	30	0	902
C_4H_3IO	2-Iodofuran	1.03	B	25	5?	1406
$C_4H_3IO_3$	α-Iodotetronic acid	5.63	D	25	0	994
C_4H_3IS	2-Iodothiophene	1.08	B	25	5	500
		1.13±0.10	B	25	0	1641
		1.14	B	30	0	902
		1.12±0.03	CCl_4	25	5	500
$C_4H_3NO_2S$	2-Nitrothiophene	4.15	B	25	-	1430
		4.23	B	30	0	902
		4.27	B	25	5?	1406
$C_4H_3NO_3$	2-Nitrofuran	4.41	B	25	5	1406

FORMULA	COMPOUND NAME	μ, D	State, or Solv.	t, °C	Method or P_a, % of P_e	Reference
$C_4H_3NO_5$	α-Nitrotetronic acid	6.15	D	25	0	995
$C_4H_4AsCl_3$	*Bis*(2-chlorovinyl)arsenic chloride	1.46	B	20	0	1348
$C_4H_4BrNO_2$	*N*-Bromosuccinimide	2.10±0.01	D	20	5	1194
$C_4H_4ClNO_2$	*N*-Chlorosuccinimide	2.86±0.01	D	20	5	1194
$C_4H_4Cl_2$	1,4-Dichloro-2-butyne	2.08 1.95±0.02 1.90-1.99	gas B CCl_4	90-160 10-50 -18.5 to 55	- 5 5	1354 1357 1357
$C_4H_4Cl_2O_2$	Succinyl chloride	3.03	B	20	0	1265
$C_4H_4Cl_2O_4$	*dl*-α,β-Dichlorosuccinic acid	2.96	B	n.s.	-	676
	meso-α,β-Dichlorosuccinic acid	2.49	B	n.s.	-	676
$C_4H_4Cl_4O_2$	2,3,5,6-Tetrachloro-1,4-dioxane (m.p. = 60°C)	0	n.s.	n.s.	-	188
	(m.p. = 100°C)	1.87	n.s.	n.s.	-	188
	(m.p. = 144°C)	1.06	n.s.	n.s.	-	188
$C_4H_4INO_2$	*N*-Iodosuccinimide	0.97±0.03	D	20	5	1194
$C_4H_4N_2$	Pyrazine	0 0 Below 1 0.66	gas B B D	n.s. 35 n.s. 35	- - - -	2012 1712 2016 1712
	Pyridazine	3.97	D	35	-	1712

FORMULA	COMPOUND NAME	μ, D	State, or Solv.	t, °C	Method or P_a, % of P_e	Reference
	Pyrimidine	2.0	B	20?	-	772
		2.44	D	35	-	1712
	Succinonitrile	3.48-3.60	gas	170	-	184
		3.8	B	25	0	2113
		3.93	B	25	-	184
		3.96	B	25,75	-	1978-9
		3.83	D	25	-	184
		2.97	Tol	-90	0	1121
		3.19	Tol	-60	0	1121
		3.39	Tol	-30	0	1121
		3.57	Tol	0	0	1121
		3.71	Tol	30	0	1121
		3.83	Tol	60	0	1121
		3.93	Tol	90	0	1121
$C_4H_4N_2O$	4-Oxypyrimidine	2.72	D	35	-	1712
	Pyridazone	2.69	D	25	-	770
$C_4H_4N_2OS$	2-Thiouracil	4.21	D	35	0	1716
	4-Thiouracil	4.47	D	35	0	1716
$C_4H_4N_2O_3$	Barbituric acid	1.04±0.01	D	35	0	1847
$C_4H_4N_2O_4$	1,4-Dinitro-1,3-butadiene	0.72	B	25	-	2031-2
$C_4H_4N_2S_2$	2,4-Dithiouracil	4.67	D	35	0	1716
$C_4H_4N_4$	Hydrocyanic acid tetramer	7.8	D	35	0	2074
C_4H_4O	Furan	0.661±0.006	gas	-	M	1779
		0.72	gas	55-155	0	666
		0.63	B	20	0	790
		0.67±0.02	B	20	-	1626
		0.71	B	25	0	666

FORMULA	COMPOUND NAME	μ, D	State, or Solv.	t, °C	Method or P_a, % of P_e	Reference
C_4H_4O (contd.)	Furan (contd.)	0.71	B	25	0	1836
		0.72	CCl_4	25	0	1076
$C_4H_4O_2$	Crotonolactone	4.62	B	25	0	685,1899
	Ketene dimer	3.56	gas	160-240	-	796
		3.19	B	25	10	39
		3.23	B	25	0	1621
		3.34	B	25	0	1428
		3.33	CCl_4	25	10	39
	Tetrolic acid	2.14	D	25	0	2128
$C_4H_4O_3$	Succinic anhydride	3.83	B	25	0	1165
		4.19	D	10	0	1585
		4.24	D	20-40	0	1585
	Tetronic acid	4.76	D	25	0	994
$C_4H_4O_4$	Fumaric acid	2.45±0.02	D	25	0	119
	Maleic acid	3.17±0.02	D	25	0	119
C_4H_4S	Thiophene	0.55±0.04	gas	20-155	0	666
		0.58	gas	55-200	6	985
		0.51_4	liq	25	5	499
		0.55	liq	15	-	1180
		0.52±0.05	B	25	0	666
	0.52 P_e = 24.61, P_a = 0	0.52_3±0.002	B	25	5	499
		0.53	B	30	0	902
		0.53-0.54	B, Hx	18	0	716
		0.54±0.02	B	20	0	1626
		0.55	B	15	0	1183
		0.54	CCl_4	25	0	1079
		0.56_2±0.003	CCl_4	25	5	500

FORMULA	COMPOUND NAME	μ, D	State, or Solv.	t, °C	Method or P_a, % of P_e	Reference
		0.52_4±0.002	cHx	25	5	499
		0.55±0.01	CS_2	25	5	499
$C_4H_4S_2$	3-Mercaptothiophene	1.07±0.01	B	25	0	1641
C_4H_4Se	Selenophene	0.41±0.03	B	20	-	1626
		0.78_1	B	25	0	1926
		0.77_2	Hx	25	0	1926
C_4H_5Br	3-Bromo-1-butyne	1.76±0.05	B	25	0	1645
C_4H_5Cl	1-Chloro-2-butyne	2.17	gas	55-110	5	306
	2-Chloro-1,3-butadiene	1.43	n.s.	n.s.	-	1062
	4-Chloro-1,2-butadiene	2.02	gas	120-220	0	661
$C_4H_5ClO_3$	3-Chloropropylene carbonate	4.72	B	25	5	897
$C_4H_5Cl_3O_2$	Ethyl trichloroacetate	2.55	B	25	0	1176
		2.57	B	25	0	1106
$C_4H_5Cl_7O_2Ti$	Ethyl trichloroacetate - titanium chloride complex	2.10	B	20	0	1447
C_4H_5N	2-Butenenitrile	3.44±0.10	B	25	0	1646
	trans-Crotononitrile	4.53	gas	135-240	-	796
	Cyclopropylnitrile	3.78	B	25	-	1650
	Methacrylonitrile	3.69	gas	120-200	0	662
	Pyrrole	1.55-1.65	gas	20-85	-	1993
		1.84±0.08	gas	15-200	-	235
		1.54_3	liq	25	5	821

FORMULA	COMPOUND NAME	μ, D	State, or Solv.	t, °C	Method or P_a, % of P_e	Reference
C_4H_5N (contd.)	Pyrrole (contd.)	1.74	B	20	-	1255
		1.80±0.01	B	25	0	931
		1.80±0.07	B	25	-	235
		1.82±0.01	B	20	-	1626
		1.85	B	20	-	341
		2.2	B	20	-	769
		1.97	D	25	-	1255
		2.15	D	25	0	1200
		1.76	CCl_4	n.s.	-	596
		1.78	CCl_4	25	0	1079
C_4H_5NO	3-Methylisoxazole	2.86±0.02	B	25	-	1858
		3.04±0.02	B	25	0	1860
		3.16	B	25	-	1932
		2.89±0.02	D	25	-	1858
		3.13±0.02	D	25	0	1860
	5-Methylisoxazole	2.86±0.01	B	25	0	1860
		2.89	B	25	-	1932
		3.04±0.02	B	25	-	1858
		2.89±0.02	D	25	0	1860
		3.13±0.02	D	25	-	1858
$C_4H_5NO_2$	Succinimide	1.55	B	20	0	345
		1.47±0.01	D	30	0	1059
C_4H_5NS	Allyl isothiocyanate	3.21	B	20	0	791
$C_4H_5O_2S$	*p*-Oxathiane oxide	2.92	B	25	0	357
C_4H_6	1,2-Butadiene	0.401±0.02	gas	-	M	1142
	1,3-Butadiene	0	gas	25-190	5	658

FORMULA	COMPOUND NAME	μ, D	State, or Solv.	t, °C	Method or P_a, % of P_e	Reference
	Butyne	0.81	gas	25-75	0	970
$C_4H_6Br_2$	*trans*-1,4-Dibromo-2-butene	1.63	B	25	0	1919
$C_4H_6Br_2Cl_4O_4Sn$	Tin chloride - di(bromoacetic acid) complex	3.59	B	n.s.	15	1441
$C_4H_6Br_2O_2$	*trans*-2,3-Dibromodioxane	1.90	B	n.s.	-	21
		1.86	CCl_4	n.s.	-	21
	trans-2,5-Dibromodioxane	0.8	B	n.s.	-	21
$C_4H_6Cl_2$	1,3-Dichloro-2-butene	2.15	B	20	0	1491
	cis-2,3-Dichloro-2-butene	2.43	B	21-30	-	1308
	trans-2,3-Dichloro-2-butene	0	B	20-26	-	1308
	1,1-Dichloro-2-methylpropane	2.03	B	30	0	394
$C_4H_6Cl_2O_2$	*cis*-2,3-Dichlorodioxane	3.06	B	n.s.	-	21
		3.00	CCl_4	n.s.	-	21
	trans-2,3-Dichlorodioxane	1.63	B	n.s.	-	21
		1.62	CCl_4	n.s.	-	21
		1.6	n.s.	n.s.	-	188
	trans-2,5-Dichlorodioxane	0.6	B	n.s.	-	21
		0.6	CCl_4	n.s.	-	21
	Ethyl dichloroacetate	2.63	B	25	0	1106
$C_4H_6Cl_2O_3$	2-Hydroxy-2-dichloromethyl-1,3-dioxolane	3.38	B	25	0	5
$C_4H_6Cl_3NO_2$	*Tris*(chloromethyl)nitromethane	2.85±0.03	B	25	0	1196

FORMULA	COMPOUND NAME	μ, D	State, or Solv.	t, °C	Method or P_a, % of P_e	Reference
$C_4H_6Cl_4I_2O_4Sn$	Tin chloride - di(iodoacetic acid) complex	4.15	B	n.s.	15	1441
$C_4H_6Cl_6O_4Sn$	Tin chloride - di(chloracetic acid) complex	3.44	B	n.s.	15	1449
$C_4H_6N_2$	1-Methylimidazole	3.73±0.03	B	20	15	2176
	4-Methylimidazole	6.2	B	70	-	769
		6.3	B	25	-	769-70
		5.1	D	20,25	-	769-70
		5.8	CCl_4	18,25	-	769
	N-Methylimidazole	3.6	B	20	-	769
		3.8	D	20	-	769
	4(5)-Methylimidazole	3.3±0.2	B	20	15	2176
	1-Methylpyrazole	2.30	B	25	15	859
	3-Methylpyrazole	1.44	B	25	15	859
$C_4H_6N_2O$	Dimethylfurazan	4.04	D	20	0	1320
	Dimethyloxydiazole	3.31	D	20	0	1320
	3-Methyl-5-pyrazolone	2.56	D	25	15	859
$C_4H_6N_2O_2$	Dimethylfurazan peroxide	4.81	B	25	0	1931
	Dimethylfuroxan	4.43	B	25	15	859
	Ethyl diazoacetate	2.045±0.015	B	22	-	2135
		2.05±0.02	B	25	15	2137
	1-Nitroso-2-pyrrolidinone	4.58	B	25	0	782

FORMULA	COMPOUND NAME	μ, D	State, or Solv.	t, °C	Method or P_a, % of P_e	Reference
$C_4H_6N_2O_4$	*cis*-2,3-Dinitro-2-butene	5.16	B,CCl_4	25	0	1293
C_4H_6O	Butenone	3.00	B	25	0	1632
	3-Butyne-1-ol	1.73±0.05	B	25	0	1645
	3-Butyne-2-ol	1.69±0.05	B	25	0	1645
	Crotonaldehyde	3.70	gas	140-245	-	796
		3.54±0.03	CCl_4	25	0	127
	trans-Crotonaldehyde	3.50	B	25	5	468
		3.58	D	25	5	468
	Cyclobutanone	2.46	liq	25	0	65
		2.76	B	25	0	65
	2,5-Dihydrofuran	1.54±0.01	B	20	-	1626
	Dimethylketene	1.85	B	25	0	788
		1.8_7	B	27	0	41
	3,4-Epoxy-1-butene	1.86	B	25	0	1632
	Ethoxyacetylene	2.00	liq	25	-	814
		1.96±0.03	B	20	0	422
	α-Methylacrolein	2.68	gas	90-190	0	662
		2.72	B	25	0	1631
	Methyl 1-propynyl ether	1.57±0.03	B	20	0	422
	Vinyl ether	1.07	B	20	0	1836

FORMULA	COMPOUND NAME	μ, D	State, or Solv.	t, °C	Method or P_a, % of P_e	Reference
$C_4H_6O_2$	Butadione	1.06-1.28	gas	55-105	0	184
		1.26	gas	55	-	2165
		1.30-1.49	gas	85-230	-	2165
		1.0_8	B	25	0	363
		1.86	B	25	0	251
		1.0_4	CCl_4	25	0	363
	Butyne-1,4-diol	2.63±0.03	B	n.s.	0	1968
		1.2[kk]	-	-	m	1984
	γ-Butyrolactone	3.82	B	25	0	1165
		4.13	B	25	0	317
		4.15	B	25	0	1259
	trans-Crotonic acid	2.13	B	30	0	1276
	Methylacrylic acid	1.65	B	0	-	953
		1.77-1.81	B,D,Hx	0-40	-	958
	Vinyl acetate	1.7_9	B	25	5	1076
		1.7_6	CCl_4	25	5	1076
		1.7_7	n.s.	n.s.	-	1062
$C_4H_6O_3$	Acetic anhydride	2.7-2.9	gas	50-270	10	2170
		3.15	B	20	-	447
		3.15	mm	20	-	447
		3.12	Hx	20	-	447
		2.85	CS_2	25	-	1505
	Cyclic ester of carbonic acid with 1,3-propanediol	5.25	B	20	5	54
	Propylene carbonate	4.98	B	25	5	897
$C_4H_6O_4$	Ethyl oxalate	2.40-2.53	liq	20-95	0	909
		2.57-2.61	B	20-75	0	907
		2.49-2.58	D	20-75	0	907

kk) Hydroxyl group positions not given. m) From potential measurements when adsorbed on Pt.
mm) Pentane.

FORMULA	COMPOUND NAME	μ, D	State, or Solv.	t, °C	Method or P_a, % of P_e	Reference
	Succinic acid	2.14±0.02	D	25	0	119
		2.15	D	35	5	1399
		2.20±0.10	D	25	0	1635
$C_4H_6O_4S_2$	Dithiodiacetic acid	3.08±0.10	D	25	0	1635
$C_4H_6O_6$	Tartaric acid	3.24±0.02	D	25	0	119
	meso-Tartaric acid	3.64±0.02	D	25	0	119
C_4H_6S	Divinyl sulfide	1.20	gas	130-195	0	661
C_4H_7Br	Bromocyclobutane	2.09	B	25	0	1622
C_4H_7BrO	1-Bromo-2-butanone	2.35	Hx	n.s.	0	1345
$C_4H_7BrO_2$	2-Bromomethyl-1,3-dioxolane	2.30	B	25	0	1459
	Ethyl bromoacetate	2.42	B	25	0	1176
	Methyl-2-bromopropionate	2.20	B	25	0	1176
C_4H_7Cl	1-Chloro-2-butene	2.01	gas	75-250	-	796
		2.10	B	20	0	1491
	3-Chloro-1-butene	1.99	B	20	0	1491
	1-Chloro-2-methylpropene	1.85	gas	105-205	0	661
C_4H_7ClO	Butyryl chloride	2.51	B	25	0	927
		2.63	B	20	0	1263
$C_4H_7ClO_2$	Chlorodioxane	2.24	B	n.s.	-	21
		2.28	CCl_4	n.s.	-	21

FORMULA	COMPOUND NAME	μ, D	State, or Solv.	t, °C	Method or P_a, % of P_e	Reference
$C_4H_7ClO_2$ (contd.)	Ethyl chloroacetate	2.64	B	25	0	1176
		2.66	B	25	0	1106
$C_4H_7ClO_3$	2-Hydroxyethyl chloroacetate	3.97	B	25	0	5
$C_4H_7Cl_2NO_2$	1,1-Di(chloromethyl)-1-nitroethane	3.24±0.03	B	25	0	1196
$C_4H_7Cl_3$	1,1,2-Trichloro-2-methylpropane	1.82	gas	10-195	5	1329
		1.99	Hp	n.s.	0	1327
$C_4H_7Cl_3O_2$	Chloroform - methyl acetate complex	2.64	n	25	-	1450
C_4H_7N	1-Butyronitrile	4.07±0.01	gas	65-170	0	623
		3.49	B	20	0	2096
		3.60	B	20	0	342
	2-Butyronitrile	3.61	B	25	0	1630
	3-Pyrroline	1.44±0.01	B	20	-	1626
C_4H_7NO	Acetone cyanhydrin	3.17	B	25	0	1630
	cis-γ-Butyrolactam	3.55	B	25	0	783
	dimer	2.2	B	25	0	783
	2-Pyrrolidinone	2.3	B	20	-	403
		3.1	B	25	0	1165
		3.7p	B	30	5	513
		3.79	D	30	0	1060
$C_4H_7NO_2$	Trimethylurethan	5.10±0.02	D	30	0	1061
$C_4H_7N_2O_3$	Nitromorpholine	3.22	D	20	-	569
$C_4H_7N_3O_2$	Ethyl azidoacetate	2.79	B	25	-	1762

n) Mutually soluble. p) Or less due to association.

FORMULA	COMPOUND NAME	μ, D	State, or Solv.	t, °C	Method or P_a, % of P_c	Reference
C_4H_8	1-Butene	0.30	gas	25-95	-	2069
		0.37	gas	0-190	-	1839
	cis-2-Butene	0.33[q]	gas	30-90	0	189
	trans-2-Butene	0	gas	25-95	-	2069
	Methylpropene	0.49	gas	25-95	-	2069
		0.503±0.009	gas	-	M	1051
$C_4H_8AlBr_3O_2$	Aluminum bromide - dioxane (1:1) complex	5.27	B	25	10	1740
$C_4H_8AlCl_3O_2$	Aluminum chloride - dioxane (1:1) complex	5.23	B	25	10	1740
$C_4H_8Al_2Br_6O_2$	Aluminum bromide - dioxane (2:1) complex	4.65	B	25	10	1740
$C_4H_8BF_3O$	Boron trifluoride - tetrahydrofuran complex	5.68±0.17	B	25	0	1452
$C_4H_8Br_2$	1,4-Dibromobutane	2.16±0.03	liq	20-70	0	907
		1.99	B	25	0	1943
		2.02	B	25	0	1835
		2.06±0.02	B	20-70	0	907
		2.067±0.004	B	25	0	692
		2.098±0.004	B	64	0	692
		2.10±0.04	D	20-70	0	907
		1.95±0.02	CCl_4	20-70	0	907
		1.91	Hx	20-70	0	907
		1.98	Hp	25	0	1835
		2.03	Hp	50	0	1835
	dl-2,3-Dibromobutane	2.20	gas	22	-	1967
		1.63	liq	25	-	2130
	meso-2,3-Dibromobutane	2.20	gas	22	-	1967
		1.75	liq	25	-	2130

q) Microwave frequencies.

FORMULA	COMPOUND NAME	μ, D	State, or Solv.	t, °C	Method or P_a, % of P_e	Reference
$C_4H_8Br_2$ (contd.)	1,2-Dibromo-2-methylpropane	1.16-1.31	gas	80-140	5	1329
		1.09	Hp	-20	0	1326
		1.18	Hp	1	0	1326
		1.26	Hp	25	0	1326
		1.30	Hp	50	0	1326
		2.5±0.3	Hp	n.s.	0	1327
$C_4H_8Br_2O_2$	Bromine - dioxane complex	1.3	D	n.s.	0	1920
$C_4H_8Br_3InO_2$	Indium bromide - dioxane complex	1.90	D	n.s.	14	1738
$C_4H_8Br_4O_2Zr$	Ethyl acetate - zirconium bromide complex	3.18	B	20	10	923
C_4H_8ClNO	Methyl 2-chloropropionamide	2.4	CCl_4	n.s.	-	558
$C_4H_8Cl_2$	1,4-Dichlorobutane	2.22	gas	160-220	0	1436
		2.13	liq	25	-	707
		2.08	B,Hx	20	RT	921
		2.10	B	25	0	1943
	1,2-Dichloro-2-methylpropane	2.58±0.08	gas	n.s.	0	1327
		1.53-1.63	gas	65-195	5	1329
		2.8±0.3	Hp	n.s.	0	1327
$C_4H_8Cl_2N_2$	*N,N'*-Dichloropiperazine	0	B	n.s.	-	33
		0.71	B	20	-	568
$C_4H_8Cl_2O$	1-Chloroethyl 2-chloroethyl ether	1.78	Hx	20	0	1350
	Bis(2-chloroethyl) ether	2.60	B	25,50	0	1835
		2.61	B	20	15	1980
		2.42	Hx	20	0	1350
		2.43	cHx,r	20	15	1980
		2.36	pXy	20	15	1980
		2.57	s	20	15	1980

r) Methylcyclohexane. s) Styrene.

FORMULA	COMPOUND NAME	μ, D	State, or Solv.	t, °C	Method or P_a, % of P_e	Reference
		2.47	t	20	-	1981
		2.60	u	20	-	1981
$C_4H_8Cl_2S$	*Bis*(2-chloroethyl) sulfide	1.78	Hx	20	22	1348-9
$C_4H_8Cl_3O_2P$	Methyl ethyl(trichloromethyl)phosphinate	3.23	B	25	5	61
$C_4H_8Cl_3O_2Tl$	Thallium chloride - dioxane complex	1.95	D	25	14	1738
$C_4H_8Cl_4O_2Ti$	Ethyl acetate - titanium chloride complex	4.95	B	20	0	1447
$C_4H_8Cl_4O_4Sn$	Tin chloride - *bis*(acetic acid) complex	6.38	B	n.s.	15	1449
$C_4H_8Cl_8OSn_2$	Tetrahydrofuran - *bis*(tin tetrachloride) complex	1.71	B	n.s.	-	592
$C_4H_8I_2O$	*Bis*(2-iodoethyl) ether	2.25	B	25,50	0	1835
$C_4H_8I_2O_2$	Iodine - dioxane complex	0.95	D	n.s.	0	1920
		1.41±0.02	D	25	-	947
		3.0	cHx	25	-	949
$C_4H_8I_4O_2Zr$	Ethyl acetate - zirconium iodide complex	3.20	B	20	-	1444,1446
$C_4H_8N_2$	Acetaldazine	1.17	Hp	25	0	2099
$C_4H_8N_2O_2$	Dimethylglyoxime	1.39	D	20	-	1319
	Nitrosomorpholine	3.05	D	20	-	569
$C_4H_8N_2O_3$	Nitromorpholine	3.22	D	20	-	569
$C_4H_8N_2O_6$	1,4-Dinitroxybutane	3.58±0.03	B	25	0	1053
$C_4H_8N_4O_2$	*N*,*N*′-Dinitrosopiperazine	2.05	B	20	-	569
		2.1_0	B	25	5	69
		2.08	D	20	-	569

t) Octane. u) 2,2,4-Trimethylpentane.

FORMULA	COMPOUND NAME	μ, D	State, or Solv.	t, °C	Method or P_a, % of P_e	Reference
$C_4H_8N_4O_4$	*N,N'*-Dinitropiperazine	2.21	D	20	-	569
		2.33	D	25	-	569
		2.46	D	30	-	569
		2.68	D	40	-	569
C_4H_8O	Butyraldehyde	2.74	gas	80-140	-	796
		2.450	liq	40	RT	1034
		2.50	B	18	0?	673
		2.59	B	20	0	329
	2-Butanone	3.2	liq	30-40	-	1497
		3.41	liq	20	-	657
		2.5	B	30	MA	2052
		2.76	B[v]	25	5	469
		2.77_7	B	22	15	2140
		2.78±0.03	B?	n.s.	-	302
		2.82	B	15	10	2141
		2.82	D[v]	25	5	469
		2.75	Ether	20	-	1448
		2.76±0.02	w	20	-	1448
	Ethyl vinyl ether	1.27±0.03	B	20	0	422
	2-Methylpropanone	2.60	B	20	0	329
	Tetrahydrofuran	1.63	gas	65-175	24	566
		1.63[x]	liq	22-55	-	824
		1.48-1.57[x]	B	25-45	-	824
		1.69±0.01	B	20	-	1626
		1.73	B	25,50	0	1836
		1.75	B	25	0	357
		1.76	B	25	0	5
		1.84	D	25,50	0	1836
		1.89	n.s.	n.s.	0	709

v) Benzene and dioxane mixtures. w) Methyl benzoate. x) Jatkar's equation.

FORMULA	COMPOUND NAME	μ, D	State, or Solv.	t, °C	Method or P_a, % of P_e	Reference
C_4H_8OS	*p*-Oxathiane	0.42[y]	B	25	0	357
		0.47	n.s.	n.s.	-	188
	Tetrahydrothiophene oxide	4.17	B	25	0	357
C_4H_8OSe	*p*-Oxaselenane	0.30	n.s.	n.s.	-	188
$C_4H_8O_2$	*cis*-2-Butene-1,4-diol	2.50±0.03	B	n.s.	0	1968
	trans-2-Butene-1,4-diol	2.47±0.03	B	n.s.	0	1968
	Butyric acid	0	liq	10-70	-	1827
		1.23	liq	25	-	1495
		1.34	liq	60	-	1495
		0.68±0.02	B	16	15	2137
		0.94	B	22	-	208
		1.9	B	30	-	1526
		1.59	D	n.s.	-	1530
	1,3-Dioxane	2.13_4±0.03	B	25	-	2058
		2.14_5±0.03	cHx	25	-	2058
	1,4-Dioxane	0	gas	65-115	-	1721
		0.43-0.49	gas	55-205	10	988
		0	liq	14-85	-	2035
		0.3	liq	25	-	444
		0.45	liq	25	0	1833
		0.47	liq	50	0	1833
		0.4	B	25	0	2115
		0.40	B	20	0	792
		0.45	B	25	-	1696
		0.21	cHx	25	-	949
		0.3	n.s.	n.s.	-	188

y) Large P_A may cause high error.

FORMULA	COMPOUND NAME	μ, D	State, or Solv.	t, °C	Method or P_a, % of P_e	Reference
$C_4H_8O_2$ (contd.)	Ethyl acetate	1.78	gas	30-195	-	2168
		1.83	liq	25	-	1497
		2.05	liq	20	-	657
		1.50	B, CCl_4	25-45	-	831
		1.83	B	25	0	1366
		1.83	B	25	0	2017-8
		1.83_5±0.03	B	22	15	2140
		1.84±0.02	B	50	0	1832
		1.88±0.02	B	25	0	1832
		1.88	B	50	-	1679
		1.76	CCl_4	25	0	966
		1.83^z	CCl_4	19.2	-	882
		1.89	CCl_4	25	0	2017-8
		1.84	Hx	25	0	2017-8
		1.89	Hp	-70	0	1810
		1.92	Hp	-50	0	1810
		1.95	Hp	-30	0	1810
		1.97	Hp	-10	0	1810
		2.02	Hp	30	0	1810
		1.88	n.s.	n.s.	-	1062
		1.81	zz	25	-	1747
	Isobutyric acid	1.09	liq	25	-	1495
		1.34	liq	60	-	1495
		1.79	B, Hx, D	0-40	-	958
	2-Methyl-1,3-dioxolane	1.22	B	25	0	1459
	Methyl propionate	1.703±0.03	B	22	15	2140
		1.75_5±0.03	CCl_4	22	15	2140
		1.66_7±0.03	Hp	22	15	2140
	Propyl formate	1.91_3±0.03	B	22	15	2140
$C_4H_8O_2S$	Methyl allyl sulfone	4.60	B	30	0	88

z) 10-cm frequency.   zz) Acetone.

FORMULA	COMPOUND NAME	μ, D	State, or Solv.	t, °C	Method or P_a, % of P_e	Reference
	Tetrahydrothiophene dioxide	4.69	B	25	0	1165
		4.81	B	25	0	357
$C_4H_8O_3$	2-Hydroxyethyl acetate	2.65	liq	30	-	1498
		2.36	B	30	0	590
$C_4H_8O_3S$	*p*-Oxathiane dioxide	3.29	B	25	0	357
C_4H_8S	1-Methyl-2-thiacyclobutane	1.79±0.02	B	25	0	1191
	Tetrahydrothiophene	1.89±0.01	B	20	-	1626
		1.90	B	25	0	357
$C_4H_8S_2$	Dithiane	~0	B	25	0	248
C_4H_8Se	Tetrahydroselenophene	1.81±0.01	B	20	-	1626
C_4H_9Br	1-Bromobutane	2.17±0.01	gas	80-200	0	623
		1.92	liq	25	-	707
		2.04	liq	20	-	657
		1.95	B	20	0	348
		1.956±0.004	B	25	0	692
		1.966±0.004	B	64	0	692
		1.99±0.10	B	10-50	0?	1477
		2.00	B	(27)	0	1469
		1.97	CCl_4	20	-	79
		2.00	CCl_4	(27)	0	1469
		2.11	Hx	(27)	0	1469
		1.83	Hp	-90 to 70	-	1828
		2.11	Hp	(27)	0	1469
		1.98	Tol	(27)	0	1469
		1.89	n.s.	n.s.	-	1806
	2-Bromobutane	2.22	gas	70	0	623
		2.14	liq	25	-	707
		2.14±0.10	B	20	0?	1477

FORMULA	COMPOUND NAME	μ, D	State, or Solv.	t, °C	Method or P_a, % of P_e	Reference
C_4H_9Br (contd.)	1-Bromo-2-methylpropane	1.92	liq	25	-	707
		1.99±0.10	B	20	0?	1477
	2-Bromo-2-methylpropane	2.21	gas	-	M	1962
		2.40	liq	25	-	707
		2.23±0.10	B	10-50	0?	1477
		2.19	CCl_4	20	0	79
C_4H_9BrHg	*n*-Butylmercury bromide	3.48	D	25	0	366
C_4H_9Cl	1-Chlorobutane	2.06	gas	40-205	0	1822
		2.13±0.01	gas	15-100	0	623
		1.94	B	20	RT	921
		1.99±0.10	B	10-50	0?	1477
		2.05	B	20	0	1969
		2.02	Hx	20	RT	921
		1.90	Hp	-90 to 70	-	1828
		1.91	n.s.	n.s.	-	1806
	2-Chlorobutane	2.14	gas	60-120	0	2132
		2.07±0.10	B	25	0	1646
		2.07	B	20	RT	921
		2.11±0.10	B	10-50	0?	1477
		2.13	Hx	20	RT	921
		2.16	Hp	-70 to 70	25	1807
	1-Chloro-2-methylpropane	2.06	gas	70-130	0	2132
		1.98±0.10	B	10-50	0?	1477
	2-Chloro-2-methylpropane	2.11	gas	25	5	1802
		2.15	gas	80	0	2132
		2.15	gas	-	M	1962
		1.96	B	(27)	0	1469
		2.1	B, CCl_4	25	-	2015
		2.13	B	25	5	1802
		2.14	B	20	RT	921

FORMULA	COMPOUND NAME	μ, D	State, or Solv.	t, °C	Method or P_a, % of P_e	Reference
		2.14±0.04	B	25	-	313
		2.17±0.10	B	20	0?	1477
		2.20	D	25	5	1802
		2.21	D	25	-	313
		1.92	CCl_4	20	0	79
		1.97	CCl_4	(27)	0	1469
		2.0	CCl_4	n.s.	-	2014
		2.06	CCl_4	20	5-25	77
		2.17	CCl_4	25	5	1802
		2.19	CCl_4	25	-	313
		2.05	Hx	(27)	0	1469
		2.17	Hx	25	5	1802
		2.18	Hx	25	-	313
		2.22	Hx	20	RT	921
		2.04	Hp	(27)	0	1469
		2.16	Hp	25	5	1802
		2.18	Hp	25	-	313
		1.94	Tol	(27)	0	1469
		1.96	a	25	5	1802
		1.98	a	25	-	313
		2.01	CS_2	25	5	1802
		2.03	CS_2	25	-	313
		2.17	b	25	5	1802
		2.16	c	25	5	1802
C_4H_9ClHg	*n*-Butylmercury chloride	3.90	B	(20)	-	1700
C_4H_9ClO	1-Chloroethyl ethyl ether	1.83	Hx	20	0	1350
	2-Chloroethyl ethyl ether	2.20	Hx	20	0	1350
$C_4H_9ClO_2S$	Butyl chlorosulfite	2.74	B	25,45	0	1974
	Isobutyl chlorosulfite	2.69	B	25,45	0	1975
C_4H_9D	2-Methyl-2-propene-d_1	0.1406±0.0012	gas	-	M	1138

a) Tetralin. b) CCl_4 plus *p*-dichlorobenzene. c) Dioxane plus *p*-dichlorobenzene.

FORMULA	COMPOUND NAME	μ, D	State, or Solv.	t, °C	Method or P_a, % of P_e	Reference
C_4H_9F	2-Fluoro-2-methylpropane	1.959±0.005	gas	-	M	1145
		2.15±0.02	gas	-	M	32
C_4H_9I	1-Iodobutane	2.10	gas	75-140	0	623
		1.90±0.10	B	10-50	0?	1477
		1.90	B	20	0	348
		1.95	B	(27)	0	1469
		1.95	CCl_4	20	5-25	78
		1.96	CCl_4	(27)	0	1469
		2.02	Hx	(27)	0	1469
		1.60	Hp	-90 to 70	-	1828
		2.00	Hp	(27)	0	1469
		1.92	Tol	(27)	0	1469
	2-Iodobutane	2.06±0.10	B	20	0?	1477
		2.12	CCl_4	20	5-25	78
	1-Iodo-2-methylpropane	1.89±0.10	B	20	0?	1477
		1.94	CCl_4	20	5-25	78
	2-Iodo-2-methylpropane	2.13	gas	-	M	1962
		2.15	B	20	0?	1477
		2.22	CCl_4	20	5-25	78
C_4H_9Li	Butyllithium	0.87	B	25	13	1637
	hexamer	1.43[d]	B	25	48	1637
		1.1	Hx	25	-	2034
		1.15-1.20	Hx	n.s.	-	1628
C_4H_9N	Pyrrolidine	1.58±0.01	B	20	-	1626
C_4H_9NO	Acetone methyloxime	1.20	liq	n.s.	5	1593
		1.172±0.005	B	20	5	1593

d) Recalculated from J. Am. Chem. Soc. 68, 2748 (1946).

FORMULA	COMPOUND NAME	μ, D	State, or Solv.	t, °C	Method or P_a, % of P_e	Reference
	n-Butyramide	3.48	B	30	0	108
		3.59	B	30	0	744
		3.86	D	30	0	108
	Isobutyramide	3.46	B	30	0	108
		3.70	B	30	0	744
		3.88	D	30	0	108
	N,*N*-Dimethylacetamide	3.81	D	30	10	1011
	N-Ethylacetamide	3.90	D	30	10	1011
	Ethyl acetimino ether	1.34	D	30	10	1011
	Methyl ethyl ketoxime	0.9_6	"pure"	19.8	0	249
	Morpholine	1.68	liq	50	-	1426
		1.75	liq	25	-	1426
		1.49	B	20	-	1473
		1.52	B	30	0	1274
		1.52	B	25	0	68
		1.59	B	25	0	1120
C_4H_9NOS	*t*-Butylthionitrite	2.63	B	20	-	969
		2.71	D	20	-	969
		2.55	CCl_4	20	-	969
		2.45	CS_2	20	-	969
	Sulfinylisobutylamine	1.63	B	25	15	855
$C_4H_9NO_2$	*t*-Butylnitrite	2.70	gas	15-55	5	305
	Ethyl ester of glycine	2.13	B	5-75	0?	2152
	1-Nitrobutane	3.61±0.01	gas	100-200	0	623
		3.32	B	20	-	793
		3.40±0.1	B	25	0	1316

FORMULA	COMPOUND NAME	μ, D	State, or Solv.	t, °C	Method or P_a, % of P_e	Reference
$C_4H_9NO_2$ (contd.)	1-Nitro-2-methylpropane	3.71	gas	n.s.	-	1805
	2-Nitro-2-methylpropane	3.74	gas	120-175	0	2132
$C_4H_9NO_3$	*n*-Butylnitrate	2.99	B	20	0	338
	2-Nitro-1-butanol	3.50±0.05	B	25	0	257
	2-Nitro-2-methyl-1-propanol	3.35±0.05	B	25	0	257
$C_4H_9NO_4$	2-Nitro-2-methyl-1,3-propanediol	3.52±0.05	B	25	0	257
		3.52±0.05	D	25	0	1196
C_4H_{10}	*n*-Butane	0	gas	25-95	-	2069
		<0.05	gas	25-160	-	643
	Isobutane	0	gas	29-95	-	2069
		0.13±0.01	gas	25-160	-	643
		0.132	gas	25	MA	1272
		0.132±0.001	gas	-	M	1145
$C_4H_{10}Al$	Diethylaluminum	0.95±0.04	B	25	5-15	694
	dimer	1.3	B	25	5-15	694
$C_4H_{10}AlBr_3O$	Aluminum bromide - ethyl ether complex	6.44	B	(20)	-	2006
		6.48	B	(20)	-	2007
		6.64	B	20	15	1415,1734
$C_4H_{10}AlCl_3O$	Aluminum chloride - ethyl ether complex	6.50	B	(20)	-	2006
		6.59	B	(20)	-	2007
		6.73	B	25	15	1415
		5.99	CS_2	20	15	1415
$C_4H_{10}AuBr$	Diethylgold bromide	~0	B	25	-	578
		<1	B	25	-	232
		1.32	CCl_4	25	-	233

FORMULA	COMPOUND NAME	μ, D	State, or Solv.	t, °C	Method or P_a, % of P_e	Reference
$C_4H_{10}BCl_3S$	Boron trichloride - ethyl sulfide complex	6.05±0.01	B	25	5	1502
$C_4H_{10}BF_3O$	Boron trifluoride - ethyl ether complex	4.96±0.38	B	25	-	1044
		4.96±0.14	B	25	0	1452
		5.34±0.03	B	25	5	1502
$C_4H_{10}Be$	Diethylberyllium	1.70	B	20	15	1887
		4.3	D	20	15	1887
		1.0	Hp	20	15	1887
$C_4H_{10}Br_2Si$	Diethyldibromosilane	2.51±0.04	B	25	5	1592
$C_4H_{10}Br_2Sn$	Diethyltin bromide	4.0	Hx	n.s.	-	875
$C_4H_{10}Br_3InO$	Ethyl ether - indium tribromide complex	1.16	B	20	10	1739
$C_4H_{10}Cd$	Diethylcadmium	0.53	B	20	10	1887
		1.14	D	20	10	1887
		0.31	Hp	20	10	1887
$C_4H_{10}ClN$	*N*-Chlorodiethylamine	2.02±0.02	B	20	0	6
$C_4H_{10}Cl_2Ge$	Diethyldichlorogermane	3.19	B	25	-	886
$C_4H_{10}Cl_2Pb$	Diethyllead dichloride	4.74±0.10	B	25	-	1118
$C_4H_{10}Cl_2Si$	Diethyldichlorosilane	2.41	B	25	0	373
	(Dichloromethyl)trimethylsilane	2.28	B	30	0	543
$C_4H_{10}Cl_2Sn$	Diethyltin dichloride	3.88	CCl_4, Hx	25	0	1852
$C_4H_{10}Cl_3OTl$	Ethyl ether - thallium chloride complex	2.06	Ether	25	14	1738
$C_4H_{10}Cl_8SSn_2$	*n*-Butylmercaptan - *bis*(tin tetrachloride) complex	1.48	B	n.s.	-	592

FORMULA	COMPOUND NAME	μ, D	State, or Solv.	t, °C	Method or P_a, % of P_e	Reference
$C_4H_{10}F_2Si$	Diethyldifluorosilane	2.25	B	25	0	373
$C_4H_{10}Fe_2N_4O_4S_2$	Di-μ-ethanethiolato-*bis*(dinitrosyliron)	1.88±0.05	B	n.s.	-	711
		1.85±0.07	Hx	n.s.	-	711
$C_4H_{10}Fe_2N_4O_4Se_2$	Di-μ-ethaneselenolato-*bis*(dinitrosyliron)	0.92±0.2	B	n.s.	-	711
$C_4H_{10}Hg$	Diethylmercury	0	B,D,Hp	20	10	1887
		0.37±0.01	B	25	5-15	694
		0.39	B	15	0	158
		0.55	B	20	-	1700
$C_4H_{10}Mg$	Diethylmagnesium	4.8	D	20	10	1887
$C_4H_{10}N_2$	Piperazine	1.47	B	20	-	568
		1.4_7	B	25	5	69
$C_4H_{10}N_2O$	*n*-Propyl urea	4.1	D	20	-	403
	Isopropyl urea	4.44±0.04	D	25	0	120
$C_4H_{10}N_2O_2$	Diethyl hyponitrite	1.5	B	20	-	793
$C_4H_{10}N_4O_4$	1,2-*Bis*(methylnitroamine)ethane	3.61	D	20	-	569
	N,N'-Methylene-*bis*(*N*-methyldiimide dioxide)	3.80	D	20	-	567
$C_4H_{10}O$	1-Butanol	1.60	gas	35-105	16	984
		1.67_0	gas	110-215	-	1311
		2.87	liq	40	10	1295
		2.96	liq	20-30	-	1498
		2.98	liq	20	10	1295
		1.53[e]	B	28	RT	1401
		1.60	B	20	-	977
		1.63	B	25	0	1239
		1.66	B	20	-	1039

e) 3.26 cm wavelength.

FORMULA	COMPOUND NAME	μ, D	State, or Solv.	t, °C	Method or P_a, % of P_e	Reference
		1.67_2±0.02	B	22	15	2140
	1.81 P_e = 21.68 P_a = 2.17	1.69	B	25	0	194
		1.71	B	20	10	1295
		1.75	B	25	5	1108
		1.76	B	20-70	0	1830
		1.77	D	50	0	1833
		1.83	D	25	0	1833
		1.75	CCl_4	25	5	1078,1108
		1.79_5	CCl_4	20	10	1295
		1.95	Hx	25	-	1150
		1.58	cHx	40	10	1295
	Isobutanol	1.64	gas	55-210	-	983
		2.96	liq	30	-	1498
		1.375	B	n.s.	0	1665
		1.59[ee]	B	28	-	1401
		1.66	B	30	-	1237
		1.71	B	25	-	724
		1.73	B	20	-	1039
		1.78	B	(20)	0	749
		1.81	B	25	0	1234
		1.42	CS_2	20	-	724
		1.8	Ether	20	-	717,723
		1.65	f	55	-	1865
	2-Butanol	1.66	B	30	-	1237
	t-Butanol	1.44[g]	B	28	RT	401
		1.56±0.02	B,Hx	7-30	-	697
		1.63	B	70	0	1807
	1.67 P_e = 21.68 P_a = 2.17	1.67±0.022	B	22	15	413
		1.67	B	30	0	1267
		1.69	B	25	-	194
		1.76	B	50	0	1807
		1.81	B	10	0	1807
		1.88	B	30	0	1807

ee) 3.26 cm wavelength. f) *p*-Dichlorobenzene. g) 3.26 cm wavelength.

FORMULA	COMPOUND NAME	μ, D	State, or Solv.	t, °C	Method or P_a, % of P_e	Reference
$C_4H_{10}O$ (contd.)	Ethyl ether	1.0_0	gas	40-160	-	1687
		1.11±0.02	gas	40-160	-	1693,1695
		1.14	gas	n.s.	-	745
	1.17	1.14	gas	30-100	0	1352
		1.15±0.01	gas	15-80	-	1896
		1.15	gas	20	-	554
		1.15_6±0.012	gas	40-160	-	1690
		1.19±0.01	gas	15-200	8	624
		1.05	liq	-17 to 20	-	463
		1.21	liq	-79 to 16	ES	706
		1.22	liq	n.s.	-	823
		1.27	liq	18	REF	1115
		1.433	liq	-186 to 18	-	809
		2.00	liq	20	-	657
		0.74	B	20	-	816
		1.16±0.02	B	7-45	-	1304
	1.30 P_e = 21.93	1.17	B	25	15	1841
	P_a = 2.19	1.2	B	25	0	1747
		1.23	B	20	-	1039
		1.23	B	25	0?	2121
		1.23	B	18	0	1653
		1.26	B	25	-	99
		1.27	B	25	-	1926
		1.28	B	18	0	671,683-4
		1.28	B	25	-	1863
		1.30±0.03	B	20	0	422
		1.53	B	20	-	2133
		1.25	CCl_4	25	0?	966
		1.28[h]	CCl_4	17.5	-	882
		1.15[h]	CS_2	17.3	-	882
		1.79	$CHCl_3$	19	0	671,683-4
	Methyl *n*-propyl ether	1.24	B	25	-	1863
$C_4H_{10}OS$	Diethyl sulfoxide	3.88	B	20	0	636

h) 10 cm wavelength.

FORMULA	COMPOUND NAME	μ, D	State, or Solv.	t, °C	Method or P_a, % of P_e	Reference
$C_4H_{10}O_2$	1,4-Butanediol	3.75	liq	70	0	907
		3.86	liq	45	0	907
		3.93	liq	20	0	907
		2.5±0.003	B	n.s.	0	1968
		2.42±0.02	D	15,30	-	2060
		2.55	D	20	0	907
	t-Butylhydroperoxide	1.80±0.01	B	50	0	1163
		1.80	B	50	0	1164
		1.82	B	30	0	1164
		1.83±0.02	B	30	0	1163
		1.87±0.05	B	25	0	1639
	1-Hydroxyethyl ethyl ether	1.63	Hx	20	0	1350
	2-Hydroxyethyl ethyl ether	2.10	B	25	0	246
		2.24	B	30	0	590
		2.12	Hx	20	0	1350
$C_4H_{10}O_2S$	Diethyl sulfoxalate	1.92±0.2	B	20	10	1705
	Diethyl sulfone	4.44	B	25	-	1907
		4.50	B	20	0	636
$C_4H_{10}O_2S_2$	Diethyl thiosulfite	2.03±0.02	B	20	10	1705
	Ethylsulfinyl ethyl sulfide	4.37	B	25	0	619
$C_4H_{10}O_3$	Diethylene glycol	5.50	liq	20	-	933
		2.69	D	25	10	1999
	Bis(2-hydroxyethyl) ether	2.33	B	20	0	1350
	Bis(oxymethoxy)methyl ether	1.41	Hx	25	10	2000
	Trimethyl orthoformate	1.90	B	25	5	60

FORMULA	COMPOUND NAME	μ, D	State, or Solv.	t, °C	Method or P_a, % of P_e	Reference
$C_4H_{10}O_3S$	Diethyl sulfite	2.99±0.02	B	24.18	5	1914
$C_4H_{10}O_6S$	Dioxane - sulfuric acid complex	4.63	D	n.s.	-	1920
$C_4H_{10}S$	1-Butanethiol	1.54	liq	25	27	1280
		1.33	B	20	0	791
		1.49	B	25,50	0	2059
		1.53	B	25	27	1280
	2-Butanethiol	1.65	liq	25	27	1280
		1.57	B	25	27	1280
	Ethyl sulfide	1.52	gas	35-200	6	985
		1.58	B	20	0	790
		1.59	B	25,50	0	2059
		1.62	B	25	-	1907
	2-Methyl-1-propanethiol	1.53	liq	25	27	1280
		1.53	B	25	27	1280
	2-Methyl-2-propanethiol	1.67	liq	25	27	1280
		1.55±0.02	B	20	0	1193
		1.59	B	25	27	1280
$C_4H_{10}S_2$	Diethyldisulfide	1.96±0.03	B	30	0	1026
		1.99	B	25	0	2102
$C_4H_{10}S_3$	Diethyltrisulfide	1.64	B	25	0	2102
$C_4H_{10}Se$	Diethylselenide	1.52±0.02	B	25	-	309
		1.52±0.02	B	20	0	310
$C_4H_{10}Se_2$	Diethyldiselenide	1.87±0.05	B	25	0	1639

FORMULA	COMPOUND NAME	μ, D	State, or Solv.	t, °C	Method or P_a, % of P_e	Reference
$C_4H_{10}Zn$	Diethylzinc	0.49	B	20	15	1887
		0.53	B	20	10	1893
		0.62±0.02	B	25	5-15	694
		1.6_9	D	20	15	1887,1893
		0.2	Hx	20	10	1893
		0.3-0.4	Hx	25	5-15	694
		0	Hp	20	15	1887
$C_4H_{11}BO_2$	*n*-Butylboric acid	1.87	B	25	-	1458
		1.96	D	25	-	1458
$C_4H_{11}ClSi$	(Chloromethyl)trimethylsilane	2.03	B	30	0	543
	Diethylchlorosilane	2.03±0.02	B	25	5	1853
$C_4H_{11}N$	*n*-Butylamine	1.00	gas	20	5	337
		1.00	gas	75-155	-	101
		1.22	liq	20	5	296
		1.13	B	25	0	1729
	1.32 P_e = 23.57, P_a = 2.36	1.20	B	25	-	101
		1.32	B	25	5	369
		1.322	B	25	0	508
		1.37	B	20	5	337
		1.40	B	20	0	336
		1.40	B	25	0	1630
		1.50	B	30	-	1579
		1.305	D	25	0	508
		1.32	D	25	5	369
		1.51	CCl_4	25	0	1729
		1.33	Hx	20	5	336
		1.36	Hx	20	0	337
		1.32	cHx	20	5	336
		1.36	cHx	20	0	337
	Isobutylamine	1.2_7	B	25	-	101

FORMULA	COMPOUND NAME	μ, D	State, or Solv.	t, °C	Method or P_a, % of P_e	Reference
$C_4H_{11}N$ (contd.)	*sec*-Butylamine	1.28	B	25	0	1630
	t-Butylamine	1.29	B	25	0	1630
		1.322	B	25	0	508
		1.322	D	25	0	508
	Diethylamine	0.91	gas	25-95	-	572
		0.92_0	gas	25	0	100
		1.11	liq	25	0	100
		1.02_6	B	25	0	100
		1.14_1	B	25	0	725
		1.21	B	25	0	530
		1.11_3	Hx	25	0	725
$C_4H_{11}NO_2$	Diethanolamine	2.84	D	25	0	1485
$C_4H_{11}O_2P$	Ethyl ethylphosphinate	3.51	B	25	5	61
$C_4H_{11}O_3P$	Diethylphosphite	3.17	B	20	-	50
		3.06	CCl_4	20	-	50
		2.85	cHx	20	-	50
		3.04	i	20	-	50
	Dimethyl ethylphosphonate	3.38	liq	30	-	950
$C_4H_{11}O_4P$	Diethyl hydrogenphosphate	2.32	B	25	-	906
$C_4H_{12}BN$	Dimethylaminodimethylborine	1.40±0.03	B	20	0	115
$C_4H_{12}B_2Cl_4N_2$	Dichlorodimethylaminoborine dimer	0.34±0.11	B	25	5	224
$C_4H_{12}Cl_2N_2Si$	*Bis*(dimethylamino)dichlorosilane	2.64±0.03	B	20	-	1702
$C_4H_{12}Cl_2Si_2$	1,2-Dichlorotetramethyldisilane	1.35±0.5	B	25	-	690
		1.75±0.1	CCl_4	25	-	690
$C_4H_{12}Cl_3FeO_4$	Diaquodioxanate of ferric chloride	3.33	D	n.s.	-	1290

i) Decalin.

FORMULA	COMPOUND NAME	μ, D	State, or Solv.	t, °C	Method or P_a, % of P_c	Reference
$C_4H_{12}Cl_4O_2Sn$	Tin tetrachloride - *bis*(ethanol) complex	6.08	B	20	0	1439
$C_4H_{12}N_2$	1,4-Butanediamine	1.95	B	25,45	0	1977,1979
		2.35	B	25	0	1763
	Diethylhydrazine	0.51	B	25	-	1555
$C_4H_{12}O_3Si$	Methyltrimethoxysilane	1.70	B	20	0	1041
$C_4H_{12}O_4Si$	Tetramethoxysilane	1.62	B	20	5	53
		1.70	B	20	0	1041
$C_4H_{12}S_2Si_2$	Tetramethylcyclodithiosiloxane	0	B	30	16	1014
$C_4H_{12}S_6Si_4$	Tetramethyltricyclotetrasilthiane	0.61±0.04	B	25	0	1191
$C_4H_{12}Si$	*n*-Butylsilane	0.76	liq	20	6	24
	Isobutylsilane	0.75	liq	20	6	24
	Diethylsilane	0.75	liq	20	6	24
	Tetramethylsilane	0	liq	20	0	24
$C_4H_{14}BN$	*t*-Butylamine - borane complex	4.64±0.05	B	25	15	1420
C_4NiO_4	Nickel carbonyl	0	CCl_4	0	-	1905
		~0.3	CCl_4	0	0	1907
C_5BrMnO_5	Manganese pentacarbonyl bromide	3.11±0.04	B	25	20	116
C_5Cl_6	Hexachlorocyclopentadiene	0.72	liq	n.s.	-	2175
		0.88	Hx	n.s.	-	2175
C_5F_{12}	Perfluoropentane	0	gas	100-230	35	577

FORMULA	COMPOUND NAME	μ, D	State, or Solv.	t, °C	Method or P_a, % of P_e	Reference
C_5F_{12} (contd.)	Perfluoroisopentane	0	gas	100-230	30	577
C_5FeO_5	Iron pentacarbonyl	0	B	25	20	715
		0.64	B	12	0	133-5
		0.82	B	20	0	608
C_5HMnO_5	Manganese pentacarbonyl hydride	0.70±0.05	B	25	0	713
$C_5H_2Br_3N$	2,3,6-Tribromopyridine	3.11	B	25	0	358
	2,4,6-Tribromopyridine	2.07	B	17-67	0	589
		2.29	B	25	0	358
	3,4,5-Tribromopyridine	0.58	B	25	0	358
$C_5H_2ClN_3O_4$	2,6-Dinitro-3-chloropyridine	5.32	B	25	25	1841
$C_5H_2O_5$	Croconic acid	9.3±0.3	D	35	0	2064
		9.5±0.5	D	25	0	2064
$C_5H_3BrO_3$	4-Bromo-2-furoic acid	1.03	B	25	5	1406
	5-Bromo-2-furoic acid	2.19	B	25	5	1406
$C_5H_3Br_2N$	2,5-Dibromopyridine	2.33	B	25	0	358
	2,6-Dibromopyridine	3.46	B	17-67	0	589
		3.54	B	25	0	358
	3,4-Dibromopyridine	1.16	B	25	0	358
	3,5-Dibromopyridine	0.99	B	17-67	0	589
		1.02	B	25	0	358
$C_5H_3ClN_2O_2$	2-Nitro-5-chloropyridine	3.96	B	25	17	1841

FORMULA	COMPOUND NAME	μ, D	State, or Solv.	t, °C	Method or P_a, % of P_e	Reference
$C_5H_3Cl_2N$	2,6-Dichloropyridine	3.65	B	25	0	358
	3,5-Dichloropyridine	0.95	B	25	0	358
$C_5H_3F_3N_2OS$	6-Trifluoromethyl-2-thiouracil	2.27	D	35	0	1716
$C_5H_3NO_4$	5-Nitro-2-furfural	3.45	B	25	0	1408
$C_5H_3NO_5$	5-Nitro-2-furoic acid	4.09	B	25	5	1406
$C_5H_3N_3OS$	5-Cyano-2-thiouracil	5.19	D	35	0	1716
$C_5H_3N_3O_4$	3,5-Dinitropyridine	1.11	B	25	25	1841
$C_5H_4BCl_4N$	4-Chloropyridine - boron trichloride complex	6.71±0.03	B	25	0	110
C_5H_4BrN	2-Bromopyridine	3.01	B	20-67	0	589
		3.21	B	25	0	358
	3-Bromopyridine	1.95	B	17-67	0	589
		2.00±0.01	B	25	0	1730
		2.02	B	25	0	358
	4-Bromopyridine	0.77±0.02	B	25	0	1730
		0.89	B	25	0	358
C_5H_4BrNO	3-Bromopyridine 1-oxide	3.69±0.01	B	25	0	1730
	4-Bromopyridine 1-oxide	2.90±0.01	B	25	0	1730
C_5H_4ClN	2-Chloropyridine	3.22	B	25	0	1640
		3.25	B	25	0	358
	3-Chloropyridine	2.02±0.01	B	25	0	1730
		2.02	B	25	0	358

FORMULA	COMPOUND NAME	μ, D	State, or Solv.	t, °C	Method or P_a, % of P_e	Reference
C_5H_4ClN (contd.)	4-Chloropyridine	0.84	B	25	10	1112
		0.84	B	25	0	358
		~0.9	B	20	-	772
C_5H_4ClNO	3-Chloropyridine 1-oxide	3.68±0.01	B	25	0	1730
	4-Chloropyridine 1-oxide	2.82±0.01	B	25	0	891
		2.83±0.01	B	25	0	1730
$C_5H_4F_8O$	2,2,3,3,4,4,5,5-Octafluoro-1-pentanol	2.88	B	20	0	1666
C_5H_4IN	3-Iodopyridine	1.93±0.01	B	25	0	1730
C_5H_4INO	3-Iodopyridine 1-oxide	3.79±0.01	B	25	0	1730
$C_5H_4N_2O_2$	2-Nitropyridine	5.06	B	25	13	1841
		5.06±0.10	B	25	0	1634
	3-Nitropyridine	3.31	B	25	13	1841
	4-Nitropyridine	1.58±0.01	B	25	0	891
		1.61±0.10	B	25	0	1634
		1.68	B	25	13	1841
		1.55	CCl_4	25	0	1729
$C_5H_4N_2O_3$	4-Nitropyridine-1-oxide	0.69±0.02	B	25	0	891
C_5H_4OS	4*H*-Pyran-4-thione	4.08±0.01	B	20	-	1660
	4*H*-1-Thiapyran-4-one	3.96±0.02	B	20	-	1660
	2-Thiophene carboxaldehyde	3.55	B	30	0	903
$C_5H_4O_2$	2-Furaldehyde	2.13	liq	35	5	1397
		3.60	B	n.s.	-	675
		3.6_3	B	25	0	249

FORMULA	COMPOUND NAME	μ, D	State, or Solv.	t, °C	Method or P_a, % of P_e	Reference
		3.63 3.54	B CCl_4	25 25	0 0	363 363
	4*H*-Pyran-4-one	3.72±0.01	B	20	-	1660
$C_5H_4O_2S$	2-Thiophenecarboxylic acid	1.26±0.02 1.30±0.02	B CCl_4	25 25	5 5	500 500
$C_5H_4O_3$	Citraconic anhydride	4.26-4.35	B	10-40	0	1585
	2-Furoic acid	1.38 2.29±0.10	B D	25 25	5 0	1406 1641
$C_5H_4S_2$	4*H*-1-Thiapyran-1-thione	4.41±0.03	B	20	-	1660
$C_5H_5BBr_3N$	Boron tribromide - pyridine complex	7.90±0.06	B	25	0	111
$C_5H_5BCl_3N$	Boron trichloride - pyridine complex	7.70±0.02	B	25	0	111
$C_5H_5BF_3N$	Boron trifluoride - pyridine complex	6.90±0.05	B	25	0	111
$C_5H_5BrO_3$	Methyl α-bromotetronate	6.24	D	25	0?	994
$C_5H_5Br_2N$	Bromine - pyridine complex	5.45	B	25	-	895
C_5H_5ClIN	Pyridine - iodine chloride complex	8.26±0.01	B	25	15	850
$C_5H_5ClN_2O$	2-Methoxy-5-chloropyrimidine	0	D	35	-	1712
C_5H_5ClS	2-Thenyl chloride	1.58	B	30	-	903
$C_5H_5Cl_2OV$	Cyclopentadienyl vanadium oxydichloride	4.91±0.09	B	25	15	520
$C_5H_5IO_3$	Methyl α-iodotetronate	6.16	D	25	0?	994

FORMULA	COMPOUND NAME	μ, D	State, or Solv.	t, °C	Method or P_a, % of P_e	Reference
$C_5H_5I_2N$	Iodine - pyridine complex	4.17	B	n.o.	0	1920
		4.5	cHx	25	-	919
		4.8	j	25	0	947
C_5H_5N	1,3-Butadiene-4-nitrile	3.90	gas	155-190	0	662
	Pyridine	2.15	gas	-	M	402
		2.25	gas	140-315	0	234
		2.23[k]	liq	25-85	0	825
		2.31	liq	20	VS	1701
		2.43	liq	20	-	657
		1.96-2.18[k]	B	25-45	-	825
		2.1	B	20	0	772
		2.11	B	20	-	819
		2.13	B	24	-	1039
		2.20	B	25	10	1112
		2.20±0.02	B	20	0	91
		2.21	B	25	0	360
		2.23	B	15	0	138
		2.25	B	17-67	0	589
		2.27	B	25	-	234
		2.28±0.01	B	10-40	-	1589
		2.28	B	20	-	1307
		2.22	D	25	10	1112
		2.22±0.02	D	25	0	92
		2.345	CCl_4	n.s.	-	596
		2.35	CCl_4	20	-	1307
		2.39	CCl_4	25	-	234
		2.23	Hx	20	-	1307
		2.22	cHx	20	-	1307
		2.28	cHx	25	-	949
		2.27	Tol	20	-	1307
		2.12	CS_2	20	-	1307
		2.26±0.04	kk	20	-	1448
		2.29	m	25	-	234

j) Pyridine. k) Jatkar's equation. kk) Dimethylaniline. m) Light petroleum.

FORMULA	COMPOUND NAME	μ, D	State, or Solv.	t, °C	Method or P_a, % of P_e	Reference
C_5H_5NO	4-Hydroxypyridine	6.0	B	50	10	1112
	Pyridine 1-oxide	4.18	B	25	0	1729
		4.24	B	25	0	1158
		4.24	B	25	0	1552
		4.32	D	25	0	1158
		4.19	cHx	25	0	1729
	2-Pyridone	2.10	B	27	0	1934
	2-Pyrrolecarboxyaldehyde	1.88	B	20	-	1255
		2.42	D	20	-	1255
$C_5H_5NO_2$	2-Furaldoxime (m.p. 75°C)	1.1_7	B	25	0	249
		0.8_2	CCl_4	25	0	249
	2-Furaldoxime (m.p. 90°C)	1.9_2	B	25	0	249
	2-Furamide	3.64±0.10	D	25	0	1641
$C_5H_5NO_3$	2-Methyl-5-nitrofuran	4.91	B	25	5	1406
	3-Methyl-5-isoxazolecarboxylic acid	3.09±0.02	D?	25	0	1859
	5-Methyl-3-isoxazolecarboxylic acid	3.35±0.02	D?	25	0	1859
C_5H_6	Cyclopentadiene	0.416±0.01	gas	-	M	1046
		0.53	gas	70-180	0	660
		0.45	B	(25)	-	1921
	2-Methyl-1-buten-3-yne	0.55	B	20	0	1492
	cis-2-Penten-4-yne	0.86	B	20	0	1492
	trans-2-Penten-4-yne	0.94	B	20	0	1492

FORMULA	COMPOUND NAME	μ, D	State, or Solv.	t, °C	Method or P_a, % of P_e	Reference
C_5H_6 (contd.)	4-Pentenyne	0.57	B	20	0	1492
$C_5H_6Br_2$	*cis*-3,5-Dibromocyclopentene	3.40	D	30	0	1007
	trans-3,4-Dibromocyclopentene	1.37	D	30	0	1007
$C_5H_6N_2$	2-Aminopyridine	2.04±0.02	B	25	0	92
		2.06±0.01	B	25	0	1634
		2.19	B	17-67	0	589
	3-Aminopyridine	3.12±0.02	B	25	0	92
		3.22	B	17-67	0	589
	4-Aminopyridine	3.82	B	17	0	589
		3.95±0.05	B	25	0	92
		3.97±0.10	B	25	0	1634
		4.36	D	25	10	1112
	Glutaronitrile	3.91	B	25	-	1978-9
		3.91	B	25	0	1943
		4.02	B	75	-	1978-9
$C_5H_6N_2OS$	2-(Methylthio)-4(*3H*)-pyrimidone	3.29	D	35	0	1716
$C_5H_6N_2S_2$	2-(Methylthio)-4-thiouracil	4.07	D	35	0	1716
C_5H_6O	1-Cyclopenten-3-one	3.44±0.05	B	n.s.	-	2061
	2-Methylfuran	0.68	liq	35	5	1397
		0.70±0.10	B	25	0	1641
		0.74	B	25	5	1406
$C_5H_6O_2$	Furfuryl alcohol	1.92	liq	35	5	1397
		1.92	B	25	5	1406
$C_5H_6O_4$	Mesaconic acid	2.39±0.02	D	25	-	119

FORMULA	COMPOUND NAME	μ, D	State, or Solv.	t, °C	Method or P_a, % of P_e	Reference
C_5H_6S	2-Methylthiophene	0.67	B	30	0	903
	3-Methylthiophene	0.82	B	30	0	903
C_5H_7Br	3-Bromo-3-methyl-1-butyne	1.94±0.05	B	25	0	1645
C_5H_7BrO	2-Bromocyclopentanone	3.47	B	30	0	1006
		3.64	D	30	0	1006
		3.49	Hp	30	0	1006
$C_5H_7BrO_2$	*cis*-3-Bromo-6-oxabicyclo[3.1.0]-hexan-2-ol	2.40	D	30	0	1007
	trans-3-Bromo-6-oxabicyclo[3.1.0]-hexan-2-ol	2.88	D	30	0	1007
$C_5H_7Br_3$	1,3-Dibromo-2-(bromomethyl)-2-methyl-propane	1.85	B	20	0	636
C_5H_7ClO	2-Chlorocyclopentanone	3.62	B	30	0	1006
		3.69	D	30	0	1006
		3.63	Hp	30	0	1006
C_5H_7N	Methylcyclopropanecarbonitrile	3.81	B	25	-	1632
	N-Methylpyrrole	1.92±0.02	B	25	0	931
	2-Methylpyrrole	1.89	B	25	-	642
C_5H_7NO	3,4-Dimethylisoxazole	3.15±0.02	B	25	-	1858
		3.18±0.02	D	25	-	1858
	3,5-Dimethylisoxazole	3.08±0.02	B	25	0	1858,1860
		3.21	B	25	0	1932
		3.10±0.02	D	25	0	1858,1860

FORMULA	COMPOUND NAME	μ, D	State, or Solv.	t, °C	Method or P_a, % of P_e	Reference
C_5H_7NO (contd.)	4,5-Dimethylisoxazole	3.23±0.02 3.34±0.02	D D	25 25	- -	1858 1858
	Pyridine - water complex	3.13	B	20	-	894
$C_5H_7NOS_2$	3-Ethylrhodanine	1.77	B	25	0	1767
$C_5H_7NO_2$	Glutarimide	2.58±0.04	D	30	0	1059
	N-Methylsuccinimide	1.61±0.02	D	30	0	1059
$C_5H_7NO_3$	*N*-Acetyloxazolidone	2.81±0.02	D	30	0	1061
	5,5-Dimethyloxazolidine-2,4-dione	1.74±0.02	D	30	0	1061
$C_5H_7N_3$	2,6-Diaminopyridine	1.46±0.03	B	25	0	92
C_5H_8	Cyclopentene	0.93 0.98	CCl_4 Hx	25 25	0 0	1540 1540
	2-Methyl-1,3-butadiene	0.15 0.38±0.1	gas gas	-75 to 25 85-205	10 5	496 658
	1,3-Pentadiene	0.50	B	25	8	497
	trans-1,3-Pentadiene	0.68±0.05	gas	115-195	5	658
	1,4-Pentadiene	0.38±0.02	B	10	0	173,1189
	Pentyne	0.86	gas	75-125	0	970
C_5H_8BN	Pyridine - borane complex	5.86±0.01	B	25	0	111
C_5H_8BrClO	3-Chloromethyl-3-bromomethyloxacyclobutane	1.23	B	25	-	533
$C_5H_8Br_2$	*trans*-1,2-Dibromocyclopentane	1.51	D	30	0	1006-7

FORMULA	COMPOUND NAME	μ, D	State, or Solv.	t, °C	Method or P_a, % of P_e	Reference
$C_5H_8Br_2O$	3,3-*Bis*(bromomethyl)oxacyclobutane	1.13-1.313	B	20-92	-	533
$C_5H_8Br_4$	1,3-Dibromo-2,2-*bis*(bromomethyl)propane	0	gas	n.s.	MB	467
		0	B	n.s.	-	429
		0	B	25	-	2112
		0.22[n]	B	25	0	1359
		0.50	B	25	0	1945
		0.59[n]	D	25	0	1359
		0.68	CCl_4	25	0	1081
		0.49	pXy	96	0	1945
		0.59	pXy	116	0	1945
$C_5H_8Cl_2$	1,1-Dichlorocyclopentane	2.37	B	25	-	1650
	trans-1,2-Dichlorocyclopentane	1.64	B	30	0	1006
		1.67	D	30	0	1006
		1.56	Hp	30	0	1006
$C_5H_8Cl_2O$	2-Chloromethyl-2-methylpropionyl chloride	3.21±0.03	B	25	0	1196
	3,3-*Bis*(chloromethyl)oxacyclobutane	1.22	B	25	-	533
$C_5H_8Cl_4$	1,3-Dichloro-2,2-*bis*(chloromethyl)propane	0	B	n.s.	-	429
		<0.2	B	25	-	431
		0.43	B	25	0	1945
		0.63±0.04	B	25	0	1196
		0.65	CCl_4	25	0	1081
$C_5H_8F_4$	1,3-Difluoro-2,2-*bis*(fluoromethyl)propane	2.2	n.s.	n.s.	-	1945
		0.77±0.03	B	25	0	1196
$C_5H_8I_2O$	3,3-*Bis*(iodomethyl)oxacyclobutane	0.93	B	20	-	533

n) "Apparent" due to atomic polarization.

FORMULA	COMPOUND NAME	μ, D	State, or Solv.	t, °C	Method or P_a, % of P_e	Reference
$C_5H_8I_4$	1,3-Diiodo-2,2-*bis*(iodomethyl)propane	0	B	n.s.	-	429
		0	B	25	0	1359
		0.7[nn]	D	25	0	1359
$C_5H_8N_2O$	Methylethylfurazan	4.091	D	20	0	1320
$C_5H_8N_2O_2$	Methylethylfurazan peroxide	4.79	B	25	0	1931
$C_5H_8N_4O_{12}$	Pentaerythritol tetranitrate	2.48	D	25	0	1359
C_5H_8O	Cyclopentanone	3.30±0.01	gas	-	M	940
		2.86	B[p]	20	5	1513
		2.89±0.03	B	20	0	630
		2.93±0.02	B	25	0	127
		3.03±0.03	B	22	15?	416
		2.96±0.03	CCl_4	20	0	630
		3.0_2	CCl_4	25	5	1074
		2.84±0.03	cHx	20	0	630
	Cyclopropyl methyl ketone	2.87	B	25	-	1632
	Dihydropyran	1.43	liq	35	5	1397
		1.39	B	35	0	1848
	Ethyl 1-propynyl ether	1.70±0.03	B	20	0	422
	Ethyl 2-propynyl ether	1.46±0.05	B	25	0	1645
	6-Oxabicyclo[3.1.0]hexane	1.8	n.s.	n.s.	-	1362
	2-Methyl-3-butenone	2.74	B[q]	25	5	469
		2.80	D[q]	25	5	469
	2-Methyl-3-butyn-2-ol	1.59±0.05	B	25	0	1645

nn) "Apparent" due to atomic polarization. p) Plus water. q) Extrapolated from mixtures of B and D.

FORMULA	COMPOUND NAME	μ, D	State, or Solv.	t, °C	Method or P_a, % of P_e	Reference
	3-Penten-2-one	3.20	B	25	5	468
		3.26	D	25	5	468
	Tiglaldehyde	3.39	B	25	0	1631
$C_5H_8O_2$	Acetylacetone	3.03	gas	50-200	10	2170
		2.81±0.08	B	16	15?	2137
	2,6-Dioxa-4-spiroheptane	0.81	n.s.	n.s.	-	701
	Methyl methacrylate	1.60	B	n.s.	-	1680
		1.97	B	18	-	1310
$C_5H_8O_3$	Butylene carbonate	5.27	B	25	0	1165
	Cyclic ester of carbonic acid with 1,3-butanediol	5.32	B	20	5	54
$C_5H_8O_4$	Glutaric acid	2.38	D	35	5	1399
		2.39±0.02	D	25	0	119
		2.64±0.10	D	25	0	1635
	Isopropyl oxalate	2.32	liq	20	0	909
		2.39	liq	45	0	909
		2.46	liq	70	0	909
		2.53	liq	95	0	909
$C_5H_8S_2$	2,6-Dithia-4-spiroheptane	1.12	n.s.	n.s.	-	701
$C_5H_9AsO_3$	1-Methyl-4-arsena-3,5,8-trioxabicyclo-[2.2.2]octane	2.36±0.05	D	25	0	230
C_5H_9Br	Bromocyclopentane	2.16	B	25	0	1622
		2.22	B	25	-	1650
		2.1_5	CCl_4	25	5	1074

FORMULA	COMPOUND NAME	μ, D	State, or Solv.	t, °C	Method or P_a, % of P_e	Reference
$C_5H_9BrO_2$	2-Bromomethyl-1,3-dioxane	2.92	B	25	0	1459
	Methyl 2-bromo-2-methylpropionate	2.34	B	25	0	1176
$C_5H_9Br_3$	1,3-Dibromo-2-bromomethyl-2-methylpropane	2.05	B	25	0	1944
C_5H_9Cl	Chlorocyclopentane	2.10	B	25	-	1650
		1.98[r]	CCl_4	25	-	1542
		2.0_6	CCl_4	25	5	1074
	1-Chloro-3-methyl-2-butene	2.24	B	20	0	1491
	3-Chloro-3-methyl-1-butene	2.09	B	20	0	1491
	2-Chloro-3-pentene	2.17	B	20	0	1491
C_5H_9ClO	Isovaleryl chloride	2.65	B	20	0	1263
	Valeryl chloride	2.63	B	20	0	1263
$C_5H_9ClO_2$	Ethyl 2-chloropropionate	2.46	B	25	0	1176
$C_5H_9Cl_3$	1-Chloro-2,2-di(chloromethyl)propane	1.79±0.02	B	25	0	1196
	1,3-Dichloro-2-methyl-2-chloromethylpropane	2.10	B	25	0	1945
$C_5H_9Cl_3O$	Trichlorohydrin of pentaerythritol	2.03±0.02	B	25	0	1196
C_5H_9F	Fluorocyclopentane	1.88	B	25	-	1650
C_5H_9I	Iodocyclopentane	2.08	B	25	-	1650
		2.0_9	CCl_4	25	5	1074
C_5H_9N	2,2-Dimethylpropanenitrile	3.65	B	25	0	1630

r) $\overline{\mu}^2$ value.

FORMULA	COMPOUND NAME	μ, D	State, or Solv.	t, °C	Method or P_a, % of P_e	Reference
	Pentanenitrile	4.12±0.01 3.60	gas B	150-250 20	5 0	622 342
C_5H_9NO	1-Azacyclohexan-2-one	2.64	B	25	0?	2152
	3-(Dimethylamino)prop-2-en-1-al	6.24±0.02	B	25	0	798
	1-Methylpyrrolidin-2-one	4.09±0.04 4.06±0.03	B B	30 30	5 0	513 1059
	5-Methylpyrrolidinone	4.08±0.04	D	30	0	1060
	cis-δ-Valerolactam dimer	3.83 2.4	B B	25 25	0 0	783 783
$C_5H_9N_3$	Cyclopentylazide	2.29±0.03	B	25	5	125
$C_5H_9N_3O_9$	2,2-*Bis*(nitroxymethyl)-1-nitroxypropane	3.39±0.03	B	25	0	1053
$C_5H_9N_3O_{10}$	2,2-*Bis*(nitroxymethyl)-3-nitroxy-1-propanol	3.73±0.03	B	25	0	1053
$C_5H_9O_3P$	1-Methyl-4-phospha-3,5,8-trioxabicyclo-[2.2.2]octane	4.15±0.05	D	25	0	230
$C_5H_9O_3PS$	1-Methyl-4-phospha-3,5,8-trioxabicyclo-[2.2.2]octane 4-sulfide	6.77±0.05	D	25	0	230
$C_5H_9O_4P$	1-Methyl-4-phospha-3,5,8-trioxabicyclo-[2.2.2]octane 4-oxide	7.10±0.05	D	25	0	230
C_5H_{10}	Cyclopentane	0 0	liq CCl_4	20 25	- 5	2013 1074
	Ethylcyclopropane	0.18	liq	20	-	2013

FORMULA	COMPOUND NAME	μ, D	State, or Solv.	t, °C	Method or P_a, % of P_e	Reference
C_5H_{10} (contd.)	2-Methyl-1-butene	0.50	liq	20	-	23
		0.54	liq	20	-	2013
	1-Pentene	0.34	liq	20	-	23
		0.47	liq	20	-	2013
$C_5H_{10}Br_2$	1,3-Dibromo-2,2-dimethylpropane	2.34	B	25	0	1944
	1,2-Dibromopentane	1.77	B	25	-	1744
	1,5-Dibromopentane	2.30-2.50	gas	-40 to 60	0	1813
		2.33±0.01	liq	20-70	0	907
		2.27	B	20	0	907
		2.27	B	25	0	1943
		2.28	B	25,50	0	1835
		2.284±0.004	B	25	0	692
		2.318±0.004	B	64	0	692
		2.36	D	20	0	907
		2.27	CCl_4	20	0	907
		2.26	Hx	20	0	907
	2,3-Dibromopentane	2.14	B	25	-	1744
	dl-erythro-2,3-Dibromopentane	1.68	liq	25	-	2130
	dl-threo-2,3-Dibromopentane	1.92	liq	25	-	2130
$C_5H_{10}Br_4O_2Zr$	Ethyl propionate - zirconium bromide complex	3.60	B	20	10	923
$C_5H_{10}ClN$	*N*-Chloropiperidine	2.20±0.03	B	20	0	1191
$C_5H_{10}Cl_2$	2,2-*Bis*(chloromethyl)propane	2.26±0.02	B	25	0	1196
	1,3-Dichloro-2,2-dimethylpropane	2.27	B	25	0	1945
	1,5-Dichloropentane	2.36	B	25	0	1943

FORMULA	COMPOUND NAME	μ, D	State, or Solv.	t, °C	Method or P_a, % of P_e	Reference
	3,3-Dichloropentane	2.24	B	20	RT	921
		2.32	Hx	20	RT	921
$C_5H_{10}Cl_2O_2$	Dichlorohydrin of pentaerythritol	2.81±0.02	D	25	0	1196
$C_5H_{10}Cl_2Si$	Trimethyl(2,2-dichlorovinyl)silane	1.64	B	25	15	887
$C_5H_{10}Cl_3O_2P$	Ethyl ethyl(trichloromethyl)phosphinate	3.18	B	25	5	61
$C_5H_{10}Cl_3O_3P$	Diethyl trichloromethylphosphonate	3.28	B	25	5	61
$C_5H_{10}Cl_4O_2Ti$	Ethyl propionate - titanium chloride complex	4.88	B	20	0	1447
$C_5H_{10}Hg$	Mercuracyclohexane	0.86	B	20	-	1700
$C_5H_{10}N_2O_2$	Methyl ethylglyoxime	1.116	D	20	-	1319
	N-Methyl-*N*-nitrosobutramide	0.92	B	25	0	782
$C_5H_{10}N_2O_6$	1,2-Dinitroxypentane	4.10±0.03	B	25	0	1053
	1,5-Dinitroxypentane	3.72±0.03	B	25	0	1053
$C_5H_{10}N_4O_2$	2-Methyl-*N*,*N*′-dinitrosopiperazine	2.10	D	20	-	569
$C_5H_{10}N_4O_4$	2-Methyl-*N*,*N*′-dinitropiperazine	2.21	D	20	-	569
$C_5H_{10}O$	Allyl ethyl ether	1.37±0.01	B	25	0	1646
	Cyclopentanol	1.7_2	CCl_4	25	5	1108
	3,3-Dimethyloxacyclobutane	1.93	B	20	-	533
	Isovaleraldehyde	2.62	B	20	0	329,1472

FORMULA	COMPOUND NAME	μ, D	State, or Solv.	t, °C	Method or P_a, % of P_e	Reference
$C_5H_{10}O$ (contd.)	2-Methyl-3-butanone	2.76	B[s]	25	5	469
	3-Methyl-2-butanone	2.77±0.02	B?	n.s.	-	302
	2-Pentanone	2.5	B	30	MA	2052
		2.72±0.03	B	22	15	2140
		2.74	B	15	10	2141
	3-Pentanone	2.5	B	30	MA	2052
		2.72	B[t]	20	15	609
		2.74	B	15	10	2141
		2.82	CCl_4	25	-	67
	Tetrahydropyran	1.63	liq	35	0	1397
		1.55	B	25	0	357
		1.63	B	35	0	1848
		1.89	B	25	0	5
	Valeraldehyde	2.59	B	20	0	329,1472
$C_5H_{10}OS$	Tetrahydrothiapyran oxide	4.19	B	25	0	357
$C_5H_{10}OS_2$	*S*-Ethyl-*O*-ethylxanthate	1.6_7	B	20	0	1243
$C_5H_{10}O_2$	2,2-Dimethyl-1,3-dioxolane	1.13	B	25	0	1459
	2,4-Dimethyl-1,3-dioxolane	1.33	B	25	0	1459
	Ethyl propionate	1.752±0.03	B	22	15	2140
		1.80	B	50	0	1832
		1.83	B	25	0	1832
		1.83	B	25,50	-	1679
	Isobutyl formate	1.89_0±0.01	B	22	15	2140
	Isopropyl acetate	1.86_1±0.03	B	22	15	413

s) Plus dioxane. t) Also in Hx, Tol, CS_2.

FORMULA	COMPOUND NAME	μ, D	State, or Solv.	t, °C	Method or P_a, % of P_e	Reference
	Isovaleric acid	1.15	liq	25	-	1495
		1.32	liq	60	-	1495
		0.63±0.01	B	25	15	2137
		0.90	B	22	-	208
	Methyl butyrate	1.71_7±0.02	B	22	15	2140
	Methyl isobutyrate	1.98	B	20	-	1309
	2-Methyl-1,3-dioxane	1.91	B	25	0	1459
	Propyl acetate	1.79_2±0.03	B	22	15	2140
		1.8_8	B	25	0	1366
		1.91_5±0.03	CCl_4	22	15	2140
		1.79_2±0.03	Hp	22	15	2140
	Tetrahydrofurfuryl alcohol	2.12	liq	35	5	1397
	Trimethyl acetic acid	1.77-1.81	B,D,Hx	0-40	-	958
		1.9	B	30	-	1526
$C_5H_{10}O_2S$	Tetrahydrothiapyran dioxide	4.82	B	25	0	357
$C_5H_{10}O_3$	Diethyl carbonate	1.07	gas	80-205	10	990
		0.91	B	25	10	1946
		0.95-1.05	B	10-50	10	2157
		0.68-0.92	Hx	-25 to 50	10	2157
		0.67-0.98	Hp	-40 to 70	10	2157
	Ethyl lactate	2.14[u]	B	28	RT	1401
		2.34	B	20	0	1719
	2-Methoxyethyl acetate	2.15	B	30	0	590
$C_5H_{10}S$	3,3-Dimethyl-1-thiacyclobutane	1.78	n.s.	n.s.	-	701

u) 3.26 cm wavelength.

FORMULA	COMPOUND NAME	μ, D	State, or Solv.	t, °C	Method or P_a, % of P_c	Reference
$C_5H_{10}S$ (contd.)	Tetrahydrothiapyran	1.71	B	25	0	357
$C_5H_{11}BCl_2$	Dichloro-*n*-pentylborine	1.55	B	25	5	370
		4.49	D	25	5	370
$C_5H_{11}BF_2$	Difluoro-*n*-pentylborine	1.64	B	25	5	370
		3.37	D	25	5	370
$C_5H_{11}Br$	1-Bromo-3-methylbutane	1.95	B	20	0	1479
	2-Bromo-2-methylbutane	2.27	B	20	0	1479
	1-Bromopentane	2.21	gas	110, 120	20-30	405
		1.99	liq	25	-	707
		2.09	liq	20	-	2055
		1.97	B	20	0	348
		2.04	B	18	-	698
$C_5H_{11}BrHg$	*n*-Pentylmercury bromide	3.50	D	25	0	366
		3.56	D	25	0	366
$C_5H_{11}Cl$	1-Chloro-2,2-dimethylpropane	1.93	B	30	0	543
	1-Chloro-3-methylbutane	1.94	B	20	0	1479
	2-Chloro-2-methylbutane	2.16	B	20	0	1479
	1-Chloropentane	2.14±0.01	gas	80-110	0	623
		1.94	B	20	RT	921
		2.01	Hx	20	RT	921
$C_5H_{11}ClSi$	Trimethyl(chlorovinyl)silane	1.71	B	25	15	887
$C_5H_{11}Cl_3O$	Chloroform - ethyl ether complex	2.51	v	25	-	1450

v) Mutual solvents.

FORMULA	COMPOUND NAME	μ, D	State, or Solv.	t, °C	Method or P_a, % of P_e	Reference
$C_5H_{11}F$	2-Fluoro-2-methylbutane	1.92	B	25	0	1630
		1.55	CCl_4	20	-	79
	1-Fluoropentane	1.85	B	25	0	1630
		1.31	CCl_4	20	-	79
$C_5H_{11}I$	1-Iodo-3-methylbutane	1.85	B	20	0	1479
	2-Iodo-2-methylbutane	2.20	B	20	0	1479
		2.21	CCl_4	20	-	78
	1-Iodopentane	1.90	B	20	0	348
		1.93	CCl_4	20	-	78
	2-Iodopentane	2.11	CCl_4	20	-	78
	3-Iodopentane	2.11	CCl_4	20	-	78
$C_5H_{11}I_2N$	Piperidine - iodine complex	7.90[w]	B	25	-	895
$C_5H_{11}Li$	*n*-Pentyllithium	1.1	Hx	25	0	2034
		1.15-1.20	Hx	n.s.	-	1628
$C_5H_{11}N$	Piperidine	1.10±0.02	B	20	0	1191
		1.18	B	10-40	-	1589
		1.19	B	25	0	68
		1.29	CCl_4	25	0	1729
		1.07±0.03	pXy	30,80	5	512
	Propylideneethylamine	1.52±0.01	B	10-40	-	1589
$C_5H_{11}NO$	2-Butanone methyloxime	1.19	liq	20	5	1593
		1.118±0.005	B	20	5	1593
	Methyl propyl ketoxime	0.9_9	"pure"	19.8	0	249

w) Lower limit to μ - not accurately determined.

FORMULA	COMPOUND NAME	μ, D	State, or Solv.	t, °C	Method or P_a, % of P_e	Reference
$C_5H_{11}NO_2$	Betaine	10.7	x	25	-	438
	Ethyl 2-aminopropionate	2.11	B	25	0	2152
	Ethyl 3-aminopropionate	2.16	B	5-75	0	2152
	Pentyl nitrite	2.29	B	25	0	2084
		2.24	CCl_4	25	0	2017-8
$C_5H_{11}NO_4$	2-Ethyl-2-nitro-1,3-propanediol	3.54±0.05	B	n.s.	-	257
$C_5H_{11}NS_2$	Ethyl dimethyldithiocarbamate	3.25	B	25	0	639
	S-Ethyl-*N*-ethyldithiocarbamate	3.1_2	B	20	0	1243
$C_5H_{11}N_5$	1-Ethyl-5-ethylaminotetrazole	7.36	B	25	0	892
C_5H_{12}	*n*-Pentane	0	gas	35-110	-	983
		<0.05	gas	25-145	-	1040
		<0.07	gas	25-160	-	643
		0	liq	-90 to 30	-	418
		0	liq	20	-	2013
		0.37	B	27	0	720
	Isopentane	0.1	gas	25-160	-	643
		0.13±0.01	gas	25-145	-	1040
	neo-Pentane	<0.05	gas	25-160	-	643
$C_5H_{12}N_2O$	*t*-Butylurea	4.25	D	25	-	562
		4.27±0.04	D	25	0	120
		5.40	y	25	-	562
	asym-Diethylurea	4.08	D	25	-	562
		4.11	D	25	0	120
		4.62	y	25	-	562

x) Benzene-ethanol. y) Acetone.

FORMULA	COMPOUND NAME	μ, D	State, or Solv.	t, °C	Method or P_a, % of P_t	Reference
	Tetramethylurea	3.92	"pure"	25	-	562
		3.47±0.04	B	25	0	120
		3.49	D	25	-	562
		3.66±0.04	CCl_4	25	0	120
		3.47±0.04	Hx	25	0	120
		3.40±0.04	cHx	25	0	120
		3.43±0.04	Hp	25	0	120
		3.34±0.04	Tol	25	0	120
		3.34±0.04	z	25	0	120
		3.29±0.04	a	25	0	120
		3.28±0.04	b	25	0	120
$C_5H_{12}N_2S$	*n*-Butylthiourea	5.698±0.002	D	25	5	275-6
	Isobutylthiourea	5.703±0.002	D	25	5	275-6
	sym-Diethylthiourea	4.9	D	20	0	792
$C_5H_{12}O$	Ethyl *n*-propyl ether	1.16	B	25	-	1863
	2-Methylbutan-2-ol	1.96	liq	30-40	-	1498
		1.66	B	30	-	1237
		1.67	B	18	0	681a
		1.7	B	25	0	1747
		1.72±0.04	B	20	0	1184
		1.85	B	24	-	1039
	3-Methylbutan-1-ol	1.63	B	10-70	-	1875
		1.65	B	30	-	1237
		1.78	B	18-64	-	1039
		1.8	B	25	0	1747
		1.84	B	26.5	0	1234
		1.87	CCl_4	25	0?	966
		1.64	c	55	-	1865
	Methyl *n*-butyl ether	1.25	B	25	-	1863

z) *trans*-Decalin. a) *m*-Xylene. b) Isopropylbenzene. c) *p*-Dichlorobenzene.

FORMULA	COMPOUND NAME	μ, D	State, or Solv.	t, °C	Method or P_a, % of P_e	Reference
$C_5H_{12}O$ (contd.)	Methyl *t*-butyl ether	1.225±0.02	B	20	0	1198
	1-Pentanol	0.89	liq	18	REF	1115
		1.37	B	n.s.	0	1665
		1.59	B	20	-	977
		1.66	B	30	-	1237
		1.66	B	20?	-	722
		1.66	B	25	5	1108
		1.67	B, Hx	18	0	716
		1.68	B	20	15	1241a
		1.71±0.02	B	20	0	1184
		1.72_5	CCl_4	25	-	1078
		1.7_3	CCl_4	25	5	1108
		1.8	Ether	20	0	717,723
$C_5H_{12}O_2$	Diethoxymethane	1.23	gas	55-80	10	987
		1.27	gas	135-200	10	987
		0.75	B	n.s.	5	334
		0.93	B	25	0	1457
	1,2-Pentanediol	2.24±0.02	D	25	-	386
	1,4-Pentanediol	2.51±0.03	D	25	-	386
	1,5-Pentanediol	3.64-3.87	liq	20-70	0	907
		2.37±0.03	D	25	-	386
		2.45	D	20	0	907
	2,3-Pentanediol	2.13±0.05	D	25	-	386
	2,4-Pentanediol	2.71±0.03	D	25	-	386
$C_5H_{12}O_3$	2-Hydroxymethyl-2-methyl-1,3-propanediol	2.79±0.10	D	15-30	-	2060
$C_5H_{12}O_4$	Pentaerythritol	~2	gas	n.s.	MB	465

FORMULA	COMPOUND NAME	μ, D	State, or Solv.	t, °C	Method or P_a, % of P_e	Reference
	Tetramethyl orthocarbonate	<0.3	gas	100-180	-	554
		>0	B	-	-	429
		0.84	B	25	5	56
$C_5H_{12}O_4S_2$	2,2-*Bis*(methylsulfonyl)propane	1.27	B	20	0	636
$C_5H_{12}S$	Methyl *t*-butyl sulfide	1.57±0.02	B	20	0	1193
	1-Pentanethiol	1.54	liq	25	0	1280
		1.51	B	25,50	0	2059
$C_5H_{12}S_4$	Tetramethyl orthothiocarbonate	0.50	liq	70	-	82
$C_5H_{12}Si$	Trimethylvinylsilane	0	B	25	15	887
$C_5H_{13}BO_2$	*n*-Pentylboric acid	1.92	B	25	-	1458
		1.94	D	25	-	1458
$C_5H_{13}ClGe$	Trimethyl-(1-chloroethyl)germane	2.34	B	25	-	886
$C_5H_{13}ClOSi$	Dimethylchloromethylethoxysilane	2.14	B	30	0	545
$C_5H_{13}N$	Amylamine	1.55	B	30	5	1578
	Isoamylamine	1.53	B	30	5	1578
$C_5H_{13}NO_2$	2-Methyl-3-azapentane-1,5-diol	2.90±0.05	n.s.	n.s.	-	1720
	N-Methyldiethanolamine	2.86±0.03	n.s.	n.s.	-	1720
$C_5H_{13}O_2P$	Isopropyl ethylphosphinate	3.37	B	25	5	61
	Propyl ethylphosphinate	3.40	B	25	5	61
$C_5H_{13}O_3P$	Diethyl methylphosphonate	3.33	liq	30	-	950

FORMULA	COMPOUND NAME	μ, D	State, or Solv.	t, °C	Method or P_a, % of P_e	Reference
$C_5H_{14}N_2$	1,5-Diaminopentane	1.93	B	25,45	0	1977,1979
$C_5H_{14}OSi$	Trimethylethoxysilane	1.18	B	30	0	545
$C_5H_{15}ClN_2Si$	*Bis*(dimethylamino)methylchlorosilane	2.26±0.02	B	20	-	1702
$C_5H_{15}NSi$	Dimethylaminotrimethylsilane	0.67±0.07	B	20	-	1702
	Trimethyl-*N*-ethylsilazane	0.72	liq	30	-	328
		0.78	B	30	-	328
C_5IMnO_5	Manganese pentacarbonyl iodide	3.16±0.04	B	25	20	116
C_6BrCl_5O	4-Bromo-2,3,4,5,6-pentachloro-2,5-cyclohexadien-1-one	1.74	B	25	25?	1745
	Pentachlorophenylhypobromite	2.05	B	25	25?	1745
$C_6Br_2I_2O_2$	Dibromodiiodo-*p*-benzoquinone	0?	B	35	0	1851
$C_6Br_3N_3O_6$	1,3,5-Tribromo-2,4,6-trinitrobenzene	0	B	20,59	45	1207
		0.66	B	20	-	1949
$C_6Br_4O_2$	Tetrabromo-*p*-benzoquinone	0	B	35	0	1851
$C_6Cl_3N_3O_6$	1,3,5-Trichloro-2,4,6-trinitrobenzene	0.64	B	20	-	1949
$C_6Cl_4O_2$	Tetrachloro-*p*-benzoquinone	0	B	35	0	1851
		0.87[d]	D	25	0	1761
		0.38	n.s.	n.s.	-	701
C_6Cl_6	Hexachlorobenzene	0.2_0	B	50	-	1949
		0.54	B	20	SPC	1300
		0.74	B	30	SPC	1300
		0.80	B	35	SPC	1300
		0.86	B	40	SPC	1300

d) If P_a = 33% P_e then $\mu = 0$.

FORMULA	COMPOUND NAME	μ, D	State, or Solv.	t, °C	Method or P_a, % of P_e	Reference
		0.95	B	60	SPC	1300
		0.21	CCl_4	20	SPC	1300
		0.50	CCl_4	30	SPC	1300
C_6Cl_6O	2,3,4,4,5,6-Hexachloro-2,5-cyclohexa-dien-1-one	1.63	B	25	29	1745
	Pentachlorophenol hypochlorite	2.09	B	25	23	1745
C_6Cl_{10}	Perchlorocyclohexene	0	B	30	5	1328,1614
C_6Cl_{12}	Perchlorocyclohexane	0	B	25,30	5	1614,1754
$C_6F_3MnO_5$	Trifluoromethylmanganese pentacarbonyl	3.57±0.03	B	25	20	116
$C_6F_4I_2$	1,4-Diiodotetrafluorobenzene	0.14	B	30	SPC	1300
C_6F_{12}	Perfluorocyclohexane	0	B	n.s.	-	1803
$C_6F_{15}N$	Perfluorotriethylamine	1.36	B	25	15	1647
$C_6HBr_3N_2O_4$	1,3,5-Tribromo-2,4-dinitrobenzene	3.10	B	25	~5	1162
C_6HCl_5	Pentachlorobenzene	0.88±0.03	B	20	0	904
		0.89	B	25	0	1814
C_6HCl_5O	Pentachlorophenol	2.14	B	25	29	1745
C_6HF_{11}	Undecafluorocyclohexane	1.64	B	n.s.	-	1803
$C_6H_2Br_2N_2O$	3,5-Dibromo-1,2-benzoquinone-2-diazide	2.9_2	B	25	0	36
$C_6H_2Br_2O_2$	Dibromo-*p*-quinone	0.70	B	25	0	648
$C_6H_2Br_3NO_2$	1,3,5-Tribromo-2-nitrobenzene	3.12	B	25	~5	1162
$C_6H_2Br_4$	1,2,4,6-Tetrabromobenzene	0.7	B	n.s.	0	678

FORMULA	COMPOUND NAME	μ, D	State, or Solv.	t, °C	Method or P_a, % of P_r	Reference
$C_6H_2ClN_3O_6$	Picryl chloride	1.1_3	B	25	0	1071
$C_6H_2Cl_2O_2$	2,5-Dichloro-1,4-benzoquinone	0	gas	-90 to -30	26	331
		0.64	B	25	0	648
$C_6H_2Cl_4$	1,2,3,4-Tetrachlorobenzene	1.92	B	25	0	1814
	1,2,4,5-Tetrachlorobenzene	0.73±0.01	B	20-40	SPC	1300
	1,2,4,5-Tetrachlorobenzene	0.65	B	n.s.	0	678
$C_6H_2F_{10}$	*trans*-1Ha:2Ha-Decafluorocyclohexane	0.88	B	n.s.	-	1803
	cis-1He:2He-Decafluorocyclohexane	2.59	B	n.s.	-	1803
	trans-1Ha:4Ha-Decafluorocyclohexane	0.2-0.15	B	(20)	-	474
	cis-1Ha:4He-Decafluorocyclohexane	2.59±0.07	B	(20)	-	474
$C_6H_2N_4O_6$	1,3-Dinitro-*o*-benzoquinone furoxan	3.01	D	25	0	1323,1930
	2,3-Dinitro-*o*-benzoquinone furoxan	2.77	D	25	0	1323,1930
$C_6H_3BrN_2O_2$	5-Bromofurazan *N*-oxide	3.65	D	25	-	1324
$C_6H_3BrN_2O_4$	1-Bromo-2,4-dinitrobenzene	3.1±0.1	B	20	35	1209
		3.39	B	25	0	1162
	1-Bromo-3,5-dinitrobenzene	2.3	B	20	35	1209
$C_6H_3Br_3$	1,3,5-Tribromobenzene	0.2	B	n.s.	0	751
		0.28	B	20	-	1949
		1.46	B	18	RT	1532
		1.55	B	25	0	598
		1.57	B	n.s.	0	678

FORMULA	COMPOUND NAME	μ, D	State, or Solv.	t, °C	Method or P_a, % of P_e	Reference
		1.98	D	25	-	451
		1.44	cHx	25	-	451
$C_6H_3ClN_2O$	3-Chloro-*o*-benzoquinone furazan	3.21	D	25	0	1930
	5-Chlorobenzofurazan	2.88	B	25	-	736
$C_6H_3ClN_2O_2$	3-Chloro-*o*-benzoquinone furoxan	3.93	D	25	0	1930
$C_6H_3ClN_2O_4$	1-Chloro-2,4-dinitrobenzene	3.0±0.1	B	20	36	1209
		3.24	B	25	0	1071
		3.275	B	20	-	1673
		3.28	B	25	0	1162
		3.31	B	n.s.	0	678
$C_6H_3ClN_2S$	5-Chloro-2,1,3-benzothiadiazole	0.84	B	25	-	736
$C_6H_3ClN_2Se$	5-Chloro-2,1,3-benzoselenadiazole	0.5	B	25	-	736
$C_6H_3Cl_2I$	Dichloroiodobenzene	2.6_2	B	25	0	1072
$C_6H_3Cl_2NO_2$	2,3-Dichloronitrobenzene	3.89	B	25	5	1947
	2,4-Dichloronitrobenzene	2.68	B	25	5	1947
	2,5-Dichloronitrobenzene	3.48	B	n.s.	0	678
		3.48	B	25	5	1947
		3.48	B	25	0	1071
	2,6-Dichloronitrobenzene	3.17	B	25	-	2028
		4.21	B	25	5	1947
	3,4-Dichloronitrobenzene	2.19	B	25	5	1947
	3,5-Dichloronitrobenzene	2.68	B	25	5	1947

FORMULA	COMPOUND NAME	μ, D	State, or Solv.	t, °C	Method or P_a, % of P_e	Reference
$C_6H_3Cl_3$	1,2,3-Trichlorobenzene	2.33±0.01	B	20	0	904
	1,2,4-Trichlorobenzene	1.26	B	n.s.	0	678
	1,3,5-Trichlorobenzene	0.28 0.3_1	B B	20 25	- 0	1949 1071
$C_6H_3Cl_3O$	2,4,5-Trichlorophenol	1.43 1.63 1.88 1.38	B B D cHx	25 n.s. 25 25	0 0 - -	598 678 451 451
$C_6H_3Cl_9O_3$	2,4,6-*Tris*(α,α,α-trichloromethyl)-1,3,5-trioxacyclohexane	1.64±0.1	n.s.	n.s.	-	1421
$C_6H_3Cl_{13}O_6Sn$	Tin chloride - *tris*(trichloroacetic acid) complex	1.89	B	n.s.	-	1449
$C_6H_3FN_2O_4$	2,4-Dinitrofluorobenzene	3.30[e] 3.55	B B	30 30	MA -	1574 1571
$C_6H_3FeNO_4$	Iron methylnitrosyltetracarbonyl	5.02±0.03	B	25	20	715
$C_6H_3IN_2O_4$	1-Iodo-2,4-dinitrobenzene	3.4±0.1	B	20	31	1209
$C_6H_3I_3$	1,3,5-Triiodobenzene	0.24	B	20	-	1949
$C_6H_3MnO_5$	Methylpentacarbonylmanganese	0.4_3	B	25	20	714
$C_6H_3N_3O_3$	1-Nitro-*o*-benzoquinone furazan	5.80	D	25	0	1323,1930
$C_6H_3N_3O_4$	1-Nitro-*o*-benzoquinone furoxan	5.51	D	25	0	1323,1930
	3-Nitro-*o*-benzoquinone furoxan	2.52	D	25	0	1323,1930

e) At 9515 Mc frequency.

FORMULA	COMPOUND NAME	μ, D	State, or Solv.	t, °C	Method or P_a, % of P_e	Reference
$C_6H_3N_3O_6$	1,3,5-Trinitrobenzene	0	B,D	25	22	272
		0.3	B	n.s.	-	750
		0.41±0.07	B	20	15	214
		~0.5	B	25,45	0	1064
		0.7	B	n.s.	0	751
		0.71	B	20	0	1949
		0.8	B	25	-	2112,2114
		0.8	B	10-50	0	1476
		0.81	B	20	0	1719
		1.09	B	25	-	2123
		0.4	D	25	0	1064
		1.26[f]	CCl_4	25	27	272
		0.56-0.78	g	85	-	213,215
		0	$CHCl_3$	25	-	841
$C_6H_3N_3O_7$	2,4,6-Trinitrophenol	1.345	B	20	-	1673
		1.50[h]	B	6-35	-	1532
		1.51	B	25	0	1597
		1.75	B	30	0	1269
		2.13	D	25	0	1597
$C_6H_3N_3O_8$	2,4,6-Trinitro-1,3-dihydroxybenzene	1.5±0.03	B	25	27	1764
$C_6H_3N_5O_4$	1-Azido-2,4-dinitrobenzene	2.66	B	25	0	1762
$C_6H_4AlBr_3ClNO_2$	Aluminum bromide - *o*-chloronitrobenzene complex	9.62	B	20	0	1734
	Aluminum bromide - *p*-chloronitrobenzene complex	7.50	CS_2	20	0	560
$C_6H_4AlCl_4NO_2$	Aluminum chloride - *o*-chloronitrobenzene complex	9.54	B	n.s.	-	1733
	Aluminum chloride - *p*-chloronitrobenzene complex	7.84	B	n.s.	-	1733

f) Low accuracy because of low solubility. g) Naphthalene.
h) In private communication the original author suggests 1.50 D as the best average in the temperature range given.

FORMULA	COMPOUND NAME	μ, D	State, or Solv.	t, °C	Method or P_a, % of P_e	Reference
$C_6H_4BCl_3N_2$	4-Cyanopyridine - boron trichloride complex	4.20±0.05	B	25	0	110
C_6H_4BrCl	*o*-Bromochlorobenzene	2.15	B	20	0	140
		2.23	B	22	15	185
		2.25	Hx	22	15	185
	m-Bromochlorobenzene	1.52	B	22	15	185
		1.54	Hx	22	15	185
	p-Bromochlorobenzene	<0.03	B	22	15	185
		0.1	B	25	-	2114
		<0.2	B	25	-	2112
$C_6H_4BrClHg$	*p*-Bromophenylmercury chloride	1.75	D	(20)	-	1700
	p-Chlorophenylmercury bromide	1.58	D	25	0	366
$C_6H_4BrClO_2$	*p*-Bromophenylsulfonyl chloride	3.26	B	20	0	1266
		3.33	B	25	0	634
C_6H_4BrF	*o*-Bromofluorobenzene	1.39	liq	(30)	-	1374
		2.29	B	22	0	140
	m-Bromofluorobenzene	1.44	liq	(30)	-	1374
		1.40	B	30	-	1367,1370
	p-Bromofluorobenzene	0.5	gas	160-250	0	794
		0.53	liq	(30)	-	1374
		0	B	25	5	1114
C_6H_4BrI	*o*-Bromoiodobenzene	1.75	B	20	0	2057
		1.88	B	19	0	140
	m-Bromoiodobenzene	1.15	B	20	0	2057

FORMULA	COMPOUND NAME	μ, D	State, or Solv.	t, °C	Method or P_a, % of P_e	Reference
	p-Bromoiodobenzene	0.49	B	20	0	2057
C_6H_4BrNO	*p*-Bromonitrosobenzene	1.94	B	20	0	1990
$C_6H_4BrNO_2$	*o*-Bromonitrobenzene	4.01±0.06	B	25,65	-	1208
		4.19±0.01	B	20	0	1186
		4.23	B	20	-	1949
	m-Bromonitrobenzene	3.21	liq	140	-	1570
		3.27	liq	55	-	1570
		3.20±0.08	B	25,65	-	1208
		3.44	B	20	-	1949
		3.44±0.02	B	20	0	1186
		3.44±0.01	B	25	0	1730
	p-Bromonitrobenzene	2.47±0.1	B	25,65	-	1208
		2.55	B	n.s.	0	751
		2.64±0.02	B	20	0	1186
	2.69 P_e = 40.00 P_a = 4.00	2.66±0.01	B	25	0	1730
		2.67	B[i]	20[i]	-	1949
		2.68	B	25	0	481
		2.68±0.02	B	30	0	170
		2.68	B	25	0	1162
		2.71	B	n.s.	-	750
		2.47±0.02	j	30	0	170
$C_6H_4BrN_3$	*o*-Azidobromobenzene	2.25	B	25	5	1854
	m-Azidobromobenzene	1.40	B	25	5	1854
	p-Azidobromobenzene	0.02	B	25	5	1854
		0.26	B	25	0	1762
		0.64	B	23	0	155-6

i) Inferred from table listing in original. j) 1-Methylnaphthalene.

FORMULA	COMPOUND NAME	μ, D	State, or Solv.	t, °C	Method or P_a, % of P_e	Reference
$C_6H_4Br_2$	*o*-Dibromobenzene	1.53	B	n.s.	14	454
		1.57	B	n.s.	-	455
	1.90 P_e = 39.98, P_a = 4.00	1.89	B	20	0	2057
		2.05	B	23	0	140
		2.13	B	20	0	514
		2.13	B	20	-	1949
		2.14	B	25	0	1773
	m-Dibromobenzene	1.10	B	n.s.	-	455
		1.47	B	20	-	1949
		1.56	B	20	0	2057
	p-Dibromobenzene	0	B	n.s.	-	455
		0	B	20	0	2057
		0.22	B	20	-	1949
$C_6H_4Br_3N$	2,4,6-Tribromoaniline	1.693	B	25	0	508
		1.82	B	n.s.	-	678
		1.972	D	25	0	508
$C_6H_4Br_4$	1,2,3,4-Tetrabromo-2,3-dimethylbutane	1.55	B	25	0	1919
C_6H_4ClF	*o*-Chlorofluorobenzene	1.71	liq	(30)	-	1374
		2.35	B	18	0	140
		2.38	B	30	-	548
		2.40	Hp	30	-	548
	m-Chlorofluorobenzene	1.49	liq	(30)	-	1374
		1.52	B	30	-	1367,1370
	p-Chlorofluorobenzene	0.95	liq	30	-	1370
		0.99	liq	(30)	-	1374
C_6H_4ClI	*o*-Chloroiodobenzene	1.95	B	19	0	140
	m-Chloroiodobenzene	1.40	B	25	-	640

FORMULA	COMPOUND NAME	μ, D	State, or Solv.	t, °C	Method or P_a, % of P_e	Reference
	p-Chloroiodobenzene	0.46	B	25	-	540
$C_6H_4ClIO_2S$	*p*-Iodophenylsulfonyl chloride	3.56	B	25	0	634
C_6H_4ClNO	*p*-Chloronitrosobenzene	1.82	B	25	0	650
		1.86	B	n.s.	-	704
C_6H_4ClNOS	*m*-Chloro-*N*-sulfinylaniline	1.95	liq	25	-	968
		1.81	B	20	-	968
		1.72	D	20	-	968
$C_6H_4ClNO_2$	*o*-Chloronitrobenzene	4.63	gas	200	5	625
		6.22	liq	33-160	-	815
		3.81	B	20	0	2057
		3.99±0.06	B	25,62	-	1208
		4.28	B	n.s.	-	750-1
		4.35	B	25	-	2028
		4.37	B	20	-	1949
		4.86	B	20	-	816
	m-Chloronitrobenzene	3.72	gas	210	5	625
		3.31	liq	55-65	-	1497
		4.29	liq	46-160	-	815
	3.40 P_e = 37.30 P_a = 3.73	3.15±0.08	B	25,65	-	1208
		3.21	B	20	5	2057
		3.26	B	25	15	1841
		3.41	B	n.s.	-	750-1
		3.43	B	20	-	1949
		3.44±0.01	B	25	0	1730
	p-Chloronitrobenzene	2.81	gas	210	5	625
		2.83	liq	84-160	-	815
	2.60 P_e = 37.30 P_a = 3.73	2.36±0.1	B	25,65	-	1208
		2.38	B	20	5	1949
		2.54	B	n.s.	-	750
		2.57	B	n.s.	-	751

FORMULA	COMPOUND NAME	μ, D	State, or Solv.	t, °C	Method or P_a, % of P_e	Reference
$C_6H_4ClNO_2$ (contd.)	*p*-Chloronitrobenzene (contd.)	2.59	B	20	-	1949
		2.59	B	25	0	1162
		2.62±0.01	B	25	0	1730
		3.15	B	20	-	816
		2.38	CS_2	20	0	560
$C_6H_4ClNO_4S$	*m*-Nitrophenylsulfonyl chloride	4.12	B	25	0	634
		4.15	B	n.s.	-	632-3
$C_6H_4ClN_3$	*o*-Azidochlorobenzene	2.37	B	25	5	1854
	m-Azidochlorobenzene	1.45	B	25	5	1854
	p-Azidochlorobenzene	0.30	B	25	5	1854
		0.33-0.35	B	25	0	1774,1902
		0.47	B	21	0	155-6
$C_6H_4Cl_2$	*o*-Dichlorobenzene	2.18	gas	150-180	5	625
		2.51	gas	170-250	0	794
		2.54	gas	80-100	0	1352
		2.56	liq	20	-	657
	2.27 P_e = 34.57, P_a = 3.46	2.0	B	n.s.	14	454-5
		2.24	B	n.s.	-	976
		2.26	B	19	0	140
		2.27	B, Hp	25	17	1826
		2.27	B	24	-	1262
		2.28	B	22	0	514
		2.32	B, Hx	0-50	-	1824
		2.35	B	20	-	1949
		2.02	Ether	20	5	728
	m-Dichlorobenzene	1.68	gas	140-185	5	625
		1.59	liq	20	-	657
		1.22	B	n.s.	14	454-5
		1.38	B	24	-	1262
		1.49	B, Hp	25	13	1826
		1.49	B	20	-	1949

FORMULA	COMPOUND NAME	μ, D	State, or Solv.	t, °C	Method or P_a, % of P_e	Reference
		1.56	B, Hx	0-50	-	1824
		1.34	Ether	20	5	728
	p-Dichlorobenzene	0	gas	160	5	625
		0	B	n.s.	12	454-5
		0	B, Hx	0-50	-	1824
		0	B, Hp	25	10	1826
		0	B	n.s.	-	2111
		0	B	25	0	1067
		0	B	24	-	1262
		0.23	B	20	-	1949
		0.36	B	20	-	977
		0	Hp	20	12	209
		0	CCl_4	25	0	1067, 1719
		1.66	NB	25	0	1067
		1.68	NB	30	0	754
		0.83	k	25	0	1067
		1.61	l	25	0	1067
		0.86	m	25	0	1067
		0.70	n	25	0	1067
$C_6H_4Cl_2Hg$	(*m*-Chlorophenyl)mercury chloride	2.91	D	20	0	1777
$C_6H_4Cl_2O$	2,4-Dichlorophenol	1.60	B	25	0	1900
		2.09	B	40	RT	1864
$C_6H_4Cl_2O_2S$	*p*-Chlorophenylsulfonyl chloride	3.23	B	25	0	634
$C_6H_4Cl_3I$	*o*-Chlorophenyl iododichloride	2.98	B	25	-	640
	m-Chlorophenyl iododichloride	2.13	B	25	-	640
	p-Chlorophenyl iododichloride	1.36	B	25	-	640
$C_6H_4Cl_3N$	2,4,6-Trichloroaniline	1.96	B	n.s.	-	678
		2.168	B	20	-	1673

k) Chlorobenzene. l) Benzonitrile. m) Ethyl benzoate. n) Dimethylaniline.

FORMULA	COMPOUND NAME	μ, D	State, or Solv.	t, °C	Method or P_a, % of P_e	Reference
$C_6H_4Cl_8$	1ea2ea3e4e5a6e-Octachlorocyclohexane	1.82	B	30	-	1606
	1ea2ea3e4e5e6a-Octachlorocyclohexane	1.43	B	30	-	1606
	1ea2ea3e4e5e6e-Octachlorocyclohexane	0.40-0.44	B	25	5	1356
	1ea2e3e4ea5e6e-Octachlorocyclohexane	0.40-0.44	B	25	5	1356
	Octachlorocyclohexane[p]	1.69	B	30	5	1328
C_6H_4FI	*o*-Fluoroiodobenzene	2.07	liq	(30)	-	1374
		2.02	B	22	0	140
	m-Fluoroiodobenzene	1.34	liq	(30)	-	1374
		1.38	B	30	-	1367,1370
	p-Fluoroiodobenzene	0.9	gas	200-220	0	794
		0.76	liq	(30)	-	1374
$C_6H_4FNO_2$	*p*-Fluoronitrobenzene	2.87	gas	215-250	0	794
		2.64	B	25	5	1114
		2.65	B	21	0	141
$C_6H_4F_2$	*o*-Difluorobenzene	2.40	B	22	0	140
	m-Difluorobenzene	1.58	gas	80-150	0	1352
C_6H_4INO	*p*-Iodonitrosobenzene	2.18	B	22	0	1990
$C_6H_4INO_2$	*o*-Iodonitrobenzene	3.69±0.07	B	25,65	-	1208
		3.95	B	22	15	1527
	m-Iodonitrobenzene	3.25±0.09	B	25,65	-	1208
		3.47	B	22	15	1527
		3.54±0.01	B	25	0	1730

p) Configuration not given.

FORMULA	COMPOUND NAME	μ, D	State, or Solv.	t, °C	Method or P_a, % of P_e	Reference
	p-Iodonitrobenzene	2.65±0.12	B	25,65	-	1208
		2.90±0.01	B	25	0	1730
		3.07_5	B	22	15	1527
$C_6H_4I_2$	*o*-Diiodobenzene	1.33	B	n.s.	14	454-5
		1.70	B	20	-	1949
		1.71	B	23	0	140
	m-Diiodobenzene	1.02	B	n.s.	14	454-5
		1.28	B	20	-	1949
	p-Diiodobenzene	0.19	B	20	-	1949
$C_6H_4N_2$	2-Cyanopyridine	5.24	B	25	0	358
	3-Cyanopyridine	3.46	B	25	0	358
	4-Cyanopyridine	1.63	B	25	0	358
	Isonicotinonitrile	1.61	B	25	10	1112
		1.65±0.02	B	25	0	92
	Nicotinonitrile	3.48±0.02	B	25	0	92
	Picolinonitrile	5.23±0.03	B	25	0	92
$C_6H_4N_2O$	Benzofurazan	4.03	B	25	-	736
	o-Benzoquinone furazan	4.40	D	25	-	1322,1930
	4-Cyanopyridine-1-oxide	1.22±0.13	B	25	0	110
	4-Quinonediazide	5.0_2	B	25	0	36
$C_6H_4N_2O_2$	*o*-Benzoquinone furoxan	5.33	D	25	0	1322,1930

FORMULA	COMPOUND NAME	μ, D	State, or Solv.	t, °C	Method or P_a, % of P_e	Reference
$C_6H_4N_2O_2S_2$	Dithionyl-*p*-phenylenediamine	1.61	B	25	0	855
$C_6H_4N_2O_3$	*p*-Nitronitrosobenzene	0.85	B	n.s.	-	704
$C_6H_4N_2O_3S$	*o*-Nitrothionylaniline	3.74	B	25	15	855
	m-Nitrothionylaniline	3.50	B	20	-	968
		3.51	B	25	15	855
	p-Nitrothionylaniline	3.09	B	25	15	855
		3.20	B	20	-	968
$C_6H_4N_2O_4$	*o*-Dinitrobenzene	2.27	B	20	-	977
		6.00	B	n.s.	-	750
		6.03	B	20	-	1949
		6.05	B	n.s.	0	751
		6.10	B	25	-	2123
		6.31	B	25	-	581
		6.87	NB	25	-	581
		6.51	q	25	-	581
		6.52	r	25	-	581
	m-Dinitrobenzene	3.65	liq	140	-	1570
		3.68	liq	92	-	1570
		3.73	B	n.s.	0	751
	3.87 P_e = 40.03, P_a = 0	3.743	B	20	-	1673
		3.81	B	20	-	1949
		3.82	B	20	-	977
		3.83	B	25	0	1162
		3.84	B	25	-	2123
		3.86	B	25	0	1071
		3.99	B	25	5	728
		4.03	B	25	-	963
		4.05	B	n.s.	-	750
		4.10	Hx	25	5	728

q) *o*-Dichlorobenzene. r) 50% *o*-Dichlorobenzene plus benzene.

FORMULA	COMPOUND NAME	μ, D	State, or Solv.	t, °C	Method or P_a, % of P_e	Reference
		3.74	s	85	-	213,215
		3.25	Ether	25	5	728
	p-Dinitrobenzene	0	gas	-80 to -20	-	331
		0	B,D	25	32	272
		<0.3	B	n.s.	-	2111
		0.3	B	n.s.	-	750
		0.32	B	25	-	2123
		0.5	B	25,45	0	1064
		0.58	B	20	-	1949
		0.69	B	40	0	1525
		0.78	B	20	0	1525
		0.8	B	n.s.	0	751
		0.2	$CHCl_3$	25	0	1064
$C_6H_4N_2O_5$	2,4-Dinitrophenol	3.02	B	25	0	1597
		3.314	B	20	-	1673
		3.51	D	25	0	1597
	2,6-Dinitrophenol	3.20	B	18	RT	1532
		3.89	B	25	0	1597
		3.47	D	25	0	1597
$C_6H_4N_2S$	Benz-1-thia-2,3-diazole	3.5_7	B	25	0	36
	Benz-2-thia-1,3-diazole	1.73	B	25	-	736
$C_6H_4N_2Se$	Benz-2-seleno-1,3-diazole	0.94	B	25	-	736
$C_6H_4N_4O_2$	*o*-Azidonitrobenzene	4.25	B	25	0	1762
		4.46	B	25	5	1854
	m-Azidonitrobenzene	3.52	B	25	0	1762
		3.56	B	25	5	1854

s) Naphthalene.

FORMULA	COMPOUND NAME	μ, D	State, or Solv.	t, °C	Method or P_a, % of P_e	Reference
$C_6H_4N_4O_2$ (contd.)	*p*-Azidonitrobenzene	2.89	B	25	0	1762
		2.90	B	25	5	1854
		2.99	B	18	0	156
$C_6H_4N_4O_6$	2,4,6-Trinitroaniline	2.87	B	25	5	1800
		3.27	D	25	5	1800
		3.28	D	25	-	2026
$C_6H_4O_2$	*o*-Benzoquinone	5.1	B	25	0	1389-90
	p-Benzoquinone[ss]	0	gas	-155 to -30	-	331
		0	gas	-	MB	467
		<0.2	gas	-	M	283
		<0.15	B	n.s.	EB	283
		0.50	B	40	-	1301
		0.63	B	30	-	1301
		0.65	B	25	0	648
		0.67	B	n.s.	0	675
		0.69	B, CCl_4	25,45	0	1065
		0.75	B, CCl_4	20	-	1301
		0.60	D	40	-	1301
		0.65	D	30	-	1301
		0.70	D	20	-	1301
		0.68	CCl_4	25	0	648
		0.66	Hx	40	0	648
$C_6H_4S_2$	*trans*-Thiophthene	0	Hp	n.s.	-	279
$C_6H_4Se_2$	Isoselenophthene	~0	B	25	0	1927
	cis-Selenophthene	1.53	B	25	0	1927
	trans-Selenophthene	1.08	B	25	0	1927

ss) 283 and 284 argue that $P_a \approx 23\%\ P_e$ and hence $\mu = 0$.

FORMULA	COMPOUND NAME	μ, D	State, or Solv.	t, °C	Method or P_a, % of P_e	Reference
$C_6H_5AlBr_3NO_2$	Aluminum bromide - nitrobenzene complex	9.14-9.19	B	(20)	-	2006-7
		9.36±0.13	B	20	-	1415
		8.57	CS_2	20	0	560
$C_6H_5AlCl_3NO_2$	Aluminum chloride - nitrobenzene complex	9.08	B	(20)	-	2006
		9.31	B	20	15	1415
$C_6H_5BCl_2$	Phenyldichloroborine	2.19	B	25	5	370
		4.72	D	25	5	370
$C_6H_5BF_2$	Phenyldifluoroborine	1.90	B	25	5	370
		3.63	D	25	5	370
C_6H_5Br	Bromobenzene	1.70±0.03	gas	85-210	-	1092
		1.72	gas	100-210	5	622
		1.79	gas	180	0	794
		1.48	liq	(30)	-	1374
		1.50	liq	20	-	657
		1.52	liq	25	-	707
		1.24t	B	20	-	882
		1.49±0.05	B	n.s.	0	1966
		1.50	B	20	0	140
		1.5	B	25	-	2112
		1.52	B	25	-	2114
		1.52	B	30	RT	1580
		1.53	B	25	0	1366
		1.53	B	30	RT	1376
		1.54	B	20	-	1949
		1.54	B	10-40	0	347
		1.55	B	25	0	226
		1.56	B	25	0	1460
		1.56	B	25	0	1068
		1.57	B	n.s.	0	750-1
		1.57	B	25	0	358
		1.72	B	n.s.	-	379
		1.59	D	30	0	223

t) 10 cm wavelength.

FORMULA	COMPOUND NAME	μ, D	State, or Solv.	t, °C	Method or P_a, % of P_e	Reference
C_6H_5Br (contd.)	Bromobenzene (contd.)	1.51	CCl_4	25	5	1073
		1.52	CCl_4	0-40	0	347
		1.54	CCl_4	10-60	-	384
		1.69	CCl_4	20	5-25	77
		1.36	Hx	-60 to 60	-	693
		1.58	Hx	-23 to 40	0	347
		1.61	Hx	25	0	1906
		1.57	cHx	10-40	0	347
		1.50±0.02	Tol	-23 to 40	0	347
		1.39±0.02	CS_2	-23 to 40	0	347
		1.49	n.s.	n.s.	-	1424
C_6H_5BrHg	Phenylmercury bromide	3.09	D	50	0	336
$C_6H_5BrN_2O_2$	4-Bromobenzoquinone dioxime	2.64	D	25	-	1324
C_6H_5BrO	*o*-Bromophenol	1.27	B	40	RT	1864
		1.37	B	25	0	2119
		1.39	B	25	0	1598
		2.39	D	25	0	46
		1.15	CCl_4	25	0	46
		1.05	cHx	25	0	1598
	p-Bromophenol	2.14_5	B	(22)	15	415
		2.25	B	25	0	46
		2.27	B	25	0	598
		2.2_8	B	25	5	1109
		2.89	B	25	0	2119
		2.75	D	25	-	451
		2.78	D	25	0	46
		2.0_{95}	CCl_4	25	5	1109
		2.15	cHx	25	0	451
C_6H_5BrS	*o*-Bromothiophenol	1.96±0.02	B	20	0	1202
	m-Bromothiophenol	1.51±0.02	B	20	0	1202

FORMULA	COMPOUND NAME	μ, D	State, or Solv.	t, °C	Method or P_a, % of P_e	Reference
	p-Bromothiophenol	1.14±0.03	B	20	0	1202
$C_6H_5Br_2N$	2,4-Dibromoaniline	2.63	B	25	5	1800
		2.98	D	25	5	1800
	3,5-Dibromoaniline	3.02	B	25	5	1800
		3.42	D	25	5	1800
$C_6H_5Br_3Si$	Phenyltribromosilane	2.36±0.04	B	25	5	1592
C_6H_5Cl	Chlorobenzene	1.62	gas	-50 to 70	0	2017
		1.70±0.01	gas	85-220	13	621
	1.75	1.70	gas	80-150	0	1352
		1.72	gas	25	-	1097
		1.72	gas	100-245	-	1285
		1.72	gas	160	0	794
		1.76	gas	90-175	0	327
		1.23	liq	25	-	1097
		1.35	liq	27-86	-	823
		1.39	liq	20	-	863
		1.52	liq	-80 to 80	-	1825
		1.54	liq	25	-	707
		1.56	liq	20	VS	1701
		1.56	liq	(30)	-	1374
		1.64	liq	20	-	657
		1.50	B	30	RT	1580
		1.5	B,D	25	0?	2116
	1.58 P_e = 29.86, P_a = 2.99	1.51	B	25	-	581
		1.54	B	n.s.	0	1384
		1.54	B	30	5	390
		1.54	B	30	RT	1376
		1.55	B	25	RT	1710
		1.55	B	25	0	1068
		1.55	B	20	0	904
		1.56	B	25	0?	2121
		1.56	B	18	0	1653

FORMULA	COMPOUND NAME	μ, D	State, or Solv.	t, °C	Method or P_a, % of P_e	Reference
C_6H_5Cl (contd.)	Chlorobenzene (contd.)	1.57	B,Hp	25	10	1826
		1.57	B	21	0	140
		1.57	B	19	0	671,683
		1.57	B	21	-	1949
		1.57	B	22	0	514
		1.57	B	20	-	977
		1.57	B	n.s.	-	976
		1.57	B	20	0,RT	921
		1.58	B	25	0	1366
		1.58	B	25	0	366
		1.58[u]	B	25	-	1542
		1.58	B	30	-	548
		1.59	B	n.s.	-	750
		1.59	B	32	0	589
		1.59	B	25	0	1082
		1.59	B	25	0	358
		1.60	B	25	0	1097
		1.60	B	25	0	226
		1.6	B	20	5	1969
		1.61	B	25	0	1861
		1.62	B	0-50	-	1824
		1.62	B	6-53	-	1304
		1.65	B	n.s.	0	751
		1.62	D	30	0	223
		1.68	D	25	0	366
		1.56-1.59	CCl_4	0-60	-	1418
		1.60	CCl_4	10-60	-	384
		1.65	CCl_4	20	5-25	77
		1.70	CCl_4	20	5	602
		1.53	Hx	-80 to 80	11	1825
		1.56	Hx	25	-	2122
		1.60	Hx	20	0	1349
		1.62	Hx	0-50	-	1824
		1.62	Hx	20	0,RT	921
		1.57	Hp,B	25	10	1826
		1.58	Hp	30	-	548

u) $\overline{\mu}^2$ value.

FORMULA	COMPOUND NAME	μ, D	State, or Solv.	t, °C	Method or P_a, % of P_e	Reference
		1.62-1.64	Hp	C-60	-	1418
		1.56-1.58	Tol	0-60	-	1418
		1.57	Tol	0-75	0	730
		1.61	Tol	18	0	671,683-4,
		1.53	v	85	-	213,215
		1.49	CS_2	0-40	-	1418
		1.53	CS_2	25	-	2122
		1.19	$CHCl_3$	25	0	1097
		1.52	$CHCl_3$	18	0	671,683-4
		1.3	Ether	25	-	717,723
		1.63	NB	25	0	1097
		1.56±0.02	w	20	-	1448
		1.67	x	20	5	1512
		1.50	y	25	5	581
C_6H_5ClHg	Phenylmercury chloride	2.99	D	20	-	1777
$C_6H_5ClN_2O_2$	4-Chloro-2-nitroaniline	4.44	B	25	0	2026
C_6H_5ClO	*o*-Chlorophenol	2.19±0.04	gas	150-325	-	1155
		1.46	liq	20	10	1295
		1.55	liq	30-40	-	1498
		1.24	B	n.s.	RT	1368
		1.3	B	25	-	2112
		1.31_9	B	(22)	15	415
		1.34	B	25	0	46
		1.43	B	25	0	2119
		2.13	D	25	0	46
		1.00	CCl_4	20	10	1295
		1.16	CCl_4	-15 to 30	0	935
		1.16	CCl_4	25	0	46
		1.13	Hx	10-40	0	935
		1.10	cHx	25	0	1598
		1.11	Hp	-30 to 70	0	935
		1.24	Tol	0-75	0	730

v) Naphthalene. w) Bromobenzene. x) Several solvents.
y) 25 mole per cent *o*-dichlorobenzene plus benzene.

FORMULA	COMPOUND NAME	μ, D	State, or Solv.	t, °C	Method or P_a, % of P_e	Reference
C_6H_5ClO (contd.)	*m*-Chlorophenol	2.12	B	(22)	15	415
		2.19	B	25	0	2119
		2.11	Tol	0-75	0	730
	p-Chlorophenol	1.82	B	n.s.	-	1368
		2.2_2	B	25	5	1109
		2.24_2	B	22	15	415
		2.27	B	25	0	598
		2.4	B	25	-	2112
		2.70	B	25	0	2119
		2.82	D	25	-	451
		2.16	CCl_4	25	5	1109
		2.25	cHx	25	-	451
		2.27	Tol	0-75	0	730
$C_6H_5ClO_2$	3-Chloro-3-cyclohexen-1,2-dione	3.14	B	30	5	1328
	2-Chloro-6-hydroxy-2,5-cyclohexadien-1-one	3.14	B	30	-	1604
$C_6H_5ClO_2S$	Benzenesulfonyl chloride	4.50	B	20	0	1266
		4.57	B	25	0	634
		4.58	B	n.s.	-	632-3
	p-Chlorobenzenesulfinic acid	2.20	B	25	0	641
		3.32	D	25	0	641
	Phenyl chlorosulfite	2.46	B	25,45	0	1975
C_6H_5ClS	*o*-Chlorothiophenol	1.98±0.02	B	20	0	1202
	m-Chlorothiophenol	1.56±0.02	B	20	0	1202
	p-Chlorothiophenol	1.16±0.01	B	20	0	1193
C_6H_5ClSe	*p*-Chloroselenophenol	0.93±0.02	B	20	0	310

FORMULA	COMPOUND NAME	μ, D	State, or Solv.	t, °C	Method or P_a, % of P_e	Reference
$C_6H_5Cl_2I$	Phenyl iododichloride	2.63	B	25	-	640
$C_6H_5Cl_2N$	2,5-Dichloroaniline	1.69	B	n.s.	0	678
$C_6H_5Cl_3OTi$	Trichlorophenoxytitanium	2.97	B	25	0	355
$C_6H_5Cl_3Si$	Phenyltrichlorosilane	2.21	B	25	15	887
		2.43	B	25	0	373
$C_6H_5Cl_3Sn$	Phenyltrichlorotin	4.24	B	n.s.	-	592
		5.81	D	n.s.	-	592
		3.99	Hx	n.s.	-	592
$C_6H_5Cl_7$	1ea2e3a4e5e6a-Heptachlorocyclohexane (η)	0.7	B	25	-	1595
		0.92	B	yy	yy	1520
	1ea2e3e4a5a6e-Heptachlorocyclohexane (δ)	2.20	B	yy	yy	1520
		2.22	B	25	5	1356
	1ea2e3e4a5e6e-Heptachlorocyclohexane (ρ)	1.2	B	25	-	1595
		1.48	B	yy	yy	1520
	1ea2e3e4e5a6a-Heptachlorocyclohexane (ε)	2.21	B	yy	yy	1520
		2.22	B	25	5	1356
	1ea2e3e4e5e6a-Heptachlorocyclohexane (γ)	1.32	B	yy	yy	1520
		1.36	B	25	5	1356
	1ea2e3e4e5e6e-Heptachlorocyclohexane (α)	1.19	B	yy	yy	1520
		1.20	B	25	5	1356
	1ea2ea4ea5e-Heptachlorocyclohexane (m.p. = 119-120°C)	0.69	B	25	5	1605,1610
C_6H_5F	Fluorobenzene	1.57	gas	~160	0	794
		1.58	gas	70-235	-	1285

yy) Not given in Chem. Abstrs. 54, 16055c.

FORMULA	COMPOUND NAME	μ, D	State, or Solv.	t, °C	Method or P_a, % of P_e	Reference
C_6H_5F (contd.)	Fluorobenzene (contd.)	1.61	gas	80-150	0	1352
		1.66±0.03	gas	-	M	401
		1.52	liq	(30)	-	1374
	1.35 P_e = 23.50	1.39	B	30	RT	1376
	P_a = 2.35	1.40	B	20	0	2057
		1.42	B	25	0	2017-8
		1.46	B	21	0	140
		1.47	B	30	0	225
		1.47	B	30	-	548
		1.48	B	25	0	1068
		1.50	B	25	0	1378
		1.54	B	25	5	1114
		1.51	D	30	0	746
		1.45	CCl_4	20	5-25	77
C_6H_5FO	*o*-Fluorophenol	1.32	B	25	0	1598
		1.84	D	25	0	672
		1.87	D	25	0	1598
		1.16	CCl_4	25	0	46
	p-Fluorophenol	2.10	B	25	5	1114
		2.17	B	25	0	598
$C_6H_5F_3Si$	Phenyltrifluorosilane	2.80	B	25	0	373
$C_6H_5F_7O_2$	Ethyl perfluorobutyrate	3.09	B	25	15	1647
C_6H_5I	Iodobenzene	1.70	gas	200-250	0	794
		1.71±0.03	gas	155-290	-	1092
		1.27	liq	(30)	-	1374
		1.26	B	20	0	2057
		1.29	B	30	RT	1376
		1.30_5	B	20	0	140
		1.39	B	20	-	1949
		1.38	CCl_4	25	0	919
		1.70	CCl_4	20	-	79

FORMULA	COMPOUND NAME	μ, D	State, or Solv.	t, °C	Method or P_a, % of P_e	Reference
		1.26	Hx	-60 to 60	-	693
		1.44	cHx	30	RT	1844
$C_6H_5IN_2O_2$	4-Iodo-2-nitroaniline	4.59	B	25	0	2026
C_6H_5IO	*o*-Iodophenol	1.54	B	25	0	1598
		2.70	D	25	0	1598
		1.25	cHx	25	0	1598
	p-Iodophenol	2.21	B	25	0	598
C_6H_5IS	*p*-Iodothiophenol	1.04±0.04	B	20	0	1202
C_6H_5NO	Pyridine-2-carboxaldehyde	3.35±0.02	B	25	0	91
	Pyridine-3-carboxaldehyde	2.37±0.02	B	25	0	91
	Pyridine-4-carboxaldehyde	1.74±0.02	B	25	0	91
	Nitrosobenzene	3.11_5	B	n.s.	-	704
		3.17	B	25	0	650
		3.25	B	n.s.	0	675
		3.25	n.s.	n.s.	-	263
C_6H_5NOS	Thionylaniline	1.95	liq	20	-	968
		1.90	B	20	-	968
		1.92	B	25	15	855
		2.6	B	20	-	159
$C_6H_5NO_2$	Isonicotinic acid	2.7	D	25	10	1112
	Nicotinic acid	4.0	z	n.s.	-	1330
	Nitrobenzene	3.98	gas	10-75	0	2017
		4.22±0.05	gas	170-275	-	1285
		4.28±0.01	gas	130-250	5	621
		4.28	gas	25	-	1097

z) 90% phenol or pyridine.

FORMULA	COMPOUND NAME	μ, D	State, or Solv.	t, °C	Method or P_a, % of P_c	Reference
$C_6H_5NO_2$ (contd.)	Nitrobenzene (contd.)	4.4	gas	n.s.	-	1898
		0.71	liq	18	-	1115
		1.74	liq	25	-	1097
		3.93	liq	20	VS	1701
		3.97	liq	20	10	1295
		4.69	liq	20	-	657
		3.85	B	25	-	281
	3.93 P_e = 32.59	3.87	B	24-65	-	1039
	P_a = 3.26	3.87	B	25	-	2122
		3.89	B	25	-	964
		3.91	B	30	RT	1580
		3.93	B	25	-	2123
		3.93±0.01	B[a]	n.s.	-	258
		3.95	B	27	-	1463
		3.96	B	22	15	1527
		3.96±0.01	B	25	0	1082
		3.96±0.01	B	25	-	2020
		3.97	B	28	MA	979
		3.97	B	20	-	1365
		3.97	B	25	0	838,840
		3.98	B	25	0	1080
		3.99-4.02	B	10-40	-	1588
		3.99	-	-	-	1424
		4.00	B	20	-	1949
		4.00	B	32	0	589
		4.01	B	19	0	683-4
		4.02	B[b]	20	5	1513
		4.04	B	25	0	2026
		4.04	B[b]	20	0	1951
		4.06	B	n.s.	0	1384
		4.06	B	25	0	1068
		4.06	B	20	-	977
		4.08	B	25	0	1097
		4.11	B	21	0	141
		4.11	B	n.s.	-	976
		4.4	B	30	RT	1375,1377

a) Saturated with water. b) Plus water.

FORMULA	COMPOUND NAME	μ, D	State, or Solv.	t, °C	Method or P_a, % of P_e	Reference
		3.93-3.97	D	10-40	-	1588
		3.96	D	25	0	839
		4.02	D	30	5	390
		4.07	D	25	0	2026
		4.07	D	14.3	0	839
		4.15	D	20	-	1777
		3.84	CCl_4	20	10	1295
		3.962	CCl_4	25	0	838
		3.97	CCl_4	10-50	-	1463
		4.00	CCl_4	25	0	2017-8
		4.05	CCl_4	10-40	-	1588
		3.92	Hx	25	-	2122
		3.99	Hx	25	-	1504
		4.0_3	Hx	25	-	1503
		4.079	Hx	25	0	838
		4.13	Hx	10-40	-	1588
		3.96	cHx	24	0	228
		4.004	cHx	25	0	838
		4.09-4.12	cHx	10-40	-	1588
		4.13	Hp	10-40	-	1588
		3.86	Tol	19	-	671,683-4
		3.87	Tol	24-100	-	1039
		4.27	pXy,c	20-40	-	491
		4.28±0.05	pXy,d	20-120	-	489
		3.960	e	25	0	838
		3.76	f	85	-	213,215, 1365
		3.688	CS_2	25	0	838
		3.72	CS_2	20	0	560
		3.76	CS_2	24	-	1039
		3.92	CS_2	25	-	2122
		3.08	$CHCl_3$	25	0	1097
		3.202	$CHCl_3$	25	0	838
		3.27	$CHCl_3$	n.s	-	684
		3.33	$CHCl_3$	18	-	671
		3.2	Ether	25	-	717,722-3

c) Decane. d) Diisoamyl. e) Decalin. f) Naphthalene.

FORMULA	COMPOUND NAME	μ, D	State, or Solv.	t, °C	Method or P_a, % of P_e	Reference
$C_6H_5NO_2$ (contd.)	Nitrobenzene (contd.)	3.96±0.04	Ether	20	-	1448
		4.06±0.03	g	20	-	1448
		4.32	h	20	5	1512
		2.51	i	25	0	1097
		3.82	i	25	-	581
		3.39	j	25	-	581
		3.77	k	25	-	581
		3.59	l	25	-	581
		4.0	n.s.	n.s.	-	1454
	p-Nitrosophenol	4.44	D	35	0	1851
		4.76	D	20	0	339
$C_6H_5NO_3$	1-(2-Furyl)-2-nitroethylene	5.04	B	25	-	2031-2
		5.07	B	25	5	1406
	o-Nitrophenol	3.13	B	25	0	2119
		3.14	B	25	0	842
		3.22	B	25	0	1597
		3.07	D	25	0	1597
		3.23	D	30	0	47
		3.01	CCl_4	20	10	1295
		3.09-3.19	Tol	0-75	0	730
	m-Nitrophenol	3.90±0.10	B	25	0	1636
		3.93	B	25	0	2119
		4.47	D	30	0	47
		3.69-3.89	Tol	0-75	10	730
	p-Nitrophenol	4.83	B	25	5	1109
		5.05_6	B	(22)	15	415
		5.07	B	25	0	598
		5.09	B	25	0	2119
		5.43	D	30	0	47
		4.76-4.64	Tol	25-75	10	730

g) Methyl benzoate. h) Several solvents. i) Chlorobenzene. j) *o*-Dichlorobenzene.
k) 50 mole per cent *o*-dichlorobenzene plus benzene. l) 75 mole per cent *o*-dichlorobenzene plus benzene.

FORMULA	COMPOUND NAME	μ, D	State, or Solv.	t, °C	Method or P_a, % of P_e	Reference
$C_6H_5NO_4S$	*m*-Nitrobenzenesulfinic acid	4.26	D	25	0	641
$C_6H_5N_3$	1,2,3-Benzotriazole	4.10	D	25	15	859
	Azidobenzene	1.44	B	25	-	1762
		1.55	B	22	0	155-6
		1.56	B	25	0	1774,1902
$C_6H_5N_3O_4$	2,3-Dinitroaniline	7.35	D	25	0	2026
	2,4-Dinitroaniline	5.85	B	25	5	1800
		6.51	D	25	5	1800
		6.53	D	25	0	2026
	2,5-Dinitroaniline	2.69	D	25	0	2026
	2,6-Dinitroaniline	1.90	B	25	0	2026
		1.93	B	25	0	1599
		1.69	D	25	0	1599
	3,4-Dinitroaniline	8.96	D	25	0	2026
	3,5-Dinitroaniline	5.45	B	25	5	1800
		5.88	D	25	5	1800
		5.96	D	25	0	2026
C_6H_6	Benzene	0	gas	20-50	-	660
		0	gas	50-205	-	1284
		0	gas	70-250	-	621
		0	gas	23-95	-	1564
		0	gas	140	-	795
		0-0.2	gas	5-80	-	1685
		0	liq	n.s.	-	809
		0	liq	25	0	1067
		0	liq	n.s.	0	1665
		0	liq	20-30	5-15	22

FORMULA	COMPOUND NAME	μ, D	State, or Solv.	t, °C	Method or P_a, % of P_e	Reference
C_6H_6 (contd.)	Benzene (contd.)	0	liq	20-60	6	1489
		0	liq	25	-	1254
		0.07	liq	20	VS	1701
		0.14	liq	20,50	-	695
		0	CCl_4	20	4	209
		0.06 [0.21]m	CCl_4	25	0	966
		0	Hx	20	0	1349
		0.08 [0.31]m	Hx	25	-	2122
		0	cHx	25	-	949
		0	Hp	20	3	209
		0.09 [0.39]m	CS_2	25	-	2122
		0	$CHCl_3$	25	0	1067
		0	$CHCl_3$	25	-	841
		0.53	$CHCl_3$	18	0	671,683-4 717,723
		0	Ether	25	-	1067
		1.53	NB	25	0	754
		1.56	NB	30	0	1067
		0.73	n	25	0	
	Fulvene	1.1	B	28	0	1939
	1,5-Hexadien-3-yne	0-0.2	B	25	-	1750
$C_6H_6AlBr_3$	Aluminum bromide - benzene complex	5.1	B	20?	15	1415
$C_6H_6AsCl_3$	*Tris*(2-chlorovinyl)arsine	0.39	B	20	0	1348
$C_6H_6BrCl_5$	1e-Bromo-1a2e4ea5e-pentachlorocyclohexane	0.3	n.s.	n.s.	-	1615
C_6H_6BrN	*o*-Bromoaniline	1.62	B	40	RT	1864
		1.79	B	20	-	1950
	m-Bromoaniline	2.61	B	40	RT	1864
		2.66	B	25	5	1799
		2.67	B	20	-	1950
		2.99	D	25	5	1799

m) Values in brackets as recalculated in Ref. 1284. n) Chlorobenzene.

FORMULA	COMPOUND NAME	μ, D	State, or Solv.	t, °C	Method or P_a, % of P_e	Reference
	p-Bromoaniline	2.88	B	25	0	1068
		2.90	B	23	0	159
		2.99	B	25	5	1799
		3.02	B	20	-	1950
		3.34	D	25	5	1799
$C_6H_6BrNO_2S$	*p*-Bromobenzenesulfonamide	4.44	D	25	0	635
$C_6H_6Br_2Cl_4$	1e2a-Dibromo-3a4a5e6e-tetrachloro-cyclohexane (γ)	2.64	B	n.s.	-	1519
		2.66	B	30	5	1609,1754
	1e2e-Dibromo-3e4e5a6a-tetrachloro-cyclohexane (α) (m.p. = 178°C)	1.84	B	30	5	304,1609
		1.87	B	n.s.	-	1519
	1e2e-Dibromo-3e4e5e6e-Tetrachloro-cyclohexane (β) (m.p. = 285°C)	0.3	B	30	5	304,1609
		0.3	B	n.s.	-	1519
$C_6H_6Br_2Si$	Phenyldibromosilane	2.33±0.04	B	25	5	1592
$C_6H_6Br_4Cl_2$	1e2e3a4a-Tetrabromo-5e6e-dichloro-cyclohexane	1.93	B	30	5	1328,1603
	1e2e4e5e-Tetrabromo-1a4a-dichloro-cyclohexane	0.3	n.s.	n.s.	-	1615
$C_6H_6Br_6$	1e2e3e4e5a6a-Hexabromocyclohexane	2.0	B	30	-	1609
		2.14	D	25	-	881
	1e2e3e4e5e6e-Hexabromocyclohexane	0.4	B	30	-	1609
C_6H_6ClN	*o*-Chloroaniline	1.72	B	40	-	1864
		1.773	B	20	-	1673
		1.79	B	20	-	1950
		1.83	B	n.s.	-	678
		1.86	B	25	0	530

FORMULA	COMPOUND NAME	μ, D	State, or Solv.	t, °C	Method or P_a, % of P_e	Reference
C_6H_6ClN (contd.)	*m*-Chloroaniline	2.50	B	40	-	1864
		2.68	B	20	-	1950
		2.73	B	n.s.	-	678
		2.94	B	25	0	530
	p-Chloroaniline	2.77	B	40	-	1864
		2.93	B	24	0	159
		2.96	B	18	-	671
		2.98	B[p]	n.s.	-	472
	3.01 P_e = 33.22, P_a = 3.32	2.99	B	25	5	369
		2.994	B	25	5	508
		3.00	B	20	-	1950
		3.03	B	25	0	530
		3.36	D	25	5	369
		3.372	D	25	5	508
$C_6H_6ClNO_2S$	*p*-Chlorobenzenesulfonamide	3.94	B	25	0	635
		4.42	D	25	0	635
$C_6H_6Cl_2$	1,4-Dichloro-1,4-cyclohexadiene	0.3	B	30	5	1328
$C_6H_6Cl_2F_{12}P_2Pt$	Dichloro*bis*(methyldi(trifluoromethyl)-phosphine)platinum	0	CCl_4	n.s.	-	118
$C_6H_6Cl_3Tl$	Benzene - thallium chloride complex	3.96	B	20	14	1738
$C_6H_6Cl_4$	3,4,5,6-Tetrachlorocyclohex-1-ene (α)	2.34	n.s.	n.s.	-	107
	3,4,5,6-Tetrachlorocyclohex-1-ene (β)	2.65	n.s.	n.s.	-	107
	3,4,5,6-Tetrachlorocyclohex-1-ene (γ)	3.62	n.s.	n.s.	-	107
	3,4,5,6-Tetrachlorocyclohex-1-ene (δ)	2.30	n.s.	n.s.	-	107
	3,4,5,6-Tetrachlorocyclohex-1-ene (ε)	2.19	n.s.	n.s.	-	107

p) Mixed with dioxane.

FORMULA	COMPOUND NAME	μ, D	State, or Solv.	t, °C	Method or P_a, % of P_e	Reference
$C_6H_6Cl_4O$	2,2,6,6-Tetrachlorocyclohexanone	4.40	B	30	-	1328,1604
$C_6H_6Cl_6$	Hexachlorocyclohexanes[q,r]					
	1ea2ea4e5e	0.65	B	25	-	1605,1754
	1ea2e4ea5e (ξ)[s]	0.46	B	25	5	1754
		0[t]	CCl_4	n.s.	-	105
	1a2a3a4e5e6e (γ) (m.p. = 112.5°C)	2.60-2.55	liq	113-159	-	830
		2.8	B	30	-	1609
		2.83	B	25	5	1356
	(m.p. = 112.8°C)	2.84	B	20	0	1654
		2.84	B	30	0	1151
	(m.p. = 113.3°C)	2.92	B	25	0	708
		2.53	D	25,40	-	830
		3.00	D	30	0	1151
		2.51	CCl_4	25,40	-	830
	1e2a3a4e5a6a (ξ)	0	B	n.s.	-	105
	1e2e3a4e5a6a (ξ) (m.p. = 88°C)	1.7±0.2	B	26	-	1711
	1e2e3a4e5e6a (ε)	0	B	25	5	1356
		0.43	B	30	0	1151
		0.45	D	30	0	1151
	1e2e3e4a5e6a (α) (m.p. = 156.3°C)	2.22	B	25	0	708
		2.24	B	25	5	1356
	(m.p. = 157.5°C)	2.12	B	20	0	1654
		2.16	B	30	0	1151
		2.22	B	n.s.	-	676

q) These compounds are often misnamed "hexachlorobenzene."
r) For earlier measurements see Ref. 676 and 2120.
s) Note these designations are not used uniformly or consistently.
t) As corrected by Ref. 252.

FORMULA	COMPOUND NAME	μ, D	State, or Solv.	t, °C	Method or P_a, % of P_e	Reference
$C_6H_6Cl_6$ (contd.)	Hexachlorocyclohexanes (contd.)					
	1e2e3e4a5e6a (α) (m.p. = 158°C)	1.73	B	n.s.	-	829
		1.70	D, CCl_4	25	-	829-30
		2.34	D	30	0	1151
	1e2e3e4e5a6a (α) (m.p. = 159°C)	2.2	B	30	-	1609
	1e2e3e4e5e6a (δ)	2.58	liq	n.s.	-	829
	(m.p. = 138°C)	0	D, CCl_4	25,40	-	830
		2.32	D	20	0	1151
	(m.p. = 138.2°C)	2.19	B	25	0	27,708
		2.24	B	25	5	1356
		2.24	B	25	0	1151
		2.58	D	n.s.	-	829
	1e2e3e4e5e6e (β) (m.p. = 308°C)	0	B	n.s.	-	676
		0	B	20	0	1654
		0	B	25	5	1356
		0	B	25	0	708
		0	B	30	-	1609
		0	B	n.s.	-	829
		0.34	B	30	0	1151
		0	D	n.s.	-	829
		0	D	25,40	-	830
		0.46	D	30	0	1151
		0	CCl_4	25,40	-	830
C_6H_6FN	*p*-Fluoroaniline	2.48	B	25	5	1114
		2.77	B	24	0	159
C_6H_6IN	*p*-Iodoaniline	2.84	B	25	0	159
$C_6H_6INO_2S$	*p*-Iodobenzenesulfonamide	4.53	D	25	0	635

FORMULA	COMPOUND NAME	μ, D	State, or Solv.	t, °C	Method or P_a, % of P_e	Reference
$C_6H_6I_2$	Benzene - iodine complex	1.8	cHx	25	-	949
$C_6H_6N_2$	Dimethylfumarodinitrile	0.7	B	30	-	117
	Dimethylmaleodinitrile	6.0	B	30	-	117
$C_6H_6N_2O$	Isonicotinamide	3.91	D	25	10	1112
	Nicotinamide	4.2	u	n.s.	-	1330
	Pyridine-2-aldoxime	2.47±0.03	D	25	0	92
	Pyridine-3-aldoxime	2.37±0.02	D	25	0	92
	Pyridine-4-aldoxime	2.46±0.03	D	25	0	92
$C_6H_6N_2O_2$	*o*-Benzoquinone dioxime	3.87	D	25	-	1322
	p-Benzoquinone dioxime	2.39	D	25	-	1322
	o-Nitroaniline	5.00	liq	90-110	-	1497
		4.04	B	20	-	816
		4.06	B	25	0	1599
		4.23	B	20	-	1673
		4.28	B	20	0	1950
		4.29	B	25	-	2026
		4.38	B	(21)	-	1216
		4.48	B	n.s.	0	751
		4.36	D	25	0	1599
		4.68	D	(21)	-	1216
	m-Nitroaniline	4.680	B	20	-	1673
	4.90 P_e = 35.95 P_a = 3.60	4.76	B	n.s.	0	751
		4.83	B	(21)	-	1216
		4.89	B	25	5	1799
		4.98	B	40	-	1950

u) 90% phenol or pyridine.

FORMULA	COMPOUND NAME	μ, D	State, or Solv.	t, °C	Method or P_a, % of P_e	Reference
$C_6H_6N_2O_2$ (contd.)	*m*-Nitroaniline (contd.)	5.04	B	30	5	390
		5.22	D	25	5	1799
		5.24	D	(21)	-	1216
		5.26	D	30	5	390
	p-Nitroaniline	5.6±0.6	gas	n.s.	MB	467,2134
		7.16	liq	160-180	-	1497
		6.14±0.06	B	20	-	1351
	5.29 P_e = 35.95	6.20	B	25	0	1799
	P_a = 3.60	6.22	B	25	-	2026
		6.30	B	20	FL	375
		6.31	B	25	5	369
		6.32	B[v]	25	5	471
		6.33	B	(21)	-	1216
		6.37	B	25	0	1068
		6.4	B	70	-	1950
		7.1	B	n.s.	0	751
		6.73	D	30	10	1011
		6.80	D	25	0	1799
	6.91 P_e = 35.95	6.86	D	25	-	2026
	P_a = 3.60	6.87	D[v]	25	5	471
		6.88	D	25	5	369
		6.92	D	(21)	-	1216
$C_6H_6N_2O_2Se$	5,6-Dihydro-7-nitroso-1,2,3,4*H*-benzoxa-selenazole	2.90±0.03	n.s.	n.s.	-	915
$C_6H_6N_2O_4S$	*m*-Nitrobenzenesulfonamide	4.97	D	25	0	635
$C_6H_6N_4O_4$	2,4-Dinitrophenylhydrazine	5.8±0.6	B	17	0	2008

v) Extrapolated from data for benzene-dioxane mixtures.

FORMULA	COMPOUND NAME	μ, D	State, or Solv.	t, °C	Method or P_a, % of P_e	Reference
C_6H_6O	Phenol	1.40±0.03	gas	20	10	xx
		1.41	gas	175	5	625
		2.22	liq	20	10	xx
		1.45	B	25	10	xx
		1.45	B	25	0	194
		1.45	B	30	0?	890
		1.46	B	n.s.	-	1368
		1.47±0.02	B	20	0	1193
		1.5	B	70	0	1261
		1.53±0.03	B	20	3	1718
		1.53	B	25	0	789
		1.54	B	20	0	1387
		1.54	B	25	0	598
		1.57	B	20	0	81
		1.59	B	(22)	15	415
		1.65	B	25	5	913
		1.72	B	25	0	2117,2121
		1.75	B	20	-	1824[w]
		1.80	D	20	0	1387
		1.86	D	25	-	451
		1.92	D	20	10	xx
		1.39	CCl_4	30	0?	890
		1.40	CCl_4	20	10	xx
	1.55 P_e = 26.61 P_a = 2.66	1.46±0.03	CCl_4	10-60	3	1718
		1.49	CCl_4	20	0	81
		1.50	CCl_4	20	0	1387
		1.5_3	CCl_4	25	5	1109
		1.55	CCl_4	27	-	1987
		1.37	Hx	30	0?	890
		1.32±0.03	cHx	20	10	xx
		1.33±0.03	cHx	20	3	1718
		1.39	cHx	30	0?	890
		1.43	cHx	25	-	451
		1.37	Hp	20	0	1387
		1.44	Hp	20	0	81
		1.86	Hp[x]	20	0	1387

w) Calculated from data in J. Chem. Soc. 1905, 998-1003. x) Plus dioxane.
xx) Ref. 1221, 1295-6.

FORMULA	COMPOUND NAME	μ, D	State, or Solv.	t, °C	Method or P_a, % of P_e	Reference
C_6H_6O (contd.)	Phenol (contd.)	1.44-1.53	Tol	75-0	0	947
		1.46	Tol	30	0?	890
		1.56	y	30	0?	890
		1.79	y	20	10	1221,1296
		1.31	z	30	0?	890
		1.43	a	30	0?	890
		1.38	CS_2	20	10	yy
		1.39	CS_2	30	0?	890
		1.64	CS_2	25	-	2122
		2.14	Ether	25	0	1387
		2.14	Ether	20	0	81
		2.29	Ether	20	10	1221,1296
		1.37	b	20	10	yy
		1.41	c	20	10	yy
		1.45	ClB	20	10	yy
C_6H_6OS	2-Acetylthiophene	3.36±0.01	B	25	5	500
		3.37	B	30	-	903
		3.41	CCl_4	25	5	500
C_6H_6OSe	Hydroxyselenophenol	1.45±0.02	B	20	0	310
$C_6H_6O_2$	Catechol	2.60	B	25	10	367
		2.64±0.03	B	27	5	1036
		2.95	D	25	10	367
	Resorcinol	2.09±0.02	B	44	5	1036
	Hydroquinone	1.4±0.10	B	44	5	1036
		2.49	Ether	n.s.	-	671
$C_6H_6O_2S$	Benzenesulfinic acid	3.00	B	25	0	641
		3.26	B	40	0	641
		3.79	D	25	0	641
$C_6H_6O_3S$	Benzenesulfonic acid	3.80	B	n.s.	-	632-3

y) Cyclohexene. yy) Ref. 1221, 1295-6. z) Decalin. a) Tetralin. b) Tetrachloroethylene.
c) Trichloroethylene.

FORMULA	COMPOUND NAME	μ, D	State, or Solv.	t, °C	Method or P_a, % of P_e	Reference
C_6H_6S	Thiophenol	1.13	liq	25	10	1281
		1.18	B	20	-	1190
		1.19±0.01	B	20	0	1198
		1.23	B	25	10	1281
		1.34	B	20	0	791
C_6H_6Se	Selenophenol	1.08±0.02	B	20	-	309
		1.10±0.02	B	20	0	310
$C_6H_7BCl_3N$	4-Methylpyridine - boron trichloride complex	8.37±0.06	B	25	0	110
$C_6H_7BCl_3NO$	4-Methoxypyridine - boron trichloride complex	8.86±0.02	B	25	0	110
$C_6H_7BO_2$	Phenylboric acid	1.65	B	25	0	1458
		1.74	D	25	0	1458
$C_6H_7BrN_2$	*p*-Bromophenylhydrazine	2.92±0.05	B	16	0	2008
C_6H_7BrSi	Phenylbromosilane	2.07±0.04	B	25	5	1592
$C_6H_7Br_2N$	Bromine - 4-picoline complex	5.98[d]	B	25	-	895
C_6H_7Cl	*m*-Chlorophenylsilane	1.42	B	25	0	322
	p-Chlorophenylsilane	0.99	B	25	0	322
$C_6H_7I_2N$	3,4-Diiodo-2,5-dimethylpyrrole	4.00±0.01	B	25	0	931
	Iodine - 4-picoline complex	5.87[d]	B	25	-	895
C_6H_7N	Aniline	1.49	gas	175	5	625
		1.19	liq	20	5	337
		1.7_9	liq	20	-	657
		1.30	B	25	RT	1710

d) Lower limit of μ.

FORMULA	COMPOUND NAME	μ, D	State, or Solv.	t, °C	Method or P_a, % of P_e	Reference
C_6H_7N (contd.)	Aniline (contd.)	1.32	B	n.s.	-	1368
		1.47	B	(21)	-	1216
	1.56 P_e = 28.51	1.505	B	25	5	507-8
	P_a = 2.85	1.51	B	25	-	1094
		1.52	B	n.s.	0	751
		1.52	B	25	0	1068
		1.52±0.01	B[e]	20	5	797
		1.53	B	20	-	1950
		1.53	B	20	0	589
		1.53	B	n.s.	RT	1373
		1.54	B	20	0	349
		1.55	B	25	0	722
		1.55	B	25	0	2026
		1.55	B	20	5	449
		1.55	B	25	RT	2051
		1.56	B	18	0	683
		1.56	B	20	0	1387
		1.56	B	n.s.	0	1384
		1.6	B	20-50	-	463
		1.750	D	25	5	507-8
		1.76	D	(21)	-	1216
		1.79	D	25	0	2026
		1.80	D	20	0	1387[f]
		1.92	D	25	0	1004
		1.458	CCl_4	25	5?	506
		1.46	CCl_4	20	5	337
		1.46±0.01	CCl_4[e]	20	5	797
		1.50	CCl_4	25	0	1729
		1.49	Hx	20	0	349
		1.51	Hx	25	0	722
		1.38	cHx	20	5	337
		1.50	cHx	20	0	349
		1.477	Hp	25	5?	506
		1.52	Hp	20	0	1387

e) Saturated with water. f) Also published in J. Chem. Soc. Japan 71, 560 (1951).

FORMULA	COMPOUND NAME	μ, D	State, or Solv.	t, °C	Method or P_a, % of P_e	Reference
		1.53	Tol	20	0	349
		1.37	CS_2	n.s.	RT	1373
		1.420	CS_2	25	5?	506
		1.44	CS_2	20	5	337
		1.68	Ether	25	0	722
		1.69	Ether	20	0	1387
		1.59	n.s.	n.s.	-	525
	2-Methylpyridine	1.93	liq	20-60	RT	1487
		1.74±0.01	B	10-40	-	1589
		1.88±0.01	B	20	-	1179
		1.92±0.01	B	30	10	547
		1.96±0.1	B	25	0	1640
		1.97	B	25	0	360
	3-Methylpyridine	2.32±0.01	B	10-40	-	1589
		2.40	B	25	0	360
		2.41	B	25	0	1729-30
		2.54	CCl_4	25	0	1729-30
	4-Methylpyridine	2.57	B	25	10	1112
		2.60_3	B	25	0	360
		2.715	CCl_4	n.s.	-	596
		2.75	CCl_4	25	0	1729
		1.92±0.01	n.s.	25	-	1179
C_6H_7NO	*N*-Acetylpyrrole	2.44	B	20	-	1256
		2.52	D	20	-	1256
	1-Acetylpyrrole	2.58	B	25	-	642
	2-Acetylpyrrole	1.52	B	20	-	1255
		1.79	D	20	-	1255
	m-Aminophenol	1.85	B	25	0	2119

FORMULA	COMPOUND NAME	μ, D	State, or Solv.	t, °C	Method or P_a, % of P_r	Reference
C_6H_7NO (contd.)	4-Methoxypyridine	2.94	B	25	10	1112
	3-Methylpyridine 1-oxide	4.33±0.01	B	25	0	1730
	4-Methylpyridine 1-oxide	4.50	B	25	0	1729-30
		4.74±0.01	B	25	0	891
		4.97	CCl_4	25	0	1729-30
	2-Pyridinylmethanol	3.03±0.03	B	25	0	91
		2.55±0.03	D	25	0	91
	3-Pyridinylmethanol	2.74±0.03	B	25	0	91
		2.61±0.03	D	25	0	91
	4-Pyridinylmethanol	2.84±0.03	B	25	0	91
		2.6±0.1	D	25	0	91
$C_6H_7NO_2$	2-Carbomethoxypyrrole	1.70	B	25	5	818
	3-Carbomethoxypyrrole	3.63	B	25	5	818
	4-Methoxypyridine-1-oxide	5.08±0.01	B	25	0	891
$C_6H_7NO_2S$	Benzenesulforamide	4.77	B	25	0	635
		4.79	B	-	-	632-3
		4.84	B[g]	25	5	470
		4.19	D	25	0	992
		5.09	D	25	0	1550
		5.12	D[g]	25	5	470
		5.13	D	25	0	635
$C_6H_7NO_3$	3,5-Dimethyl-4-isoxazolecarboxylic acid	3.31±0.02	D	25	0	1857
$C_6H_7N_3O_2$	1-Nitro-2,4-diaminobenzene	7.16	D	25	0	2026
	1-Nitro-3,5-diaminobenzene	5.90	D	25	0	2026

g) Extrapolated from benzene-dioxane mixtures.

FORMULA	COMPOUND NAME	μ, D	State, or Solv.	t, °C	Method or P_a, % of P_e	Reference
	p-Nitrophenylhydrazine	7.2±0.5	B	16	0	2008
C_6H_8	1,3-Cyclohexadiene	0.38±0.02	B	20	0	173,1189
	1,3,5-Hexatriene	0.2	B	36.8	5	2146
	5-Hexen-3-yne	0.62	B	20	0	1492
	2-Methyl-1-penten-3-yne	0.65	n.s.	n.s.	0	1490
$C_6H_8BrCl_3$	1e-Bromo-1a4ea-trichlorocyclohexane	0.2	n.s.	n.s.	-	1615
	1e-Bromo-2e4a5a-trichlorocyclohexane	2.66	B	25	-	1615
$C_6H_8Br_2$	3,6-Dibromo-1-cyclohexene	0	CCl_4	17,47	0	1675
$C_6H_8Br_2Cl_2$	1a2e-Dibromo-3e4e-dichlorocyclohexane (m.p. = 128° C)	3.81	n.s.	n.s.	-	106
	1e2e-Dibromo-4a5a-dichlorocyclohexane	2.69	B	26	-	1615
	1,2-Dibromo-4,5-dichlorocyclohexane (α) (m.p. = 173° C)	1.69	B	25	-	1611
	1,2-Dibromo-4,5-dichlorocyclohexane (β) (m.p. = 242° C)	0.2	B	26	-	1611
$C_6H_8Br_4$	1a2e3e4e-Tetrabromocyclohexane (m.p. = 156° C)	3.66	n.s.	n.s.	-	106
	1e2e3e4e-Tetrabromocyclohexane (m.p. = 142° C)	2.89	n.s.	n.s.	-	106
	1,2,3,4-Tetrabromocyclohexane (m.p. = 90° C)	2.82	n.s.	n.s.	-	106

FORMULA	COMPOUND NAME	μ, D	State, or Solv.	t, °C	Method or P_a, % of P_e	Reference
$C_6H_8Br_4$ (contd.)	1,2,4,5-Tetrabromocyclohexane	2.2_4	B	~25	0	646
$C_6H_8Cl_2$	4,5-Dichlorocyclohexene	1.61-1.70	B	10-40	0	1674
		1.52-1.60	CCl_4	10-40	0	1674
$C_6H_8Cl_2O$	*cis*-Dichlorocyclohexanone	4.60±0.06	CCl_4	n.s.	-	1554
	trans-Dichlorocyclohexanone	3.80±0.04	CCl_4	n.s.	-	1554
	gem-Dichlorocyclohexanone	3.84±0.04	CCl_4	n.s.	-	1554
$C_6H_8Cl_2O_4$	*dl*-Dimethyl α, β-dichlorosuccinate	2.96	B	n.s.	-	676
	meso-Dimethyl α, β-Dichlorosuccinate	2.49	B	n.s.	-	676
$C_6H_8Cl_4$	Tetrachlorocyclohexanes					
	1ea2a4e	0.48	n.s.	n.s.	-	1605
	1ea3e5e	0.11	n.s.	n.s.	-	1605
	1e2a4e5a	0	n.s.	n.s.	-	1605
	1e2a4e5e	0	n.s.	n.s.	-	1605
	1e2e4e5e (α)[h] (m.p. = 174°C)	2.45	B	26	-	1615
	(β) (m.p. = 228°C)	0	n.s.	n.s.	-	1613
		0	n.s.	n.s.	-	1605
	1,2,3,4 (m.p. = 112°C)	2.95	n.s.	n.s.	-	106
	1ea4ea (m.p. = 125°C)	0	n.s.	n.s.	-	1615

h) Note these designations not used uniformly or consistently.

FORMULA	COMPOUND NAME	μ, D	State, or Solv.	t, °C	Method or P_a, % of P_e	Reference
	No positions given (m.p. = 124°C)	0	n.s.	n.s.	-	1613
		2.38	B	25	0	708
	(m.p. = 110°C)	3.19	B	25	5	1605,1754
$C_6H_8N_2$	2-Amino-3-methylpyridine	2.17	B	(30)	-	1372
		2.17	B	25	0	92
		4.23	B	30	-	1367
	2-Amino-4-methylpyridine	2.27±0.01	B	25	0	92
		2.65	B	(30)	-	1372
		2.94	B	30	-	1367
	2-Amino-5-methylpyridine	2.02±0.01	B	25	0	92
		2.35	B	(30)	-	1372
	2-Amino-6-methylpyridine	1.65±0.01	B	25	0	92
		1.77	B	30	-	1367,1372
	2-(Aminomethyl)-pyridine	2.25±0.02	B	25	0	91
		2.30±0.02	D	25	0	91
	3-(Aminomethyl)-pyridine	2.52±0.03	B	25	0	91
	4-(Aminomethyl)-pyridine	2.84±0.03	B	25	0	91
	1,4-Butanedicarbonitrile	3.57	B	25	-	1978
		3.69	B	75	-	1978
		3.76	B	25	0	1943
		3.79	B	25	-	1979
		3.84	B	25	0	1763
		3.90	B	75	-	1979
	2,5-Dimethylpyrazine	0	B	n.s.	-	2012
		0	B	20-50	-	2016
	2,6-Dimethylpyrazine	0.5	B	n.s.	-	2012
		0.53	B	20-50	-	1840,2016

FORMULA	COMPOUND NAME	μ, D	State, or Solv.	t, °C	Method or P_a, % of P_e	Reference
$C_6H_8N_2$ (contd.)	*o*-Phenylenediamine	1.49±0.08	gas	150-325	-	1155
		1.45	B	20-30	-	1950
		1.46	B	25	0	2118
	m-Phenylenediamine	1.72	gas	150-325	-	1155
		1.79	B	20	-	1950
		1.8	B	25	0	2118
	p-Phenylenediamine	1.47	gas	150-325	-	1155
		<0.3	B	25	-	2112,2125
	1.51 P_e = 31.87, P_a = 3.19	0.3	B	25	-	2088
		~1.5	B	25	0	2118
		1.52	B	25	-	206
		1.57	B	40	-	1950
		1.58	B	25	5	70
		1.60	B	25	0	475
	Phenylhydrazine	1.65-1.81	B	18	15	76
		1.67	B	25	5	70
		1.76	B	15	0	2008
$C_6H_8N_2OS$	1-Ethyl-2-thiouracil	4.77	D	35	0	1716
	3-Ethyl-2-thiouracil	3.61	D	35	0	1716
	5-Ethyl-2-thiouracil	5.07	D	35	0	1716
	6-Ethyl-2-thiouracil	4.68	D	35	0	1716
$C_6H_8N_2O_2S$	*m*-Aminobenzenesulfonamide	5.67	D	25	0	1004
	p-Aminobenzenesulfonamide (sulfanilamide)	6.22	B[i]	25	5	471
		5.37	D	25	-	992
		6.5±0.1	D	25	-	860
		6.58	D[i]	25	5	471
		6.60	D	25	0	1550

i) From data on benzene-dioxane mixtures.

FORMULA	COMPOUND NAME	μ, D	State, or Solv.	t, °C	Method or P_a, % of P_e	Reference
		6.65	D	n.s.	-	632-3
		6.67	D	25	0	635
		6.68	D	25	0	1004
$C_6H_8N_2S$	5,6-Dihydro-3-methylimidazo[2,1-b]thiazole	4.2±0.2	B	n.s.	-	2129
$C_6H_8N_4O_2$	*N*-Nitro-*bis*(2-cyanoethyl)amine	4.02	D	20	0	569
$C_6H_8O_2$	1,2-Cyclohexanedione	2.80±0.02	B	(25)	5	1915
	1,4-Cyclohexanedione	1.41±0.1	gas	195, 217	0	1642
		1.2_3	B	25	5	1074
		1.26-1.31	B	18-50	0	18
		1.30	B	25	-	1065
	3-Hexene-2,5-dione	2.85	n.s.	n.s.	-	1706
	Methyl ketene dimer	3.30	B	25	0	1621
$C_6H_8O_4$	Dimethyl fumarate	2.25	CCl_4	25	5	1075
	Dimethyl maleate	2.48	CCl_4	25	5	1075
$C_6H_8O_6$	*l*-Ascorbic acid	3.96	D	25	0	994
C_6H_8S	2,5-Dimethylthiophene	0.51	B	30	0	282
C_6H_9Br	3-Bromocyclohexene	2.53±0.04	B	25	5	125
	1-Bromo-1-hexyne	1.07	B	25	0	1494
C_6H_9BrO	2-Bromocyclohexanone	3.489	B	25	0	15
		3.50	B	30	0	1006
		3.51±0.02	B	30	0	1522
		3.62±0.02	D	30	0	1522
		3.64	D	30	0	1006
		3.33±0.02	CCl_4	30	0	1522
		3.37	Hp	30	0	1006
		3.38±0.02	Hp	30	0	1522

FORMULA	COMPOUND NAME	μ, D	State, or Solv.	t, °C	Method or P_a, % of P_e	Reference
C_6H_9Cl	1-Chloro-1-hexyne	1.24	B	25	0	1494
C_6H_9ClO	2-Chlorocyclohexanone	3.66	B	52	0	2156
		3.71±0.02	B	25	0	13
		3.77±0.02	B	30	0	1522
		3.78	B	30	0	1006
		3.96	B	27.4	0	2156
		3.91	D	30	0	1006
		3.94±0.02	D	30	0	1522
		3.41±0.07	CCl_4	n.s.	-	1554
		3.45±0.02	CCl_4	30	0	1522
		3.45	Hp	30	0	1006
		3.45±0.02	Hp	30	0	1522
$C_6H_9Cl_7O_2Ti$	*n*-Butyl trichloroacetate - titanium chloride complex	2.12	B	20	0	1447
$C_6H_9Cl_7O_6Sn$	Tin chloride - *tris*(chloroacetic acid) complex	3.67	B	n.s.	15	1449
	Isobutyl trichloroacetate - titarium chloride complex	2.23	B	20	0	1447
C_6H_9FO	2-Fluorocyclohexanone	3.86±0.10	B	n.s.	-	898
C_6H_9I	1-Iodo-1-hexyne	0.76	B	25	0	1494
C_6H_9N	Cyclopentanecarbonitrile	3.74	B	25	-	1650
	2,4-Dimethylpyrrole	1.75	B	25	-	642
	2,5-Dimethylpyrrole	2.03	B	25	-	642
		2.08±0.03	B	25	0	931

FORMULA	COMPOUND NAME	μ, D	State, or Solv.	t, °C	Method or P_a, % of P_e	Reference
C_6H_9NO	3,4,5-Trimethylisoxazole	3.39±0.02	B	25	0	1858,1860
		3.45	B	25	0	1932
		3.40±0.01	D	25	0	1858,1860
$C_6H_9NO_2$	*N*-Methylglutarimide	2.70±0.04	D	30	0	1059
$C_6H_9N_3S$	5,6-Dihydro-3-methyl-6-aminoimidazo-[2,1-b]thiazole	1.8±0.2	B	n.s.	-	2129
$C_6H_9O_3P$	1-Phospha-2,8,9-trioxaadamantine	4.7±0.2	D	25	0	230
$C_6H_9S_3$	α-Trithioacetaldehyde trimer	2.14	n.s.	n.s.	-	682
	β-Trithioacetaldehyde trimer	2.14	n.s.	n.s.	-	682
C_6H_{10}	Cyclohexene	0.61	gas	35-105	-	989
		0.28	liq	20	NRA	23
		0.63	CCl_4	25	0	1540
		0.76	Hx	25	0	1540
	2,3-Dimethyl-1,3-butadiene	0	gas	-75 to 50	9	496
		0.52±0.07	gas	100-210	5	658
		0	B	25	8	497
	2-Ethyl-1,3-butadiene	0.45	gas	110-205	0	660
	2,4-Hexadiene 'low boiling <80°C"	0.36	B	25	8	497
	"high boiling ≈ 80°C"	0.31	B	25	8	497
	1-Hexyne	0.88	gas	25-125	0	970
	2-Methyl-1,3-pentadiene	0.65	gas	125-225	0	660
		0.59	B	25	8	497
	2-Methyl-2,3-pentadiene	0.52	gas	-75 to 50	19	496

FORMULA	COMPOUND NAME	μ, D	State, or Solv.	t, °C	Method or P_a, % of P_e	Reference
C_6H_{10} (contd.)	3-Methyl-1,3-pentadiene	0.63	gas	125-215	0	660
		0.53	B	25	8	497
$C_6H_{10}BrCl$	1-Bromo-1-chlorocyclohexane	2.46	B	25	0	122
		2.47	B	30	0	1033
		2.50	CCl_4	30	0	1033
	cis-1-Bromo-2-chlorocyclohexane	3.08	B	30	0	1033
		3.15	B	25	0	122
		3.08	CCl_4	30	0	1033
		3.16	CCl_4	25	0	122
	trans-1-Bromo-2-chlorocyclohexane	2.19±0.01	gas	175, 194	0	1642
		2.43	B	30	0	1033
		2.48	B	25-50	5	1521
		2.49	B	25	0	122
		2.04	CCl_4	30	0	1033
		2.07	CCl_4	25-50	5	1521
$C_6H_{10}Br_2$	1,1-Dibromocyclohexane	2.43	B	30	0	1033
		2.44	B	25	0	122
		2.45	CCl_4	30	0	1033
	cis-1,2-Dibromocyclohexane	3.12	B	25	0	122
		3.15	B	30	0	1033
		3.06	CCl_4	25	0	122
		3.13	CCl_4	30	0	1033
	trans-1,2-Dibromocyclohexane	2.00±0.01	gas	175, 194	0	1642
		2.11	B	25	0	122
		2.15	B	10-50	0	936
		2.15	B	25-50	5	1521
		2.16	B	30	0	1033
		1.73-1.79	CCl_4	10-40	0	936
		1.74	CCl_4	25	0	122

FORMULA	COMPOUND NAME	μ, D	State, or Solv.	t, °C	Method or P_a, % of P_e	Reference
		1.76	CCl_4	30	0	1033
		1.77	CCl_4	25-50	5	1521
	cis-1,4-Dibromocyclohexane	2.93	B	30	0	1033
		2.89	CCl_4	30	0	1033
	trans-1,4-Dibromocyclohexane	0	B	18	0	645
		0	B	30	0	1033
		0	CCl_4	30	0	1033
		0	CCl_4	25	-	960
	cis-1,4-Dibromo-2,3-dimethyl-2-butene	2.49	B	25	0	1919
	trans-1,4-Dibromo-2,3-dimethyl-2-butene	1.72	B	25	0	1919
$C_6H_{10}Cl_2$	1,1-Dichlorocyclohexane	2.46	B	30	0	1033
		2.49	B	25	0	122
		2.48	CCl_4	30	0	1033
	cis-1e2a-Dichlorocyclohexane	3.11	gas	-37	0	1992
		3.10	B	30	0	937
		3.13	B,CCl_4	30	0	1033
		3.13	B	40	0	1992
		3.12	CCl_4	40	0	937
		3.13	CCl_4	25	0	122
		3.15-3.13	Hp	30-45	0	937
	trans-1,2-Dichlorocyclohexane	2.29±0.01	gas	168, 195	0	1642
		2.31	gas	-34	0	1992
		2.54	B	30	0	936
		2.63	B	30	0	1033
		2.65	B	25-50	5	1521
		2.66	B	40	0	1992
		2.67	B	25	0	122
		2.17	CCl_4	30	0	936

FORMULA	COMPOUND NAME	μ, D	State, or Solv.	t, °C	Method or P_a, % of P_e	Reference
$C_6H_{10}Cl_2$ (contd.)	*trans*-1,2-Dichlorocyclohexane (contd.)	2.25	CCl_4	25-50	5	1521
		2.27	CCl_4	30	0	1033
	cis-1,4-Dichlorocyclohexane	2.89	B, CCl_4	30	0	1033
	trans-1,4-Dichlorocyclohexane	0	B	18	0	645
		0	B	30	0	1033
		0	CCl_4	25,30	0	960,1033
		0	n.s.	n.s.	-	938
$C_6H_{10}I_2$	*cis*-1,4-Diiodocyclohexane (m.p. = 57.5°C)	2.45	B	18	0	645
		2.48	B	30	0	1033
		2.45	CCl_4	30	0	1033
	trans-1,4-Diiodocyclohexane (m.p. = 142°C)	0	B	18	0	645
		0	B, CCl_4	30	0	1033
$C_6H_{10}I_2Ni$	Allyl - nickel iodide π-complex	1.62±0.07	B	25	15	516
		1.48±0.03	cHx	25	15	516
$C_6H_{10}N_2O_2$	*N,N'*-Diformylpiperazine	2.72±0.03	B	20	-	568
$C_6H_{10}N_2O_4$	*cis*-3,4-Dinitro-3-hexene	5.76	B	25	0	1293
		5.79	CCl_4	26	0	1293
	Diethyl azidodiformate	2.58	CCl_4	25	5	1075
$C_6H_{10}O$	Butoxyacetylene	2.05	liq	25	-	814
	Cyclohexanone	2.75	B	15	-	2136
		2.8	B	25	0	2115
		2.94	B[j]	20	5	1513
	3.08 P_e = 27.06, P_a = 0	3.01±0.02	B	25	0	127
		3.04±0.01	B?	n.s.	-	302
		3.07±0.1	B	25	0	1113

j) Plus water.

FORMULA	COMPOUND NAME	μ, D	State, or Solv.	t, °C	Method or P_a, % of P_e	Reference
		3.11±0.03	B	20	0	630
		2.93	D	25	0	647
		2.98±0.03	CCl_4	n.s.	-	1554
		3.14	CCl_4	25	5	1074
		3.17±0.03	CCl_4	20	0	630
		3.07±0.03	cHx	20	0	630
	Epoxycyclohexane	1.7	n.s.	n.s.	-	1362
	2-Methyl-2-penten-4-one	2.79	B	25	5	469
		2.84±0.03	B	25	0	127
		2.83	D	25	5	469
		2.85±C.03	CCl_4	25	0.	127
	4-Methyl-3-penten-2-one	3.20	B	25	5	468
		3.28	D	25	5	468
$C_6H_{10}O_2$	Ethyl methacrylate	2.15	B	15	-	1310
$C_6H_{10}O_2S$	Ethyl 3-thioacetoacetate	2.40	B	25	0	503
		2.24	CCl_4	25	0	503
		2.30	Hx	25	0	503
		2.18	CS_2	25	0	503
$C_6H_{10}O_2S_2$	Allylsulfinyl allyl sulfide	3.99	B	25	0	619
$C_6H_{10}O_3$	Ethyl acetoacetate	2.9_6	gas	120-160	10	2170
	enol form	2.06	CS_2	-80	-	174
	keto form	3.25	B	Rm. Temp.	-	174
$C_6H_{10}O_4$	Adipic acid	2.32±0.02	D	25	0	119
		2.41	D	35	5	1399
		2.50	D	35	-	1581
		2.60±0.10	D	25	0	1635
		4.07	D	25	0	1983

FORMULA	COMPOUND NAME	μ, D	State, or Solv.	t, °C	Method or P_a, % of P_e	Reference
$C_6H_{10}O_4$ (contd.)	Diethyl oxalate	2.40-2.54	liq	20-95	0	907
		2.22	B	25-45	-	831
		2.51	B	25	0	1832
		2.51	B	25,50	-	1679
		2.54	B	50	0	1832
		2.57	B	20,45	0	907
		2.61	B	75	0	907
		2.49	D	20	0	907
		2.54	D	45	0	907
		2.58	D	75	0	907
		2.22	CCl_4	25-45	-	831
		2.41	CCl_4	25	5	1075
	Ethylene glycol diacetate	2.34	B	30	0	590
	cis-2,5,7,10-Tetraoxabicyclo[4.4.0]decane[k]	1.92	B	20	-	700-1
	trans-2,5,7,10-Tetraoxabicyclo[4.4.0]decane m.p. = 136°C	0.77	B	20	-	700
		0.72	n.s.	n.s.	-	701
		0	n.s.	n.s.	-	382
$C_6H_{10}O_6$	*dl*-Dimethyl tartrate	2.9_5	B	25	0	2082
		2.96	B	25	-	2139
	d-Dimethyl tartrate	2.9_6	B	25	0	2082,2083
$C_6H_{10}S_2$	Diallyl disulfide	1.76	B	25	-	2030
$C_6H_{10}S_4$	2,2'-Di(1,3-dithiacyclopentyl)	1.16	n.s.	n.s.	-	701
$C_6H_{11}Br$	Bromocyclohexane	2.08	liq	25	-	707
		2.13	B	60	0	680
		2.20	B	25	0	835
		2.24	B	25	0	1646
		2.3±0.1	B	25	0	2115

k) Reference 382 claims the compound is actually 2,2'-di(1,3-dioxacyclopentyl).

FORMULA	COMPOUND NAME	μ, D	State, or Solv.	t, °C	Method or P_a, % of P_e	Reference
		2.31	B	25	0	1622
		2.2_6	CCl_4	25	5	1074
$C_6H_{11}BrHgO$	α-(2-Methoxycyclopentyl)mercuric bromide	3.4	B,D	20	-	221
	β-(2-Methoxycyclopentyl)mercuric bromide	4.7	B,D	20	-	221
$C_6H_{11}BrO_2$	*dl-erythro*-2-Acetoxy-3-bromobutane	2.25	liq	25	-	2130
	dl-threo-2-Acetoxy-3-bromobutane	2.28	liq	25	-	2130
	Propyl 2-bromopropionate	2.36	B	25	0	1176
$C_6H_{11}Cl$	Chlorocyclohexane	2.09-2.12	B	60-18	0	680,681a
		2.18±0.10	B	25	0	1646
		2.24	B	25	0	122
		2.3±0.1	B,D	25	0	2115-6
		2.18	CCl_4	25	5	1074
		2.2[m]	CCl_4	25	-	1542
	Chloromethylcyclopentane	2.35[m]	CCl_4	25	-	1542
$C_6H_{11}ClHgO$	α-(2-Methoxycyclopentyl)mercuric chloride	2.9	B,D	20	-	221
	β-(2-Methoxycyclopentyl)mercuric chloride	4.1	B,D	20	-	221
$C_6H_{11}HgIO$	α-(2-Methoxycyclopentyl)mercuric iodide	3.2	B,D	20	-	221
	β-(2-Methoxycyclopentyl)mercuric iodide	3.9	B,D	20	-	221
$C_6H_{11}I$	Iodocyclohexane	2.00	B	60	-	680
		2.2_3	CCl_4	25	5	1074
$C_6H_{11}N$	Isohexanenitrile	3.53	B	25	0	1630

m) $\overline{\mu}^2$ value.

FORMULA	COMPOUND NAME	μ, D	State, or Solv.	t, °C	Method or P_a, % of P_e	Reference
$C_6H_{11}NO$	*cis*-ε-Caprolactam	3.88	B	25	0	783
	dimer	2.6	B	25	0	783
	Cyclopentanone methyloxime	1.262±0.005	B	20	5	1593
	4-Dimethylamino-3-buten-2-one	5.10	B	25	0	1749
	5,5-Dimethylpyrrolidinone	4.05±0.03	D	30	0	1059
	N-Methyl-2-piperidone	4.04	B	25	0	1922
$C_6H_{11}NO_2$	Nitrocyclohexane	3.6±0.1	B	20	0	256
$C_6H_{11}N_3$	Azidocyclohexane	2.39±0.03	B	25	5	125
$C_6H_{11}O_2$	2-Hydroxycyclohexanone	2.90±0.05	B	(25)	5	1915
C_6H_{12}	Cyclohexane	0.61	gas	35-105	-	989
		0	B	25	0	2115
		0.2	B	25	-	1696
		0	CCl_4	25	5	1074
		1.78	NB	30	0	754
	Ethylcyclobutane	0.05	liq	20	-	2013
	1-Hexene	0.34	liq	20	NRA	23
		0.45	liq	20	0	1189
	cis-3-Hexene	0.34	liq	20	NRA	23
	trans-3-Hexene	0	liq	25	NRA	23
	Methylcyclopentane	0	liq	20	-	2013
	n-Propylcyclcpropane	0.75	B	20	-	187

FORMULA	COMPOUND NAME	μ, D	State, or Solv.	t, °C	Method or P_a, % of P_e	Reference
$C_6H_{12}BNO_3$	*triptych*-Boroxazolidine	8.8±0.12	D	25	-	553
$C_6H_{12}Br_2$	1,6-Dibromohexane	2.38±0.03	liq	20-95	0	907
		2.32	B	20	0	907
		2.328±0.004	B	25	0	692
		2.379±0.004	B	64	0	692
		2.40	B	25	0	1943
		2.41	D	20	0	907
		2.38	CCl_4	20	0	907
		2.29	Hx	20	0	907
		2.38	cHx	20	0	907
		2.42±0.01	Hp	20	0	1837
	dl-3,4-Dibromohexane	2.08	liq	25	-	2130
	meso-3,4-Dibromohexane	1.58	liq	25	-	2130
		1.24	B	25	0	1355
	2,3-Dibromo-2,3-dimethylbutane	0.88	CCl_4	-15	0	1355
		0.97	CCl_4	6	0	1355
		1.03	CCl_4	25	0	1355
		1.15	CCl_4	55	0	1355
		0.77	Hp	-20	0	1355,1326
		0.84	Hp	1	0	1355,1326
		0.96	Hp	25	0	1355,1326
		1.02	Hp	50	0	1355,1326
		2.6±0.3	Hp	n.s.	0	1327
$C_6H_{12}Br_2Cl_6NiP_2$	*Bis*(tri(chloromethyl)phosphine)nickel bromide	1.5±0.2	B	25	15	335
$C_6H_{12}Br_4O_4Zr$	*Bis*(ethyl formate) - zirconium bromide complex	7.91	B	20	10	923

FORMULA	COMPOUND NAME	μ, D	State, or Solv.	t, °C	Method or P_a, % of P_e	Reference
$C_6H_{12}Cl_2$	1,6-Dichlorohexane	2.27	B	25	0	1943
		2.35	B	20	RT	921
		2.47	Hx	20	RT	921
	2,3-Dichloro-2,3-dimethylbutane	1.75	B	25	0	1355
		1.24	CCl_4	-20	0	1355
		1.30	CCl_4	0	0	1355
		1.37	CCl_4	25	0	1355
		1.47	CCl_4	55	0	1355
		1.14	Hp	-20	0	1355,1326
		1.23	Hp	1	0	1355,1326
		1.31	Hp	25	0	1355,1326
		1.41	Hp	50	0	1355,1326
		3.1±0.3	Hp	n.s.	0	1327
$C_6H_{12}Cl_2F_6P_2Pt$	Dichloro*bis*(dimethyl(trifluoromethyl)-phosphine)platinum	9.0±0.05	$CHCl_3$	n.s.	-	118
$C_6H_{12}Cl_2N_2$	*N,N'*-Dichloro-2,5-dimethylpiperazine	0.47	B	20	0	568
$C_6H_{12}Cl_3O_2P$	Propyl ethyl(trichloromethyl)phosphinate	3.31	B	25	-	61
$C_6H_{12}Cl_4O_2Sn$	Stannic chloride - *bis*(acetone) complex	7.7	B	15	0	2003
$C_6H_{12}Cl_4O_2Ti$	Titanium chloride - *n*-butyl acetate complex	4.86	B	20	0	1447
	Titanium chloride - ethyl butyrate complex	4.52	B	20	15	1438
$C_6H_{12}Cl_4O_4Zr$	Zirconium chloride - *bis*(ethyl formate) complex	7.63	B	20	15	1442
$C_6H_{12}Cl_4O_6Sn$	Stannic chloride - *tris*(acetic acid) complex	7.23	B	n.s.	15	1449
$C_6H_{12}Cl_8O_4Zr_2$	Zirconium chloride - ethyl formate complex dimer	4.18	B	20	0?	924

FORMULA	COMPOUND NAME	μ, D	State, or Solv.	t, °C	Method or P_a, % of P_e	Reference
$C_6H_{12}Hg$	Mercuracycloheptane	0.90	CS_2	20	-	1700
$C_6H_{12}Hg_2O_2$	1,7-Dioxa-4,10-dimercuracyclodecane	1.17	B	20	-	1700
$C_6H_{12}I_4O_4Zr$	Zirconium iodide - *bis*(ethyl formate) complex	4.24	B	(0)	-	1446
$C_6H_{12}N_2$	Dimethylketazine	1.52	gas	75-230	10	184
$C_6H_{12}N_2O_2$	Methyl-*n*-propylglyoxime	1.26	D	20	-	1319
$C_6H_{12}N_2O_3$	Glycylglycine ethyl ester	3.23	B	50	0	2152
$C_6H_{12}N_2O_6$	1,6-Dinitroxyhexane	3.76±0.03	B	25	0	1053
$C_6H_{12}N_2S_3$	*Bis*(dimethylthiocarbamoyl) sulfide	5.33	B	25	0	639
$C_6H_{12}N_2S_4$	*Bis*(dimethylthiocarbamoyl) disulfide	2.23	B	25	0	638,639
$C_6H_{12}N_2S_5$	*Bis*(dimethylthiocarbamoyl) trisulfide	4.84	B	25	0	639
$C_6H_{12}N_2S_6$	*Bis*(dimethylthiocarbamoyl) tetrasulfide	5.82	B	25	0	639
$C_6H_{12}N_4$	Hexamethylenetetramine	0	$CHCl_3$	25-45	5	1093
$C_6H_{12}N_4O_2$	2,5-$\uparrow\uparrow$-Dimethyl-*N*,*N*′-dinitrosopiperazine	2.73	D	20	-	569
	2,5-$\downarrow\uparrow$-Dimethyl-*N*,*N*′-dinitrosopiperazine	1.61	D	20	-	569
$C_6H_{12}N_4O_4$	2,5-$\uparrow\uparrow$-Dimethyl-*N*,*N*′-dinitropiperazine	4.32	D	20	-	569
	2,5-$\downarrow\uparrow$-Dimethyl-*N*,*N*′-dinitropiperazine	1.90	D	20	-	569
$C_6H_{12}O$	Cyclohexanol	1.3	liq	n.s.	-	594
		1.4	B, CCl_4	n.s.	-	594
		1.70	B	18	0	681a
		1.9	B	25	0	2115
		1.84	D	25	0	647
		1.8_6	CCl_4	25	5	1108

FORMULA	COMPOUND NAME	μ, D	State, or Solv.	t, °C	Method or P_a, % of P_e	Reference
$C_6H_{12}O$ (contd.)	2,2-Dimethyl-3-butanone	2.81	B	15	10	2141
	2-Hexanone	2.5	B	30	RT	2052
		2.68_8±0.02	B	22	15	2140
		2.75	B	15	10	2141
	Vinyl *n*-butyl ether	1.25	B	25	0	1631
	Vinyl isobutyl ether	1.20	B	25	0	1631
$C_6H_{12}O_2$	Amyl formate	1.92	gas	100-245	-	2168
	n-Butyl acetate	1.82	liq	30-40	-	1497
		1.84	liq	25	-	1254
		1.86±0.01	B	22	15	2140
		1.8_7	B	25	0	1366
	Isobutyl acetate	1.87_4±0.01	B	22	15	2140
		1.8_9	B	25	0	1366
	t-Butyl acetate	1.93_2±0.025	B	22	15	413
	Caproic acid	1.13	liq	25	-	1495
		1.31	liq	60	-	1495
	cis-1,2-Cyclohexanediol	2.33±0.02	B	25	5	1915
	trans-1,2-Cyclohexanediol	2.39±0.03	B	25	5	1915
	1,2-Cyclohexanediol (monoenol)	2.80±0.02	B	25	5	1915
	1,2-Cyclohexanediol (dienol)	2.78±0.02	B	25	5	1915
	cis-1,4-Cyclohexanediol	1.80±0.03	B	25	5	1915
	trans-1,4-Cyclohexanediol	2.50±0.02	B	25	5	1915

FORMULA	COMPOUND NAME	μ, D	State, or Solv.	t, °C	Method or P_a, % of P_e	Reference
	cis-2,3-Dimethyl-2-butene-1,4-diol	2.52	B	25	0	1919
	trans-2,3-Dimethyl-2-butene-1,4-diol	1.93	B	25	0	1919
	Ethyl butyrate	1.75_8±0.02	B	22	15	2140
	Ethyl isobutyrate	2.07	B	20	-	1309
	4-Hydroxy-4-methyl-2-pentanone	3.24	B	20	-	977
	Methyl valerate	1.61_6±0.03	B	22	15	2140
	4-Methylvaleric acid	1.13	liq	25	-	1495
		1.30	liq	60	-	1495
	Propyl propionate	1.78_7±0.03	B	22	15	2140
$C_6H_{12}O_2S$	Tetrahydrodimethylthiophene 1,1-dioxide	4.80	B	25	0	1165
$C_6H_{12}O_3$	3-Oxapentyl acetate	2.32	liq	30-50	-	1497
		2.24	B	30	0	590
	Paraldehyde	1.44	gas	n.s.	-	1086
		1.73	liq	n.s.	-	1086
		1.91	liq	25	-	1098
		1.87	B	25	-	1086
		1.94	B	18	0?	673
		1.95	B	n.s.	0	671,684
		2.05	B	25	-	1098
		1.90	D	25	-	1086
		1.98	CCl_4	25	-	1086
		2.14	CCl_4	25	-	1098
		2.01	Tol	n.s.	0	671,683-4
		1.77	mm	25	-	1086
		2.14	$CHCl_3$	25	-	1086
		2.28	$CHCl_3$	25	-	1098

mm) Light petroleum.

FORMULA	COMPOUND NAME	μ, D	State, or Solv.	t, °C	Method or P_a, % of P_e	Reference
$C_6H_{12}O_3$ (contd.)	Paraldehyde (contd.)	1.79	ClB	25	-	1098
		1.69	n	25	-	1098
		1.96	p	25	-	1098
		1.99	NB	25	-	1098
		1.78	q	25	-	1098
		1.82	r	25	-	1098
$C_6H_{12}S$	Cyclohexanethiol	1.64	liq	25	10	1281
		1.73	B	25	10	1281
$C_6H_{13}BCl_2$	Dichloro-*n*-hexylborine	1.55	B	25	5	370
		4.46	D	25	5	370
$C_6H_{13}BF_2$	Difluoro-*n*-hexylborine	1.61	B	25	5	370
		3.30	D	25	5	370
$C_6H_{13}Br$	1-Bromohexane	1.99[s]	liq	25	-	707
		2.06	liq	20	-	657
		1.99	B	18	-	698
$C_6H_{13}Cl$	1-Chlorohexane	1.94	B	20	RT	921
	2-Chlorohexane	2.06	B	20	RT	921
		2.11	Hx	20	RT	921
	3-Chlorohexane	2.03	B	20	RT	921
		2.10	Hx	20	RT	921
	2-Chloro-3-methylpropane	2.16	B	20	RT	921
		2.18	Hx	20	RT	921
	3-Chloro-3-methylpropane	2.17	B	20	RT	921
		2.20	Hx	20	RT	921
$C_6H_{13}ClO$	Butyl 1-chloroethyl ether	2.09	B	25	0	1763

n) Dimethylaniline. p) Benzonitrile. q) Ethyl benzoate. r) *o*-Dichlorobenzene.

FORMULA	COMPOUND NAME	μ, D	State, or Solv.	t, °C	Method or P_a, % of P_e	Reference
$C_6H_{13}ClO_2S$	Hexanechlorosulfite	2.75	B	25,45	0	1974
$C_6H_{13}I$	1-Iodohexane	1.94	CCl_4	20	-	78
$C_6H_{13}N$	*N*-*n*-Butylethylidenamine	1.61±0.03	B	25	-	478
	Cyclohexylamine	1.22	liq	20	5	337
		1.26	B	20	5	337
		1.33	B	25	0	1120
		1.34	B	25	5	70
		1.31	Hx	20	5	337
		1.32	Hp	25	-	920
	N-Ethylbutylidenamine	1.67±0.05	B	25	0	478
	N-Methylpiperidine	0.80±0.01	B	25	0	1113
		0.92±0.01	B	10-40	-	1589
		0.92	B	25	0	68
$C_6H_{13}NO$	*N*,*N*-Diethylacetamide	3.75	D	30	10	1011
	N-Ethylbutyramide	3.87	B	25	0	1165
$C_6H_{13}NO_2$	Ethyl α-aminobutyrate	2.15	B	25	0	2152
	Ethyl β-aminobutyrate	2.13	B	25	0	2152
	Methyl α-aminovalerate	1.6	B	20	-	463
	Methyl β-aminovalerate	2.7	B	20-50	-	463
$C_6H_{13}NO_4$	2-Propyl-2-nitro-1,3-propanediol	3.49±0.05	B	n.s.	-	257
C_6H_{14}	*n*-Hexane	0	gas	80, 110	-	983
		0	gas	65-285	-	1819
		0	liq	-90 to 50	-	418

FORMULA	COMPOUND NAME	μ, D	State, or Solv.	t, °C	Method or P_a, % of P_e	Reference
C_6H_{14} (contd.)	*n*-Hexane (contd.)	0	liq	-90 to 70	-	1825
		0	liq	-60 to 60	-	693
		0	liq	20	VS	1701
		0.05	B	25	-	2122
		0.08	CS_2	25	-	2122
$C_6H_{14}Hg$	Di-*n*-propylmercury	0.54	B	20	-	1700
$C_6H_{14}N_2$	1,4-Dimethylpiperazine	0.44	B	25	5	69
		0.53	B	25	0	1253
		0.59	B	25	-	568
	2,5-Dimethylpiperazine	1.55	D	20	-	568
$C_6H_{14}N_2O_2$	Di-*n*-propylnitramine	4.76	D	20	-	569
	Diisopropylnitramine	4.91	D	20	-	569
$C_6H_{14}O$	Ethyl *n*-butyl ether	1.22	B	25	-	1863
		1.25	B	25	0	1128
		1.20	t	25	0	1128
		1.2	n.s.	n.s.	-	1128
	1-Hexanol	1.55	B	20	-	977
		1.65	B	25	0	1239
		1.68	B	25	5	1108
		1.73±0.04	B	20	0	1184
		1.77	CCl_4	25	-	1078
		1.7_7	CCl_4	25	5	1108
		1.92	Hx	25	-	1150
	n-Propyl ether	1.03±0.05	gas	95-175	-	1695
		1.21±0.01	gas	60-200	2	624
		1.13	B	25	-	99
		1.31	B	25	-	1863
		1.17±0.05	Hx	-5 to 50	0	1305

t) Kerosene.

FORMULA	COMPOUND NAME	μ, D	State, or Solv.	t, °C	Method or P_a, % of P_e	Reference
	Isopropyl ether	1.13	gas	80-160	-	99
		1.26	B	25	-	99
$C_6H_{14}O_2$	1,1-Diethoxyethane	1.08	gas	50-135	10	987
		1.22	gas	200	10	987
	2-Hydroxyethyl *n*-butyl ether	2.10	B	24	0	246
	1,6-Hexanediol	2.50	D	25,50	0	1833
	2-Methyl-2,4-pentanediol	2.9	D	n.s.	-	2105
		2.1	Hp	n.s.	-	2105
$C_6H_{14}O_2S$	Dipropylsulfone	4.47±0.03	B	20	0	1201
$C_6H_{14}O_4$	Triethylene glycol	5.58	liq	20	-	933
		2.99	B	55	10	1999
$C_6H_{14}S$	1-Hexanethiol	1.55	liq	25	27	1280
	n-Propyl sulfide	1.56	B	20	0	791
$C_6H_{14}S_2$	*n*-Propyl disulfide	1.98±0.01	B	30	0	1026
		1.98	B	25	-	2030
$C_6H_{14}Si$	Methylethylallylsilane	0.55	B	25	15	887
$C_6H_{15}Al$	Triethylaluminum	0.6	B	20	15	1888-9
		3.5	D	20	15	1888
		0.5	Hp	20	15	1888-9
$C_6H_{15}As$	Triethylarsine	1.05	B	20	0	1028
$C_6H_{15}AsCl_2Pt$	*cis*-Triethylarsine platinum dichloride complex	10.9	B	25	0	296

FORMULA	COMPOUND NAME	μ, D	State, or Solv.	t, °C	Method or P_a, % of P_e	Reference
$C_6H_{15}AuClP$	Chloro(triethylphosphine)gold	8.4	B	25	-	262
$C_6H_{15}B$	Triethylboron	0.3	D	20	15	1888
		0	Hp	20	15	1888
$C_6H_{15}BO_3$	Triethyl borate	0.75	B	20	0	343
		0.75	CCl_4	25	0	72
$C_6H_{15}BrPb$	Triethyllead bromide	4.49	B	25	-	1118
		4.91	Tol	23	-	1250
$C_6H_{15}BrSi$	Triethylbromosilane	2.42±0.04	B	25	5	1592
$C_6H_{15}BrSn$	Triethyltin bromide	3.35	CCl_4	25	0	1852
		3.15	Hx	n.s.	-	875
$C_6H_{15}ClPb$	Triethyllead chloride	4.42	B	25	-	1118
		4.69	CCl_4	22	-	1250
$C_6H_{15}ClO_3Ti$	Chlorotriethoxytitanium	2.87	B	25	-	355
		2.51	Hx	25	0?	274
$C_6H_{15}ClSi$	Triethylchlorosilane	2.09	B	25	0	373
		2.09±0.01	CCl_4	20	0	1248
$C_6H_{15}ClSn$	Triethyltin chloride	3.47	CCl_4	25	0	1852
$C_6H_{15}Cl_2PPt$	*cis*-Triethylphosphine platinum dichloride complex	10.9	B	25	0	296
$C_6H_{15}Cl_2PtSb$	*cis*-Triethylstibine platinum dichloride complex	10.5	B	25	0	296
$C_6H_{15}FSi$	Triethylfluorosilane	1.74	B	25	0	373

FORMULA	COMPOUND NAME	μ, D	State, or Solv.	t, °C	Method or P_a, % of P_e	Reference
$C_6H_{15}Ga$	Triethylgallium	0.7	B	20	15	1888
		2.1	D	20	15	1888
		0	Hp	20	15	1888
$C_6H_{15}I_2N$	Iodine - triethylamine complex	9.30[u]	B	25	-	895
		11.3	D	20	-	1988
$C_6H_{15}In$	Triethylindium	1.4	B	20	15	1888
		1.3	D	20	15	1888
		0.6	Hp	20	15	1888
$C_6H_{15}N$	Di-*n*-propylamine	1.01	liq	20	5	337
		1.03	B	20	5	337
		1.08	B	20	0	336
		0.99	Hx	20	5	337
		1.04	Hx	20	0	336
	Hexylamine	1.32	B	30	-	1579
		1.59	B	30	5	1578
	Isohexylamine	1.60	B	30	5	1578
	Triethylamine	0.61	gas	-	M	802
		0.66_0	gas	25	0	100
		0.83	gas	25-105	-	572
		0.70	liq	n.s.	ABS	423
		0.75_8	liq	25	0	100
		0.79_9	B	25	0	725
		0.80	B	25	0	195
		0.87	B	25	0	68
		0.91	B	25	0	530
		0.91_3	B	25	0	100
		0.98	CCl_4	25	0	1729
		0.74_9	Hx	25	0	725
$C_6H_{15}NO_2S$	Triethylamine - sulfur dioxide complex	4.95	B	25	5	1344

u) Lower limit only.

FORMULA	COMPOUND NAME	μ, D	State, or Solv.	t, °C	Method or P_a, % of P_e	Reference
$C_6H_{15}NO_3$	Triethanolamine	3.60	D	25	0	1485
		3.43±0.05	n.s.	n.s.	-	1720
$C_6H_{15}NO_3S$	Triethylamine - sulfur trioxide complex	6.84	B	25	5	1344
		7.3	D	25	5	1344
$C_6H_{15}O_2P$	*n*-Butyl ethylphosphinate	3.44	B	25	5	61
	Isobutyl ethylphosphinate	3.41	B	25	5	61
$C_6H_{15}O_2PS$	*O,O′*-Diethyl ethylphosphonothioate	3.25	B	25	5	61
$C_6H_{15}O_3P$	Diethyl ethylphosphonate	2.95±0.02	liq	32	-	950-1
		2.91	CCl_4	20	-	50
		2.84	Hp	32	-	951
	Dipropylphosphite	3.15	CCl_4	20	-	50
	Diisopropylphosphite	3.08	CCl_4	20	-	50
	Triethylphosphite	1.82±0.05	D	25	0	230
		1.96	CCl_4	20	-	50
$C_6H_{15}O_3PS$	Triethylthiophosphate	2.82	B	25	5	473
		2.94	B	20	0	59
$C_6H_{15}O_4P$	Triethylphosphate	3.00	B	25	-	906
		3.08	B	25	5	473
		3.10±0.03	B	24.3	5	1914
		3.07	CCl_4	20	-	50
		2.84	cHx	20	-	50
$C_6H_{15}P$	Triethylphosphine	1.84	cHx	25	-	920

FORMULA	COMPOUND NAME	μ, D	State, or Solv.	t, °C	Method or P_a, % of P_e	Reference
$C_6H_{15}Tl$	Triethylthallium	0.5	B	20	15	1888
		0.7	D	20	15	1888
		0	Hp	20	15	1888
$C_6H_{16}Cl_2Si_2$	1,2-*Bis*(chloromethyl)-1,1,2,2-tetramethyl-disilane	2.23	B	30	0	544
$C_6H_{16}N_2$	1,6-Diaminohexane	1.94	B	25	-	1979
$C_6H_{16}OSi$	Triethylsilanol	0.62±0.08	liq	20	-	626
		1.51+0.05	CCl_4	22	0	1248
$C_6H_{16}OSn$	Triethyltin hydroxide	1.93±0.01	B	22	0	1249
$C_6H_{16}O_2Si$	Dimethyldiethoxysilane	1.36	B	30	0	545
$C_6H_{16}O_3Si$	Triethoxysilane	1.80±0.06	B	25	5	1853
$C_6H_{16}Si$	Triethylsilane	0.60	liq	20	6	24
		0.76±0.07	B	25	5	1853
		0.50	CCl_4	25	15	1247
$C_6H_{17}NO$	Triethylamine - water complex	2.24	B	20	-	894
$C_6H_{17}NSi$	Trimethyl-*N*-propylsilazane	0.79	B	30	-	328
		0.75	cHx	30	-	328
	Trimethyl-*N*-isopropylsilazane	0.77	B	30	-	328
		0.76	cHx	30	-	328
$C_6H_{18}BN$	Trimethylborine - trimethylamine complex	3.92±0.03	B	20	0	57
$C_6H_{18}B_3N_3$	Hexamethylborazole	0.72	B	25	5	2065
$C_6H_{18}Cl_2O_6P_2Pt$	Dichloro*bis*(dimethyl methylphosphate)-platinum	9.23	B	n.s.	-	57

FORMULA	COMPOUND NAME	μ, D	State, or Solv.	t, °C	Method or P_a, % of P_e	Reference
$C_6H_{18}Cl_2P_2Pt$	Dichloro*bis*(trimethylphosphine)platinum	1.31±0.5	B, $CHCl_3$	n.s.	-	118
$C_6H_{18}N_2Si$	*Bis*(dimethylamino)dimethylsilane	0.81±0.06	B	20	-	1702
$C_6H_{18}OSi$	Hexamethyldisiloxane	0.66±0.05	gas	47-247	16	755
	[For higher polymers of the type	0.43	liq	25	-	83
	$(CH_3)_3Si[OSi(CH_3)_2]_nCH_3$, see the	0.46	liq	20,40	16	755
	entry near the end of the section	0.7	B	20	0	1041
	on polymers, page 574.]	0.79	B	30	0	545
		0.81	B	20	10	1014
		0.78	Hx	25	0	627-8
$C_6H_{18}O_3Si_3$	Hexamethylcyclotrisiloxane	0	B	30	20	1014
$C_6H_{18}S_3Si_3$	Hexamethylcyclotrithiosiloxane	1.04	B	20	13	1014
$C_6H_{19}BN$	Di-*n*-propylamine - borane complex	4.55±0.05	B	25	15	1420
$C_6H_{21}Ga_3O_3$	Dimethylgallium hydroxide trimer	1.8	B	25	0	900
$C_7F_3MnO_6$	Trifluoroacetylmanganese pentacarbonyl	3.47±0.05	B	25	20	116
C_7HF_{15}	1-Hydroperfluoroheptane	1.75	liq	25	0	1292
$C_7H_2BrCl_2F_3$	3-Bromo-4,5-dichlorotrifluoromethylbenzene	0.87	B	25	0	1378
	5-Bromo-2,4-dichlorotrifluoromethylbenzene	1.59	B	25	0	1378
$C_7H_2Br_3N_3$	*cis*-2,4,6-Tribromobenzenediazocyanide	2.5	B	25	0	1103
	trans-2,4,6-Tribromobenzenediazocyanide	4.0	B	25	0	1103
$C_7H_2Cl_3N$	2,4,6-Trichlorobenzonitrile	3.70	B	25	-	2028
		3.91	B	n.s.	-	678

FORMULA	COMPOUND NAME	μ, D	State, or Solv.	t, °C	Method or P_a, % of P_e	Reference
$C_7H_3BrClF_3$	2-Bromo-5-chlorotrifluoromethylbenzene	2.29	B	25	0	1378
	5-Bromo-2-chlorotrifluoromethylbenzene	2.41	B	25	0	1378
$C_7H_3Br_3O$	2,4,6-Tribromobenzaldehyde	2.53	B	25	0	1796
$C_7H_3Br_3O_2$	3,5,7-Tribromo-2-hydroxy-2,4,6-cyclohepta-trien-1-one	1.84 2.46±0.05	B B	25 25	0 -	991,1019 1013
$C_7H_3ClN_2O_5$	3,5-Dinitrobenzoyl chloride	1.21	B	20	0	1264
$C_7H_3Cl_2F_3$	3,4-Dichlorotrifluoromethylbenzene	1.51	B	25	0	1378
$C_7H_3Cl_2N$	2,5-Dichlorobenzonitrile	3.82	B	n.s.	-	678
$C_7H_3Cl_5$	2,3,4,5,6-Pentachlorotoluene	1.56	B	25	0	1814
$C_7H_3CrNO_5$	Chromium pentacarbonyl - acetonitrile complex	6.9	B	20	-	1891
$C_7H_3MnO_6$	Acetylmanganese pentacarbonyl	2.16±0.05	B	25	20	116
$C_7H_3O_6Re$	Acetylrhenium pentacarbonyl	2.47±0.06	B	25	20	116
C_7H_4BrClO	*p*-Bromobenzoyl chloride	2.05	B	20	0	1264
C_7H_4BrN	*p*-Bromobenzonitrile	2.66	B	21	0	141
C_7H_4BrNS	*p*-Bromophenylisothiocyanate	1.55	B	20	0	159
$C_7H_4BrN_3$	*cis*-*o*-Bromobenzenediazocyanide	3.82	B	25	0	1103
	trans-*o*-Bromobenzenediazocyanide	5.36	B	25	0	1103
	cis-*p*-Bromobenzenediazocyanide	2.93	B	25	0	1103

FORMULA	COMPOUND NAME	μ, D	State, or Solv.	t, °C	Method or P_a, % of P_e	Reference
$C_7H_4BrN_3$ (contd.)	*trans*-*p*-Bromobenzenediazocyanide	3.81	B	25	0	1103
$C_7H_4Br_2O$	2,7-Dibromo-2,4,6-cycloheptatrien-1-one	3.27±0.03	B	25	0	1013,1015
$C_7H_4Br_2O_2$	3,7-Dibromo-2-hydroxy-2,4,6-cyclohepta-trien-1-one	3.57±0.10	B	25	0	1021
	1,2-Dibromo-4,5-methylenedioxybenzene	2.63	B	25	0	1862
$C_7H_4ClF_3$	2-Chlorotrifluoromethylbenzene	3.46	gas	120-240	0	327
		3.27	B	25	0	1378
	3-Chlorotrifluoromethylbenzene	2.14	B	25	0	226
		2.22	B	30	-	548
	4-Chlorotrifluoromethylbenzene	1.58	gas	120-240	0	327
		1.15	B	30	-	548
		1.22	B	25	0	1378
		1.22	Hp	30	-	548
C_7H_4ClN	2-Chlorobenzonitrile	4.73	B	25	-	1856
		4.77	B	n.s.	-	678
		4.80	B	21	0	141
	3-Chlorobenzonitrile	3.42	B	22	0	2142
	4-Chlorobenzonitrile	2.10	B	22	0	2142
		2.50	B	25	-	1856
		2.63	B	23	0	161
	2-Chlorobenzoisonitrile	2.09	B	25	0	649
		2.52	B	22	0	2142
C_7H_4ClNO	2-Chlorobenzonitrile *N*-oxide	4.78	B	25	-	1856
	4-Chlorobenzonitrile *N*-oxide	2.62	B	25	-	1856

FORMULA	COMPOUND NAME	μ, D	State, or Solv.	t, °C	Method or P_a, % of P_e	Reference
	4-Chlorophenylisocyanate	0.85	CCl_4	25	-	1774
$C_7H_4ClNO_3$	*p*-Nitrobenzoyl chloride	1.12	B	20	0	1264
C_7H_4ClNS	*p*-Chlorophenylisothiocyanate	1.56	B	19	-	159
	p-Chlorophenylthiocyanate	2.96	B	21	0	141
	2-Chlorobenzothiazole	2.30 3.36	B B	25 40	0 0	2029 2029
C_7H_4ClNSe	*p*-Chlorophenylselenocyanate	3.31±0.3	B	25	0	1638
$C_7H_4ClN_3$	*cis*-*p*-Chlorobenzenediazocyanide	2.96	B	25	0	1103
	trans-*p*-Chlorobenzendiazocyanide	3.76	B	25	0	1103
$C_7H_4Cl_2O$	*p*-Chlorobenzoyl chloride	2.02	B	20	0	1264
$C_7H_4Cl_3F$	3-Fluorotrichloromethylbenzene	1.78 1.84	B Hp	30 30	- -	548 548
	4-Fluorotrichloromethylbenzene	0.68	B	30	-	548
$C_7H_4Cl_4$	4-Chlorotrichloromethylbenzene	0.78	B	30	-	548,1274
$C_7H_4F_4$	3-Fluorotrifluoromethylbenzene	2.19 2.24	B Hp	30 30	- -	548 548
$C_7H_4F_{12}O$	2,2,3,3,4,4,5,5,6,6,7,7-Dodecafluoro-1-heptanol	2.95	B	20	0	1666
C_7H_4IN	*p*-Iodobenzonitrile	2.84	B	23	0	161
$C_7H_4N_2O_2$	*o*-Nitrobenzonitrile	6.24	B	25	0	1903

FORMULA	COMPOUND NAME	μ, D	State, or Solv.	t, °C	Method or P_a, % of P_e	Reference
$C_7H_4N_2O_2$ (contd.)	*m*-Nitrobenzonitrile	3.81	B	25	0	1903
	p-Nitrobenzonitrile	0	gas	208-250	29	331
		0.66	B	25	0	649
		0.72	B	18	-	439,671
$C_7H_4N_2O_2S$	*p*-Nitrophenylthiocyanate	3.10	B	25	0	268
$C_7H_4N_2O_2Se$	*p*-Nitrophenylselenocyanate	3.58	B	25	0	268
$C_7H_4N_2O_3$	*o*-Nitrophenylisocyanate	5.22±0.10	B	25	0	1633
	m-Nitrophenylisocyanate	3.42±0.10	B	25	0	1633
	p-Nitrophenylisocyanate	3.42±0.10	B	25	0	1633
$C_7H_4N_4O_2$	*cis*-*p*-Nitrobenzenediazocyanide	2.06	B	25	0	1103
	trans-*p*-Nitrobenzenediazocyanide	1.48	B	25	0	1103
$C_7H_5AlCl_4O$	Aluminum chloride - benzoyl chloride complex	8.93-8.98	B	(20)	-	2006-7
		9.10	B	20	15	1415
		7.99±0.3	CS_2	20	15	1415
C_7H_5BrO	Benzoyl bromide	3.40	B	20	0	1264
	p-Bromobenzaldehyde	2.21±0.02	D	25	0	1485
		2.22	D	25	0	1483
	2-Bromo-2,4,6-cycloheptatrien-1-one	4.94±0.02	B	30	0	1013,1015
$C_7H_5BrO_2$	*o*-Bromobenzoic acid	2.73±0.02	B	15-45	10	1464
	dimer	1.91±0.03	B	15-45	10	1464
		2.52	D	30	0	746

FORMULA	COMPOUND NAME	μ, D	State, or Solv.	t, °C	Method or P_a, % of P_e	Reference
	m-Bromobenzoic acid	2.17	D	30	0	223
	p-Bromobenzoic acid	2.10	D	30	0	223
	3-Bromo-2-hydroxy-2,4,6-cyclohepta-trien-1-one	3.91±0.02	B	25	0	1013
	5-Bromo-2-hydroxy-2,4,6-cyclohepta-trien-1-one	2.07	B	25	0	1013,1021
$C_7H_5Br_3$	3,5-Dibromobenzoyl bromide	1.66	B	30	0	1274
	2,4,6-Tribromotoluene	0.73	B	30	0	1274
C_7H_5ClO	Benzoyl chloride	3.16 3.26 3.36	B B B	25 20 20	0 15 -	927 1415 1264
	p-Chlorobenzaldehyde	2.05	B	20	0	329
$C_7H_5ClO_2$	*o*-Chlorobenzoic acid dimer	2.49±0.06 2.45 2.03±0.03	B D B	15-45 30 15-45	10 0 10	1464 746 1464
	m-Chlorobenzoic acid	2.22	D	30	0	223
	p-Chlorobenzoic acid	2.02	D	30	0	223
	4-Chlorosalicylaldehyde	2.31	B	(25)	-	433
	5-Chlorosalicylaldehyde	1.37	B	(30)	-	433
$C_7H_5ClO_4S$	*m*-Sulfochlorobenzoic acid	3.87 3.88	B B	25 n.s.	0 -	634 632-3

FORMULA	COMPOUND NAME	μ, D	State, or Solv.	t, °C	Method or P_a, % of P_e	Reference
$C_7H_5Cl_3$	(Trichlormethyl)benzene	2.04	B	30	-	548
		2.09	B	25	-	1903
		2.17	B	20	-	1480
		2.14	Hp	30	-	548
	2,4,6-Trichlorotoluene	0.54	B	25	0	1814
		0.57	B	30	0	1274
		0.6_6	B	25	0	1071
$C_7H_5Cl_3GaN$	Gallium trichloride - benzonitrile complex	8.71	B	Rm. Temp.	0	2005
$C_7H_5Cl_3O$	2,4,6-Trichloroanisole	1.40	B	20	-,RT	1241b
$C_7H_5Cl_4GaO$	Gallium trichloride - benzoyl chloride complex	6.90	CCl_4	Rm. Temp.	0	2005
$C_7H_5Cl_4NTi$	Titanium tetrachloride - benzonitrile complex	6.21	B	15	0	2003
$C_7H_5CoO_2$	Cyclopentadienylcobalt dicarbonyl	2.87±0.1	B	25	10	715
$C_7H_5CrNO_3$	Cyclopentadienylchromium dicarbonyl nitrosyl	3.19±0.05	B	25	10	715
C_7H_5FO	*p*-Fluorobenzaldehyde	1.98	B	25	5	1114
$C_7H_5FO_2$	*o*-Fluorobenzoic acid	2.46±0.04	B	15-45	10	1464
		2.12	D	30	0	746
	dimer	2.21±0.03	B	15-45	10	1464
	m-Fluorobenzoic acid	2.18	D	30	0	746
	p-Fluorobenzoic acid	2.01	D	30	0	746

FORMULA	COMPOUND NAME	μ, D	State, or Solv.	t, °C	Method or P_a, % of P_e	Reference
$C_7H_5F_3$	(Trifluoromethyl)benzene	2.86	gas	80 to 120	0	327
		2.54	B	25	0	226
		2.56	B	25	-	195,548, 964
		2.60	B	25	0	1625
		2.63	Hp	30	-	548
C_7H_5HgN	Phenylmercury cyanide	3.92	D	20	-	1777
$C_7H_5IO_2$	o-Iodobenzoic acid	2.27±0.03	B	15-45	10	1464
	dimer	1.83±0.04	B	15-45	10	1464
C_7H_5N	Benzonitrile	4.14±0.05	gas	-	M	1134
		4.3	gas	n.s.	-	1898
		4.40±0.02	gas	110-250	5	621-2
		3.16	liq	20	-	657
		3.77	B	20	0	2097
	3.93 P_e = 29.47, P_a = 2.95	3.82	B	40	0	2097
		3.85	B	60	0	2097
		3.87	B	20	0	2096
		3.91	B	60	-	671,674
		3.94	B	21	0	141
		3.95	B	40	-	671,674
		3.96	B	18	-	671,674
		3.97	B	22	0-14	1528,2142
		3.97	B	15	-	1856
		4.05	B	20	0	346
		4.05	B	25	0	475
		4.05	B	25	0	358
		4.02	CCl_4	25	5	1073
		4.07	CCl_4	0-40	0	346
		4.08	Hx	24	0	228
		4.12	Hx	-23	0	346
		4.17	Hx	20	0	346
		4.13	cHx	20	0	346
		3.84	Tol	-79	0	346

FORMULA	COMPOUND NAME	μ, D	State, or Solv.	t, °C	Method or P_a, % of P_e	Reference
C_7H_5N (contd.)	Benzonitrile (contd.)	3.94	Tol	-23	0	346
		3.97	Tol	0	0	346
		3.98	Tol	20	0	346
		3.64	CS_2	-79	0	346
		3.75	CS_2	-23	0	346
		3.78	CS_2	0	0	346
		3.80	CS_2	20	0	346
		4.44	v	20	5	1512
	Benzoisonitrile	3.56	B	22	15	1528
C_7H_5NO	Anthranil	3.09	B	25	15	859
	Benzonitrile *N*-oxide	4.00	B	15	-	1856
	Benzoxazole	1.48	B	25	15	859
	4,5-Benzoisoxazole	3.06	B	25	15	859
	Phenol-*p*-carbonitrile	4.95	B	25	0	598
		5.00	B	25	-	2028
	Phenylisocyanate	2.29	B	20	MA	870
		2.30	B	20	0	344,1475
		2.35	B	18	-	671
		2.25	CCl_4	25	0	1774
	Salicylonitrile	4.38	B	25	5	368
		5.03	D	25	5	368
		3.2	CCl_4	25	5	368
C_7H_5NOS	2-Benzothiazol	3.36	B	25	0	2029
$C_7H_5NO_3$	*o*-Nitrobenzaldehyde	4.26	B	n.s.	5	1212
		4.30	B	25	0	249
		4.35	D	n.s.	5	1212

v) Several solvents.

FORMULA	COMPOUND NAME	μ, D	State, or Solv.	t, °C	Method or P_a, % of P_e	Reference
	m-Nitrobenzaldehyde	3.32	B	n.s.	5	1212
		3.39	D	n.s.	5	1212
	p-Nitrobenzaldehyde	2.4	B	25	-	2113-4
		2.43	B	20	0	329
		2.47	B	n.s.	5	1212
		2.65	D	n.s.	5	1212
$C_7H_5NO_4$	1,2-Methylenedioxy-4-nitrobenzene	4.84	B	25	0	1862
	o-Nitrobenzoic acid	3.82	B	n.s.	5	1212
		4.07	D	n.s.	5	1212
	m-Nitrobenzoic acid	3.50	B	n.s.	5	1212
		4.03±0.10	D	25	0	1633
		4.11	D	n.s.	5	1212
	p-Nitrobenzoic acid	3.5	B	25	-	2113-4
		4.05	D	25	0	2128
		4.18	D	n.s.	5	1212
C_7H_5NS	Benzothiazole	1.46	B	25	0	1430
	Phenylthiocyanate	3.62	B	23	0	141
	Phenylisothiocyanate	2.79	B	24	0	141
		2.85	B	20	MA	870
		3.03	B	20	-	159
$C_7H_5NS_2$	2-Mercaptobenzothiazole	4.03	B	25	0	1430
		4.05	B	40	0	2029
		4.70	B	25	0	2029
$C_7H_5N_3$	*trans*-Benzenediazocyanide	5.0	B	25	0	1087
$C_7H_5N_3O_2$	5(6)-Nitrobenzimidazole	6.00	D	25	0	1451

FORMULA	COMPOUND NAME	μ, D	State, or Solv.	t, °C	Method or P_a, % of P_e	Reference
$C_7H_5N_3O_6$	2,4,6-Trinitrotoluene	1.16±0.04	B	25	28	1764
		1.16	B	25	22	272
		1.3_7	B	25	0	1071
		1.0	D	25	22	272
$C_7H_5N_3O_7$	2,4,6-Trinitroanisole	2.0_3	B	25	0	1071
C_7H_6BrCl	*p*-Bromobenzyl chloride	1.73	Hp	25,50	0	1834
	p-Chlorobenzyl bromide	1.75	Hp	25,50	0	1834
$C_7H_6BrCl_3Si$	Trichloro(*p*-bromobenzyl)silane	1.83	B	25	15	887
C_7H_6BrNO	*m*-Bromobenzamide	3.40	B	30	0	108,744
		3.66	D	30	0	108
	p-Bromobenzamide	3.16	B	30	0	108
		3.25	B	30	0	744
		3.68	D	30	0	108
$C_7H_6BrNO_2$	*p*-Bromo(nitromethyl)benzene	2.87	B	25	-	535
	p-Nitrobenzyl bromide	3.60	B	25,50	0	1834
$C_7H_6BrN_3O$	1-(*p*-Bromophenylazo)formamide	~4[w]	B	25	0	542
$C_7H_6Br_2$	3,5-Dibromotoluene	1.86	B	30	0	1274
$C_7H_6Br_3N$	2,4,6-Tribromo-*N*-methylaniline	1.68	B	25	0	1801
		1.81	D	25	0	1801
C_7H_6ClNO	*m*-Chlorobenzamide	3.57	B	30	0	744
		3.64	B	30	0	108
		3.67	D	30	0	108

w) Measured in absence of light.

FORMULA	COMPOUND NAME	μ, D	State, or Solv.	t, °C	Method or P_a, % of P_e	Reference
	p-Chlorobenzamide	3.38	B	30	0	108
		3.49	B	30	0	744
		3.73	D	30	0	108
$C_7H_6ClNO_2$	2-Chloro-4-nitrotoluene	4.05	B	30	-	1575
	4-Chloro-2-nitrotoluene	3.63	B	30	-	1575
	4-Chloro-3-nitrotoluene	4.82	B	30	-	1575
	6-Chloro-2-nitrotoluene	2.95	B	30	-	1575
	6-Chloro-3-nitrotoluene	3.11	B	30	-	1575
	o-Nitrobenzyl chloride	4.13	B	30	0	395
		3.94	CCl_4	30	0	395
		3.98±0.2	pXy	20-120	0	488
	m-Nitrobenzyl chloride	3.85	B	30	0	395
		3.92	CCl_4	30	0	395
		3.79±0.04	pXy	20-120	0	488
	p-Nitrobenzyl chloride	3.63	B	25,50	0	1834
		3.66	B	20	0?	136
		3.48±0.02	pXy	20-120	0	488
$C_7H_6ClNO_4S$	2-Methyl-5-nitrophenylsulfonyl chloride	4.86	B	25	0	634
		4.89	B	-	-	632-3
$C_7H_6Cl_2$	*o*-Chlorobenzyl chloride	2.41	B	30	0	395
		2.32	CCl_4	30	0	395
		2.27	Hp	30	0	395
	m-Chlorobenzyl chloride	2.07	B	30	0	395
		2.07	CCl_4	30	0	395
		2.11	Hp	30	0	395

FORMULA	COMPOUND NAME	μ, D	State, or Solv.	t, °C	Method or P_a, % of P_e	Reference
$C_7H_6Cl_2$ (contd.)	*p*-Chlorobenzyl chloride	1.73	B	30	0	395
		1.76	B	25	-	2085
		2.13	B	20	0	136
		1.74	CCl_4	30	0	395
		1.71	Hp	30	0	395
	α,α-Dichlorotoluene	2.05	B	25	0	1903
		2.07	B	20	-	1480
	2,4-Dichlorotoluene	1.95	B	30	-	1575
	2,6-Dichlorotoluene	1.11	B	30	-	1575
	3,4-Dichlorotoluene	2.95	B	30	-	1575
	3,5-Dichlorotoluene	1.90	B	30	0	1274
$C_7H_6Cl_2O$	2,6-Dichloroanisole	2.63	B	20	RT	1241b
$C_7H_6F_3N$	*p*-Trifluoromethyltoluidine	4.28	B	25	0	1625
$C_7H_6N_2$	*p*-Aminobenzonitrile	5.02	B	25	-	2028
		5.96	B	25	5	369
		6.46	D	25	5	369
	Benzimidazole	3.96	D	25	15	859
		4.06	D	25	0	1451
		4.11	D	25	0	1922
	Indazole	1.85	B	25	15	859
$C_7H_6N_2O$	1-Methyl-*o*-benzoquinone furazan	4.54	D	25	0	1322,1930
	3-Methyl-*o*-benzoquinone furazan	4.95	D	25	0	1322,1930

FORMULA	COMPOUND NAME	μ, D	State, or Solv.	t, °C	Method or P_a, % of P_e	Reference
$C_7H_6N_2O_2$	1-Methyl-*o*-benzoquinone furoxan	4.97	D	25	0	1322,1930
	3-Methyl-*o*-benzoquinone furoxan	5.42	D	25	0	1322,1930
$C_7H_6N_2O_3$	5-Methoxybenzofurazan *N*-oxide	4.29	D	25	-	1324
	p-Nitrobenzamide	4.60[x]	B	25	5	471
		4.84[x]	D	25	5	471
		4.9	D	20	-	403
$C_7H_6N_2O_4$	2,3-Dinitrotoluene	5.81	B	n.s.	-	663
	2,4-Dinitrotoluene	3.78	B	n.s.	-	663
		4.33	B	25	28	272
		4.38	D	25	28	272
		3.90	CCl_4	n.s.	-	663
	2,5-Dinitrotoluene	0.58	B	25	28	272
		0.95	B	n.s.	-	663
		0.57	D	25	28	272
	2,6-Dinitrotoluene	2.81	B	25	28	272
		2.98	B	n.s.	-	663
		2.85	D	25	28	272
		2.77	CCl_4	n.s.	-	663
	3,4-Dinitrotoluene	6.38	B	n.s.	-	663
		6.26	CCl_4	n.s.	-	663
	3,5-Dinitrotoluene	4.09	B	n.s.	-	663
		4.33	B,D	25	28	272
$C_7H_6N_2O_5$	2,6-Dinitro-*p*-cresol	4.12[y]	B	5-33	RT	1532
$C_7H_6N_2S$	*p*-Aminophenylthiocyanate	5.22	B	25	0	268

x) Extrapolated from data on benzene-dioxane mixtures.
y) In a private communication the original author suggests this average for the temperature range given.

FORMULA	COMPOUND NAME	μ, D	State, or Solv.	t, °C	Method or P_a, % of P_e	Reference
$C_7H_6N_2Se$	*p*-Aminophenylselenocyanate	5.16	B	25	0	268
C_7H_6O	Benzaldehyde	2.724	liq	40	RT	1034
		2.77	liq	20	10	1295
		2.97	liq	27-72	-	823
		2.78	B	25	-	2112,2114
		2.80	B	25	0	587
		2.9_8	B	25	0	249
		2.99	B	20	0	329
		2.89	CCl_4	20	10	1295
	2,4,6-Cycloheptatrien-1-one	4.17±0.03	B	25	0	1022
		4.30±0.1	B	40	0	407
C_7H_6OS	2-Mercapto-2,4,6-cycloheptatrien-1-one	4.36±0.01	B	25	0	1013,1015
$C_7H_6O_2$	Benzoic acid	0.56	B	22	-	208
		1.0	B	25	0	2117
		1.65	B	30	-	1526
		1.72[z]	B,D	25	5	470
		2.12±0.04	B	15-45	10	1464
		1.73	D	25	0	2128
		1.80	D	30	0	223
		1.86±0.02	D	25	0	119
	dimer	0.97±0.03	B	15-45	10	1464
	Bicyclo-[4.3.0]-1,2-dioxa-4,6,8-nonatriene	1.0	n.s.	n.s.	-	186
	o-Hydroxybenzaldehyde	2.80	liq	20	10	1295
		3.09	liq	32-87	-	823
		3.13	liq	30-40	-	1497
		2.86±0.01	B	20	0	1186,1203
		2.91	B	25	0	367
		3.02	D	25	0	367
		2.78	CCl_4	20	0	1295

z) Extrapolated from data on benzene-dioxane mixtures.

FORMULA	COMPOUND NAME	μ, D	State, or Solv.	t, °C	Method or P_a, % of P_e	Reference
	p-Hydroxybenzaldehyde	4.21	D	25	-	1483
		4.22	D	25	0	1485
		4.23±0.02	D	20	0	1186
		4.66	n.s.	25	-	1484
	2-Hydroxy-2,4,6-cycloheptatrien-1-one	3.53±0.02	B	25	0	1022
		3.64	B	35	0	991,1019
		3.71	B	25	0	991,1019
	1,2-Methylenedioxybenzene	0.81	B	25	0	1862
$C_7H_6O_3$	*o*-Hydroxybenzoic acid	2.65	D	25	0	2128
	m-Hydroxybenzoic acid	2.39	D	25	0	2128
	p-Hydroxybenzoic acid	2.76	D	25	0	2128
$C_7H_6O_5$	Dimethyl croconate	5.85±0.15	D	25	0	2064
$C_7H_7AlBr_3NO_2$	Aluminum bromide - *o*-nitrotoluene complex	9.36	B	20	-	1734
	Aluminum bromide - *p*-nitrotoluene complex	9.82	B	20	-	1734
$C_7H_7AlCl_3NO_2$	Aluminum chloride - *o*-nitrotoluene complex	8.98	n.s.	n.s.	-	1733
	Aluminum chloride - *p*-nitrotoluene complex	9.74	n.s.	n.s.	-	1733
$C_7H_7BCl_2$	Dichloro-*p*-tolylborine	2.68	B	25	5	370
		4.63	D	25	5	370
$C_7H_7BF_2$	Difluoro-*p*-tolylborine	2.48	B	25	5	370
		3.58	D	25	5	370
C_7H_7Br	Benzyl bromide	1.87	B	25,50	0	1834
		1.90	Hx	20	9	1345
		1.89	Hp	25,50	0	1834

FORMULA	COMPOUND NAME	μ, D	State, or Solv.	t, °C	Method or P_a, % of P_e	Reference
C_7H_7Br (contd.)	*o*-Bromotoluene	1.45	B	20	-	1949
	m-Bromotoluene	1.77	B	20	-	1949
	p-Bromotoluene	1.95	B	20	-	1949
		1.98	B	25	0	654
		2.17	B	n.s.	-	379
C_7H_7BrHg	*p*-Tolylmercury bromide	3.42	D	25,50	0	366
C_7H_7BrHgO	*o*-(Methoxyphenyl)mercury bromide	3.90	D	20	0	1777
	m-(Methoxyphenyl)mercury bromide	3.49	D	20	0	1777
	p-(Methoxyphenyl)mercury bromide	3.85	D	20	0	1777
$C_7H_7BrHg_2$	1-Bromomercuri-4-methylmercuribenzene	1.94	CS_2	(40)	-	1700
C_7H_7BrO	*o*-Bromoanisole	2.47	B	25	0	46
		2.56	D	25	0	46
	p-Bromoanisole	2.12	liq	30-40	-	1497
		2.26	B	25,50	0	1836
		2.29	B	20	0	160
		2.33	Hp	25,50	0	1836
$C_7H_7BrO_2S$	*p*-Bromophenyl methyl sulfone	3.83±0.03	B	20	0	1201
C_7H_7BrS	*o*-Bromophenyl methyl sulfide	2.53±0.03	B	20	0	1202
	m-Bromophenyl methyl sulfide	1.85±0.03	B	20	0	1202
	p-Bromophenyl methyl sulfide	1.83±0.02	B	20	0	1202
$C_7H_7Br_2N$	2,4-Dibromo-*N*-methylaniline	2.86	B	25	0	1801
		3.10	D	25	0	1801

FORMULA	COMPOUND NAME	μ, D	State, or Solv.	t, °C	Method or P_a, % of P_e	Reference
$C_7H_7Br_3OSi$	Tribromo-(*p*-methoxyphenyl)silane	3.34±0.04	B	25	5	1592
C_7H_7Cl	Benzyl chloride	1.75	B	30	-	1579
		1.84	B	25	0	1908
		1.86_5	B	22	0	141
		1.87	B	20	-	1480
		1.8_9	B	25	-	2081
		1.80	D	20	0	1969
		1.74	a	20-160	0	488
	o-Chlorotoluene	1.57	gas	140-160	0	327
		1.41	liq	20	VS	1701
		1.33_4±0.02	B	(22)	15	2143
		1.36	B	30	-	1579
		1.40	B	20	0	2057
		1.44	B	20	-	1949
	m-Chlorotoluene	1.77	liq	20	VS	1701
		1.61	B	20	0	2057
		1.79	B	20	-	1949
		1.81_4±0.04	B	(22)	15	2143
	p-Chlorotoluene	2.21	gas	120-150	0	327
		1.60	liq	20-60	RT	1487
		1.92	liq	20	VS	1701
		1.76	B	20	0	2057
		1.90±0.015	B	22	15	2143
		1.96	B	20	-	1949
		1.96	B	20	0	1969
		1.86	cHx	30	RT	1844
C_7H_7ClHg	Benzylmercury chloride	3.08	B	25	0	1427
	o-Tolylmercury chloride	2.78±0.07	B	30.4	0?	1245

a) Decalin.

FORMULA	COMPOUND NAME	μ, D	State, or Solv.	t, °C	Method or P_a, % of P_e	Reference
C_7H_7ClHgO	*o*-(Methoxyphenyl)mercury chloride	3.84	D	20	0	1777
	m-(Methoxyphenyl)mercury chloride	3.35	D	20	0	1777
	p-(Methoxyphenyl)mercury chloride	3.65	D	20	0	1777
$C_7H_7ClN_2O_2$	4-Chlorobenzeneazomethane dioxide	2.69	B	20	-	567
C_7HClO	*o*-Chloroanisole	2.44	B	20	RT	1241b
		2.49	B	10-50	0	1336
		2.50	B	25	0	46
		2.49	Hp	-15 to 60	0	1336
	p-Chloroanisole	1.73	liq	20-60	RT	1618
		2.26	B	22	0?	136
		2.32	B	20	RT	1241b
$C_7H_7ClO_2S$	Benzylsulfonyl chloride	3.88	B	25	0	634
	m-(Methylsulfonyl)chlorobenzene	4.36±0.10	B	25	0	1636
		4.39±0.03	B	20	0	1201
	p-(Methylsulfonyl)chlorobenzene	3.82±0.03	B	20	0	1201
	p-Toluenesulfonyl chloride	5.04±0.02	B	25	0	634
		5.05	B	20	-	1266
		5.05	B	n.s.	-	632-3
C_7H_7ClS	*o*-Chlorophenyl methyl sulfide	2.56±0.03	B	20	0	1202
	m-Chlorophenyl methyl sulfide	1.89±0.02	B	20	0	1202
	p-Chlorophenyl methyl sulfide	1.83±0.02	B	20	0	1193
C_7H_7ClSe	*p*-Chlorophenyl methyl selenide	1.72±0.02	B	20	0	310

FORMULA	COMPOUND NAME	μ, D	State, or Solv.	t, °C	Method or P_a, % of P_e	Reference
$C_7H_7Cl_2I$	*o*-Tolyliododichloride	2.57	B	25	-	640
	m-Tolyliododichloride	2.85	B	25	-	640
	p-Tolyliododichloride	3.05	B	25	-	640
$C_7H_7Cl_3GaNO_2$	Gallium trichloride - *p*-nitrotoluene complex	9.22	B	20	0	2005
$C_7H_7Cl_3Si$	Benzyltrichlorosilane	1.78	B	25	15	887
C_7H_7F	Benzyl fluoride	1.77±0.05	B	25	0	1630
	o-Fluorotoluene	1.35 1.26 1.30	gas liq B	80-150 (30) 30	0 - -	1352 1374 1367,1369
	m-Fluorotoluene	1.85 1.66 1.78	gas liq B	80-150 (30) 30	0 - -	1352 1374 1367,1369
	p-Fluorotoluene	2.01 1.76 1.68 1.82 1.88	gas liq B B B	80-150 (30) 25 25 30	0 - 5 0 -	1352 1374 1114 1378 1367,1369
C_7H_7FO	*o*-Fluoroanisole	2.31	B	25	0	46
	p-Fluoroanisole	2.04 2.11	B B	25 20	5 0	1114 160
C_7H_7HgIO	*o*-(Methoxyphenyl)mercury iodide	4.19	D	20	0	1777
	p-(Methoxyphenyl)mercury iodide	3.55	D	20	0	1777

FORMULA	COMPOUND NAME	μ, D	State, or Solv.	t, °C	Method or P_a, % of P_e	Reference
C_7H_7I	*o*-Iodotoluene	1.22 1.30	B CCl_4	22 25	15 0	1527 919
	m-Iodotoluene	1.58	B	22	15	1527
	p-Iodotoluene	1.73	B	22	15	1527
C_7H_7IO	*p*-Iodoanisole	2.14	B	20	0	160
$C_7H_7IO_2S$	*p*-Iodophenyl methyl sulfone	4.05±0.03	B	20	0	1201
C_7H_7IS	*p*-Iodophenyl methyl sulfide	1.80±0.03	B	20	0	1202
C_7H_7N	2-Vinylpyridine	1.92	liq	20-60	RT	1487
	4-Vinylpyridine	2.00	liq	20-60	RT	1487
C_7H_7NO	2-Acetylpyridine	2.85	B	30	0	1367,1372
	3-Acetylpyridine	2.53	B	30	0	1367,1372
	4-Acetylpyridine	2.41±0.01	B	25	0	891
	2-Amino-2,4,6-cycloheptatrien-1-one	3.78±0.05	B	25	0	1013,1017
	Benzaldoxime (m.p. = 35°C)	1.2_3 0.6_8	B $CHCl_3$	25 25	0 0	249 249
	Benzaldoxime (m.p. = 128°C)	1.5_5 1.2_5	B $CHCl_3$	25 25	0 0	249 249
	Benzaldoxime (isomer not given)	0.85 0.87±0.04 0.77±0.04	"pure" B D	20 25 25	0 0 0	249 92 92

FORMULA	COMPOUND NAME	μ, D	State, or Solv.	t, °C	Method or P_a, % of P_e	Reference
	Benzamide	3.55	B	30	0	744
		3.65	B	30	0	108
		3.77[b]	B	25	5	470
		3.6	D	20	-	403
		3.80	D	30	5	390
		3.84	D	30	0	108
		3.88[b]	D	25	5	470
	Formanilide	3.37	CCl_4	25	0	1909
	Methyl 2-pyridyl ketone	2.84±0.03	B	25	0	91
	Methyl 3-pyridyl ketone	2.33±0.03	B	25	0	91
	Methyl 4-pyridyl ketone	2.37±0.03	B	25	0	91
	p-Nitrosotoluene	3.82	B	20	0	1990
C_7H_7NOS	*m*-Methyl-*N*-sulfinylaniline	2.16	liq	20	-	968
		2.14	B	20	-	968
	p-Methyl-*N*-sulfinylaniline	2.39	liq	20	-	968
		2.40	B	20	-	968
$C_7H_7NO_2$	4-Acetyl pyridine-l-oxide	3.19±0.02	B	25	0	891
	o-Aminobenzoic acid	1.52	D	25	0	2019
	m-Aminobenzoic acid	2.73	D	25	0	2019
	p-Aminobenzoic acid	3.10[b]	B	25	5	471
		3.39[b]	D	25	5	471
		3.54	D	25	0	2019
	α-Nitrotoluene	3.33	B	25	-	535

b) Extrapolated from data on benzene-dioxane mixtures.

FORMULA	COMPOUND NAME	μ, D	State, or Solv.	t, °C	Method or P_a, % of P_e	Reference
$C_7H_7NO_2$ (contd.)	*o*-Nitrotoluene	4.12	liq	20	-	657
		3.59	B	n.s.	-	750
		3.67	B	n.s.	0	751
	3.63 P_e = 37.13	3.69	B	22	15	1527
	P_a = 3.71	3.72	B	20	-	1949
		3.78	B	25	-	2123
		3.82	B	30	0	1466
		4.26	B	20	-	816
		3.84	CCl_4	30	0	1466
		3.94	Hx	30	0	1466
		3.97	Hp	30	0	1466
		3.74	Tol	30	0	1466
		3.64	CS_2	30	0	1466
		3.04	$CHCl_3$	30	0	1466
	m-Nitrotoluene	4.17	B	22	15	1527
		4.20	B	20	-	1949
		4.23	B	25	-	2123
		4.32	B	30	0	1466
		4.26	CCl_4	30	0	1466
		4.36	Hx	30	0	1466
		4.41	Hp	30	0	1466
		4.15	Tol	30	0	1466
		4.04	CS_2	30	0	1466
		3.35	$CHCl_3$	30	0	1466
	p-Nitrotoluene	4.34	B	n.s.	-	751
		4.36	B	n.s.	-	750
	4.39 P_e = 37.13	4.45	B	22	15	1527
	P_a = 0	4.47	B	20	-	1949
		4.47	B	25	0	1460
		4.54	B	25	-	2123
		4.56	B	30	0	1466
		4.48	CCl_4	30	0	1466
		4.72	Hx	30	0	1466

FORMULA	COMPOUND NAME	μ, D	State, or Solv.	t, °C	Method or P_a, % of P_e	Reference
		4.46	cHx	24	0	228
		4.64	Hp	30	0	1466
		4.44	Tol	30	0	1466
		4.32	CS_2	30	0	1466
		3.62	$CHCl_3$	30	0	1466
$C_7H_7NO_2S$	*p*-Methoxy-*N*-sulfinylaniline	2.90	B	20	-	968
	Methyl *p*-nitrophenyl sulfide	4.36±0.03	B	20	0	310
$C_7H_7NO_2Se$	Methyl *p*-nitrophenyl selenide	4.38±0.03	B	20	0	310
$C_7H_7NO_3$	*p*-Aminosalicylic acid	3.20	D	25	5	822
	4-Amino-3-hydroxybenzoic acid	2.78	D	25	5	820
	1-(2-Furyl)-2-nitro-1-propene	4.63	B	25	-	2031
	o-Nitroanisole	4.82	gas	205	5	625
		4.84	B	n.s.	0	751
		4.85_4	B	(22)	15	415
		4.87	B	20	-	340
	m-Nitroanisole	4.54	gas	200	5	625
		3.89	B	20	-	340
	p-Nitroanisole	5.26	gas	105	5	625
		4.36	B	n.s.	0	751
		4.78	B	20	-	340
		4.79_3	B	(22)	15	415
		4.8_2	B	25	0	1071
		4.86±0.03	B	20	0	310
$C_7H_7NO_4S$	Methyl *p*-nitrophenyl sulfone	3.47±0.05	B	20	0	1201
C_7H_7NS	*cis*-Thioformanilide	3.02±0.15	CCl_4	25	0	1910

FORMULA	COMPOUND NAME	μ, D	State, or Solv.	t, °C	Method or P_a, % of P_t	Reference
$C_7H_7N_3$	*α*-Azidotoluene	2.60	B	25	0	1762
	o-Azidotoluene	1.39	B	25	5	1854
	m-Azidotoluene	1.75	B	25	5	1854
	p-Azidotoluene	1.90	B	25	5	1854
		1.98	B	25	-	1774,1902
	1-Methyl-1*H*-benzotriazole	4.1_6	B	25	0	1083
$C_7H_7N_3O$	1-Phenylazoformamide	3.86	B	25	0	542
$C_7H_7N_3O_2$	*cis*-Nitroformaldehyde phenylhydrazone	3.37	B	20	-	854
$C_7H_7N_3O_3$	*o*-Nitro-*N*-nitrosomethylaniline	5.83	B	25	0	1218
		6.14	D	25	0	1218
	m-Nitro-*N*-nitrosomethylaniline	3.66	B	25	0	1218
		3.71	D	25	0	1218
	p-Nitro-*N*-nitrosomethylaniline	3.52	B	25	0	1218
		3.79	D	25	0	1218
$C_7H_7N_3O_4$	2,4-Dinitro-*N*-methylaniline	6.36	B	25	0	1801
		6.58	D	25	0	1801
	o-Nitrobenzeneazomethane dioxide	5.19	B	20	-	567
	m-Nitrobenzeneazomethane dioxide	3.74	B	20	-	567
	p-Nitrobenzeneazomethane dioxide	3.50	B	20	-	567
C_7H_8	2-Methyl-1,5-hexadien-3-yne	0.51	B	25	0	1750
	6-Methylfulvene	1.1	B	25.7	0	1940

FORMULA	COMPOUND NAME	μ, D	State, or Solv.	t, °C	Method or P_a, % of P_e	Reference
	Toluene	0.37	gas	85-210	-	1284
		0.37	gas	75-180	-	84
		0.31±0.01	liq	20-60	RT	1487,1489
		0.31	liq	20-30	13	22
		0.37	liq	25	-	1254
		0.38	liq	20	VS	1701
		0.63	liq	18	-	1115
		0.34	B	25	0	1080
		0.34	B	25	6	359
	0.43 P_e = 29.68, P_a = 0	0.39	B,Hx	20	-	1949
		0.40	B	20	-	1824
		<0.49	B	22	0	1528
		0.50	B	n.s.	-	379
		0.52	B	25	0?	2121
		0.55	B	20	-	2133
		0.35	CCl_4	25	5	1073
		0.40	CCl_4	25	0?	966
		0.53	Hx	20	0	1349
		0.3	Ether	25	0	717,723
$C_7H_8AlBr_3O$	Aluminum bromide - anisole complex	6.63	B	20	0	1734
$C_7H_8AlCl_3O$	Aluminum chloride - anisole complex	6.59	n.s.	n.s.	-	1733
C_7H_8BrN	*p*-Bromo-*N*-methylaniline	3.31	B	25	0	1801
		3.56	D	25	0	1801
$C_7H_8Br_2$	*exo-cis*-5,6-Dibromo-2-norbornene	2.97±0.02	B	25	-	1056
$C_7H_8Br_2O$	3,3-Dibromoendocamphor	2.40	B	25	0	972
	3,3-Dibromonorcamphor	4.30	B	25	0	972
	2,3-Dibromo-1,1-dimethyl-2-cyclopentene-4-one	3.68	B	20	-	1994

FORMULA	COMPOUND NAME	μ, D	State, or Solv.	t, °C	Method or P_a, % of P_e	Reference
C_7H_8ClN	2-Chloro-4-toluidine	2.35	B	30	MA	1576
	2-Chloro-6-toluidine	2.81	B	30	MA	1576
	3-Chloro-2-toluidine	2.08	B	30	MA	1576
	4-Chloro-2-toluidine	3.07	B	30	MA	1576
	5-Chloro-2-toluidine	3.26	B	30	MA	1576
$C_7H_8Cl_2$	*endo-cis*-5,6-Dichloronorbornene	3.92	B	25	0	1623
	trans-5,6-Dichloronorbornene	1.97	B	25	0	1623
$C_7H_8Cl_2Si$	Methylphenyldichlorosilane	2.51	Hx	20	-	1964
$C_7H_8FeN_3O_2S_3$	Iron (III) thiocyanate - dioxane complex	8.2±1	D	30	-	311
$C_7H_8N_2$	Ethylmethylfumarodinitrile	1.4	B	30	-	117
	Ethylmethylmaleodinitrile	5.5	B	30	-	117
$C_7H_8N_2O$	*p*-Aminobenzamide	4.65[c]	B	25	5	471
		4.7	D	20	-	403
		4.98[c]	D	25	5	471
	Benzoyl hydrazide	2.72	B	25	15	549
	N-Nitroso-*N*-methylaniline	3.61	B	25	-	1218
		3.65	B	20	0	339
		3.91	D	25	-	1218
	p-Nitroso-*N*-methylaniline	7.43	B	25	0	1100

c) Extrapolated from data on benzene-dioxane mixtures.

FORMULA	COMPOUND NAME	μ, D	State, or Solv.	t, °C	Method or P_a, % of P_e	Reference
$C_7H_8N_2O_2$	Phenylurea	3.8±0.2	D	17	0	166
		4.31±0.04	D	25	0	120
		4.525	D	30	0	1282
	Benzeneazomethane dioxide	3.26	B	20	-	567
	2-Methyl-4-nitroaniline	6.40	B	25	5	1800
		6.98	D	25	5	1800
	3-Nitro-4-aminotoluene	4.40	B	25	0	2026
	2-Nitro-*N*-methylaniline	4.36	B	(21)	-	1216
		4.45	D	(21)	-	1216
	3-Nitro-*N*-methylaniline	5.25	B	(21)	-	1216
		5.51	D	(21)	-	1216
	4-Nitro-*N*-methylaniline	6.66	B	(21)	-	1216
		6.82	B	25	0	1801
		7.12	D	25	0	1801
		7.22	D	(21)	-	1216
$C_7H_8N_2O_3$	4-Methoxy-*o*-benzoquinone dioxime	3.68	D	25	-	1324
$C_7H_8N_2O_4S$	2-Methyl-5-nitrobenzenesulfonamide	4.51	D	25	0	635
	p-Nitrotoluene-*o*-sulfonamide	4.51	D	n.s.	-	632-3
$C_7H_8N_2S$	Phenylthiourea	5.12	D	30	0	1282
		5.16_4±0.002	D	25	5	275-6
$C_7H_8N_2Se$	Phenylselenourea	5.49	D	30	0	1282
$C_7H_8N_4O_2$	Theophylline	4.6	d	n.s.	-	1330

d) 90% phenol or pyridine.

FORMULA	COMPOUND NAME	μ, D	State, or Solv.	t, °C	Method or P_a, % of P_e	Reference
C_7H_8O	Anisole	1.36	gas	130	5	625
		1.36	gas	n.s.	-	1071
		0.8	liq	20	-	463
		1.06	liq	20-60	RT	1618
		1.18	liq	27-86	-	823
		1.22	liq	30-40	-	1497
		1.49	liq	20	-	657
		0.8	B	20-50	-	463
		1.17	B	n.s.	-	751
	1.25 P_e = 31.41, P_a = 3.14	1.24±0.01	B	20	15	416
		1.24	B	20	-,RT	1241b
		1.25	B	20	RT	922
		1.30	B	25	0	1071
		1.32	B	20	0	1387
		1.4C	B	20	0	81
		1.26±0.10	D	25	0	1633
		1.35	D	20	0	1387
		1.50	D	20	0	81
		1.245	CCl_4	25	5	71
		1.25	CCl_4	25	0	919
		1.28	CCl_4	20	0	1387
		1.32	CCl_4	20	0	81
		1.25	Hp	20	0	1387
		1.30	Hp	20	0	81
		1.15	Ether	25	0	1387
		1.20	Ether	25	0	81
		1.35	e	20	0	1387
	Benzyl alcohol	1.70±0.01	gas	178-190	0	236
		1.59	B	n.s.	RT	1368
		1.67	B	25	0	194
		1.67_5±0.005	B	25	0	236
		1.70	B	(22)	15	185
		1.70	B,Hx	18	0	716
		1.70	B	25	5	913
		1.71	B	n.s.	RT	1373

e) Heptane plus dioxane.

FORMULA	COMPOUND NAME	μ, D	State, or Solv.	t, °C	Method or P_a, % of P_c	Reference
		1.79±0.02	D	20,60	-	771
		1.23	cHx	n.s.	RT	1373
		1.23	CS_2	n.s.	RT	1373
		1.6	f	120-160	0	488
	o-Cresol	1.60	liq	20	10	1295
		2.32	liq	25-30	-	1498
		1.42_1	B	(22)	15	415
		1.45	B	25	-	2112
		1.45	B	40	RT	1864
		1.55	B	20	-	1824
		1.29	CCl_4	20	10	1295
		1.36	Tol	0-50	0	730
	m-Cresol	2.39	liq	25-30	-	1498
		1.55_3	B	(22)	15	415
		1.61	B	25	-	2112
		1.78	B	20	-	1824
		1.59-1.67	Tol	50-0	0	730
	p-Cresol	2.35	liq	20	10	1295
		1.53±0.02	B	20	0	1193
		1.57	B	25	0	598
		1.58_3	B	22	15	415
		1.65	B	25	-	2112
		1.83	B	20	-	1824
		1.43	D	20	10	1295
		1.49	D	25	5	1109
		1.83	CCl_4	25	0	451
		1.44	cHx	25	0	451
		1.55-1.63	Tol	75-0	0	947
C_7H_8OS	2,6-Dimethyl-4*H*-pyran-4-thione	5.12±0.01	B	20	-	1660
	2,6-Dimethyl-4*H*-1-thiapyran-4-one	4.30	B	20	-	1660
$C_7H_8OS_2$	Epidithio-2,4-heptadien-6-one	4.00±0.01	B	20,50	0	1684

f) Decalin.

FORMULA	COMPOUND NAME	μ, D	State, or Solv.	t, °C	Method or P_a, % of P_e	Reference
$C_7H_8OSe_2$	Epidiseleno-2,4-heptadiene-6-one	3.30±0.01	B	20,50	0	1684
$C_7H_8O_2$	2,6-Dimethyl-4*H*-pyran-4-one	4.09	B	20	0	792
		4.51-4.55	B	10-40	-	1584
		4.58±0.02	B	20	-	1660
		4.62±0.02	B	15-65	15	2023
		4.6_9	B	25	0	1070
	Guaiacol	2.33	liq	20	10	1295
		2.37	B	25	0	1596
		2.41	B,D	25	5	367
		2.44	B	25	0	1152
		2.40	D	25	0	1596
		2.35	CCl_4	20	10	1295
	o-Hydroxybenzyl alcohol	2.64	B	25	5	367
		2.78	D	25	5	367
$C_7H_8O_2S$	2-Carbethoxythiophene	1.91	B	30	0	903
	Methyl phenyl sulfone	4.65±0.03	B	20	0	1201
		4.73±0.10	B	25	0	1636
		4.80	B	30	0	88
		4.58±0.05	D	20	0	1201
		4.77±0.10	D	25	0	1636
	Toluene-2-sulfinic acid	3.05	B	25	0	641
	Toluene-4-sulfinic acid	3.35	B	25	0	641
		4.01	B	40	0	641
		4.12	D	25,40	0	641
$C_7H_8O_3S$	*m*-(Methylsulfonyl)phenol	4.10±0.10	B	25	0	1636
	p-(Methylsulfonyl)phenol	5.32±0.10	B	25	0	1636
C_7H_8S	Methyl phenyl sulfide	1.38	B	20	-	1190

FORMULA	COMPOUND NAME	μ, D	State, or Solv.	t, °C	Method or P_a, % of P_e	Reference
	α-Toluenethiol	1.32	liq	25	10	1281
		1.44	B	25	10	1281
	p-Methylthiophenol	1.46±0.02	B	20	-	1190,1198
	Thioanisole	1.28	B	21	0	141
		1.38±0.02	B	20	0	1198
$C_7H_8S_2$	2,6-Dimethyl-4*H*-1-thiapyran-1-thione	4.90±0.01	B	20	-	1660
$C_7H_8S_3$	Epidithio-2,4-heptadien-6-thione	3.52±0.01	B	20,50	0	1684
C_7H_8Se	Methyl phenyl selenide	1.31±0.02	B	20	-	309,310
	p-Tolueneselenol	1.45±0.02	B	20	-	309
$C_7H_9BCl_3NO$	4-Ethoxypyridine - boron trichloride complex	7.74±0.07	B	25	0	110
C_7H_9BrO	*endo*-3-Bromonorcamphor	3.97	B	25	0	971
	exo-3-Bromonorcamphor	3.75	B	25	0	971
C_7H_9ClO	*n*-Butylpropiolyl chloride	3.09	B	25	0	927
	anti-7-Chloronorcamphor	1.55	B	25	0	1623
	syn-7-Chloronorcamphor	4.29	B	25	0	1623
C_7H_9ClSi	*m*-Chlorophenylmethylsilane	1.61	B	25	0	322
	p-Chlorophenylmethylsilane	1.45	B	25	0	322
$C_7H_9Cl_2NPt$	*trans*-Dichloro(ethylene)(pyridine)-platinum (II)	3.75±0.2	B	25	15	287

FORMULA	COMPOUND NAME	μ, D	State, or Solv.	t, °C	Method or P_a, % of P_e	Reference
C_7H_9N	Benzylamine	1.15	liq	20	5	337
		1.31	B	20	5	337
		1.38±0.10	B	25	0	1646
		1.26	Hx	20	5	337
		1.28	cHx	20	5	337
		1.22	Tol	20	5	337
	n-Butylpropiolnitrile	4.24	B	25	0	371
	2,3-Dimethylpyridine	2.07±0.01	liq	30	-	1179
		2.20	B	25	0	360
	2,4-Dimethylpyridine	2.24	liq	20-60	MA	1566
		2.30	B	25	0	360
	2,5-Dimethylpyridine	2.16±0.01	liq	20	-	1179
		2.15	B	25	0	360
	2,6-Dimethylpyridine	1.78	liq	20-60	MA	1566
		1.64	B	25	0	1729
		1.66	B	25	0	589
		1.66	B	25	0	360
		1.87±0.01	B	25	0	1640
		1.78	CCl_4	25	0	1729
	3,4-Dimethylpyridine	1.87±0.01	liq	20	-	1179
	3,5-Dimethylpyridine	2.58	B	25	0	360
	2-Ethylpyridine	1.96	liq	20-60	RT	1487
	3-Ethylpyridine	2.41±0.01	B	25	0	1730
	4-Ethylpyridine	2.25	B	20-60	RT	1487
		2.65_5	B	25	0	360

FORMULA	COMPOUND NAME	μ, D	State, or Solv.	t, °C	Method or P_a, % of P_e	Reference
	N-Methylaniline	1.643	B	25	5	507-8
	1.67 P_e = 33.17, P_a = 3.32	1.65	B	25	0	530
		1.67	B	25	0	101
		1.68	B	20	0	336
		1.71	B	(21)	-	1216
		1.77±0.02	B	25	0	524
		1.76	D	(21)	-	1216
		1.833	D	25	5	507-8
		1.73	Hx	20	0	336
	Methylphenylamine	1.22	liq	20	5	337
		1.65	B	20	5	337
		1.70	Hx	20	5	337
	o-Toluidine	1.45	B	25	0	530
		1.58_6	B	(22)	15	415
		1.59	B	20	-	1950
		1.60	B	25	0	101
		1.61±0.02	B	25	0	524
		1.60	n.s.	n.s.	-	525
	m-Toluidine	1.44_2	B	(22)	15	415
		1.45	B	20	-	1950
		1.45	B	25	0	101
		1.49±0.02	B	25	0	524
		1.58	B	25	0	530
	p-Toluidine	1.28_3	B	(22)	15	415
	1.64 P_e = 33.05, P_a = 3.30	1.30	B	25	5	1799
		1.32	B	20	-	1950
		1.36±0.02	B	25	0	524
		1.52	B	25	0	101
		1.61	B	40	RT	1864
		1.66	B	25	0	530
		1.47	D	25	5	1799
		1.35	n.s.	n.s.	-	525

FORMULA	COMPOUND NAME	μ, D	State, or Solv.	t, °C	Method or P_a, % of P_e	Reference
C_7H_9NO	4-Ethylpyridine 1-oxide	4.54±0.01	B	25	0	1730
	o-Methoxyaniline	1.63	gas	150-325	-	1155
		1.46_9	B	(22)	15	415
		1.51	B	25	0	2119
	p-Methoxyaniline	1.82	B	25	0	2119
		1.89_4	B	(22)	15	415
$C_7H_9NO_2S$	Benzylsulfonamide	4.05	B	25	0	635
		4.66	D	25	0	635
	m-(Methylsulfonyl)aniline	5.11±0.10	B	25	0	1636
	p-(Methylsulfonyl)aniline	6.09±0.10	B	25	0	1636
		6.41±0.03	B	20	0	1201
	p-Toluenesulfonamide	5.04	B	25	0	635
		5.06	B	n.s.	-	632-3
		5.41	D	25	0	635
C_7H_9NS	Methyl *p*-aminophenyl sulfide	2.58±0.02	B	25	0	1202
$C_7H_9N_3O$	Phenylsemicarbazide	4.11	D	30	0	1282
$C_7H_9N_3S$	Phenyl-3-thiosemicarbazide	4.67	B	30	0	1282
		5.15	D	30	0	1282
$C_7H_9N_3Se$	2-Phenyl-3-selenosemicarbazide	5.58	D	30	0	1282
C_7H_{10}	Bicyclo[2.2.1]hept-2-ene	0.40±0.05	Hp	25	-	12
	6-Hepten-4-yne	0.65	B	20	0	1492
$C_7H_{10}Br_2$	*exo-cis*-2,3-Dibromonorbornane	3.21±0.02	B	25	-	1056

FORMULA	COMPOUND NAME	μ, D	State, or Solv.	t, °C	Method or P_a, % of P_c	Reference
$C_7H_{10}Cl_2O_4$	Diethyl dichloromalonate	3.40	B	25	0	1176
$C_7H_{10}N_2$	2-Amino-4,6-dimethylpyridine	2.20	B	30	-	1367,1372
	4-Dimethylaminopyridine	4.31±0.01	B	25	0	891
	1-Methyl-1-phenylhydrazine	1.81 1.84	B B	19 15	15 0	76 2008
	1,5-Pentanedicarbonitrile	4.10 4.13 4.29	B B B	25 25 75	0 - -	1943 1978-9 1978-9
	5,6,7,8-Tetrahydrobenzimidazole	3.98	D	25	0	1451
	o-Tolylhydrazine	1.55±0.08	B	18	0	2008
	m-Tolylhydrazine	1.65±0.06	B	15	0	2008
	p-Tolylhydrazine	2.06±0.08	B	18	0	2008
$C_7H_{10}N_2O$	4-Dimethylaminopyridine-1-oxide	6.76±0.04	B	25	0	891
$C_7H_{10}N_2OS$	2-(Ethylthio)-3-methyl-4-pyrimidone	3.63	D	25	0	1716
	5-Ethyl-2-(methylthio)-4-pyrimidone	3.06	D	25	0	1716
	1-Methyl-2-(ethylthio)-4-pyrimidone	6.47	D	25	0	1716
$C_7H_{10}N_2O_2S$	3-Methylsulfanilamide	6.06	D	25	0	1550
$C_7H_{10}N_3S$	3,5-Dimethyl-6-imino-5,6-dihydroimidazo-[2,1-b]thiazole	3.3±0.2	B	n.s.	-	2129
$C_7H_{10}O$	*n*-Butylpropynylaldehyde	3.20	B	25	0	587

FORMULA	COMPOUND NAME	μ, D	State, or Solv.	t, °C	Method or P_a, % of P_e	Reference
$C_7H_{10}O$ (contd.)	2-Methyl-5-hexen-3-yn-2-ol	1.67	B	25	-	1750
	Norcamphor	3.11	B	25	0	971
$C_7H_{10}O_4$	*cis*-Dimethyl citraconate	2.7_0	B	20	0	216
	trans-Dimethyl mesaconate	2.0_8	B	20	0	216
$C_7H_{10}O_7$	*cis*-Dimethyl citraconate ozonide	2.8	B	20	0	216
	trans-Dimethyl mesaconate ozonide	2.5	B	220	0	216
$C_7H_{10}Zn$	Ethyl cyclopentadienylzinc	0.6 1.9 0.6	B D Hp	20 20 20	15 15 15	1894 1894 1894
$C_7H_{11}Br$	1-Bromo-1-heptyne	1.06	B	25	0	1494
$C_7H_{11}BrO$	*trans*-2-Bromo-5-methylcyclohexanone	3.96±0.02 4.09±0.02 3.74±0.02	B D Hp	(25) (25) (25)	- - -	14 14 14
	exo-3-Bromonorborneol	2.26	B	25	0	971
$C_7H_{11}Cl$	1-Chloro-1-heptyne	1.28	B	25	0	1494
$C_7H_{11}ClO$	*trans*-2-Chloro-5-methylcyclohexanone	4.14±0.02	B	25	0	9
$C_7H_{11}ClO_4$	Diethyl chloromalonate	3.26	B	25	0	1176
$C_7H_{11}Cl_7O_2Ti$	Isopentyl trichloroacetate - titanium chloride complex	2.23	B	20	0	1447
$C_7H_{11}F_3$	(Trifluoromethyl)cyclohexane	2.40	B	25	0	1625
$C_7H_{11}I$	1-Iodo-1-heptyne	0.81	B	25	0	1494

FORMULA	COMPOUND NAME	μ, D	State, or Solv.	t, °C	Method or P_a, % of P_e	Reference
$C_7H_{11}N$	1,2,5-Trimethylpyrrole	2.07±0.01	B	25	0	931
		2.08	B	25	-	642
$C_7H_{11}NO$	Cyclohexane-1,3-lactam	3.73±0.02	D	30	0	1060
	Cyclohexane-1,4-lactam	4.24±0.05	D	30	0	1060
	5-(Dimethylamino)-2,4-pentadienal	7.67±0.05	B	25	0	798
$C_7H_{11}NO_2$	Cyclohexane-1,3-carbamate	5.60±0.04	D	30	0	1061
	Cyclohexane-1,4-carbamate	5.64±0.02	D	30	0	1061
	Cyclohexenylnitromethane	3.6±0.1	B	20	0	256
	Cyclohexylidenenitromethane	4.3±0.1	B	20	0	256
$C_7H_{11}NO_3$	5-Ethyl-5,*N*-dimethyloxazolidine-2,4-dione	1.69±0.03	D	30	0	1061
$C_7H_{11}N_3S$	3-Methyl-6-methylamino-5,6-dihydroimidazo-[2,1-b]thiazole	1.6±0.2	B	n.s.	-	2129
C_7H_{12}	1-Heptyne	0.87	gas	75, 125	0	970
	Norbornane	0.58	B	25	0	971
$C_7H_{12}O$	Cycloheptanone	3.07±0.03	B	n.s.	-	630
		3.1_2	CCl_4	25	0	1077
		3.13±0.03	CCl_4	n.s.	-	630
		3.05±0.03	cHx	n.s.	-	630
	3,4-Dimethylpenten-2-one	2.56[g]	B,D	25	5	469
	1-Oxaspiro[5.2]octane	1.8	n.s.	n.s.	-	1362
	1-Methyl-7-oxabicyclo[4.1.0]heptane	1.8	n.s.	n.s.	-	1362

g) Extrapolated from data on benzene-dioxane mixtures.

FORMULA	COMPOUND NAME	μ, D	State, or Solv.	t, °C	Method or P_a, % of P_e	Reference
$C_7H_{12}O$ (contd.)	4-Methyl-7-oxabicyclo[4.1.0]heptane	1.6	n.s.	n.s.	-	1362
	endo-Norbornanol	1.79	B	25	0	971
	exo-Norborranol	1.90	B	25	0	971
$C_7H_{12}O_2$	Cyclohexanecarboxylic acid	0.9	B	25	0	2115
	exo-7-*sym*-Dihydroxynorbornane	3.18	B	25	0	971
	1,4-Dioxaspiro[4.4]nonane	1.24±0.02	B	30	0	151,1514
	Heptanolide	3.70	B	n.s.	-	781
	Methyl ester of cyclopentane carboxylic acid	1.64	n.s.	n.s.	-	1361
	Propyl methacrylate	2.12	B	18	-	1310
$C_7H_{12}O_4$	Diethyl malonate	2.49±0.02	liq	20-70	0	907
		2.56	liq	25-30	-	1497
		2.1	B,CCl_4	25-45	-	831
		2.54	B	25	0	1747
		2.57	B	25,50	0	1832
		2.61	B	20-75	0	907
		2.63	D	20-75	0	907
	Cyclic dioxenetrimethylene ether (m.p. = 111°C)	0.86	n.s.	n.s.	-	701
	(m.p. = 158°C)	0.97	n.s.	n.s.	-	701
	α,α-Dimethylglutaric acid	2.35±0.02	D	25	-	119
	O,O′,O′′,O′′′-Dimethylenepentaerythritol	2.6_9	CCl_4	25	0	1081

FORMULA	COMPOUND NAME	μ, D	State, or Solv.	t, °C	Method or P_a, % of P_e	Reference
	Pimelic acid	2.39±0.02	D	25	-	119
	5,5'-Spiro*bis*(1,3-dioxane)	2.66	B	25	0	1312
$C_7H_{12}O_5$	Glycerol diacetate	2.90	D	25	-	946
$C_7H_{12}Si$	5,5-Dimethyl-5-silica-1-hexen-3-yne	0.44	B	20	-	1493
$C_7H_{13}BrHgO$	α-(2-Methoxycyclohexyl)mercury bromide	4.5	B,D	20	-	221
	β-(2-Methoxycyclohexyl)mercury bromide	>4	B,D	20	-	221
$C_7H_{13}BrO_2$	Ethyl 2-bromovalerate	2.53	B	25	0	1176
$C_7H_{13}Cl$	2-(Chloroethyl) cyclopentane	2.38[h]	CCl_4	25	-	1542
	1-Methyl-1-chlorocyclohexane	2.13	B	25	0	835
$C_7H_{13}ClHgO$	α-(2-Methoxycyclohexyl)mercury chloride	3.8	B,D	20	-	221
	β-(2-Methoxycyclohexyl)mercury chloride	>4	B,D	20	-	221
$C_7H_{13}HgIO$	α-(2-Methoxycyclohexyl)mercury iodide	3.7	B,D	20	-	221
	β-(2-Methoxycyclohexyl)mercury iodide	>4	B,D	20	-	221
$C_7H_{13}N$	*N,N*-Diethyl-2-propylamine	0.80±0.05	B	25	0	1645
$C_7H_{13}NO$	Cyclohexanone methyloxime	1.25	liq	n.s.	5	1593
		1.278±0.005	B	20	5	1593
	1-Ethyl-4-piperidone	2.95±0.1	B	25	0	1113
	cis-2-Oxoheptamethylenimine	3.86	B	25	0	783
	dimer	2.5	B	25	0	783

h) $\bar{\mu}^2$ value.

FORMULA	COMPOUND NAME	μ, D	State, or Solv.	t, °C	Method or P_a, % of P_e	Reference
$C_7H_{13}NO_2$	Nitromethylcyclohexane	3.5±0.1	B	20	0	256
C_7H_{14}	1-Heptene	0.34	liq	20	NRA	23
	Methylcyclohexane	0 0	gas B	100-185 25	- 0	84 2115
	2-Methyl-2-hexene	0.92	liq	20	-	1180
$C_7H_{14}AuN$	Di-*n*-propylgold cyanide	1.48	CCl_4	25	-	233
$C_7H_{14}BF_3O$	Boron trifluoride - diisopropyl ketone complex	6.64	B	20	0	278
$C_7H_{14}Br_2$	1,2-Dibromoheptane	1.78	D	25	-	1744
	1,7-Dibromoheptane	2.441±0.004 2.480±0.004	B B	25 64	0 0	692 692
	2,3-Dibromoheptane	2.15	B	25	-	1744
	3,4-Dibromoheptane	2.15	B	25	-	1744
$C_7H_{14}Cl_4O_2Ti$	Isopentyl acetate - titanium chloride complex	4.48	B	20	0	1447
$C_7H_{14}N_2$	Diisopropylcarbodiimide	2.10	D	35	0	1713
	Diisopropylcyanamide	4.80	D	35	0	1713
$C_7H_{14}O$	Cyclohexyl methyl ether	1.30	gas	135-200	5	625
	2,4-Dimethyl-3-pentanone	2.66 2.73±0.02 2.730	B B(?) B	20 n.s. 25	0 - 10	278 302 726

FORMULA	COMPOUND NAME	μ, D	State, or Solv.	t, °C	Method or P_a, % of P_e	Reference
		2.83	CCl_4	25	0	67
		2.756	Hx	25	10	726
		2.65	CS_2	25	10	726
	n-Heptaldehyde	2.266	liq	40	RT	1034
		2.58	B	22	0	462
	2-Heptanone	2.61	B	22	0	462
	3-Heptanone	2.81	B	22	0	462
	4-Heptanone	2.5	B	30	RT	2052
		2.74	B	22	0	462
		2.76	B	15	10	2141
		2.70	CCl_4	25	-	67
	2-Methylcyclohexanol	2.39	liq	40	-	1498
		2.51	liq	35	-	1498
		2.58	liq	30	-	1498
		1.95	B	25	0	2115
	3-Methylcyclohexanol	2.51	liq	30-35	-	1498
		1.9	B	25	0	2115
	cis-3-Methylcyclohexanol	1.77	n.s.	n.s.	-	1781
	trans-3-Methylcyclohexanol	1.93	n.s.	n.s.	-	1781
	4-Methylcyclohexanol	2.70	liq	30-35	-	1498
		1.73	B	18	0	681a
		1.9	B	25	0	2115
$C_7H_{14}OS$	Cyclohexyl methyl sulfoxide	4.11	B	25	-	356
		4.14	B	25	0	361

FORMULA	COMPOUND NAME	μ, D	State, or Solv.	t, °C	Method or P_a, % of P_e	Reference
$C_7H_{14}O_2$	*n*-Amyl acetate	1.72	gas	100-245	-	2168
		1.9_3	B	25	0	1366
	Isoamyl acetate	1.76	liq	30-40	-	1497
		1.8	B	25	0	1747
		$1.84_3 \pm 0.02$	B	22	15	2140
	n-Butyl propionate	$1.79_8 \pm 0.02$	B	22	15	2140
	2,2-Diethyl-1,3-dioxolane	1.08 ± 0.02	B	30	0	151
	Ethyl 3-methylbutyrate	1.97[i]	B	28	RT	1401
	Ethyl valerate	1.76[i]	B	28	RT	1401
	Propyl butyrate	1.75	liq	25	-	1495
		1.81	liq	60	-	1495
		2.12	B	20	-	1309
	Isopropyl isobutyrate	2.09	B	20	-	1309
$C_7H_{14}O_2S$	Cyclohexyl methyl sulfone	4.63	B	25	0	356,361
$C_7H_{14}S$	Cyclohexyl methyl sulfide	1.66	B	25	0	356,361
$C_7H_{15}Br$	1-Bromoheptane	2.17	gas	100-160	0	1822
		1.90	liq	25	-	707
		2.02	liq	20	-	657
		1.8	B	20	-	461
		1.87	B	22	0	462
		1.85	Hp	-90 to 70	-	1828
		1.88	n.s.	n.s.	-	1806
	2-Bromoheptane	2.08	B	22	0	461-2
	3-Bromoheptane	2.06	B	22	0	461-2

i) 3.26 cm frequency.

FORMULA	COMPOUND NAME	μ, D	State, or Solv.	t, °C	Method or P_a, % of P_e	Reference
	4-Bromoheptane	2.06	B	22	0	461-2
$C_7H_{15}BrO$	1-Bromo-2-ethoxypentane	2.32	B	25	0	1744
	2-Bromo-3-ethoxypentane	2.07	B	25	0	1744
	3-Bromo-2-ethoxypentane	2.15	B	25	-	1744
$C_7H_{15}Cl$	1-Chloroheptane	1.86	B	22	0	462
		1.91	B	20	RT	921
		2.01	Hx	20	RT	921
	2-Chloroheptane	2.05	B	22	0	462
	3-Chloroheptane	2.06	B	22	0	462
	4-Chloroheptane	2.06	B	22	0	462
$C_7H_{15}I$	1-Iodoheptane	1.86	B	22	0	462
		1.91	CCl_4	20	0?	78
	3-Iodoheptane	1.95	B	22	0	462
$C_7H_{15}N$	*N*-*n*-Butylpropylidenamine	1.61±0.02	B	25	0	478
	N-*n*-Propylbutylidenamine	1.55±0.02	B	25	0	478
	N-Isopropylbutylidenamine	1.61±0.02	B	25	0	478
$C_7H_{15}NO$	Pinacolone methyloxime	0.98	liq	n.s.	5	1593
		0.953±0.005	B	20	5	1593
$C_7H_{15}NO_2$	Ethyl α-aminoisovalerate	2.13	B	25	0?	2152
	Ethyl α-aminovalerate	2.15	B	25	0?	2152

FORMULA	COMPOUND NAME	μ, D	State, or Solv.	t, °C	Method or P_a, % of P_c	Reference
$C_7H_{15}NO_2$ (contd.)	3-(3-Methylpentyl) carbamate	2.64±0.06	D	30	0	1061
$C_7H_{15}NO_4$	2-*n*-Butyl-2-nitro-1,3-propanediol	3.57±0.05	B	n.s.	-	257
$C_7H_{15}NS_2$	*S*-Ethyl-*N*-isobutyldithiocarbamate	3.1_5	B	22	0	1243
	S-Ethyl-*N*,*N*-diethyldithiocarbamate	3.2_1	B	20	0	1243
C_7H_{16}	2,2-Dimethylpentane	0	liq	-120 to 90	0	1829
	2,3-Dimethylpentane	0	liq	20,30	0	1829
	2,4-Dimethylpentane	0	liq	20,30	0	1829
	3,3-Dimethylpentane	0	liq	20,30	0	1829
	3-Ethylpentane	0	liq	-120 to 90	0	1829
	n-Heptane	0	gas	110	-	983
		0	gas	65-275	-	1819
		0	liq	-120 to 90	0	1829
		~0	liq	-90 to 90	-	418
	2-Methylhexane	0	liq	20,30	0	1829
	3-Methylhexane	0	liq	20,30	0	1829
	2,2,3-Trimethylbutane	0	liq	20,30	0	1829
$C_7H_{16}N_2O$	1,3-Dimethyl-1,3-diethylurea	3.89	"pure"	25	-	562
		3.48±0.04	B	25	0	120
		3.46	D	25	-	562
		3.52±0.04	D	25	0	120

FORMULA	COMPOUND NAME	μ, D	State, or Solv.	t, °C	Method or P_a, % of P_e	Reference
		3.62±0.04	CCl_4	25	0	120
		3.39±0.04	Hp	25	0	120
		4.07	j	25	-	562
		3.97	k	25	-	562
		4.63	l	25	-	562
$C_7H_{16}O$	Amyl ethyl ether	1.2	B	20	-	463
	Isoamyl ethyl ether	1.16	B	20-50	-	463
	3-Ethyl-3-pentanol	1.14±0.05	liq	20	-	626
	1-Heptanol	1.67	B	30	-	1237
		1.73	B	20	0	462
		1.7_8	CCl_4	25	5	1108
	2-Heptanol	1.73	B	20	0	462
	3-Heptanol	1.73	B	20	0	462
	4-Heptanol	1.72	B	20	0	462
	n-Propyl *n*-butyl ether	1.17	B	25	-	1863
$C_7H_{16}O_3$	Triethyl orthoformate	1.92	B	25	5	60
		0.76	Hx	25	0	274
$C_7H_{16}S$	1-Heptanethiol	1.55	liq	25	27	1280
$C_7H_{17}BO_4$	Trimethyl borate - tetrahydrofuran complex	1.68	Hp	20	15	1890
$C_7H_{17}N$	Heptylamine	1.60	B	30	5	1578
$C_7H_{17}O_3P$	Diethyl propylphosphonate	2.94	liq	30	-	950
	Diethyl isopropylphosphonate	2.85	liq	30	-	950

j) Ethanol. k) Acetone. l) Methylcellosolve.

FORMULA	COMPOUND NAME	μ, D	State, or Solv.	t, °C	Method or P_a, % of P_e	Reference
$C_7H_{17}O_3P$ (contd.)	Diisopropyl methylphosphonate	2.83	liq	30	-	950
$C_7H_{18}AuP$	Methyl(triethylphosphine)gold(I)	5.5	B	25	-	262
$C_7H_{18}O_3Si$	Methyltriethoxysilane	1.72	B	20	0	1041
$C_7H_{19}NSi$	Trimethyl-*N*-*n*-butylsilazane	0.72 0.78 0.73	liq B cHx	30 30 30	- - -	328 328 328
$C_7H_{20}Si_2$	1,1,1,3,3,3-Hexamethyl-1,3-disilicapropane	0.60	Hx	25	0?	627
C_8F_{16}	Decafluoro-1,3-*bis*(trifluoromethyl)-cyclohexane	0.61	cHx	25	0	226
	Decafluoro-1,4-*bis*(trifluoromethyl)-cyclohexane	0.83	cHx	25	0	226
	Hendecafluoro(pentafluoroethyl)cyclohexane	0.73	cHx	25	0	226
$C_8HCl_3F_6$	2,3,5-Trichloro-1,4-*bis*(trifluoromethyl)-benzene	1.07	B	25	0	1378
	4,5,6-Trichloro-1,3-*bis*(trifluoromethyl)-benzene	1.31	B	25	0	1378
$C_8H_2Cl_2F_6$	4,5-Dichloro-1,2-*bis*(trifluoromethyl)-benzene	2.10	B	25	0	1378
	4,5-Dichloro-1,3-*bis*(trifluoromethyl)-benzene	1.42 1.51	B B	25 30	0 -	1378 548
$C_8H_3BrF_6$	2,5-*Bis*(trifluoromethyl)bromobenzene	1.27	B	25	0	226
	3,5-*Bis*(trifluoromethyl)bromobenzene	1.43	B	25	0	226

FORMULA	COMPOUND NAME	μ, D	State, or Solv.	t, °C	Method or P_a, % of P_e	Reference
$C_8H_3ClF_6$	2,4-*Bis*(trifluoromethyl)chlorobenzene	2.01	B	25	0	1378
	2,5-*Bis*(trifluoromethyl)chlorobenzene	1.14	B	30	-	548
		1.44	B	25	0	226
	3,5-*Bis*(trifluoromethyl)chlorobenzene	1.29	B	30	-	548
		1.34	B	25	0	1378
$C_8H_3Cl_6F$	2,4-*Bis*(trichloromethyl)fluorobenzene	1.91	B	30	-	548
$C_8H_3MnO_6$	Acryloylmanganese pentacarbonyl	2.30±0.04	B	25	20	116
$C_8H_4BrNO_2$	*p*-Nitrophenylbromoacetylene	3.30	B	25	0	2127
$C_8H_4ClF_5$	3-Pentafluoroethyl)chlorobenzene	2.25	B	25	0	1378
C_8H_4ClI	*o*-Chlorophenyliodoacetylene	1.43	B	25	0	2127
	p-Chlorophenyliodoacetylene	1.07	B	25	0	2127
$C_8H_4ClNO_2$	3-Chlorophthalimide	2.28±0.03	D	20	0	1192
	4-Chlorophthalimide	1.43±0.03	D	20	0	1192
$C_8H_4Cl_2N_2$	2,3-Dichlorcquinoxaline	3.2	B	n.s.	-	2016
$C_8H_4Cl_2N_2O_2$	*C*-Chloro-*N*-(*p*-chlorophenyl)sydnone	4.88±0.02	B	25	0	737
$C_8H_4Cl_2O_2$	*sym*-Phthalyl dichloride	5.16	B	20	0	1265
$C_8H_4Cl_6$	1,3-*Bis*(trichloromethyl)benzene	1.99	B	30	-	548
$C_8H_4F_6$	1,2-*Bis*(trifluoromethyl)benzene	4.09	cHx	25	0	226
	1,3-*Bis*(trifluoromethyl)benzene	2.10	B	25	0	226
		2.43	B	30	-	548

FORMULA	COMPOUND NAME	μ, D	State, or Solv.	t, °C	Method or P_a, % of P_e	Reference
$C_8H_4F_6$ (contd.)	1,4-*Bis*(trifluoromethyl)benzene	0.64	B	25	0	226
$C_8H_4HgI_2S_2$	2,2'-Mercuri*bis*(5-iodothiophene)	3.21±0.1	D	30	-	874
$C_8H_4INO_2$	*p*-Nitrophenyliodoacetylene	3.83	B	25	0	2127
$C_8H_4N_2$	*p*-Dicyanobenzene	0 0	gas B	200-250 25	- -	331 2086
	p-Diisocyanobenzene	0.6-0.9	B	25	0	1416
$C_8H_4N_2O_4$	4-Nitrophthalimide	2.60±0.03	D	20	0	1192
$C_8H_4N_2SSe$	*p*-Selenocyanophenylthiocyanate	4.02	B	25	0	268
$C_8H_4N_6O_{12}$	1,3,5-Trinitro-2(2,2,2-trinitroethyl)-benzene	3.38	B	n.s.	5	80
$C_8H_4O_3$	Phthalic anhydride	5.29±0.03	B	10-40	0	1585
C_8H_5Br	*o*-Bromophenylacetylene	1.81	B	n.s.	-	1461
	m-Bromophenylacetylene	1.36	B	n.s.	-	1461
	p-Bromophenylacetylene	0.96	B	n.s.	-	1461
	1-Bromo-2-phenylacetylene	0.86	B	25	0	2127
$C_8H_5BrN_2O_2$	*C*-Bromo-*N*-phenylsydnone	6.42 6.5_2	B B	25 25	- 0	735-6 427-8
	N-*p*-Bromophenylsydnone	5.13	B	25	0	428
$C_8H_5Br_3O$	2,4,6-Tribromoacetophenone	2.58	B	25	0	1796

FORMULA	COMPOUND NAME	μ, D	State, or Solv.	t, °C	Method or P_a, % of P_e	Reference
$C_8H_5Br_3O_2$	3,5,7-Tribromo-2-hydrcxy-4-methyl-2,4,6-cycloheptatrien-1-one	3.02±0.04	B	25	0	1017
C_8H_5Cl	*o*-Chlorophenylacetylene	1.70	B	n.s.	-	1461
	m-Chlorophenylacetylene	1.39	B	n.s.	-	1461
	p-Chlorophenylacetylene	0.97	B	n.s.	-	1461
	1-Chloro-2-phenylacetylene	1.11	B	25	0	2127
		1.22	B	19	-	129
$C_8H_5ClN_2O_2$	*C*-Chloro-*N*-phenylsydnone	6.49	B	25	0	735
	N-(*p*-Chlorophenyl)sydnone	5.01±0.02	B	25	0	735-6
		5.08	B	25	-	87
$C_8H_5Cl_5$	Ethylpentachlorobenzene	0.89	B	25	0	1814
C_8H_5I	1-Iodo-2-phenylacetylene	0.55	B	25	0	2127
		0.63	B, CCl_4	30	0	599
$C_8H_5MnO_3$	Cyclopentadienylmanganese tricarbonyl	3.2	B	20	-	1894
		3.30±0.05	B	25	10	715
		3.3	D	20	-	1894
		3.1	Hp	20	-	1894
$C_8H_5MnO_6$	Propionylmanganese pentacarbonyl	2.03±0.03	B	25	20	116
$C_8H_5NO_2$	Isatin	5.76	B	20	0	345
	p-Nitrophenylacetylene	3.45	B	25	-	1461
		3.66	B	13	0	161
	Phthalimide	2.12	B	20	0	345
		2.14±0.02	D	20	0	1192

FORMULA	COMPOUND NAME	μ, D	State, or Solv.	t, °C	Method or P_a, % of P_e	Reference
$C_8H_5N_3O_4$	*C*-Nitro-*N*-phenylsydnone	6.18±0.02	B	25(?)	0	737
C_8H_6	Phenylacetylene	0.72±0.03	B	20	0	1195
		0.73	B	20	0	1181,1189
		0.79	B	25	-	814
		0.84	Hp	10-70	0	1808
C_8H_6BrN	α-Bromobenzylnitrile	3.40	B	20	16	1345
$C_8H_6BrNO_2$	2-Bromo-2-nitro-1-phenylethylene	4.08	B	25	-	2031
$C_8H_6BrN_4$	1-(*p*-Bromophenyl)-5-methyltetrazole	4.45	B	25	0	893
$C_8H_6Br_2O_2$	6,7-Dibromo-1,4-benzodioxan	4.03	B	25	0	1862
	3,7-Dibromo-2-hydroxy-4-methyl-2,4,6-cycloheptatrien-1-one	4.19±0.02	B	30	0	1017
	3,7-Dibromo-2-hydroxy-5-methyl-2,4,6-cycloheptatrien-1-one	4.27±0.13	B	30	-	1017,1422
	5,7-Dibromo-2-hydroxy-3-methyl-2,4,6-cycloheptatrien-1-one	3.27	B	25	-	1422
$C_8H_6ClN_4$	1-(*p*-Chlorophenyl)-5-methyltetrazole	4.38	B	25	0	893
$C_8H_6Cl_2O_4$	2,3-Dichloro-5,6-dimethoxybenzoquinone	3.02	B	30	5	390
	3,6-Dichloro-2,5-dimethoxybenzoquinone	2.11	B	30	5	390
$C_8H_6Cl_4$	Tetrachloro-*o*-xylene	2.67	B	25	0	1814
	Tetrachloro-*p*-xylene	0.77	B	20	SPC	1300
		0.90	B	30	SPC	1300
		1.04	B	40	SPC	1300

FORMULA	COMPOUND NAME	μ, D	State, or Solv.	t, °C	Method or P_a, % of P_e	Reference
$C_8H_6HgS_2$	2,2'-Mercuridithiophene	1.15±0.05	D	30	-	874
$C_8H_6I_2$	*cis*-1,2-Diiodobenzocyclobutane	2.51	B	25	0	843
	trans-1,2-Diiodobenzocyclobutane	1.84	B	25	0	843
$C_8H_6N_2$	Cinnoline	4.14	B	25	0	1640
	Cycloheptimidazole	4.03	B	25	-	1423
$C_8H_6N_2O$	Phenyl furazan	4.027	D	20	0	1320
	Quinoxaline *N*-oxide	2.53	B	25	0	1552
$C_8H_6N_2O_2$	*p*-Nitrobenzylnitrile	3.98	B	25,50	0	1834
	N-Phenylsydnone	6.48±0.02	B	25	0	735-6
		6.53	B	25	0	428
		6.55	B	25	0	87
		6.6_6	B	25	0	427
	Quinoxaline *N,N'*-dioxide	2.27	B	25	0	1552
$C_8H_6N_2O_4$	1-Nitro-2-*p*-nitrophenylethylene	1.01	B	25	24	2032
		1.01	B	25	-	2031
$C_8H_6N_4O_8$	1-Nitro-4-(2,2,2-trinitroethyl)benzene	3.48	B	n.s.	5	80
C_8H_6O	Phenoxyacetylene	1.42	liq	25	-	814
$C_8H_6O_2$	*o*-Phthalic aldehyde	4.54	B	n.s.	0	675
	m-Phthalic aldehyde	2.89	B	n.s.	0	675
	p-Phthalic aldehyde	2.37	B	n.s.	0	675

FORMULA	COMPOUND NAME	μ, D	State, or Solv.	t, °C	Method or P_a, % of P_e	Reference
$C_8H_6O_3$	Piperonal	3.5_2	B	25	-	1089
$C_8H_6O_4$	Phthalic acid	2.30 2.60±0.02	D D	35 25	5 -	1398 119
	Isophthalic acid	2.27±0.02	D	25	-	119
C_8H_6S	Benzothiophene	0.62	B	30	0	282
C_8H_7Br	4-Bromostyrene	1.36 1.52	B B	25 25	0 0	1462 475
	1-Bromo-2-phenylethylene	1.52	B	18	-	129
	low melting	1.13 1.39	B CCl_4	30 30	0 0	599 599
	high melting	1.55 1.54	B CCl_4	30 30	0 0	599 599
C_8H_7BrO	*p*-Bromoacetophenone	2.31	n.s.	n.s.	0	681
	ω-Bromoacetophenone	3.14	B	20	15	1345
$C_8H_7BrO_2$	3(7?)-Bromo-2-hydroxy-4-methyl-2,4,6-cycloheptatrien-1-one	4.48±0.05	B	25	-	1013
	3-Bromo-2-hydroxy-5-methyl-2,4,6-cycloheptatrien-1-one	4.51±0.19	B	30	-	1017,1422
	5-Bromo-2-hydroxy-4-methyl-2,4,6-cycloheptatrien-1-one	2.68±0.12	B	25	0	1017
	7-Bromo-2-hydroxy-3-methyl-2,4,6-cycloheptatrien-1-one	4.05±0.13	B	25.6	0	1017

FORMULA	COMPOUND NAME	μ, D	State, or Solv.	t, °C	Method or P_a, % of P_e	Reference
	7-Bromo-2-hydroxy-4-methyl-2,4,6-cyclo-heptatrien-1-one	4.42±0.05	B	25	0	1017
	7-Bromo-2-hydroxy-5-methyl-2,4,6-cyclo-heptatrien-1-one	4.05	B	30	-	1422
	3-Bromo-2-methoxy-2,4,6-cyclohepta-trien-1-one	3.31±0.08	B	25	0	1022
	7-Bromo-2-methoxy-2,4,6-cyclohepta-trien-1-one	5.51±0.01	B	25	0	1022
	Methyl *p*-bromobenzoate	1.84	B	14	-	129
C_8H_7Cl	α-Chlorostyrene	1.93±0.02	B	20	5	1205
	β-Chlorostyrene	1.41	B	19	-	129
	4-Chlorostyrene	1.29	B	25	0	1462
		1.50	B	25	0	475
$C_8H_7ClHgO_2$	Methyl *p*-(chloromercuri)benzoate	2.82	D	20	-	1777
C_8H_7ClO	*p*-Chloroacetophenone	2.40	B	(30)	-	433
		2.29±0.02	B	(22)	0-15	555
		2.31	n.s.	n.s.	0	681
	ω-Chloroacetophenone	3.29	B	20	10	1345
		3.25	CCl_4	55	-	1392
		3.28	CCl_4	30	-	1392
		3.29	CCl_4	1	-	1392
		3.43	Hp	25	-	1392
	Phenylacetyl chloride	2.56	B	20	0	1264
	p-Toluoyl chloride	3.84	B	20	0	1264

FORMULA	COMPOUND NAME	μ, D	State, or Solv.	t, °C	Method or P_a, % of P_e	Reference
$C_8H_7ClO_2$	5-Chloro-2-hydroxyacetophenone	1.79	B	(30)	0	433
$C_8H_7Cl_3$	3,4,5-Trichloro-*o*-xylene	2.48	B	25	0	1814
$C_8H_7F_3O$	2-Methoxy(trifluoromethyl)benzene	3.54	cHx	25	0	226
	4-Methoxy(trifluoromethyl)benzene	3.28	B	25	0	1378
C_8H_7IO	*p*-Iodoacetophenone	2.25	n.s.	n.s.	0	681
C_8H_7N	Benzylnitrile	3.50	B	25	0	1834
		3.59	Hp	25	0	1834
	Indole	2.07	B	20	0	345
		2.13	B	25	-	817
		2.38	D	25	0	1200
	o-Toluisonitrile	3.38	B	22	15	1528
	p-Toluisonitrile	3.98_5	B	22	15	1528
		4.01±0.03	B	25	0	649
	o-Tolunitrile	3.80_5±0.01	B	(22)	15	2143
	m-Tolunitrile	4.21	B	22	15	1528
	p-Tolunitrile	4.40	B	22	15	1528
		4.40	B	25	-	1856
		4.63	D	24	0	228
		4.42	cHx	24	0	228
		2.85	$CHCl_3$	24	0	228
C_8H_7NO	*p*-Methoxybenzonitrile	4.82	B	25	5	369
		4.97	B	25	5	368
	p-Tolunitrile *N*-oxide	4.58	B	25	-	1856

FORMULA	COMPOUND NAME	μ, D	State, or Solv.	t, °C	Method or P_a, % of P_e	Reference
	p-Tolylisocyanate	2.68	B	20	RT	870
$C_8H_7NOS_2$	*N*-(Hydroxymethylene)benzothiazoline-2-thione	4.58	B	25	-	637
C_8H_7NOSe	*p*-Methoxyphenylselenocyanate	4.42	B	25	0	268
$C_8H_7NO_2$	β-Nitrostyrene	4.47	B,D	25	0	1596
		4.50	B	25	0	475
		4.51	B	17	-	129
		4.54	B	25	-	2031-2
		4.30	D	25	0	588
	4-Nitrostyrene	4.23	B	25	0	475
	p-Methoxyphenylisocyanate	2.98±0.10	B	25	0	1633
$C_8H_7NO_3$	2-Nitroacetophenone	4.59	B	n.s.	5	1212
		4.77	D	n.s.	5	1212
	3-Nitroacetophenone	3.57	B	n.s.	5	1212
		3.80	D	n.s.	5	1212
	4-Nitroacetophenone	3.29	B	(30)	-	433
		3.41	B	n.s.	5	1212
		3.64	D	n.s.	5	1212
	anti-Piperonaldoxime	1.7_5	B	25	0	1088
	syn-Piperonaldoxime	1.5_5	B	25	0	1088
$C_8H_7NO_4$	6-Nitro-1,4-benzodioxane	5.44	B	25	0	1862
C_8H_7NS	*p*-Methylphenylisothiocyanate	3.35	B	21	-	159

FORMULA	COMPOUND NAME	μ, D	State, or Solv.	t, °C	Method or P_a, % of P_e	Reference
$C_8H_7NS_2$	2-Methylmercaptobenzothiazole	1.33 1.43	B B	25 25	- 0	637 1430
	4-Methyl-2-mercaptobenzothiazole	4.03	B	25	0	1430
	6-Methyl-2-mercaptobenzothiazole	4.33	B	25	0	1430
	N-Methylbenzothiazolethione	4.88	B	25	0	1430
C_8H_7NSe	Benzyl selenocyanate	4.01	B	25	0	1638
	p-Tolyl selenocyanate	4.38	B	25	0	1638
$C_8H_7N_3$	1-Phenyl-1,2,3-triazole	4.11	B	25	15	859
	1-Phenyl-1,2,4-triazole	2.91	B	25	15	859
	2-Phenyl-1,2,3-triazole	0.98	B	25	15	859
	4-Phenyl-1,2,4-triazole	5.67	B	25	15	859
	trans-*p*-Toluenediazocyanide	4.4	B	25	0	1087
$C_8H_7N_3O_6$	1,3-Dimethyl-2,4,6-trinitrobenzene	1.85±0.03	B	25	26	1764
$C_8H_7N_5O_2$	1-Methyl-5-(*p*-nitrophenyl)tetrazole	3.87	B	25	0	893
	1-(*p*-Nitrophenyl)-5-methyltetrazole	3.20	B	25	0	893
C_8H_8	Cycloöctatetraene	0.069	n.s.	n.s.	-	1594
	Styrene	0 0.13±0.01 0.37[m] 0	gas liq liq B	170-190 20-60 25,75 10-70	0 RT - 0	660 1487,1489 1462 1808

m) Reference 1489 suggests this value wrong because of impurity.

FORMULA	COMPOUND NAME	μ, D	State, or Solv.	t, °C	Method or P_a, % of P_e	Reference
		<0.3	B	25	0	475
		0.37	B, CCl_4	30	0	599
		0.56	B	25	0	561
		0.31	n.s.	n.s.	-	1062
C_8H_8BrNO	*p*-Bromoacetanilide	4.36	B	25	0	1797
$C_8H_8Br_2$	*o*-*Bis*(bromomethyl)benzene	2.04	B	25	0	935
		2.10	B	50	0	935
		1.81	CCl_4	-15	0	935
		1.85	CCl_4	0	0	935
		1.89	CCl_4	15	0	935
		1.90	CCl_4	30	0	935
		1.94	CCl_4	45	0	935
		2.00	Hx	20	0	935
		2.06	Hx	40	0	935
	p-*Bis*(bromomethyl)benzene	2.06	B	22	0	514
	cis-7,8-Dibromobicyclo[4.2.0]octa-2,4-diene	2.4	n.s.	n.s.	-	570
	4,5-Dibromo-1,2-dimethylbenzene	2.89	B	25	-	1773
$C_8H_8Br_2O_2$	2,5-Dibromo-1,4-dimethoxybenzene	1.01	B	25	0	479
$C_8H_8Br_3N$	2,4,6-Tribromo-*N*,*N*-dimethylaniline	1.048	B	25	0	508
		1.029	D	25	0	508
C_8H_8ClNO	*p*-Chloroacetanilide	4.32	B	25	0	1797
$C_8H_8Cl_2$	*cis*-7,8-Dichlorobicyclo[4.2.0]octa-2,4-diene	2.5	n.s.	n.s.	-	570
	p-*Bis*(chloromethyl)benzene	2.19	B	22	0	514
		2.2_5	B	25	0	2080

FORMULA	COMPOUND NAME	μ, D	State, or Solv.	t, °C	Method or P_a, % of P_e	Reference
$C_8H_8Cl_2$ (contd.)	4,5-Dichloro-1,2-dimethylbenzene	3.04	B	25	0	1814
$C_8H_8Cl_2O_2$	1,4-Dichloro-2,5-dimethoxybenzene	0 1.03±0.02	gas B	140-245 25	- 0	331 477
$C_8H_8I_2O_2$	1,4-Diiodo-2,5-dimethoxybenzene	1.01	B	20	0	1186
$C_8H_8I_2Pt$	Platinum iodide - cycloöctatetraene complex	7.3	n.s.	n.s.	-	853
$C_8H_8N_2$	*N*-Methylbenzimidazole	4.07	D	25	0	1451
$C_8H_8N_2O$	1,3-Diazaazulene monohydrate	4.39	B	25	-	1016
	1,3-Dimethyl-*o*-benzoquinone furazan	4.86	D	25	0	1930
$C_8H_8N_2O_2$	1,3-Dimethyl-*o*-benzoquinone furoxan	5.44	D	25	0	1930
	5-Ethoxybenzofurazan	4.11	D	25	-	1324
	α-Phenylgloxime	1.406	D	20	0	1319
	β-Phenylgloxime	1.722	D	20	0	1319
$C_8H_8N_2O_3$	5-Ethoxybenzofurazan *N*-oxide	4.84	D	25	-	1324
	p-Nitrobenzaldoxime-*N*-methyl ether	6.4±0.1	B	25	0	1908
	α-*p*-Nitrobenzaldoxime-*O*-methyl ether	3.42_5	B	25	0	1938
	β-*p*-Nitrobenzaldoxime-*O*-methyl ether	3.91_5	B	25	0	1938
	N-Nitrosophenylglycine	3.18	B	25	0	428
$C_8H_8N_2O_4$	Methyl *N*-(*o*-nitrophenyl) carbamate	1.76 2.42	B D	25 25	- -	1217 1217

FORMULA	COMPOUND NAME	μ, D	State, or Solv.	t, °C	Method or P_a, % of P_e	Reference
	Methyl *N*-(*m*-nitrophenyl) carbamate	5.04	B	25	0	1217
		5.26	D	25	0	1217
	Methyl *N*-(*p*-nitrophenyl) carbamate	7.09	D	25	0	1217
$C_8H_8N_4$	1-Methyl-5-phenyltetrazole	5.70	B	25	0	892
	1-Phenyl-5-methyltetrazole	5.64	B	25	0	892
C_8H_8O	Acetophenone	3.03	gas	135-220	5	622
		3.06	gas	25	0	1071
		3.00	liq	30-40	-	1497
		3.40	liq	20	-	657
		2.6	B	30	RT	1375,1377
		2.80	B	25	0	587
		2.8_9	B	25	0	1071
		2.91±0.01	B[n]	n.s.	-	258
		2.92±0.02	B	(22)	15	555
		2.92	B	n.s.	RT	1373
		2.95	B	20	0	773
		2.96±0.02	B	(25)	-	301
		2.96±0.02	B	25	0	127
		2.97	B[n]	20	5	1513
		2.99	B	18	0	671-2, 683-4
		3.00	CCl_4	20	0	773
		2.63	Hx	n.s.	RT	1373
		2.65	cHx	n.s.	RT	1373
		2.98	cHx	20	0	773
		2.89	Tol	19	0	671,683-4
		3.18	CS_2	n.s.	RT	1373
		2.83	$CHCl_3$	19	0	671,683-4
		2.88	n.s.	n.s.	-	681
		3.02	p	20	5	1512
	1,2-Epoxyethylbenzene	1.65	B	12	0	157

n) Plus water. p) Several solvents.

FORMULA	COMPOUND NAME	μ, D	State, or Solv.	t, °C	Method or P_a, % of P_e	Reference
C_8H_8O (contd.)	Phenylacetaldehyde	2.50	B	20	0	329
	Phenyl vinyl ether	1.1	B	25	0	1763
	trans-*p*-Tolualdehyde	3.24	B	25	5	468
		3.33	B	25	0	1485
		3.30	D	25	-	1483
		3.29	n.s.	25	-	1484
C_8H_8OS	2-Methylthio-2,4,6-cycloheptatrien-1-one	4.64±0.02	B	25	0	1013,1015
$C_8H_8O_2$	Anisaldehyde (isomer not given)	3.513	liq	40	RT	1034
		3.90±0.02	B	20	0	1186
	m-Anisaldehyde	4.19±0.02	B	20	0	1186
	p-Anisaldehyde	3.26	B	32-47	-	823
	1,4-Benzodioxan	1.43	B	25	0	1862
	2,5-Dimethyl-1,4-benzoquinone	0	gas	-130 to -27	23	331
	1,4-Dioxa-2,3-hydronaphthalene	1.41	n.s.	n.s.	-	701
	o-Hydroxyacetophenone	3.10	liq	20	10	1295
		2.81	B	25	0	1211
		3.19	B	25	10	367
		3.26	D	25	10	367
		2.86	CCl_4	20	10	1295
	m-Hydroxyacetophenone	2.96	B	25	0	1211
	p-Hydroxyacetophenone	3.61±0.10	B	25	0	1636
		3.95	B	25	0	1211

FORMULA	COMPOUND NAME	μ, D	State, or Solv.	t, °C	Method or P_a, % of P_e	Reference
	2-Hydroxy-3-methyl-2,4,6-cyclohepta-trien-1-one	3.27±0.01	B	25	0	1017
	2-Hydroxy-4-methyl-2,4,6-cyclohepta-trien-1-one	3.88±0.01	B	30	0	1017,1422
	2-Hydroxy-5-methyl-2,4,6-cyclohepta-trien-1-one	3.94±0.05	B	25	0	1017,1422
	2-Hydroxy-6-methyl-2,4,6-cyclohepta-trien-1-one	3.9	B	n.s.	-	1998
	o-Methoxybenzaldehyde	4.19±0.02	B	20	0	1203
		4.24	B	25	10	367
	p-Methoxybenzaldehyde	3.26	B	n.s.	-	823
		3.88	B	25	0	1485
		3.73	D	25	-	1483-4
	2-Methoxy-2,4,6-cycloheptatrien-1-one	4.72±0.01	B	25	0	1022
	Methyl benzoate	1.83	liq	30-40	-	1497
		1.7	B,CCl_4	25-45	-	831
		1.8	B	20-50	-	463
		1.85_4±0.02	B	22	15	413
		1.9±0.5	B	26	0	2092
		2.53	B	25	0	1912
	5-Methyl-2-hydroxybenzaldehyde	3.32	B	(25)	-	433
	Phenyl acetate	1.50	B,CCl_4	25-40	-	831
		1.53_7±0.015	B	22	15	413
		1.6_5	CCl_4	25	5	71
	Phenylacetic acid	1.66	B	30	0	1276
		1.77	D	25	0	2128

FORMULA	COMPOUND NAME	μ, D	State, or Solv.	t, °C	Method or P_a, % of P_e	Reference
$C_8H_8O_2$ (contd.)	*o*-Toluic acid	1.70	B	30	0	1276
	m-Toluic acid	2.05	B	30	0	1276
$C_8H_8O_2S$	Phenylvinylsulfone	5.29	B	30	0	88
$C_8H_8O_3$	*o*-Anisic acid	5.52	B	25	0	945
	Anisic acid (isomer not given)	2.37±0.10	D	25	0	1633
	2-Hydroxy-3-methoxybenzaldehyde	3.67	B	25	0	1152
		4.14	D	25	0	1152
	4-Hydroxy-3-methoxybenzaldehyde	2.45[q]	B	25-45	-	824
		2.87	B	25	0	1152
		3.39	D	25	0	1152
	Methyl *m*-hydroxybenzoate	2.39±0.03	B	20	0	1186
	Methyl *p*-hydroxybenzoate	2.87	B	n.s.	0	334
		3.01±0.03	D	20	0	1186
	Methyl salicylate	2.23	liq	20	10	1295
		2.40	B	25	0	1747
		2.43	B	13,40	-	766,876
		2.47	B	25	0	944
		2.51±0.02	B	20	0	1186,1203
		2.53	B	n.s.	5	334
		2.57	D	25	0	1596
		2.40	CCl_4	20	10	1295
$C_8H_8O_4$	3-Acetyl-6-methyl-1,2-pyran-2,4(3)-dione	2.8_6	n.s.	25	0	1070
$C_8H_8O_4S$	*m*-(Methylsulfonyl)benzoic acid	4.53±0.10	D	25	0	1636
	p-(Methylsulfonyl)benzoic acid	4.95±0.10	D	25	0	1636

q) Calculated from Jatkar's equation.

FORMULA	COMPOUND NAME	μ, D	State, or Solv.	t, °C	Method or P_a, % of P_e	Reference
C_8H_9Br	*o*-Bromoethylbenzene	1.79	liq	(30)	-	1374
		2.06	B	30	-	1367,1371
	m-Bromoethylbenzene	1.79	liq	(30)	-	1374
		2.03	B	30	-	1367,1371
	p-Bromoethylbenzene	1.74	liq	(30)	-	1374
		2.08	B	30	-	1367,1371
	4-Bromo-*o*-xylene	2.09	B	25	-	1773
	α-Bromo-*p*-xylene	2.09	Hx	20	15	1349
	2-Bromo-*p*-xylene	1.72	B	32	-	31
		1.86	D	32	-	31
		1.70	CCl_4	32	-	31
		1.74	cHx	32	-	31
		1.76	pXy	32	-	31
C_8H_9BrO	*p*-Bromophenetole	2.42	B	25,50	0	1836
		2.49	Hx	25,50	0	1836
$C_8H_9BrO_2S_2$	Ethylsulfinyl *p*-bromobenzene sulfide	4.08	B	25	0	619
C_8H_9Cl	(Chloroethyl)benzene	1.92[r]	CCl_4	25	-	1542
	o-Chloroethylbenzene	1.44	liq	(30)	-	1374
		1.57	B	30	-	1367,1371
	m-Chloroethylbenzene	1.69	liq	(30)	-	1374
		1.81	B	30	-	1367,1371
	p-Chloroethylbenzene	1.68	liq	(30)	-	1374
		2.00	B	30	-	1367,1371
C_8H_9ClO	*o*-Chlorophenetole	2.54	B	25	0	46

r) $\overline{\mu}^2$ value.

FORMULA	COMPOUND NAME	μ, D	State, or Solv.	t, °C	Method or P_a, % of P_e	Reference
$C_8H_9ClO_2$	7-*syn*-Chloronorcamphorquinone	5.33±0.01	B	30	0	1009
$C_8H_9Cl_3Si$	Trichloro(phenethyl)silane	1.81	B	25	15	887
C_8H_9NO	Acetanilide	3.55	B	22,44	-	992
		3.65	B	25	0	1797
		3.72	B	25	-	1042
		4.04	B	25	0	1068
		3.77	D	22,42	-	992
		4.02	D	n.s.	-	1388
		3.52	CCl_4	25	0	1909
	o-Aminoacetophenone	2.05	B	25	0	1213
	m-Aminoacetophenone	3.59	B	25	0	1213
		5.4	B	18	-	2090
	p-Aminoacetophenone	4.45	B	25	0	1213
		4.48	B	25	5	369
		4.94	D	25	5	369
		4.88	s	25	5	369
		4.32	n.s.	n.s.	0	681
	O-Methylbenzaldoxime	0.83	liq	n.s.	5	1593
		0.630±0.005	B	20	5	1593
		0.87	B	18	0	672
$C_8H_9NO_2$	2,5-Diacetylpyrrole	3.93	D	20	-	1255
	Ethyl isonicotinate	2.49	B	25	10	1112
	Methyl *o*-aminobenzoate	1.0	gas	100	MB	464
		1.0	B	20,50	-	463
	Methyl *m*-aminobenzoate	2.4	gas	100	MB	464
		2.4	B	20,50	-	463

s) Dioxane-*trans*-dichloroethylene.

FORMULA	COMPOUND NAME	μ, D	State, or Solv.	t, °C	Method or P_a, % of P_e	Reference
	Methyl *p*-aminobenzoate	3.3	gas	100	MB	464
		3.3	B	20,50	-	463
	Methyl *N*-phenylcarbamate	3.69	B	25	0	1217
		4.11	D	25	0	1217
	2-Nitro-1,3-dimethylbenzene	3.39	B	30	-	1577
		3.70±0.01	B	25	0	932
	2-Nitro-1,4-dimethylbenzene	3.91±0.01	B	25	0	932
		4.06	B	30	-	1577
		4.15	B	32	-	31
		4.12	D	32	-	31
		4.10	CCl_4	32	-	31
		4.17	cHx	32	-	31
		4.19	pXy	32	-	31
	4-Nitro-1,2-dimethylbenzene	4.82	B	30	-	1577
	4-Nitro-1,3-dimethylbenzene	4.40	B	30	-	1577
$C_8H_9NO_2S$	*N*-Sulfinylphenetidine	3.02	B	20	-	968
$C_8H_9NO_3$	4-Ethoxycarbonylpyridine 1-oxide	3.80±0.01	B	25	0	891
$C_8H_9NO_3S$	*N*-Phenylsulfonylacetamide	4.96	D	25	0	1550
$C_8H_9NO_4$	1,4-Dimethoxy-3-nitrobenzene	4.60	B	18	0?	673
C_8H_9NS	*trans*-Thioacetanilide	4.54±0.10	CCl_4	25	0	1910
C_8H_{10}	6,6-Dimethylfulvene	1.48±0.03	B	23.7	5?	2104

FORMULA	COMPOUND NAME	μ, D	State, or Solv.	t, °C	Method or P_a, % of P_c	Reference
C_8H_{10} (contd.)	Ethylbenzene	0.58	gas	75-180	5	84
		0.35	liq	20	-	2013
		0.36	liq	20-30	15	22
		0.37±0.01	liq	20-60	5,RT	1487,1489
		0.35	B	25	0	359
	5-Methyl-1,5-heptadien-3-yne	0.61	B	25	0	1750
	o-Xylene	0.62	gas	140-240	0	794
		0.44	liq	-20 to 130	4.5	693
		0.45±0.01	liq	20-60	10,RT	1489
		0.50	liq	20-30	20	22
		0.51	liq	15-45	-	1553
		0.52	liq	20-60	RT	1487
		0.54	liq	20	VS	1701
		0.52	B	25	0?	2121
		0.54	B	25	5	1567
		0.58	B	20	-	1949
		0.58	B	20	-	1824
	m-Xylene	0.30±0.01	liq	20-60	6,RT	1489
		0.30	liq	20-30	11	22
		0.31	liq	20-60	RT	1487
		0.33	liq	20	VS	1701
		0.34	liq	-40 to 120	4	693
		0.36	liq	14-45	-	1553
		0.37	B	20	-	1949
		0.46	B	20	-	1824
	p-Xylene	0	liq	20	-	2013
		0	liq	20	VS	1701
		0	liq	20-30	6	22
		0.02±0.01	liq	20-60	RT	1489
		0.17	liq	15-45	-	1553

FORMULA	COMPOUND NAME	μ, D	State, or Solv.	t, °C	Method or P_a, % of P_e	Reference
		0.06	B	25	0	2121
		<0.1	B	n.s.	-	2111
		0.1_2	B	20	-	1949
		0.23	B	20	-	1824
		0	Hp	20	12	209
$C_8H_{10}BeO_6$	Beryllium acetoacetate	3.10	B	25	0	2041
$C_8H_{10}BrClO_2$	2-Bromo-2-chloro-5,5-dimethyl-1,3-cyclohexadione	4.42	CCl_4	19	-	1994
$C_8H_{10}BrN$	*p*-Bromo-*N*-dimethylaniline	3.40	B	25	0	1258
		3.45	B	25	15	703
$C_8H_{10}BrN_3$	1-(*p*-Bromophenyl)-3,3-dimethyltriazene	3.9_1	B	25	0	1083
$C_8H_{10}Br_2O_2$	2,2-Dibromo-5,5-dimethyl-1,3-cyclohexadione	4.32	CCl_4	18	-	1994
$C_8H_{10}ClN$	*p*-Chloro-*N*-dimethylaniline	3.32	B	25	0	1258
$C_8H_{10}ClN_3$	1-(*o*-Chlorophenyl)-3,3-dimethyltriazene	2.8_5	B	25	0	1083
	1-(*m*-Chlorophenyl)-3,3-dimethyltriazene	3.5_2	B	25	0	1083
	1-(*p*-Chlorophenyl)-3,3-dimethyltriazene	4.0_1	B	25	0	1083
$C_8H_{10}Cl_2$	1,6-Dichloro-1,5-cycloöctadiene	2.60	B	25	0	1620
$C_8H_{10}Cl_2O_2$	2,2-Dichloro-5,5-dimethyl-1,3-cyclohexadione	4.52	B	21	-	1994
$C_8H_{10}Cl_3NPt$	*trans*-Ethylene-*p*-chloroaniline - platinum dichloride complex	1.79±0.03	B	25	20	285
$C_8H_{10}FN$	*p*-Fluorodimethylaniline	2.69	B	25	5	1114

FORMULA	COMPOUND NAME	μ, D	State, or Solv.	t, °C	Method or P_a, % of P_e	Reference
$C_8H_{10}Hg_2$	1,4-*Bis*(methylmercuri)benzene	0.91	CS_2	(10)	-	1700
$C_8H_{10}IN$	*p*-Iodo-*N*-dimethylaniline	3.25±0.02	B	25	0	1258
$C_8H_{10}N_2$	*cis*-Acetaldehyde phenylhydrazone	2.63	B	20	-	854
	trans-Acetaldehyde phenylhydrazone	2.60	B	20	-	854
$C_8H_{10}N_2O$	1-Acetyl-1-phenylhydrazine	2.74	B	25	-	1042
	2-Acetyl-1-phenylhydrazine	4.61	B	25	-	1042
	p-Nitroso-*N*-dimethylaniline	6.80	B	25	15	703
		6.81	B	n.s.	-	704
		6.90	B	20	FL	375
		6.94	B	25	0	1100
		6.96	B	45	0	1100
		6.38	CCl_4	25	0	1100
		6.45	Hx	20	FL	375
		6.70	cHx	20	FL	375
		6.80	t	25	5	369
	N-Nitroso-*N*-ethylaniline	3.64	B	20	0	339
	o-Tolylurea	4.12±0.04	D	25	0	120
	p-Tolylurea	4.19±0.04	D	25	0	120
$C_8H_{10}N_2O_2$	5-Amino-2-nitro-1,3-dimethylbenzene	5.08	B	25	-	805a
	N,*N*-Dimethyl-*o*-nitroaniline	4.01	B	(21)	-	1216
		4.30	D	(21)	-	1216
	N,*N*-Dimethyl-*m*-nitroaniline	5.23	B	(21)	-	1216
		5.36	D	(21)	-	1216

t) *trans*-Dichloroethylene.

FORMULA	COMPOUND NAME	μ, D	State, or Solv.	t, °C	Method or P_a, % of P_e	Reference
	N,*N*-Dimethyl-*p*-nitroaniline	6.20	B	25	15	703
		6.84±0.07	B	20	-	1351
		6.85	B	20	FL	375
		6.92	B	25	0	1258
		6.96	B	(21)	-	1216
		7.27	D	(21)	-	1216
$C_8H_{10}N_2O_3$	4-Ethoxy-*o*-benzoquinone dioxime	3.9	D	25	-	1324
$C_8H_{10}N_2O_3S$	*N*-Sulfanilylacetamide	6.88±0.02	D	25	0	1550
$C_8H_{10}N_2S$	*o*-Tolylthiourea	5.00_6±0.002	D	25	5	275-6
	m-Tolylthiourea	5.39_6±0.002	D	25	5	275-6
	p-Tolylthiourea	5.62_4±0.002	D	25	5	275-6
$C_8H_{10}N_4O_2$	Caffeine	3.4	u	n.s.	-	1330
	3,3-Dimethyl-1-(*o*-nitrophenyl) triazene	4.3_1	B	25	0	1083
	3,3-Dimethyl-1-(*m*-nitrophenyl) triazene	5.7_6	B	25	0	1083
	3,3-Dimethyl-1-(*p*-nitrophenyl) triazene	7.0_4	B	25	0	1083
$C_8H_{10}N_4O_7$	Ethylamine picrate	10.37	D	35	0	1559
$C_8H_{10}Ni$	π-Allyl - nickel - cyclopentadienyl complex	0.78±0.06	B	25	15	516
$C_8H_{10}O$	2,6-Dimethylphenol	1.41±0.02	B	20	0	1193
	o-Methoxytoluene	1.0	B	25	-	2112
	m-Methoxytoluene	1.18	B	25	-	2112

u) 90% phenol or pyridine.

FORMULA	COMPOUND NAME	μ, D	State, or Solv.	t, °C	Method or P_a, % of P_e	Reference
$C_8H_{10}O$ (contd.)	*p*-Methoxytoluene	1.21	B	25	-	2112
	o-Methylarisole	1.00	liq	32-82	-	823
	m-Methylarisole	1.08	liq	32-82	-	823
		1.24	B	20	RT	922
	p-Methylarisole	1.18	liq	32-82	-	823
		1.24±0.02	B	20	0	1193
	p-Methylbenzyl alcohol	1.80±0.06	B	20	0	771
		1.85±0.07	B	60	0	771
	Phenetole	1.41	gas	140, 200	5	625
		0.7	liq	20-50	-	463
		1.27	liq	30-40	-	1497
		1.43	Several	20	5	1512
		1.0	B	20-5	-	463
	1-Phenylethanol	1.51	B	20	0	773
		1.53	B	10	0	773
		1.61	B	22	15	185
		1.62	B	40	0	773
		1.55	Hx	22	15	185
	2-Phenylethanol	1.59	B	20	0	773
		1.65	B	25	0	1234
		1.82	B	40	0	773
	2,4-Xylencl	1.98	liq	n.s.	-	771
		1.48±0.12	B	20	-	771
		1.98±0.03	B	60	-	771
	2,5-Xylencl	1.43±0.03	B	20	-	771
		1.52±0.05	B	60	-	771

FORMULA	COMPOUND NAME	μ, D	State, or Solv.	t, °C	Method or P_a, % of P_e	Reference
	3,4-Xylenol	1.73±0.01	B	60	-	771
		1.77±0.04	B	20	-	771
	3,5-Xylenol	1.76±0.05	B	20,60	-	771
$C_8H_{10}OS$	*p*-Methylthioanisole	1.97±0.02	B	20	0	1202
$C_8H_{10}O_2$	Bicyclo[2.2.2]octan-2,3-dione	4.73	B	25	0	973
	o-Dimethoxybenzene	0.99^{v}	liq	25	RT	1618
		1.15-1.31	liq	25-85	-	824
		1.30	B	20	RT	922
		1.32	B	25	0	2080
		1.23	D	25	10	367
		1.3_6	CCl_4	25	5	71
		1.3_8	CCl_4	25	0	919
		1.19-1.33	n.s.	10-40	0	1336
	m-Dimethoxybenzene	1.35	liq	20-60	RT	1618
		1.49^{v}	liq	25-85	-	824
		1.5_9	B	25	0	2080
		1.60	B	20	RT	922
		1.5_9	CCl_4	25	5	71
		1.62	CCl_4	25	0	919
		1.59±0.01	n.s.	10-40	0	1336
	p-Dimethoxybenzene	1.68	gas	18-60	-	674
		1.46	liq	60-80	RT	1618
		1.60^{v}	liq	25-85	-	824
	1.70 P_e = 37.68, P_a = 3.77	1.4	B	21	RT	1531
		1.68	B	20	RT	922
		1.71	B	25	0	1387
		1.73	B	25	0	477
		1.74	B	18	0	673
		1.75	B	6-52	-	1304
		1.83	B	25	0	2080

v) Calculated from Jatkar's equation.

FORMULA	COMPOUND NAME	μ, D	State, or Solv.	t, °C	Method or P_a, % of P_c	Reference
$C_8H_{10}O_2$ (contd.)	*p*-Dimethoxybenzene (contd.)	1.78	D	25	0	1387
		1.75	CCl_4	25	0	919
		1.78	CCl_4	25	5	71
		1.82	CCl_4	25	0	1387
		1.86	x	25	0	1387
		1.6	w	23	RT	1923
		1.71	n.s.	10-40	0	1336
$C_8H_{10}O_3$	2,6-Dimethoxyphenol	2.50	B	25	0	1152
		2.43	D	25	0	1152
		2.2	y	18	RT	1531
$C_8H_{10}O_3S$	*o*-Ethoxybenzenesulfinic acid	4.02	B	25	0	641
		4.43	D	25	0	641
	m-(Methylsulfonyl)anisole	4.77±0.10	B	25	0	1636
	p-(Methylsulfonyl)anisole	5.27±0.10	B	25	0	1636
	Methyl *p*-toluenesulfonate	4.89	B	20	10	171
		5.22	B	25	0	1105
$C_8H_{10}O_4$	2-Butyne-1,4-diol acetate	2.45±0.03	B	n.s.	0	1968
$C_8H_{10}O_4S$	(Ethylbenzene)sulfate	4.82	B	25	0	619
	p-Tolylsulfocarbinol	3.90	B	25	0	619
$C_8H_{10}S$	Methyl *p*-tolyl sulfide	1.52±0.02	B	20	0	1190,1198
$C_8H_{10}S_2$	1,4-Dithiomethoxybenzene	1.85±0.02	B	20	0	1202
$C_8H_{10}Se$	Ethyl phenyl selenide	1.50±0.01	B	20	-	309
	Methyl *p*-tolyl selenide	1.46±0.02	B	20	-	309

w) Decalin. x) Petroleum benzine. y) Tetrachloroethylene.

FORMULA	COMPOUND NAME	μ, D	State, or Solv.	t, °C	Method or P_a, % of P_e	Reference
$C_8H_{11}BrO_2$	2-Bromo-5,5-dimethyl-1,3-cyclohexadione	3.84	z	22	-	1994
$C_8H_{11}ClO$	*n*-Amylpropiolyl chloride	3.11	B	25	0	927
$C_8H_{11}ClO_2$	Methyl ester of *exo*-5-chlorobicyclo-[2.1.1]hexane-*endo*-carboxylic acid	2.06±0.01	Hp	30	0	1009
	Methyl ester of *exo*-5-chlorobicyclo-[2.1.1]hexane-*exo*-carboxylic acid	2.68±0.01	Hp	30	0	1009
$C_8H_{11}ClSi$	*m*-Chlorophenyldimethylsilane	1.74	B	25	0	332
	p-Chlorophenyldimethylsilane	1.69	B	25	0	332
$C_8H_{11}Cl_2NPt$	*trans*-Dichloro(ethylene)(4-methylpyridine)-platinum(III)	4.35±0.1	B	25	15	287
$C_8H_{11}N$	*n*-Amylpropiolnitrile	4.25	B	25	0	371
	N,N-Dimethylaniline	1.61	gas	180	5	625
		1.40	B	n.s.	-	750
	1.57 P_e = 37.83, P_a = 3.78	1.55±0.01	B	25	0	524
		1.55	B	(21)	-	1216
		1.56	B	20	5	337
		1.577	B	25	5	507-8
		1.59	B	25	0	530
		1.59	B	25	0	1253
		1.60	B	20	0	336
		1.61	B	20	0	1387
		1.61	B	25	0	101
		1.633	D	25	5	507-8
		1.64	D	(21)	-	1216
		1.67	D	20	0	1387
		1.59	Hx	20	0	336
		1.55	Hp	20	0	1387
		1.30	Ether	20	0	1387
		1.55	n.s.	n.s.	-	525

z) Ethanol.

FORMULA	COMPOUND NAME	μ, D	State, or Solv.	t, °C	Method or P_a, % of P_e	Reference
$C_8H_{11}N$ (contd.)	*N*-Ethylaniline	1.32	liq	20	5	337
		1.69	B	20	0	339
		1.70	B	25	0	475
		1.67	Hx	20	5	337
		1.71	Hx	20	0	336
	N-Methyl-*o*-toluidine	1.78±0.04	B	25	0	524
	N-Methyl-*m*-toluidine	1.58±0.03	B	25	0	524
	N-Methyl-*p*-toluidine	1.41±0.02	B	25	0	524
	2-*n*-Propylpyridine	1.91	B	25	0	360
	3-*n*-Propylpyridine	2.45	B	25	0	360
	4-*n*-Propylpyridine	2.70_3	B	25	0	360
	3-Isopropylpyridine	2.43	B	25	0	1729-30
		2.55	CCl_4	25	0	1729-30
	4-Isopropylpyridine	2.70_9	B	25	0	360
	2,3,4-Trimethylpyridine	2.13±0.01	liq	20	-	1179
	2,3,6-Trimethylpyridine	2.11±0.01	liq	20	-	1179
	2,4,6-Trimethylpyridine	1.92-2.00[a]	liq	25-85	-	825
		1.95±0.01	B	10-40	-	1589
		2.24	CCl_4	n.s.	-	596
	2,4-Xylidine	1.34	n.s.	n.s.	-	525
		1.40±0.03	B	25	0	524
	2,6-Xylidine	1.63±0.02	B	25	0	524
		1.65	n.s.	n.s.	-	525

a) Calculated from Jatkar's equation.

FORMULA	COMPOUND NAME	μ, D	State, or Solv.	t, °C	Method or P_a, % of P_e	Reference
	Xylidine	1.21	liq	20	5	337
$C_8H_{11}NO$	*N*-Dimethylaniline oxide	4.79 4.85	B D	25 25	0 0	1158 1158
	4-Propylpyridine 1-oxide	4.56±C.01	B	25	0	1730
	4-Isopropylpyridine 1-oxide	4.59±0.01	B	25	0	1730
$C_8H_{11}NO_2$	Cyclohexane-1,2-imide	1.74±0.02	D	30	0	1059
	Cyclohexane-1,3-imide	2.89±0.05	D	30	0	1059
	2,5-Dimethoxyaniline	0.85 1.37[b]	B CCl_4	32 32	0 0	30 30
	2-Methyl-3-carbethoxypyrrole	3.43	B	25	5	818
	5-Methyl-3-carbethoxypyrrole	3.84	B	25	5	818
$C_8H_{11}NO_2S$	*N*-Methyl-*p*-toluenesulfonamide	5.4	B	25	0	1105
$C_8H_{11}NO_3$	Ethyl 3,5-dimethyl-4-isoxazolecarboxylate	3.13±0.03	D	25	0	1857
$C_8H_{11}N_3$	3,3-Dimethyl-1-phenyltriazene	2.2_8	B	25	0	1083
$C_8H_{11}P$	Dimethylphenylphosphine	1.22	cHx	25	-	920
C_8H_{12}	5,5-Dimethyl-1-hexen-3-yne	0.57	B	20	-	1493
	2-Isobutyl-1-buten-3-yne	0.69	B	20	0	1492
	7-Octen-5-yne	0.63	B	20	0	1492
$C_8H_{12}Br_2$	*cis*-3,4-Dibromo-1,2,3,4-tetramethyl-cyclobutene-1	2.35	B	n.s.	-	353

b) Compound formation observed.

FORMULA	COMPOUND NAME	μ, D	State, or Solv.	t, °C	Method or P_a, % of P_e	Reference
$C_8H_{12}Br_2O_2$	*cis*-3,4-Dibromotetrahydro-2,2-dimethyl-3a*H*-cyclopenta-1,3-dioxole	2.17	D	30	0	1007
	trans-3,4-Dibromotetrahydro-2,2-dimethyl-3a*H*-cyclopenta-1,3-dioxole	3.02	B	30	0	1007
		3.21	D	30	0	1007
	cis-3,5-Dibromotetrahydro-2,2-dimethyl-3a*H*-cyclopenta-1,3-dioxole	2.36	B	30	0	1007
	trans-3,5-Dibromotetrahydro-2,2-dimethyl-3a*H*-cyclopenta-1,3-dioxole	1.78	B	30	0	1007
		2.16	D	30	0	1007
$C_8H_{12}N_2$	*p*-Amino-*N*,*N*-dimethylaniline	1.43	B	25	0	1258
	Tetramethylpyrazine	0	B	n.s.	-	2012
		<0.45	B	n.s.	-	1840,2016
$C_8H_{12}N_2OS$	1,3-Diethyl-2-thiouracil	4.23	D	35	0	1716
	4-Ethoxy-2-(ethylthio)pyrimidine	2.66	D	25	0	1716
	3-Ethyl-2-ethylthio-4(3*H*)-pyrimidone	3.29	D	25	0	1716
$C_8H_{12}N_2O_2$	*N*-Cyclohexylsydnone	6.67	B	25	0	87
		6.7±0.02	B	25	0	735-6
$C_8H_{12}N_2O_3$	5,5-Diethylbarbituric acid	1.1	D	25	-	992
		1.13±0.01	D	35	0	1847
$C_8H_{12}N_3S$	3,5-Dimethyl-6-methylimino-5,6-dihydro-imidazo[2,1-b]thiazole	3.3±0.2	B	n.s.	-	2129
$C_8H_{12}O$	*n*-Amylpropiolaldehyde	3.21	B	25	0	587
	Bicyclo[2.2.2]octan-2-one	3.18	B	25	0	973

FORMULA	COMPOUND NAME	μ, D	State, or Solv.	t, °C	Method or P_a, % of P_e	Reference
	3,5-Dimethyl-2-cyclohexen-1-one	3.82	B	18	0	681a
		4.00±0.03	B	25	0	127
	3-Octyn-2-one	3.23	B	25	0	587
$C_8H_{12}O_2$	5,5-Dimethyl-1,3-cyclohexandione	3.49	B	25	0	786
		3.60	c	21	-	1994
	Ethyl sorbate	2.07	gas	235	0	662
	Tetramethylcyclobutane-1,3-dione	0	gas	-185 to -70	-	331
		0.72	B	25	0	648
		0	CCl_4	25	10	39
$C_8H_{12}O_2S_2$	Ethylsulfinyl benzyl sulfide	5.12	B	25	0	619
$C_8H_{12}O_4$	*cis*-2-Butene-1,4-diol diacetate	2.46±0.03	B	20	0	1968
	trans-2-Butene-1,4-diol diacetate	2.34±0.03	B	20	0	1968
	3-Butene-1,2-diol diacetate	2.42±0.03	B	20	0	1968
	Diethyl fumarate	2.40	B	25	0	1832
		2.42	B	50	0	1832
		2.25	CCl_4	23	0	218
		2.40	CCl_4	25	5	1075
		2.40	n.s.	25,50	-	1679
	Diethyl maleate	2.56	B	25	0	1832
		2.58	B	50	0	1832
		2.53	CCl_4	23	0	218
		2.59	CCl_4	25	5	1075
		2.56	n.s.	25,50	-	1679

c) Ethanol.

FORMULA	COMPOUND NAME	μ, D	State, or Solv.	t, °C	Method or P_a, % of P_e	Reference
$C_8H_{12}O_7$	*trans*-Diethyl fumarate ozonide	2.31	CCl_4	23	5	218
	cis-Diethyl maleate ozonide	2.55	CCl_4	23	5	218
$C_8H_{12}S$	2-*t*-Butylthiophene	0.93±0.10	B	25	0	1641
$C_8H_{13}BrO$	2-Bromocycloöctanone	3.36 3.426 3.295	B D Hp	25 25 25	0 - -	7 8 1710
$C_8H_{13}BrO_4$	Diethyl bromosuccinate	2.59	B	25	0	1176
$C_8H_{13}Cl$	1-Chloro-2,6-octadiene	2.08	B	20	0	1491
	3-Chloro-1,6-octadiene	2.00	B	20	0	1491
	1-Chloro-4-octyne	1.92±0.05	B	25	0	1645
$C_8H_{13}ClHgO$	3-Methoxy-2-norcampharylmercuric chloride	4.22	n.s.	n.s.	-	1936
$C_8H_{13}Cl_3O_2Sn$	*Bis*(methanol) - phenyltrichlorotin complex	6.36	B	n.s.	-	592
$C_8H_{13}NO_2$	3-Ethyl-3-methylglutarimide	2.84 2.92±0.05	B D	25 30	0 5	1529 1059
	Cycloheptenylnitromethane	3.6±0.1	B	20	0	256
	1-Cyclohexenyl-1-nitroethane	3.6±0.1	B	20	0	256
	4-Methyl-1-nitromethylcyclohexene	3.5±0.1	B	20	0	256
C_8H_{14}	*cis*-Cycloöctene	0.43±0.06	Hp	25	-	11
	trans-Cycloöctene	0.82±0.03	Hp	25	-	11
$C_8H_{14}BNO_3$	Trimethyl borate - pyridine complex	2.22	Hp	20	15	1890

FORMULA	COMPOUND NAME	μ, D	State, or Solv.	t, °C	Method or P_a, % of P_e	Reference
$C_8H_{14}Br_2$	*trans*-3,4-Dibromo-1,2,3,4-tetramethylcyclobutane	1.63	B	n.s.	-	353
$C_8H_{14}Cl_2Pd_2$	Isobutene - palladium chloride complex dimer	2.22±0.04	n.s.	n.s.	-	799
$C_8H_{14}INSn$	Trimethyltin iodide - pyridine complex	4.21±0.04	B	25	-	313
$C_8H_{14}N_2O_2$	Acetylproline *N*-methylamide	3.3	CCl_4	25.5	0	1338,1391
$C_8H_{14}N_2O_4$	Propyl azodiformate	2.77	CCl_4	25	5	1075
$C_8H_{14}NiO_2S_4$	*Bis*(*O*-isopropylxantho)nickel	2.6_2	B	30	0	1243
$C_8H_{14}NiO_4S_2$	*Bis*(ethylthioglycolato)nickel	2.38±0.1	B	25	0	420
$C_8H_{14}O$	Bicyclo[2.2.2]octan-2-ol	1.57	B	25	0	973
	Cycloöctanone	2.96±0.03	B	n.s.	0	630
	3,3-Dimethylcyclohexanone	2.95	B	18	0	681a
	3,4-Dimethylcyclohexanone	2.86	B	18	0	681a
	3,5-Dimethylcyclohexanone	2.92	B	18	0	681a
	3-Octyn-1-ol	1.78±0.05	B	25	0	1645
$C_8H_{14}O_2$	Butyl methacrylate	2.15	B	18	-	1310
	Cyclohexyl acetate	1.92	B	18	0	681a
	1,4-Dioxaspiro[4.5]decane	1.07±0.02	B	30	0	151
	Ethyl cyclopentanecarboxylate	1.29	n.s.	n.s.	-	1361
	Methyl cyclohexanecarboxylate	1.41	n.s.	n.s.	-	1361

FORMULA	COMPOUND NAME	μ, D	State, or Solv.	t, °C	Method or P_a, % of P_e	Reference
$C_8H_{14}O_2$ (contd.)	2-Methyl-1,4-dioxaspiro[4.4]nonane	1.22±0.02	B	30	0	151
	Octanolide	2.25	n.s.	n.s.	-	781
	Tetrahydro-2,2-dimethyl-3a*H*-cyclopenta-1,3-dioxole	0.81	B	30	0	1007
		1.37	D	30	0	1007
$C_8H_{14}O_2S_4$	Diisopropyldithio*bis*(thionoformate)	3.86	B	25	0	639
$C_8H_{14}O_4$	*dl*-2,3-Diacetoxybutane	1.97	liq	25[d]	-	2130
	meso-2,3-Diacetoxybutane	2.37	liq	25	-	2130
	1,4-Diacetoxybutane	2.47±0.03	B	n.s.	0	1968
	Diethyl succinate	2.3_2	gas	155-235	10	2166
		2.18-2.25	liq	20-70	0	907
		2.37	liq	30-40	-	1497
		2.16	B	25	0	1832
		2.21	B	20	0	907
		2.23	B	50	0	1832
		2.40	B	30	0	590
		2.64	B	45,75	0	907
		2.24	D	20	0	907
		2.31	D	45	0	907
		2.39	D	75	0	907
		2.03-2.49	e	0-180	0	1809-10
	Diisopropyl oxalate	2.32-2.53	liq	20-90	0	907
		2.60	B	20-75	0	907
		2.44	D	20-75	0	907
	Suberic acid	2.36±0.02	D	25	-	119
$C_8H_{14}O_6$	Diethyl racemate	3.4_5	B	25	0	1084

d) Extrapolated to 25°C. e) Kerosene.

FORMULA	COMPOUND NAME	μ, D	State, or Solv.	t, °C	Method or P_a, % of P_e	Reference
	dl-Diethyl tartrate	3.13	B	25	-	2139
		3.15±0.025	B	22	-	2138
	meso-Diethyl tartrate	3.69±0.04	B	12,22	-	2138-9
	Diethyl tartrate[f]	3.22	B	25	5	432
		3.4_6	B	25	0	1084
	dl-Dimethyl 2,3-dimethoxytartrate	3.23	n.s.	n.s.	-	2139
	meso-Dimethyl 2,3-dimethoxytartrate	2.93	n.s.	n.s.	-	2139
$C_8H_{14}Si$	6,6-Dimethyl-6-silica-2-hepten-4-yne	0.57	B	20	-	1493
	Trimethylcyclopentadienylsilicon	0.4	D,Hp	20	-	1894
	2,5,5-Trimethyl-5-silica-1-hexen-3-yne	0.39	B	20	-	1493
$C_8H_{15}NO$	Cyclohexanone ethyloxime	1.22	liq	20	5	1593
		1.206±0.005	B	20	5	1593
	4-Diethylamino-3-buten-2-one	5.71	B	25	0	1749
	1-Dimethylamino-1-hexen-3-one	4.87	B	25	0	1749
	2-Oxoöctamethyleneimine	3.85	B	25	0	783
	cis-Tropine	2.20	B	25	-	2173
	pseudo-Tropine	1.68	n.s.	n.s.	-	317
	trans-Tropine	1.59	B	25	-	2173
	Tropine	<0.4	n.s.	n.s.	-	317

f) Configuration not given.

FORMULA	COMPOUND NAME	μ, D	State, or Solv.	t, °C	Method or P_a, % of P_e	Reference
C_8H_{16}	2,5-Dimethyl-2-hexene	0.77	liq	20	-	1180,1189
	3,5-Dimethyl-2-hexene	0.91	liq	20	-	1180
	Ethylcyclohexane	0	gas	100-185	0	84
	3-Methyl-2-heptene	0.72	liq	20	-	1180,1189
	1-Octene	0.34	liq	20	NRA	23
	cis-3-Octene	0.29	liq	20	NRA	23
	trans-3-Octene	0	liq	25	NRA	23
	cis-4-Octene	0.26	liq	20	NRA	23
	trans-4-Octene	0	liq	25	NRA	23
$C_8H_{16}Br_2$	1,8-Dibromoöctane	2.480±0.004 2.522±0.004	B B	25 64	0 0	692 692
$C_8H_{16}Br_4O_4Zr$	*Bis*(ethyl acetate) - zirconium bromide complex	3.43	B	20	10	923
$C_8H_{16}Cl_3NO_2$	Trichloroacetic acid - triethylamine complex	9.6	D	20	-	1987
$C_8H_{16}Cl_3O_4Tl$	Thallium trichloride - *bis*(dioxane) complex	0	B	20	10	1739
$C_8H_{16}Cl_4O_2Ti$	Titanium tetrachloride - butyl butyrate complex	4.46	B	20	15	1438
$C_8H_{16}Cl_4O_4Sn$	Stannic chloride - *bis*(ethyl acetate) complex	7.3	B	18	0	962

FORMULA	COMPOUND NAME	μ, D	State, or Solv.	t, °C	Method or P_a, % of P_e	Reference
$C_8H_{16}Cl_4O_4Zr$	Zirconium tetrachloride - *bis*(ethyl acetate) complex	6.75	B	20	15	1442
	Zirconium tetrachloride - *bis*(propyl formate) complex	7.76	B	20	15	1442
$C_8H_{16}Cl_8O_4Zr_2$	Zirconium tetrachloride - ethyl acetate complex dimer	3.91	B	20	0?	924
	Zirconium tetrachloride - propyl formate complex dimer	4.21	B	20	0?	924
$C_8H_{16}I_4O_4Zr$	Zirconium tetraiodide - *bis*(ethyl acetate) complex	3.52	B	20	-	1444,1446
$C_8H_{16}N_2$	1,4-Diazaspiro[4.5]decane	1.34±0.04 1.37±0.02 1.52±0.03 1.57±0.07	B pXy pXy pXy	30 30 80 100	0 5 5 5	151 512 512 512
$C_8H_{16}O$	3-*cis*-5-*cis*-Dimethylcyclohexan-1-*cis*-ol	1.32	n.s.	20	-	1781
	3-*cis*-5-*cis*-Dimethylcyclohexan-1-*trans*-ol	1.84	n.s.	20	-	1781
	3-*cis*-5-*trans*-Dimethylcyclohexan-1-*cis*-ol	1.87	n.s.	20	-	1781
	2-Ethylcaproaldehyde	2.66	B	25	-	1485
	2-Octanone	2.72	B	15	10	2141
$C_8H_{16}O_2$	Butyl isobutyrate	2.12	B	20	-	1309
	2-Isobutyl-2-methyl-1,3-dioxalane	1.17±0.02	B	30	0	151,1514
	2-*t*-Butyl-2-methyl-1,3-dioxalane	1.06±0.02	B	30	0	151

FORMULA	COMPOUND NAME	μ, D	State, or Solv.	t, °C	Method or P_a, % of P_e	Reference
$C_8H_{16}O_2$ (contd.)	Caprylic acid	1.15	liq	25	-	1495
		1.31	liq	60	-	1495
	Ethyl hexanoate	1.80[g]	B	28	-	1401
	2-Ethyl hexanoic acid dimer	0.82	CCl_4	20	0	1719
	2,4,4,5,5-Pentamethyl-1,3-dioxolane	1.30	B	25	0	1459
$C_8H_{16}Si$	Diallyldimethylsilane	0.54	B	25	15	887
$C_8H_{17}Br$	1-Bromoöctane	1.88	liq	25	-	707
		1.99	liq	20	-	657
		1.98	B	14	-	698
		1.96	Hp	0-55	-	448
$C_8H_{17}Cl$	1-Chloroöctane	2.14	liq	25	-	707
$C_8H_{17}Cl_2NO_2$	Dichloroacetic acid - triethylamine complex	7.76	D	25	-	1987
$C_8H_{17}I$	1-Iodoöctane	1.80	liq	25	-	707
		1.90	CCl_4	20	0	78
	2-Iodoöctane	2.07	CCl_4	20	0	78
$C_8H_{17}N$	*N*-*n*-Butylbutylidenamine	1.61±0.03	B	25	0	478
	N-Isobutylbutylidenamine	1.55±0.03	B	25	0	478
$C_8H_{17}NO$	2-Nitroso-2,5-dimethylhexane	2.53	B	25	0	650
		1.00±0.10	CCl_4	0	0	651
	2-(1,3-Dimethylbutylideneamino)ethanol	2.50±0.05	B	30	0	151
$C_8H_{17}NO_2$	Ethyl α-aminocaproate	2.15	B	25	0	2152

g) 3.26 cm frequency.

FORMULA	COMPOUND NAME	μ, D	State, or Solv.	t, °C	Method or P_a, % of P_e	Reference
	Ethyl α-aminoisocaproate	2.05	B	25	0	2152
C_8H_{18}	*n*-Octane	0	gas	160	-	795
		0	liq	-50 to 110	-	418
	2,2,4-Trimethylpentane	0	liq	-100 to -10	0	1829
$C_8H_{18}Br_2Si$	1,1-Dibromoethyltriethylsilane	2.26	B	20	0	1041
	1,2-Dibromoethyltriethylsilane	2.85	B	20	0	1041
$C_8H_{18}ClNO_2$	Chloroacetic acid - triethylamine complex	7.15	D	25	-	1987
$C_8H_{18}Cl_2Sn$	Di-*n*-butyldichlorotin	4.72	B	n.s.	-	592
		5.11	D	n.s.	-	592
		4.25	Hx	n.s.	-	592
$C_8H_{18}Cl_8SSn_2$	*n*-Butylsulfide - *bis*(tin tetrachloride) complex	1.57	B	n.s.	-	592
$C_8H_{18}N_2O$	Di-*n*-butylnitrosamine	4.32	D	20	-	569
$C_8H_{18}N_2O_2$	Di-*n*-butylnitramine	4.82	D	20	-	569
$C_8H_{18}O$	*n*-Butyl ether	1.18	gas	110-180	6	624
		1.19	liq	20	0	1297
		1.09	B	25	-	99
		1.14	B	25	0	1297
		1.23	B	25	0	1128
		1.26	B	25	-	1863
		1.13	Tol	25	0	1128
		1.18	h	25	0	1128
	2-Ethyl-1-hexanol	1.76±0.02	B	25	0	1485

h) Kerosene.

FORMULA	COMPOUND NAME	μ, D	State, or Solv.	t, °C	Method or P_a, % of P_e	Reference
$C_8H_{18}O$ (contd.)	2-Methyl-3-heptanol	1.63	B	20-70	0	1831
	1-Octanol	1.60	liq	-15 to 49	0	381
		2.71	liq	20	10	1295
		1.65	B	25	0	1239
		1.72	B	20-70	0	1830
		1.72±0.02	B	20	0	1184
		1.7_6	CCl_4	25	5	1108
		1.80	CCl_4	20	10	1295
	2-Octanol	1.53	liq	49	0	381
		1.61	liq	25	0	381
		1.65	liq	-36 to 0	0	381
		1.65	B	20	-	977
$C_8H_{18}OS$	Isobutyl sulfoxide	3.93	B	25	0	652
$C_8H_{18}O_2$	Butyl *t*-butylperoxide	1.19	B	30	0	1164
	Di-*t*-butylperoxide	0.92±0.05	B	25	0	1639
		0.94	B	30,50	0	1163-4
$C_8H_{18}O_2S$	Dibutylsulfone	4.46±0.03	B	20	0	1201
$C_8H_{18}O_2Zn$	Diethylzinc - dioxane complex	1.69	D	20	10	1893
$C_8H_{18}O_5$	Tetraethylene glycol	5.84	liq	20	-	933
		3.25	D	25	10	1999
$C_8H_{18}PbS_2$	Lead *bis*(1-butanethiol)	4.61	B	25	0	639
$C_8H_{18}S$	*n*-Butyl sulfide	1.57	B	20	-	844
		1.58	B	20	0	791
		1.61	B	25	0	195
	Isobutyl sulfide	1.57±0.02	B	20	0	1193

FORMULA	COMPOUND NAME	μ, D	State, or Solv.	t, °C	Method or P_a, % of P_e	Reference
$C_8H_{18}S_2$	*n*-Butyl disulfide	1.99	B	25	-	2030
		2.06±0.05	B	25	0	1639
	Isobutyl disulfide	2.00±0.05	B	25	0	1639
	t-Butyl disulfide	1.86±0.05	B	25	0	1639
$C_8H_{18}Se_2$	*t*-Butyl diselenide	1.83±0.05	B	25	0	1639
$C_8H_{19}BrSi$	(1-Bromoethyl)triethylsilane	2.23	B	20	0	1041
$C_8H_{19}ClSi$	(1-Chloroethyl)triethylsilane	2.09	B	20	0	1041
$C_8H_{19}N$	1-Amino-5,5-dimethylhexane	1.22	B	30	5	1578
	Di-*n*-butylamine	1.06	liq	20	5	337
		1.05	B	20	5	337
		1.10	B	20	0	336
		1.01	Hx	20	5	337
		1.06	Hx	20	0	336
	Octylamine	1.42	B	30	5	1578
$C_8H_{19}NO_2$	Acetic acid - triethylamine complex	3.96	D	25	-	1987
$C_8H_{19}O_2P$	Hexyl ethylphosphinate	3.39	B	25	5	61
$C_8H_{19}O_2PS$	*O*,*O*′-Dipropyl ethylphosphonothioate	2.98	B	25	5	61
$C_8H_{19}O_3P$	Dibutylphosphite	3.17	CCl_4	20	-	50
$C_8H_{20}BCl_3O_2$	Boron trichloride - di(ethyl ether) complex	6.02	B	(20)	15?	2006-7
		6.09	B	20	15?	1415
$C_8H_{20}BeBr_2O_2$	Beryllium bromide - di(ethyl ether) complex	7.57	B	(20)	15?	2007
		7.62	B	20	15?	1415,2006

FORMULA	COMPOUND NAME	μ, D	State, or Solv.	t, °C	Method or P_a, % of P_e	Reference
$C_8H_{20}BeCl_2O_2$	Beryllium chloride - di(ethyl ether) complex	6.76	B	20	15?	1415
		6.79	B	(20)	15?	2007
		6.89	B	20	-	2006
$C_8H_{20}BrO_3P$	Ethyl bromide - ethyl diethyl phosphate complex	2.38	B	20	-	52
		2.69	B	40	-	52
		2.75	B	60	-	52
	Ethyl bromide - triethylphosphite	3.5	B	0-180	-	52
$C_8H_{20}Br_2PtS_2$	α-*Bis*(ethyl sulfide)platinum dibromide	2.28	B	20	-	844
	β-*Bis*(ethyl sulfide)platinum dibromide	8.9	B	20	-	844
$C_8H_{20}Cl_2PdS_2$	α-*Bis*(ethyl sulfide)palladium dichloride	2.21	B	20	-	844
$C_8H_{20}Cl_2PtS_2$	α-*Bis*(ethyl sulfide)platinum dichloride	2.43	B	20	-	844
	β-*Bis*(ethyl sulfide)platinum dichloride	9.9	B	20	-	844
$C_8H_{20}Cl_2PtSe_2$	α-*Bis*(ethyl selenide)platinum dichloride	2.43	B	20	-	844
	β-*Bis*(ethyl selenide)platinum dichloride	9.1	B	20	-	844
$C_8H_{20}Cl_4O_2Sn$	Tin chloride - di(ethyl ether) complex	3.63	B	15	0	2003
	Tin chloride - di(*n*-butanol) complex	6.05	B	20	0	1439
	Tin chloride - di(isobutanol) complex	6.06	B	20	0	1439
$C_8H_{20}Ge$	Tetraethylgermanium	0	B, D, Hp	20	15	1892
$C_8H_{20}I_2PtS_2$	α-*Bis*(ethyl sulfide)platinum diiodide	2.43	B	20	-	844
$C_8H_{20}N_2$	1-Amino-4-diethylaminobutane	1.20±0.02	B	25	0	942

FORMULA	COMPOUND NAME	μ, D	State, or Solv.	t, °C	Method or P_a, % of P_e	Reference
	1,8-Diaminoöctane	2.00	B	25-75	-	1977,1979
	Tetraethylhydrazine	1.62	B	25	-	1555
$C_8H_{20}N_2OS$	Sulfurous acid *bis*(diethyl amide)	3.03	B	25	15	855
$C_8H_{20}N_2O_4PtS_2$	β-*Bis*(ethylsulfide)platinum dinitrite	13.8	B	20	-	844
$C_8H_{20}NiO_4P_2S_4$	*Bis*(*O*,*O*-diethyldithiophosphono)nickel	2.2	B	31	0	1246
$C_8H_{20}NiP_2S_4$	α-*Bis*(diethyldithiophosphino)nickel	1.67	B	31	0	1246
	β-*Bis*(diethyldithiophosphino)nickel	1.72	CCl_4	21.5	0	1246
$C_8H_{20}O_4Si$	Tetraethylorthosilicate	1.72	B	20	5	53
		1.72±0.02	B	32	5	1914
		1.84	B	20	0	1041
$C_8H_{20}O_4Ti$	Tetraethylorthotitanate	1.41	B	25	-	355
		1.50	Hx	25	0	274
$C_8H_{20}O_7P_2$	Tetraethylpyrophosphate	3.88	B	(25)	5	1855
$C_8H_{20}Pb$	Tetraethyllead	0.3	B	20	15	1892
		0.4	D	20	15	1892
		0	Hp	20	15	1892
$C_8H_{20}PtS_2$	Dimethyl-1,2-*bis*(ethylthio)ethane-platinum(II)	7.1±0.1	B	25	15	293
$C_8H_{20}Si$	Dimethyldipropylsilane	0	liq	20	-	24
	Tetraethylsilane	0	liq	20	-	24
		0	B,D,Hp	20	15	1892
		0.4	B	20	0	1041
		0	CCl_4	20	15	1247

FORMULA	COMPOUND NAME	μ, D	State, or Solv.	t, °C	Method or P_a, % of P_e	Reference
$C_8H_{20}Si_2$	*trans*-1,2-*Bis*(trimethylsilyl)ethylene	0	B	25	15	885
$C_8H_{20}Sn$	Tetraethyltin	0	B	22	-	1249
		0	B,D,Hp	20	15	1892
		0-0.35	CCl_4	25	-	1852
$C_8H_{22}Cl_6O_4Sn_2$	Ethoxytrichlorotin - ethanol dimer	3.53	B	20	0	1439
$C_8H_{22}Si_2$	1,1,1,4,4,4-Hexamethyl-1,4-disilicabutane	0.56	B	25	0	627
$C_8H_{24}O_4Si_4$	Octamethylcyclotetrasiloxane	0.67	B	25	20	1014
$C_8H_{24}P_2Pt$	*cis*-*Bis*(trimethylphosphine)dimethylplatinum	5.7	B	25	15	294
$C_9H_2Cl_2F_8$	2,5-Dichloro-4-pentafluoroethyltrifluoromethylbenzene	0.90	B	25	0	1378
$C_9H_3F_9O$	1,3-*Bis*(trifluoromethyl)-4-trifluoromethoxybenzene	2.12	cHx	25	0	226
$C_9H_4F_8$	4-Pentafluoroethyltrifluoromethylbenzene	0.69	B	25	0	226
$C_9H_4F_{16}O$	2,2,3,3,4,4,5,5,6,6,7,7,8,8,9,9-Hexadecafluoro-1-nonanol	3.06	B	25	-	1666
$C_9H_5ClCrO_3$	Chlorobenzene - chromium tricarbonyl complex	5.08	B	n.s.	-	1568
C_9H_5ClINO	5-Chloro-7-iodo-8-hydroxyquinoline	2.10	B	25-40	-	992
	7-Chloro-5-iodo-8-hydroxyquinoline	3.23	B	20-40	-	992
$C_9H_5ClO_2$	*o*-Chlorophenylpropiolic acid	2.65	D	25	0	2128
	p-Chlorophenylpropiolic acid	1.97	D	25	0	2128

FORMULA	COMPOUND NAME	μ, D	State, or Solv.	t, °C	Method or P_a, % of P_e	Reference
$C_9H_5Cl_2N$	2,6-Dichloro-1-azaazulene	3.25	B	30	-	1016
$C_9H_5Cl_2NO$	5,7-Dichloro-8-hydroxyquinoline	4.23	B	22,40	-	992
$C_9H_5CrFO_3$	Fluorobenzene - chromium tricarbonyl complex	4.75±0.03	B	25	15	520
		4.91	B	n.s.	-	1568
$C_9H_5I_2NO$	5,7-Diiodo-8-hydroxyquinoline	6.36	B	20,46	0	992
C_9H_5N	Phenylpropiolnitrile	4.55	B	25	0	371
$C_9H_5NO_4$	*o*-Nitrophenylpropiolic acid	4.05	D	25	0	2128
	p-Nitrophenylpropiolic acid	3.81	D	25	0	2128
$C_9H_5O_4V$	Cyclopentadienylvanadium tetracarbonyl	3.17±0.05	B	25	10	715
C_9H_6ClN	2-Chloro-1-azaazulene	4.47	B	35	-	1016
	2-Chloroquinoline	3.26±0.1	B	25	0	1640
$C_9H_6ClNO_2$	4-Chloro-*N*-methylphthalimide	1.31+0.03	D	20	0	1192
$C_9H_6CrO_3$	Benzene - chromium tricarbonyl complex	4.92±0.05	B	25	15	520
		5.08	B	n.s.	-	1568
		5.37	D	20	15	1894
		4.31	cHx	25	15	520
$C_9H_6CrO_4$	Phenol - chromium tricarbonyl complex	5.13±0.07	B	25	15	520
$C_9H_6F_6O$	2-Methoxy-1,3-*bis*(trifluoromethyl)benzene	3.73	B	25	0	1378
	2-Methoxy-1,4-*bis*(trifluoromethyl)benzene	1.72	B	25	0	1378
	4-Methoxy-1,3-*bis*(trifluoromethyl)benzene	3.99	B	25	0	1378

FORMULA	COMPOUND NAME	μ, D	State, or Solv.	t, °C	Method or P_a, % of P_e	Reference
$C_9H_6F_6O$ (contd.)	5-Methoxy-1,3-*bis*(trifluoromethyl)benzene	3.22	B	25	0	1378
$C_9H_6N_2O_2$	Tolylene-2,4-diisocyanate	2.52	B	20	RT	870
$C_9H_6N_2O_3$	1*H*-2,4-Benzodiazepine-1,3,5(2*H*,4*H*)-trione	4.65	n.s.	n.s.	-	901
$C_9H_6N_2O_4$	3-Nitro-*N*-methylphthalimide	3.75±0.04	D	25	0	1192
	4-Nitro-*N*-methylphthalimide	2.56±0.03	D	20	0	1192
C_9H_6O	Phenylpropiolaldehyde	3.39	B	25	0	587
$C_9H_6O_2$	Coumarin	3.89[i]	liq	70-160	-	827
		3.82[i]	B	25-40	-	827
		4.5$_2$	B	25	0	1070
		4.54±0.02	B	15-65	15	2023
		4.55	B	10-40	-	1583-4
		3.85[i]	D	30-40	-	827
	1,3-Indandione	2.74	B	25	-	1766
	1-Oxaazulan-2-one	5.64	B	30	-	1016
	Phenylpropiolic acid	2.31	D	25	0	2128
C_9H_7Br	1-Bromo-2-*p*-tolylacetylene	0.98	B	25	0	2127
C_9H_7BrO	1-Bromo-2-indanone	3.43±0.06	B	30	0	1010
		3.76±0.03	Hx	30	0	1010
$C_9H_7Br_2N$	4-Quinoline - bromine complex	5.80[j]	B	25	-	895
C_9H_7Cl	1-Chloro-2-*p*-tolylacetylene	1.28	B	25	0	2127

i) Calculated from Jatkar's equation. j) Lower limit.

FORMULA	COMPOUND NAME	μ, D	State, or Solv.	t, °C	Method or P_a, % of P_e	Reference
C_9H_7ClO	1-Chloro-2-indanone	3.73±0.05	B	30	0	1010
		4.11±0.02	D	30	0	1010
		3.92±0.06	Hx	30	0	1010
	Cinnamoyl chloride	3.66	B	25	0	927
$C_9H_7CrNO_3$	Aniline - chromium tricarbonyl complex	5.40±0.05	B	25	15	520
$C_9H_7F_3$	*o*-Trifluoromethylstyrene	2.40	B	25	-	964
		2.45	Hp	25	-	964
	m-Trifluoromethylstyrene	2.48	B	25	-	964
		2.52	Hp	25	-	964
	p-Trifluoromethylstyrene	2.44	B	25	-	964
C_9H_7I	1-Iodo-2-*p*-tolylacetylene	1.02	B	25	-	2127
$C_9H_7I_2N$	Quinoline - iodine complex	4.45[k]	B	25	-	895
$C_9H_7MnO_3$	Tricarbonylmethylcyclopentadienyl-manganese	3.4	B	20	-	1894
		3.6	D	20	-	1894
		3.4	Hp	20	-	1894
C_9H_7N	β-Cyanostyrene	4.17	D	25	0	588
	Quinoline	2.19	gas	n.s.	0	2017
		2.31	gas	105-200	0	234
		2.13	liq	20	VS	1701
		2.16	liq	20	-	657
		2.16	B	14	0	138
		2.18	B	10-40	-	1589
		2.20	B	25	0	1101
		2.20	B	25	-	1953
		2.26	B	25	0	234
		2.27	B	25	0	1653

k) Lower limit.

FORMULA	COMPOUND NAME	μ, D	State, or Solv.	t, °C	Method or P_a, % of P_e	Reference
C_9H_7N (contd.)	Quinoline (contd.)	2.29	CCl_4	25	0	234
		2.19	l	25	0	234
		2.13±0.05	m	20	-	1448
	Isoquinoline	2.75	gas	80-20	0	234
		2.49±0.01	B	30	10	546
		2.54	B	25	0	1101
		2.55	B	14	0	138
		2.56	B	10-40	-	1589
		2.63	B	25	0	234
		2.67	CCl_4	25	0	234
		2.62	l	25	0	234
C_9H_7NO	1-Azaazulan-2-one	2.73	B	30	-	1016
	2-Hydroxyquinoline	1.66±0.10	B	25	0	1634
		3.45±0.10	D	25	0	1634
	8-Hydroxyquinoline	2.28	liq	100, 120	-	992
		2.23	B, CCl_4	25-40	-	992
		2.68	B	25	0	1596
		2.72	B, Hp	25	0	716
		2.51	D	25	0	1596
		1.68	CCl_4	25	0	1729
	3-Phenylisoxazole	3.16	B	n.s.	-	1515
		3.37	B	25	0	1932
	4-Phenylisoxazole	2.96	B	n.s.	-	1515
	5-Phenylisoxazole	2.80	B	n.s.	-	1515
		3.10	B	25	0	1932
	Quinoline *N*-oxide	4.00	B	25	0	1552
		4.07±0.02	B	25	0	1985

l) Light petroleum. m) Dimethylaniline.

FORMULA	COMPOUND NAME	μ, D	State, or Solv.	t, °C	Method or P_a, % of P_e	Reference
	Isoquinoline *N*-oxide	4.47±0.03	B	25	0	1985
$C_9H_7NO_2$	*N*-Methylphthalimide	2.24±0.02	D	20	0	1192
	3-Phenyl-5(4*H*)-isoöxazolone	5.0_1	B	30	0	227
$C_9H_7NO_2S_2$	2-(2'-Benzothiazolethio)acetic acid	4.44	B	25	-	637
C_9H_8	Indene	0.67	B	25	-	1921
		0.85	CCl_4, CS_2, cHx	n.s.	-	1572
		0.44	n.s.	n.s.	-	1062
	(*p*-Tolyl)acetylene	1.02	B	25	0	1461
$C_9H_8BrN_4$	1-*p*-Bromophenyl-5-ethyltetrazole	4.66	B	25	0	893
$C_9H_8Br_2$	5,6-Dibromoindane	2.50	B	25	-	1773
$C_9H_8Br_2O_2$	2,2-Dimethyl-5,6-dibromobicyclo[4.3.0]-1,3-dioxa-4,6,8-nonatriene	3.31	n.s.	n.s.	-	186
$C_9H_8ClN_4$	1-Ethyl-5-*p*-chlorophenyltetrazole	4.78	B	25	0	893
$C_9H_8Cl_2O_2$	2,2-Dimethyl-5,6-dichlorobicyclo[4.3.0]-1,3-dioxa-4,6,8-nonatriene	3.42	n.s.	n.s.	-	186
$C_9H_8N_2$	*cis*-*p*-Aminocinnamonitrile	4.95	B	25	0	412
	trans-*p*-Aminocinnamonitrile	6.30	B	25	0	412
		6.98	D	25	0	412
$C_9H_8N_2$	4-Aminoquinoline	4.38	D	25	0	45
	4-Methylcinnoline	4.53±0.1	B	25	-	1640

FORMULA	COMPOUND NAME	μ, D	State, or Solv.	t, °C	Method or P_a, % of P_e	Reference
$C_9H_8N_2$ (contd.)	2-Methylquinazoline	1.6	B	n.s.	-	2012
		2.2	B	25	-	1840
		2.4	B	n.s.	-	2012
	2-Methylquinoxaline	2.2	B	n.s.	-	2016
$C_9H_8N_2O$	3-Methyl-5-phenylazoxime	1.64	D	(20)	0	1320
	5-Methyl-3-phenylazoxime	1.42	D	(20)	0	1320
	Methylphenylfurazan	4.17	D	(20)	0	1320
	Methylphenyloxydiazole	3.46	D	(20)	0	1320
	p-Tolylfurazan	4.20	D	(20)	0	1320
$C_9H_8N_2O_2$	*N*-Benzylsydnone	6.0_9	B	25	0	427-8
		6.27±0.02	B	25	0	735-6
	Methylphenylfurazan peroxide	4.80	B	25	0	1931
	Methylphenylfuroxan	4.96	B	25	0	1931
	C-Methyl-*N*-phenylsydnone	6.56±0.02	B	25	0	736
		6.57	B	25	0	735
		6.6_0	B	25	0	427-8
	N-*p*-Tolylsydnone	6.89±0.02	B	25	0	735-6
		6.98	B	25	0	427-8
$C_9H_8N_2O_4$	β-Methyl-*p*,β-dinitrostyrene	0.41	B	25	15	2033
	cis-2-Nitro-1-(4-nitrophenyl)propene	0.88±0.04	B	20	5-10	421
		0.92	B	25	-	2031

FORMULA	COMPOUND NAME	μ, D	State, or Solv.	t, °C	Method or P_a, % of P_e	Reference
$C_9H_8N_2S_2$	2-Methyl-3-phenyl-2,5-*endo*thio-1,3,4-thiadiazoline	8.8	D	25	15	859
C_9H_8O	Cinnamaldehyde	3.74	gas	18	-	674
		3.30	liq	30-80	-	823
		3.468	liq	40	RT	1034
		3.63±0.03	B	25	0	127
	1-Indanone	3.31	B,cHx	20	0	773
		3.51	CCl_4	20	0	773
	2-Indanone	2.45	B	20	0	773
		2.40	CCl_4	20	0	773
		2.46	cHx	20	0	773
$C_9H_8O_2$	*cis*-Cinnamic acid (three isomers, m.p. = 42°C, 58°C, 68°C)	1.13	B	20	0	440
	trans-Cinnamic acid	1.31	B	n.s.	0	440
	Cinnamic acid (configuration not given)	2.26	B	30	0	1276
		1.80	D	25	0	588
$C_9H_8O_4$	Acetylsalicylic acid	2.01	D	35	-	992
		2.09	D	25	-	992
		4.36	D	25	-	944
C_9H_9Br	6-Bromohydrindane	2.17	B	25	0	1773
	o-Bromo-α-methylstyrene	1.88	B	27	-	165
	p-Bromo-α-methylstyrene	1.46	B	22	-	165
$C_9H_9BrO_2$	6-Bromo-2,2-dimethylbicyclo[4.3.0]-1,3-dioxa-4,6,8-nonatriene	2.72	n.s.	n.s.	-	186

FORMULA	COMPOUND NAME	μ, D	State, or Solv.	t, °C	Method or P_a, % of P_e	Reference
$C_9H_9BrO_2$ (contd.)	Ethyl *p*-bromobenzoate	2.31	B	30	0	1501
$C_9H_9BrO_3$	Methyl 5-bromo-2-methoxybenzoate	2.80	B	20	0	1204
$C_9H_9Br_3$	2,4,6-Tribromo-1,3,5-trimethylbenzene	0.35	B	40	-	1949
		0	CCl_4	20,54	-	1207
C_9H_9Cl	1-Chloro-1-phenyl-2-propene	2.09	D	20	0	1969
	m-Chloro-α-methylstyrene	1.91	B	27	0	165
	3-Chloropropenylbenzene	1.9	D	20	0	1969
	Cinnamyl chloride	1.92	D	25	0	588
$C_9H_9ClO_2$	6-Chloro-2,2-dimethylbicyclo[4.3.0]-1,3-dioxa-4,6,8-nonatriene	2.76	n.s.	n.s.	-	186
	Ethyl *p*-chlorobenzoate	2.02	B	20	0	136
		2.24	B	30	0	1501
$C_9H_9Cl_3$	2,4,6-Trichloro-1,3,5-trimethylbenzene	~0.1	B	25	0	1814
		0.38	B	40	-	1949
	3,5,6-Trichloro-1,2,4-trimethylbenzene	1.85	B	25	0	1814
C_9H_9F	*o*-Fluoro-α-methylstyrene	1.55	B	26	0	165
$C_9H_9F_3O$	2-Ethoxytrifluoromethylbenzene	3.54	cHx	25	0	226
	4-Ethoxytrifluoromethylbenzene	3.52	B	25	0	1378
C_9H_9I	*o*-Iodo-α-methylstyrene	1.49	B	26	0	165
C_9H_9N	3,4-Dihydroisoquinoline	1.87	liq	40	-	1709

FORMULA	COMPOUND NAME	μ, D	State, or Solv.	t, °C	Method or P_a, % of P_e	Reference
	p-Ethylbenzonitrile	4.67	D	24	0	228
		2.90	$CHCl_3$	24	0	228
		4.53	cHx	24	0	228
	1-Methylindole	2.18	B	25	-	817
	2-Methylindole	2.49	B	25	-	817
	3-Methylindole	2.10	B	20	0	345
C_9H_9NO	Cinnamoylamide	3.64	D	25	0	588
$C_9H_9NOS_2$	2-(2'-Hydroxyethylene)thiobenzothiazole	2.33	B	25	-	637
$C_9H_9NO_2$	β-Methyl-β-nitrostyrene	4.17	B	25	15	2033
	p-Methyl-β-nitrostyrene	4.81	D	25	0	588
	2-Nitro-1-phenyl-1-propene	4.17	B	25	-	2031
	1-Nitro-2-*p*-tolylethylene	5.04	B	25	-	2031
$C_9H_9NO_3$	β-Methoxy-β-nitrostyrene	5.47	B	25	15	2032
	1-Nitro-2-(*p*-methoxyphenyl)ethylene	5.47	B	25	-	2031
	3-Nitro-4-methylacetophenone	2.7_2	B	25	0	1071
$C_9H_9NO_4$	Ethyl 2-nitrobenzoate	3.94	B	n.s.	5	1212
		4.12	D	n.s.	5	1212
	Ethyl 3-nitrobenzoate	3.74	B	n.s.	5	1212
		3.96	D	n.s.	5	1212

FORMULA	COMPOUND NAME	μ, D	State, or Solv.	t, °C	Method or P_a, % of P_e	Reference
$C_9H_9NO_4$ (contd.)	Ethyl 4-nitrobenzoate	3.87	B	13	-	129
		4.05	B	30	0	1501
		4.07	B	n.s.	5	1212
		4.00	D	n.s.	5	1212
$C_9H_9NO_5$	Methyl 5-nitro-2-methcxybenzoate	3.77	B	20	0	1204
$C_9H_9N_3O_2S_2$	Sulfathiazole	7.0±0.2	D	25	-	860
$C_9H_9N_3O_4S_2$	*S*-(2,4-Dinitrophenyl)dimethyldithio-carbamate	6.36	B	25	0	639
$C_9H_9N_3O_6$	2,4,6-Trinitro-1,3,5-trimethylbenzene	0	B	20,51	37	1207
		≈0.5	B	25,45	0	1064
		0.80	B	20	-	1949
		0.6	$CHCl_3$	25	0	1064
		0.8	$CHCl_3$	25	-	841
C_9H_{10}	Allylbenzene	0.5	B	20	0	219
	Hydrindene	0.53	B	25	0	1773
	Isopropenylbenzene	0.76	B	20	0	219
	p-Methylphenylethylene	0.63	B	25	-	1462
	m-Methylstyrene	0.36±0.1	B	25	-	476
	p-Methylstyrene	0.38±0.1	B	25	-	476
	Phenylcyclopropane	0.49	B	25	0	1632
	Propenylbenzene	0.72	B	20	0	219
$C_9H_{10}ClNO$	*N*-(2-Chloroethyl)benzamide	3.04	B	30	0	2148

FORMULA	COMPOUND NAME	μ, D	State, or Solv.	t, °C	Method or P_a, % of P_e	Reference
	p-Chloro-*N*-ethylbenzamide	3.42	B	30	0	2148
$C_9H_{10}F_3N$	Trifluoromethyl-*p*-(*N*,*N*-dimethyl)toluidine	4.62	B	25	0	1625
$C_9H_{10}N_2$	*p*-Cyano-*N*,*N*-dimethylaniline	6.60	B	25	-	2028
	p-Dimethylaminobenzonitrile	5.94	B	25	15	703
	Pyrrole - pyridine complex	3.57±0.1	CCl_4	n.s.	-	596
$C_9H_{10}N_2O$	1-Phenyl-3-pyrazolidone	1.98	CCl_4	25	-	1025
$C_9H_{10}N_2O_2$	α-Methylphenylglyoxime	0.79	D	20	0	1319
	β-Methylphenylglyoxime	1.17	D	20	0	1319
	α-*p*-Tolylglyoxime	1.10	D	20	0	1319
	β-*p*-Tolylglyoxime	1.67	D	20	0	1319
$C_9H_{10}N_2O_3$	Ethyl *p*-nitrobenzimidate	3.80	B	25	5	482
$C_9H_{10}N_2O_4$	*anti*-(Ethyl *p*-nitrobenzoate oxime)	4.78	B	25	5	482
	syn-(Ethyl *p*-nitrobenzoate oxime)	4.48	B	25	5	482
$C_9H_{10}N_2S$	*p*-Dimethylaminophenylthiocyanate	5.70	B	25	0	268
$C_9H_{10}N_2Se$	*p*-Dimethylaminophenylselenocyanate	5.64	B	25	0	268
$C_9H_{10}N_4O_2S_2$	5-Methyl-2-sulfanilamido-1,3,4-thiadiazole	6.8±0.2	D	25	-	860
$C_9H_{10}O$	Cinnamyl alcohol	1.81	D	25	0	588
	1-Indanol (m.p. = 40°C)	1.76±0.02 1.87	B CCl_4	10,20 20	0 0	773 773

FORMULA	COMPOUND NAME	μ, D	State, or Solv.	t, °C	Method or P_a, % of P_e	Reference
$C_9H_{10}O$ (contd.)	1-Indanol (contd.) (m.p. = 55°C)	1.75±0.03	B	10-60	0	773
		1.70	CCl_4	40	0	773
		1.82	CCl_4	20	0	773
	2-Indanol	1.79	B	20	0	773
		1.85	B	60	0	773
		1.74	CCl_4	20	0	773
	2-Methylacetophenone	2.60	B	25	0	1796
	4-Methylacetophenone	3.20	B	25	5	468
		3.2_3	B	25	0	1071
		3.23±0.02	B	25	0	127
		3.27	D	25	5	468
	1-Phenyl-1-propanone	2.88±0.02	B	(25)	-	301
	1-Phenyl-2-propanone	2.72	B	30	0	20
	2-Phenylpropionaldehyde	2.81	B	20	0	329
	3-Phenylpropionaldehyde	2.33	B	20	0	329
		2.36	Hx	20	0	329
$C_9H_{10}OS$	Methyl *o*-methylmercaptobenzoate	2.49	B	25	0	503
$C_9H_{10}O_2$	Benzyl acetate	1.8_2	B	25	0	1366
	Bicyclo[5.4.0]-1,5-dioxa-6,8,10-undeca-triene	1.95	n.s.	n.s.	-	701
	o-Cresyl acetate	1.68_8±0.022	B	22	15	413
	m-Cresyl acetate	1.60_8±0.015	B	22	15	413
	p-Cresyl acetate	1.53_6±0.015	B	22	15	413

FORMULA	COMPOUND NAME	μ, D	State, or Solv.	t, °C	Method or P_a, % of P_e	Reference
	2,2-Dimethylbicyclo[4.3.0]-1,3-dioxa-4,6,8-nonatriene	1.03	n.s.	n.s.	-	186
	Ethyl benzoate	1.95	gas	130-230	5	622
		1.87	liq	30-40	-	1497
		1.80[n]	B	28	-	979
		1.84_9±0.02	B	22	15	413
		1.85	B	25	0	1747
		1.90	B	30	5	390
		1.93	B	25	-	163
		1.93	B	30	0	1501
		1.99	B	n.s.	5	334
		1.9_5	CCl_4	25	5	71
		1.97	CCl_4	20	5	602
	o-Methoxyacetophenone	4.02	B	25	0	1211
	m-Methoxyacetophenone	2.88	B	25	0	1211
	p-Methoxyacetophenone	3.51±0.10	B	25	0	1636
		3.56	B	25	0	1211
	2-Methoxy-6-methyl-2,4,6-cycloheptatrien-1-one	4.9	B	n.s.	-	1998
		5.06±0.10	B	25	0	1022
	Methyl *o*-toluate	1.60_8±0.015	B	22	15	413
	Methyl *m*-toluate	1.94_2±0.015	B	22	15	413
	Methyl *p*-toluate	2.06_9±0.023	B	22	15	413
	Phenyl propionate	1.53_7±0.016	B	22	15	413
	β-Phenyl propionic acid	1.65	B	30	-	1276

n) Microwave frequency.

FORMULA	COMPOUND NAME	μ, D	State, or Solv.	t, °C	Method or P_a, % of P_e	Reference
$C_9H_{10}O_2S$	Phenylallylsulfone	5.24	B	30	0	88
	trans-Methylstyrylsulfone	5.27	B	30	0	88
$C_9H_{10}O_3$	Allylbenzene ozonide	1.19	B	20	0	219
	Ethyl *p*-hydroxybenzoate	2.81	B	25	-	43
		2.90	B	n.s.	5	334
	Ethyl salicylate	1.7	B, CCl_4	25-45	-	831
		2.69	B	n.s.	5	334
		2.71	B	25	-	43
		2.85	B	25	0	944
		2.91	B	40	-	766, 876
	d-(Methyl mandelate)	2.4_9	B	25	-	2083
	dl-(Methyl mandelate)	2.4_7	B	25	-	2083
	Methyl *o*-methoxybenzoate	2.65±0.02	B	20	0	1186, 1203
		2.68	B	25	0	945
		2.84	B	n.s.	5	334
	Methyl *p*-methoxybenzoate	2.62	B	n.s.	5	334
	Methyl methoxybenzoate	2.63±0.03	B	20	0	1186
$C_9H_{11}Br$	Bromomesitylene	1.53	B	30	0	225
	1-Bromo-3-phenylpropane	1.80	D	25	0	588
$C_9H_{11}Cl$	Chloromesitylene	1.56	B	30	0	225
	1-Chloro-3-phenylpropane	1.78	D	25	0	588
		1.78	D	20	0	1969

FORMULA	COMPOUND NAME	μ, D	State, or Solv.	t, °C	Method or P_a, % of P_e	Reference
$C_9H_{11}F$	Fluoromesitylene	1.37	B	30	0	225
$C_9H_{11}I$	Iodomesitylene	1.43	B	30	0	225
$C_9H_{11}NO$	Acetophenone methyloxime	0.93	liq	n.s.	5	1593
		0.790±0.005	B	20	5	1593
	N-Acetyl-*o*-toluidine	3.71	B	25	0	1797
	N-Acetyl-*m*-toluidine	3.69	B	25	0	1797
	N-Acetyl-*p*-toluidine	3.74	B	25	0	1797
	p-Dimethylaminobenzaldehyde	5.58	B	25	5	369
		5.6	B	18	-	2090
		5.63	p	25	5	369
	N-Ethylbenzamide	3.60	B	30	0	2143
	N-Ethylbenzimidate	1.70	B	25	5	482
	Nitrosomesitylene	1.38±0.13	B	8	-	651
		1.64±0.09	B	25	-	651
$C_9H_{11}NO_2$	*N*-Dimethylanthranilic acid	6.36	B	25	-	438
	Ethyl *p*-aminobenzoate	3.28	B	25	5	369
		4.0	B	19	-	2090
		3.67	D	25	5	369
		3.41	n.s.	30	0	1501
	anti-(Ethyl benzoate oxime)	1.41	B	25	5	482
	syn-(Ethyl benzoate oxime)	2.52	B	25	5	482
	Methyl *N*-(*p*-tolyl)carbamate	3.78	B	25	-	1217
		4.02	D	25	-	1217

p) *trans*-Dichloroethylene.

FORMULA	COMPOUND NAME	μ, D	State, or Solv.	t, °C	Method or P_a, % of P_e	Reference
$C_9H_{11}NO_2$ (contd.)	Nitromesitylene	3.68	B	30	0	225
		3.70±0.01	B	25	-	651
		3.7_1	B	25	0	1071
		3.65	CCl_4	30	0	225
	Phenyl urethan	2.58±0.03	B	26.8	0	2092
$C_9H_{11}NS_2$	*S*-Ethylphenyldithiocarbamate	3.23	B	25	0	639
C_9H_{12}	*p*-Ethyltoluene	0	B	25	0	1080
	1-Methyl-2-ethylbenzene	0.56	liq	20-30	23	22
	1-Methyl-3-ethylbenzene	0.33	liq	20-30	10	22
	1-Methyl-4-ethylbenzene	0	liq	20-30	5	22
	Propylbenzene	0.35	liq	20-30	13	22
		0.37	B	25	6	359
	Isopropylbenzene	0.65	gas	95-180	5	84
		0.37	liq	20-30	13	22
		0.39±0.01	liq	20-60	5	1487,1489
		0.38	B	25	6	359
	1,2,3-Trimethylbenzene	0.56	liq	20-30	23	22
	1,2,4-Trimethylbenzene	0.30	liq	20-30	10	22
	1,3,5-Trimethylbenzene	0	liq	20	VS	1701
		0	liq	20-30	5	22
		0.0_7	B	20	-	1949
		0.1	B	25	-	2114
		<0.2	B	25	-	2112
		0.2_4	B	25	0	1071

FORMULA	COMPOUND NAME	μ, D	State, or Solv.	t, °C	Method or P_a, % of P_e	Reference
$C_9H_{12}I_2$	Iodine - 1,3,5-trimethylbenzene complex	1.1	cHx	n.s.	RT	1411
$C_9H_{12}N_2$	Acetonephenylhydrazone	2.70	B	20	-	854
$C_9H_{12}O$	2,6-Dimethylanisole	1.24±0.02	B	20	0	1193
	3-Phenylpropanol	1.64	B	25	0	1234
		1.73	D	25	0	588
	2,4,6-Trimethylphenol	1.37	B	30	0	225
		1.43	B	25	0	598
		1.66	D	25	-	451
		1.30	cHx	25	-	451
$C_9H_{12}O_2$	Cumene hydroperoxide	1.76-1.81	B	30	0	1163-4
		1.8	q	27	0	2153
$C_9H_{12}O_2S_2$	Ethylsulfinyl tolyl sulfide	5.32	B	25	0	619
$C_9H_{12}O_3$	Phloroglucinol trimethyl ether	1.8	B	25	-	2112
	1,2,3-Trimethoxybenzene	2.26	B	20	RT	922
$C_9H_{12}O_8$	Tetracarboxymethane tetramethyl ester	>0	B	n.s.	-	429
$C_9H_{12}Se$	Ethyl *p*-tolyl selenide	1.64±0.02	B	20	-	309
$C_9H_{13}BrO_2$	2-Bromo-2,5,5-trimethyl-1,3-cyclohexane-dione	2.92	B	21	-	1994
$C_9H_{13}BrSi$	(*p*-Bromophenyl)trimethylsilane	1.79	B	25	0	1845
$C_9H_{13}ClO_2$	2-Chloro-2,5,5-trimethyl-1,3-cyclohexane-dione	3.03	CCl_4	23	-	1994
$C_9H_{13}ClSi$	(Chloromethyl)dimethylphenylsilane	1.95	B	30	0	544

q) Cumene.

FORMULA	COMPOUND NAME	μ, D	State, or Solv.	t, °C	Method or P_a, % of P_e	Reference
$C_9H_{13}ClSi$ (contd.)	(*m*-Chlorophenyl) trimethylsilane	1.83	B	25	0	322
		1.85	B	30	0	544
	(*p*-Chlorophenyl) trimethylsilane	1.70	B	25	0	1624
		1.84	B	25	0	322
$C_9H_{13}Cl_2NPt$	*trans*-Ethylene-*p*-aminotoluene - platinum dichloride complex	2.35±0.02	B	25	-	285
$C_9H_{13}FSi$	(*p*-Fluorophenyl) trimethylsilane	1.69	B	25	0	1624
$C_9H_{13}N$	2-*n*-Butylpyridine	1.90	B	25	0	360
	3-*n*-Butylpyridine	2.47	B	25	0	360
	4-*n*-Butylpyridine	2.72_0	B	25	0	360
	4-Isobutylpyridine	2.73_6	B	25	0	360
	4-*sec*-Butylpyridine	2.72_0	B	25	0	360
	4-*t*-Butylpyridine	2.87	CCl_4	25	0	1729
	N,*N*-Dimethyl-*o*-toluidine	0.88±0.01	B	25	0	524
	N,*N*-Dimethyl-*m*-toluidine	1.51±0.02	B	25	0	524
	N,*N*-Dimethyl-*p*-toluidine	1.30	B	25	0	1258
		1.29±0.02	B	25	0	524
		1.30	n.s.	n.s.	-	525
		1.32	n.s.	25	15	703
	N-Methyl-2,4-xylidine	1.54±0.02	B	25	0	524
	N-Methyl-2,6-xylidine	1.28±0.01	B	25	0	524

FORMULA	COMPOUND NAME	μ, D	State, or Solv.	t, °C	Method or P_a, % of P_e	Reference
	N-*n*-Propylaniline	1.37	liq	20	5	337
		1.66	B	20	5	337
		1.69	B	20	0	336
		1.67	Hx	20	5	337
		1.70	Hx	20	0	336
	2,4,6-Trimethylaniline	1.41	B	25	-	805a
		1.45	B	25	0	1795
		1.57	D	25	0	1795
$C_9H_{13}NO$	4-*t*-Butylpyridine 1-oxide	4.63±0.01	B	25	0	1730
	7-(Dimethylamino)-2,4,6-heptatrienal	8.24±0.02	B	25	0	798
	Dimethylaniline methyl ether	1.62	gas	182	5	625
	N,*N*-Dimethyl-*p*-anisidine	1.71	B	25	15	703
$C_9H_{13}NO_2$	*N*-Acetylcyclohexane-1,3-lactam	3.63±0.05	D	30	0	1060
	N-Acetylcyclohexane-1,4-lactam	2.88±0.02	D	30	0	1060
	N-Methylcyclohexane-1,3-imide	2.88±0.05	D	30	0	1059
$C_9H_{13}NO_2S$	*N*,*N*-Dimethyl-*p*-toluenesulfonamide	5.52	B	25	0	1105
	p-(Methylsulfonyl)-*N*,*N*-dimethylaniline	6.65	B	25	-	2028
$C_9H_{13}NO_2Si$	Trimethyl(*m*-nitrophenyl)silane	4.31	B	30	0	544
$C_9H_{13}N_3$	1-(*p*-Tolyl)-3,3-dimethyltriazene	1.9_5	B	25	0	1083
C_9H_{14}	2-*t*-Butyl-1-penten-3-yne	0.67	n.s.	n.s.	0	1490
	8-Nonen-6-yne	0.65	B	20	-	1492
	2,5,5-Trimethyl-1-hexen-3-yne	0.68	B	20	-	1493

FORMULA	COMPOUND NAME	μ, D	State, or Solv.	t, °C	Method or P_a, % of P_e	Reference
$C_9H_{14}ClNO$	*p*-Chlorophenol - trimethylamine complex	4.30±0.08	B	25	-	789
$C_9H_{14}Cl_2$	1,3-Dichloro-5-methyl-2,6-octadiene	2.24	B	20	0	1491
$C_9H_{14}N_2$	1,7-Heptanedicarbonitrile	4.39	B	25	0	1943
		4.42	B	25	-	1978-9
		4.50	B	75	-	1978-9
$C_9H_{14}O$	2,6-Dimethyl-2,5-heptadiene-4-one	2.36±0.03	B	25	0	127
		2.40±0.03	CCl_4	25	0	127
	2-Nonen-2-al	3.23	B	25	0	587
	3,5,5-Trimethyl-2-cyclohexen-1-one	3.99	D	25	-	1002
$C_9H_{14}O_3Si$	Phenyltrimethoxysilane	1.62	B	20	0	1041
$C_9H_{14}O_4$	Diethyl cyclopropane-1,1-dicarboxylate	2.4	B	n.s.	-	1499
		2.42	B	25	0	495
	Monomethyl *cis*-1,2-cyclohexanedicarboxylate	1.70	B	25	-	1751
	Monomethyl *trans*-1,2-cyclohexane-dicarboxylate	1.91	B	25	-	1751
$C_9H_{14}Si$	Phenyltrimethylsilane	0.42	B	30	-	544
		0.44	B	25	0	1624
$C_9H_{15}AsO_3S_6$	Arsenic *O,O,O*-tri(ethylxanthate)	1.5_3	B, CS_2	20	15	1242
$C_9H_{15}Cl$	1-Chloro-5-methyl-2,6-octadiene	2.06	B	20	0	1491
	3-Chloro-5-methyl-1,6-octadiene	2.00	B	20	0	1491
$C_9H_{15}FeO_3S_6$	Iron *O,O,O*-tri(ethylxanthate)	2.8_1	B	25	0	1243

FORMULA	COMPOUND NAME	μ, D	State, or Solv.	t, °C	Method or P_a, % of P_e	Reference
$C_9H_{15}NO$	*pseudo*-Pelletierine	2.75±0.1	B	25	0	1113
	Phenol - trimethylamine complex	3.24±0.03	B	25	0	789
	4-Piperidyl-3-buten-2-one	5.27	B	25	0	1749
	Pyridine - *n*-butanol complex	3.37	B	25	0	318
	Pyridine - *t*-butanol complex	3.16	B	25	0	318
$C_9H_{15}NO_2$	1-Cyclohexenyl-1-nitropropane	3.6±0.1	B	20	0	256
	Cycloöctenylnitromethane	3.6±0.1	B	20	0	256
$C_9H_{15}NSi$	*p*-(Trimethylsilyl)aniline	1.65	B	25	0	1845
$C_9H_{16}NO$	Spiro-(cyclohexane-1,5'-pyrrolidone)	4.56±0.04	D	30	0	1060
$C_9H_{16}O$	Cyclononanone	2.85±0.03	B	20	-	630
$C_9H_{16}O_2$	Ethyl ester of cyclohexane carboxylic acid	1.28	n.s.	n.s.	-	1361
	2-Methyl-1,4-dioxaspiro[4.5]decane	1.06±0.02	B	30	0	151
	Methyl ester of 3-methylcyclohexane carboxylic acid	1.26	n.s.	n.s.	-	1361
	n-Propyl ester of cyclopentane carboxylic acid	1.23	n.s.	n.s.	-	1361
$C_9H_{16}O_4$	Azelaic acid	2.35±0.02 2.71	D D	25 35	- -	119 1581
	dl-erythro-2,3-Diacetoxypentane	2.50	liq	25	-	2130
	dl-threo-2,3-Diacetoxypentane	2.09	liq	25	-	2130

FORMULA	COMPOUND NAME	μ, D	State, or Solv.	t, °C	Method or P_a, % of P_e	Reference
$C_9H_{16}O_4$ (contd.)	Diethyl dimethylmalonate	2.34	B	n.s.	-	1499
	Diethyl glutarate	2.46	liq	30-40	-	1497
		2.43	B	25,50	0	1832
$C_9H_{17}N$	Quinolizidine	0.74	B	25	0	434,1989
$C_9H_{17}NO$	1-Dimethylamino-5-methyl-1-hexen-3-one	4.83	B	25	0	1749
	4-Methyl-1-oxa-4-azaspiro[4.5]decane	1.85±0.03	B	25	0	151
	trans-2-Oxanonamethylenimine	3.79	B	25	0	783
$C_9H_{17}NO_4$	Diethyl glutamate	2.58	B	25	0	2152
C_9H_{18}	2,5-Dimethyl-4-heptene	0.69	liq	20	0	1189
	2,6-Dimethyl-2-heptene	0.81	liq	20	0	1180,1189
	3,6-Dimethyl-3-heptene	0.69	liq	20	0	1180
	4-Ethyl-3-heptene	0.70	liq	20	0	1180,1189
	Isopropylcyclohexane	0	gas	120-185	-	84
$C_9H_{18}AsN_3S_6$	Arsenic *tris*(*N*,*N*-dimethyldithiocarbamate)	4.6_5±0.08	B	39	0	1242
$C_9H_{18}Br_2$	1,9-Dibromononane	2.50±0.03	liq	20-70	0	907
		2.53	B	20	0	907
		2.543±0.004	B	25	0	692
		2.57	B	25	0	1943
		2.576±0.004	B	64	0	692
		2.60	D	20	0	907
		2.55	CCl_4	20	0	907
		2.39	Hx	20	0	907
		2.57-2.60	Hp	25,50	0	1837

FORMULA	COMPOUND NAME	μ, D	State, or Solv.	t, °C	Method or P_a, % of P_e	Reference
$C_9H_{18}Cl_2$	1,9-Dichlorononane	2.51	Hx	20	RT	921
$C_9H_{18}Cl_4O_2Ti$	Titanium tetrachloride - amyl butyrate complex	4.37	B	20	15	1438
$C_9H_{18}NNaS_2$	Sodium diisobutyldithiocarbamate	3.89 4.9	B D	25 25	0 0	639 639
$C_9H_{18}N_2$	6-Methyl-1,4-diazaspiro[4.5]decane	1.52±0.04	B	30	0	151
	7-Methyl-1,4-diazaspiro[4.5]decane	1.51±0.04	B	30	0	151
	8-Methyl-1,4-diazaspiro[4.5]decane	1.41-1.58	pXy	30-90	5	512
$C_9H_{18}O$	2,6-Dimethyl-4-heptanone	2.66	CCl_4	25	-	67
	5-Nonanone	2.69	CCl_4	25	-	67
	2,2,4,4-Tetramethyl-3-pentanone	2.48±0.02 2.79	n.s. B	n.s. 15	- 10	302 2141
	1,3,5-Trimethylcyclohexanol	1.88	B	18	0	681a
$C_9H_{18}O_2$	2-Isobutyl-2,4-dimethyl-1,3-dioxolane	1.17±0.02	B	30	0	151
	2-*t*-Butyl-2,4-dimethyl-1,3-dioxolane	1.16±0.02	B	30	0	151
	3,5,5-Trimethylhexanoic acid dimer	0.86	CCl_4	20	0	1719
$C_9H_{18}O_2S$	Ethyl 3-(*n*-butylthio)propionate	2.26	B	25	0	356,361
$C_9H_{18}O_3$	Perpelargonic acid	2.38±0.04	B	30	0	1616
$C_9H_{18}O_4S$	Ethyl 3-(*n*-butylsulfonyl)propionate	4.25	B	25	0	356,361

FORMULA	COMPOUND NAME	μ, D	State, or Solv.	t, °C	Method or P_a, % of P_e	Reference
$C_9H_{19}Br$	1-Bromononane	1.89	liq	25	-	707
		1.95	liq	20	-	657
		1.91	B	n.s.	-	699
$C_9H_{19}BrO$	1-Bromo-2-ethoxyheptane	2.29	B	20	-	1744
	2-Bromo-3-ethoxyheptane	2.10	B	25	-	1744
	3-Bromo-4-ethoxyheptane	2.13	B	25	-	1744
$C_9H_{19}N$	*N*-Isoamylbutylidenamine	1.57±0.03	B	25	0	478
$C_9H_{19}N_5$	1-Butyl-5-butylaminotetrazole	7.12	B	25	0	892
C_9H_{20}	3,3-Diethylpentane	0	B,D	20	15	1892
$C_9H_{20}NO_4P$	Diethyl ester of diethylcarbamoyl phosphoric acid	2.93	B	25	5	58
$C_9H_{20}N_2O$	Tetraethylurea	3.83	"pure"	25	-	562
		3.3	B	20	-	403
		3.42±0.04	B	25	0	120
		3.45±0.04	D	25	-	120,562
		3.58±0.04	CCl_4	25	0	120
		3.28±0.04	Hp	25	0	120
		4.00	r	25	-	562
		4.08	s	25	-	562
$C_9H_{20}N_2S$	*N,N'*-Di-*n*-butylthiourea	5.68_6±0.002	D	25	5	275-6
$C_9H_{20}O$	1-Nonanol	1.61	B	25	0	1239
		1.72±0.02	B	20	0	1184
		1.7_6	CCl_4	25	5	1108
$C_9H_{20}O_2$	2,2-Dimethoxyheptane	0.91	B	25	-	1457

r) Acetone. s) Ethanol.

FORMULA	COMPOUND NAME	μ, D	State, or Solv.	t, °C	Method or P_a, % of P_e	Reference
	1,9-Nonanediol	2.46	D	25	0	907
$C_9H_{20}O_3$	1,1,1-Triethoxypropane	1.95	B	25	5	60
$C_9H_{20}O_4$	Pentaerythritol tetramethyl ether	0.8	B	25	-	431
		1.91±0.01	B	25	0	1312
	Tetraethyl orthocarbonate	>0	B	n.s.	-	429
		1.1	B	25	5	56
$C_9H_{20}Si$	Triethylallylsilane	0.2	B	25	15	887
$C_9H_{21}As$	Tripropylarsine	1.01	B	20	0	1028
$C_9H_{21}BO_3$	Tri-*n*-propyl borate	0.78	B	25	0	1122
		0.7_7	CCl_4	25	0	72
	Triisopropyl borate	0.8_6	CCl_4	25	0	72
$C_9H_{21}N$	Tri-*n*-propylamine	0.58	liq	20	5	337
		0.68	B	20	5	337
		0.74	B	25	0	68
		0.76	B	20	0	336
$C_9H_{21}NO_3$	Triisopropanolamine	3.48±0.05	n.s.	n.s.	-	1720
$C_9H_{21}O_2P$	Heptyl ethyl phosphinate	3.34	B	25	5	61
$C_9H_{21}O_3P$	Dipropyl propyl phosphonite	2.92	CCl_4	20	-	50
	Diisopropyl isopropyl phosphonite	2.91	CCl_4	20	-	50
	Tri-*n*-propyl phosphite	1.99	CCl_4	20	-	50
	Triisopropyl phosphite	1.98	CCl_4	20	-	50

FORMULA	COMPOUND NAME	μ, D	State, or Solv.	t, °C	Method or P_a, % of P_e	Reference
$C_9H_{21}O_3P$ (contd.)	Diisopropyl isopropyl phosphite	1.98	B	n.s.	-	57
$C_9H_{21}O_3PS$	Tripropyl thiophosphate	3.02	B	20	0	59
$C_9H_{21}O_4P$	Tripropyl phosphate	3.09	CCl_4	20	-	50
	Triisopropyl phosphate	2.85	CCl_4	20	-	50
$C_9H_{21}O_4V$	Tri-*n*-propyl vanadate	1.15	B	25	0	271
	Triisopropyl vanadate	1.23	B	25	0	271
$C_9H_{22}IOP$	Ethoxydiethylphosphine - propyliodide complex	5.03	n.s.	n.s.	-	1591
$C_9H_{22}N_2$	2-Amino-5-(diethylamino)pentane	1.36±0.02	B	25	0	942
$C_9H_{22}O_6P_2$	Tetraethyl methylidenediphosphonate	4.23	liq	30	-	952
$C_9H_{23}NO_4Si_4$	2,4,4,6,6,8,8-Heptamethylcyclotetrasiloxane-2-acetonitrile	3.54	liq	10-50	18	1534
$C_9H_{24}B_3N_3$	*B*-Triethyl-*N*-trimethylborazole	0.83	B	25	5	2065
$C_9H_{27}Cl_6O_9P_3Pt_3$	Dimethyl methylphosphate - platinum dichloride complex trimer	4.02	B	n.s.	-	57
$C_9H_{27}NSi_3$	*tris*(Trimethylsilyl)amine	0.51	B	n.s.	7.5	603
$C_{10}Cl_{12}$	Unnamed compound	0	CCl_4	n.s.	-	2175
$C_{10}H_3ClF_{10}$	1-Chloro-3,5-*bis*(pentafluoroethyl)benzene	1.56	B	25	0	1378
$C_{10}H_4ClCrNO_5$	Chromium pentacarbonyl - 4-chloropyridine complex	5.3	B	20	-	1891

FORMULA	COMPOUND NAME	μ, D	State, or Solv.	t, °C	Method or P_a, % of P_e	Reference
$C_{10}H_4Cl_6O_4$	Hydroquinone *bis*(trichloroacetate)	1.5±0.2	B	25	5-30	387
$C_{10}H_5CrNO_5$	Chromium pentacarbonyl - pyridine complex	6.3	B	20	-	1891
$C_{10}H_5MoNO_5$	Molybdenum pentacarbonyl - pyridine complex	6.4	B	20	-	1891
$C_{10}H_5NO_5W$	Tungsten pentacarbonyl - pyridine complex	7.0	B	20	-	1891
$C_{10}H_5N_3O_3$	4-Nitro-1,2-naphthoquinone 1-diazide	4.6_8	B	25	0	36
$C_{10}H_6BrCl$	1-Bromo-8-chloronaphthalene	2.66	B	14	0	153
$C_{10}H_6BrF$	1-Bromo-2-fluoronaphthalene	2.36	B	20	0	1394-5, 1454
$C_{10}H_6BrI$	1-Bromo-2-iodonaphthalene	1.82	B	20	0	1394-5, 1454
$C_{10}H_6BrNO_2$	1-Bromo-5-nitronaphthalene	2.51	B	22	0	1394-5, 1454
$C_{10}H_6Br_2$	1,3-Dibromoazulene	2.52±0.03	B	25	0	34
$C_{10}H_6ClF$	1-Chloro-8-fluoronaphthalene	2.89	B	19	0	153
$C_{10}H_6ClI$	1-Chloro-8-iodonaphthalene	2.57	B	14	0	153
$C_{10}H_6Cl_2$	1,3-Dichloroazulene	2.45±0.04	B	25	0	34
	1,2-Dichloronaphthalene	2.4_6	B	25	-	2087
		2.49	B	25	-	655
	1,3-Dichloronaphthalene	1.80	B	25	-	665
		1.8_6	B	25	-	2087

FORMULA	COMPOUND NAME	μ, D	State, or Solv.	t, °C	Method or P_a, % of P_e	Reference
$C_{10}H_6Cl_2$ (contd.)	1,4-Dichloronaphthalene	0	B	19	0	1394-5, 1454
		0.5_0	B	25	-	655,2087
	1,5-Dichloronaphthalene	0	B	25	-	655,2087
	1,6-Dichloronaphthalene	1.45	B	25	-	655
	1,7-Dichloronaphthalene	2.57	B	25	-	655
		2.60	B	20	0	1394-5
	1,8-Dichloronaphthalene	2.81	B	15	0	153
		2.85	B	25	-	655
		2.91_1	B	25	-	2087
	2,3-Dichloronaphthalene	2.57	B	25	-	655
	2,6-Dichloronaphthalene	0	B	25	-	655,2120
		0.2_0	B	25	-	2085
		0.60	B	14	0	1394-5, 1454
	2,7-Dichloronaphthalene	1.54	B	25	-	655,2087
$C_{10}H_6Cl_2O_4S_2$	1,5-(Disulfochloro)naphthalene	1.56	B	25	0	634
		1.67	B	n.s.	-	632-3
$C_{10}H_6F_2$	1,5-Difluoronaphthalene	0	B	22	0	1394-5, 1454
$C_{10}H_6I_2$	1,3-Diiodoazulene	2.42±0.03	B	25	0	34
$C_{10}H_6N_2$	Heptafulvene-8,8-dinitrile	7.49±0.05	D	25	5	334a
		7.54±0.05	D	25	5	2154

FORMULA	COMPOUND NAME	μ, D	State, or Solv.	t, °C	Method or P_a, % of P_e	Reference
	Phenylfumarodinitrile	1.9	B	30	-	117
		2.74	B	35	0	1714
	Phenylmaleodinitrile	6.1	B	30	-	117
		8.04	B	35	0	1714
$C_{10}H_6N_2O$	1,2-Naphthoquinone 1-diazide	4.0_7	B	25	0	36
	1,2-Naphthoquinone 2-diazide	3.5_6	B	25	0	36
	1,4-Naphthoquinone diazide	4.2_0	B	25	0	36
	1,2-Naphthoquinone dioxime furazan	4.3	B	25	0	1922
		4.88	D	20	0	1322
$C_{10}H_6N_2O_2$	1,2-Naphthoquinone dioxime peroxide	5.37	D	20	0	1322
$C_{10}H_6N_2O_4$	1,2-Dinitronaphthalene	6.48	B	25	-	1755
	1,3-Dinitronaphthalene	4.79	B	25	-	1755
	1,4-Dinitronaphthalene	1.42	B	25	-	1755
	1,5-Dinitronaphthalene	0	B	25	-	1755
		0.6	B	n.s.	0	751
	1,6-Dinitronaphthalene	3.69	B	25	-	1755
	1,7-Dinitronaphthalene	6.45	B	25	-	1755
	1,8-Dinitronaphthalene	7.1	B	n.s.	0	751
		7.27	B	25	-	2026
		7.66	B	25	-	1755
		7.92	B	14	0	153

FORMULA	COMPOUND NAME	μ, D	State, or Solv.	t, °C	Method or P_a, % of P_e	Reference
$C_{10}H_6N_2O_5$	2,4-Dinitro-1-naphthol	4.19	B	25	0	1597
		4.40	D	25	0	1597
$C_{10}H_6N_4O_2$	1-Nitro-2-azidonaphthalene	4.44	B	25	5?	1762
	1-Nitro-4-azidonaphthalene	3.12	B	25	5?	1762
	8-Nitro-2-azidonaphthalene	4.59	B	25	5?	1762
$C_{10}H_6O_2$	1,2-Naphthoquinone	5.65	B	35	0	1851
		5.72	B	25	0	1389-90
	1,4-Naphthoquinone	1.2	B	35	0	1851
		1.21	B	20	-	1301
$C_{10}H_6O_3$	2-Hydroxy-1,4-naphthoquinone	2.78	B	25	0	1596
		2.88	D	25	0	1596
$C_{10}H_6O_4$	1,4-Dihydroxy-5,8-naphthoquinone	0.48	B	20	0	183
		0.63	D	20	0	183
	Furil	3.15	B	25	5	1406
		3.19	B	25	0	363
$C_{10}H_6S_2$	Naphtho[1,8]-1,2-dithiole	1.49	B	20	0	1197
$C_{10}H_7Br$	1-Bromonaphthalene	1.29	liq	25	-	707
		1.49	B	20	0	1394-5
		1.58	B	20	0	1197
		1.59	B	20	0	1478
	2-Bromonaphthalene	1.70	B	19	0	1394-5, 1454
		1.73	B	20	0	1478
$C_{10}H_7BrN_2O_2$	4-Bromo-2-nitro-1-naphthylamine	5.64	D	25	0	2026

FORMULA	COMPOUND NAME	μ, D	State, or Solv.	t, °C	Method or P_a, % of P_e	Reference
	5-Bromo-2-nitro-1-naphthylamine	5.09	D	25	0	2026
	6-Bromo-1-nitro-2-naphthylamine	5.17	D	25	0	2026
$C_{10}H_7BrO_2S$	1-Naphthalenesulfonyl bromide	4.98	B	25	0	634
$C_{10}H_7BrS$	4-Bromo-1-thionaphthalene	1.06±0.02	B	20	0	1198
$C_{10}H_7Cl$	2-Chloroazulene	2.69	B	35	-	1016
	1-Chloronaphthalene	1.33	liq	25	-	707
		1.50	B	19	0	1394-5, 1454
		1.52	B	25	-	655
		1.60	B	20	0	1478
	2-Chloronaphthalene	1.57	B	20	0	1394-5
		1.66	B	25	-	655
		1.74	B	20	0	1478
$C_{10}H_7ClCrO_3$	*p*-Chlorotoluene - chromium tricarbonyl complex	5.19	B	n.s.	-	1563
$C_{10}H_7ClO_2S$	1-Naphthalenesulfonyl chloride	4.80	B	25	0	634
	2-Naphthalenesulfonyl chloride	5.00	B	25	0	634
		5.02	B	n.s.	-	632-3
$C_{10}H_7F$	1-Fluoronaphthalene	1.42	B	20	0	1478
		1.43	B	20	0	1394-5, 1454
	2-Fluoronaphthalene	1.50	B	19	0	1394-5
		1.57	B	20	0	1478

FORMULA	COMPOUND NAME	μ, D	State, or Solv.	t, °C	Method or P_a, % of P_e	Reference
$C_{10}H_7I$	1-Iodonaphthalene	1.44	B	20	0	1478
	2-Iodonaphthalene	1.57	B	20	0	1478
$C_{10}H_7N$	*p*-Tolylpropiolnitrile	4.94	B	25	0	371
	1-Nitroazulene	6.06±0.03	B	25	0	34
	1-Nitronaphthalene	3.31	liq	20-60	MA	1566
		3.65	B	n.s.	0	751
		3.68	B	20	-	1673
		3.87	B	25	-	1755
		3.91	B	20	-	1394-5, 1454
		3.98	B	25	0	300
		4.00	B	20-60	MA	1566
		3.90	D	25	0	300
		4.00	CCl_4	25	0	300
		3.87	Tol	25	0	300
	2-Nitronaphthalene	4.39	B	25	-	1755
		4.4	B	25	-	1148
		4.40	B	25	-	2026
	1-Nitroso-2-naphthol	4.22	B	35	0	1851
		4.33	B	25	0	63
		4.37	B	(20)	5	1215
		4.42	B, Hx	20	0	716
		4.45	D	(20)	5	1215
	1-Nitroso-4-naphthol	4.26	D	(20)	5	1215
	2-Nitroso-1-naphthol	4.24	B	35	0	1851
		4.39	B, Hx	20	0	716
		4.45	B	(20)	5	1215

FORMULA	COMPOUND NAME	μ, D	State, or Solv.	t, °C	Method or P_a, % of P_e	Reference
$C_{10}H_7NO_3$	1-Nitro-2-naphthol	3.68	B	25	0	1597
		3.91	B	(20)	5	1215
		3.79	D	25	0	1597
		4.07	D	(20)	5	1215
	1-Nitro-4-naphthol	5.27	B	(20)	5	1215
		5.88	D	(20)	5	1215
	2-Nitro-1-naphthol	3.79	B	25	0	1597
		3.89	B	(20)	5	1215
		3.62	D	25	0	1597
		4.11	D	(20)	5	1215
	3-Phenyl-5-isoxazolecarboxylic acid	2.88±0.04	D(?)	25	0	1859
	5-Phenyl-3-iscxazolecarboxylic acid	3.28±0.03	D(?)	25	0	1859
$C_{10}H_7N_3$	1-Azidonaphthalene	1.36	B	25	-	1762
	2-Azidonaphthalene	1.60	B	25	-	1762
$C_{10}H_7N_7O_{14}$	1-Nitro-3,5-*bis*(2',2',2'-trinitroethyl)-benzene	3.80	B	n.s.	5	80
$C_{10}H_8$	Azulene	1.0±0.05	B	23.7	-	2104
		1.08±0.08	B	25	0	34
	Naphthalene	0	B	20	0	1478
		0	B	25	0	2120
		0	B	20	0	1537
		0	B	20	-	1673
		0.72	Hx	25	-	2122
		0	cHx	25	-	949
		0	Hp	20	7	209
		0.69	CS_2	25	-	2122
		1.72	NB	30	-	754

FORMULA	COMPOUND NAME	μ, D	State, or Solv.	t, °C	Method or P_a, % of P_e	Reference
$C_{10}H_8$ (contd.)	1-Phenyl-1-buten-3-yne	0.97±0.03	B	25	0	1195
	3-Phenyl-1-buten-3-yne	0.27±0.1	B	20	0	1195
$C_{10}H_8BeF_6O_4$	*Bis*(trifluoroacetylacetonato)beryllium(II)	4.25	B	29.8	50	757
$C_{10}H_8BrN$	2-Bromo-2-aminoazulene	3.34	B	25	-	1148
	2-Bromo-6-aminonaphthalene	3.33	B	(25)	-	1148
	6-Bromo-2-aminonaphthalene	3.72	B	25	-	1016
$C_{10}H_8Br_2O_2$	Methyl α,β-dibromocinnamate (m.p. = 100°C)	2.70	B	21	-	129
	(m.p. = 134°C)	2.00	B	21	-	129
$C_{10}H_8ClN$	4'-Chloro-1-phenylpyrrole	0.03	B	25	0	931
$C_{10}H_8ClNO_2$	*N*-*o*-Chlorophenylsuccinimide	1.78±0.02	D	25	0	62
	N-*m*-Chlorophenylsuccinimide	2.92±0.03	D	20	0	62
	N-*p*-Chlorophenylsuccinimide	3.30±0.03	D	20	0	62
$C_{10}H_8Cl_4Sn$	Naphthalene - stannic chloride complex	1.2	liq	18	-	1986
$C_{10}H_8CoF_6O_4$	*Bis*(trifluoroacetylacetonato)cobalt(II)	3.45	B	29.8	50	757
$C_{10}H_8CrO_3$	Cycloheptatriene - chromium tricarbonyl complex	4.52	B	n.s.	-	1568
	Toluene - chromium tricarbonyl complex	5.20±0.04 5.26	B B	25 n.s.	15 -	520 1568
$C_{10}H_8CrO_4$	Methoxybenzene - chromium tricarbonyl complex	5.26±0.03 5.43	B B	25 n.s.	15 -	520 1568

FORMULA	COMPOUND NAME	μ, D	State, or Solv.	t, °C	Method or P_a, % of P_e	Reference
$C_{10}H_8CuF_6O_4$	*Bis*(trifluoroacetylacetonato)copper(II)	3.06±0.15	B	25	-	1286,1288
		3.19	B	29.8	50	757
$C_{10}H_8F_6O$	4-Ethoxy-1,3-*bis*(trifluoromethyl)benzene	4.16	B	25	0	1378
	5-Ethoxy-1,3-*bis*(trifluoromethyl)benzene	3.22	B	25	0	1378
$C_{10}H_8I_2$	Naphthalene - iodine complex	2.6	cHx	25	-	949
$C_{10}H_8N_2$	2,2'-Bipyridine	<0.68	B	17-67	0	589
		0.91	B	25	0	510
	4,4'-Bipyridine	<0.55	B	17-67	0	589
	2-Tropylpropanedinitrile	4.15±0.05	B	25	5	2154
$C_{10}H_8N_2O$	4-Cinnolinyl methyl ketone	2.53±0.1	B	25	0	1640
$C_{10}H_8N_2O_2$	1,2-Naphthoquinone dioxime	3.83	D	25	0	1322
	1-Nitro-2-naphthylamine	4.45	B	25	0	1599
		4.50	B	40	0	2024-6
		4.67	B	25	-	1148
		4.74	D	25	0	1599
	1-Nitro-4-naphthylamine	6.43	B	25	-	1148
	1-Nitro-5-naphthylamine	5.00	B	25	-	1148
	2-Nitro-1-naphthylamine	4.92	B	40	0	2024-6
		4.96	B	25	-	1148
		4.92	D	25	0	1599
	2-Nitro-6-naphthylamine	5.18	B	25	-	1148
	3-Nitro-1-naphthylamine	5.18	B	25	-	2026

FORMULA	COMPOUND NAME	μ, D	State, or Solv.	t, °C	Method or P_a, % of P_e	Reference
$C_{10}H_8N_2O_2$ (contd.)	3-Nitro-2-naphthylamine	3.82	B	25	0	1599
		4.09	D	25	0	1599
	4-Nitro-1-naphthylamine	7.02	D	25	-	2026
	4-Nitro-2-naphthylamine	4.66	B	25	-	2026
	5-Nitro-1-naphthylamine	5.26	B	25	-	2024-6
	5-Nitro-2-naphthylamine	5.07	B	25	-	2024-6
	6-Nitro-2-naphthylamine	7.15	D	25	-	2026
	8-Nitro-1-naphthylamine	3.15	B	25	-	2026
	8-Nitro-2-naphthylamine	4.50	B	25	-	2026
		3.11	D	25	0	1599
$C_{10}H_8N_2O_4$	*N*-*p*-Nitrophenylsuccinimide	5.77±0.01	D	(20)	0	62
	1,3-*Bis*(2-nitrovinyl)benzene	3.68	B	25	-	2031-2
	1,4-*Bis*(2-nitrovinyl)benzene	1.06	B	25	-	2031-2
$C_{10}H_8N_4$	*cis*-2,2'-Azodipyridine	4.0	B	n.s.	-	266
	trans-2,2'-Azodipyridine	1.77	B	n.s.	-	266
		1.8	B	25	0	1110
	cis-3,3'-Azodipyridine	2.85	B	n.s.	-	266
	trans-3,3'-Azodipyridine	2.40	B	n.s.	-	266
	4,4'-Azodipyridine	2.0	B	n.s.	-	266

FORMULA	COMPOUND NAME	μ, D	State, or Solv.	t, °C	Method or P_a, % of P_e	Reference
$C_{10}H_8O$	6-Furyl-(2)fulvene	1.68	n.s.	n.s.	-	1708
	1-Naphthol	1.0	B	21	0	1537
		1.387	B	20	-	1673
		1.41-1.44	B	20-60	0	719
		1.46	B	25	5	1214
		1.91	D	25	5	1214
		1.8	Ether	20	-	717
		1.9	Ether	20-26	-	723
		1.93	Ether	26	-	719
	2-Naphthol	1.3	B	18	0	1537
		1.412	B	20	-	1673
		1.54	B	25	0	719
		1.56	B	25	5	1214
		2.01	D	25	5	1214
	1-Phenyl-1-butyn-3-one	3.26	B	25	0	587
$C_{10}H_8O_2S$	Naphthalene-1-sulfinic acid	3.91	D	25	0	641
	Naphthalene-2-sulfinic acid	4.01	D	25	0	641
$C_{10}H_8O_4$	Furoin	3.27	B	35	-	1849
$C_{10}H_8S$	Thionaphthalene	1.17	B	20	0	1197
$C_{10}H_8S_2$	1,8-Dithionaphthalene	1.69±0.05	B	20	0	1378
$C_{10}H_9BrO_2$	Methyl β-bromocinnamate	2.65	B	22	-	129
$C_{10}H_9Br_3O_2$	3,5,7-Tribromo-2-hydroxy-4-isopropyl-2,4,6-cycloheptatrien-1-one	3.05±0.13	B	25	0	1021
$C_{10}H_9IO_4$	Diacetoxyphenyliodide	4.9_2	B	25	0	1072

FORMULA	COMPOUND NAME	μ, D	State, or Solv.	t, °C	Method or P_a, % of P_e	Reference
$C_{10}H_9N$	2-Aminoazulene	2.09	B	30	-	1016
	1-Aminonaphthalene	1.45	B	23	0	164
		1.45	B	25	0	1148
		1.50	B	25	-	2026
		1.55	B	20	0	349
		1.78	B	20	-	1673
		1.48	cHx	20	0	349
		1.49	Tol	20	0	349
	2-Aminonaphthalene	1.76	B	23	0	164
		1.78	B	20	0	349
		1.79	B	25	-	2026
		1.86	B	25	0	1148
		1.904	B	20	-	1673
		2.12	D	25	-	2026
		1.75	Tol	20	0	349
	2-Methylquinoline	1.71	liq	20-60	MA	1566
		1.88±0.01	B	10-40	-	1589
		1.95±0.1	B	25	-	1640
	4-Methylquinoline	1.96	liq	20-60	MA	1566
		2.52±0.1	B	25	-	1640
	6-Methylquinoline	1.84	liq	20-60	MA	1566
	8-Methylquinoline	1.58	liq	20-60	MA	1566
	1-Phenylpyrrole	1.32±0.04	B	25	0	931
		1.61	B	25	-	642
	Pyridine cyclopentadienilide	13.3	B	25	-	1748
		17.8	D	25	-	1748

FORMULA	COMPOUND NAME	μ, D	State, or Solv.	t, °C	Method or P_a, % of P_e	Reference
$C_{10}H_9NO$	*p*-Methoxycinnamonitrile	5.14	B	25	5	369
		5.25	u	25	5	369
	3-Methyl-5-phenylisoxazole	3.21	B	25	0	1932
	5-Methyl-3-phenylisoxazole	3.4	B	25	0	1932
	1-Methyl-2-quinoline	3.72±0.10	B	25	0	1634
$C_{10}H_9NO_2$	Phenylsuccinimide	1.68±0.02	D	20	0	62
$C_{10}H_9NO_2S$	1-Sulfonamidonaphthalene	5.16	D	25	0	635
		5.18	D	n.s.	-	632-3
	2-Sulfonamidonaphthalene	5.31	D	25	0	635
		5.40	D	n.s.	-	632-3
$C_{10}H_{10}$	1,2-Dihydronaphthalene	1.4	B	25	-	1539
	1,4-Dihydronaphthalene	1.4	Hx	25	-	1539
	p-Ethylphenylacetylene	1.06	B	n.s.	-	1461
$C_{10}H_{10}BCl_3N_2$	Trichloro*bis*(pyridine)boron	7.50±0.04	B	25	10	1707
$C_{10}H_{10}Be$	*Bis*(cyclopentadienyl)beryllium	2.46±0.06	B	25	15	520
		2.6	D	20	15	1894
		2.24±0.09	cHx	25	15	520
$C_{10}H_{10}Br_2$	*trans*-1,2-Dibromotetralin	1.34±0.03	B	25	0	556
		1.35±0.03	Hp	1-50	0	556
	trans-2,3-Dibromotetralin	1.13±0.03	B	25	0	556
		1.17±0.03	B	40	0	556
		0.95	Hp	1	0	556

u) *trans*-Dichloroethylene.

FORMULA	COMPOUND NAME	μ, D	State, or Solv.	t, °C	Method or P_a, % of P_e	Reference
$C_{10}H_{10}Br_2$ (contd.)	*trans*-2,3-Dibromotetralin (contd.)	1.04	Hp	25	0	556
		1.12	Hp	50	0	556
	6,7-Dibromotetralin	2.84	B	25	-	1773
$C_{10}H_{10}Br_2O_2$	3,7-Dibromo-2-hydroxy-4-isopropyl-2,4,6-cycloheptatrien-1-one	4.16±0.06	B	30	-	1013
		4.30	B	33	0	991, 1018-9
	5,7-Dibromo-2-hydroxy-3-isopropyl-2,4,6-cycloheptatrien-1-one	2.82±0.02	B	25	0	1021
	5,7-Dibromo-2-hydroxy-4-isopropyl-2,4,6-cycloheptatrien-1-one	3.1	B	20	-	1013
		3.14	B	20	-	991,1019
$C_{10}H_{10}Cl_2$	*trans*-1,2-Dichlorotetralin	1.52±0.02	B	25	0	556
		1.40±0.02	Hp	1	0	556
		1.48±0.02	Hp	25	0	556
		1.55±0.02	Hp	50	0	556
	trans-2,3-Dichlorotetralin	1.61±0.02	B	10	0	556
		1.66±0.02	B	25	0	556
		1.72±0.02	B	40	0	556
		1.44±0.02	Hp	1	0	556
		1.55±0.02	Hp	25	0	556
		1.61±0.02	Hp	50	0	556
$C_{10}H_{10}Cl_2N_2Pt$	*Bis*(pyridine) - platinum dichloride complex	0	B	20	-	845
$C_{10}H_{10}Cl_2N_2Zn$	*Bis*(pyridine) - zinc dichloride complex	9.20±0.04	D	25	10	1707
$C_{10}H_{10}Cl_2O$	3,7-Dichloro-2-hydroxy-4-isopropyl-2,4,6-cycloheptatrien-1-one	4.43±0.05	B	25	-	1013
		4.46	B	25	0	1020
$C_{10}H_{10}Cl_2Ti$	Dichloro*bis*(cyclopentadienyl)titanium	6.26±0.39	B	30	-	579
$C_{10}H_{10}Cl_2Zr$	Dichloro*bis*(cyclopentadienyl)zirconium	5.90±0.38	B	30	-	579

FORMULA	COMPOUND NAME	μ, D	State, or Solv.	t, °C	Method or P_a, % of P_c	Reference
$C_{10}H_{10}Co$	*Bis*(cyclopentadienyl)cobalt	0	B	25	15	520
$C_{10}H_{10}Cr$	*Bis*(cyclopentadienyl)chromium	0	B	25	15	520
		0.3	B	20	15	1894
		0.6	D	20	15	1894
$C_{10}H_{10}Fe$	*Bis*(cyclopentadienyl)iron	0	B	25	0	2076
		0.2	D	20	15	1894
		0	CCl_4	25	4.1	74
$C_{10}H_{10}Mg$	*Bis*(cyclopentadienyl)magnesium	0	B	25	15	520
		0.7	B	20	-	1894
$C_{10}H_{10}Mn$	*Bis*(cyclopentadienyl)manganese	1.0	B	20	-	1894
		3.3	D	20	-	1894
		1.0	Hp	20	-	1894
$C_{10}H_{10}N_2$	1,4-Dihydro-4-imino-1-methylquinoline	5.11	B	30	0	45
	2,3-Dimethylquinoxaline	<0.3	B	n.s.	-	2016
$C_{10}H_{10}N_2O$	6-Methoxy-8-aminoquinoline	1.91	B	25	-	1953
$C_{10}H_{10}N_2O_2$	Methyl-*p*-anisyl furazan	4.72	B	25	0	1931
	4-Nitro-2,3-dimethylindole	6.56±0.05	D	25	0	1200
	5-Nitro-2,3-dimethylindole	7.37±0.05	D	25	0	1200
	6-Nitro-2,3-dimethylindole	6.58±0.05	D	25	0	1200
	7-Nitro-2,3-dimethylindole	4.00±0.05	D	25	0	1200
$C_{10}H_{10}N_2O_2S$	4-Aminonaphthalenesulfonamide	6.53±0.02	D	25	0	1550

FORMULA	COMPOUND NAME	μ, D	State, or Solv.	t, °C	Method or P_a, % of P_e	Reference
$C_{10}H_{10}N_2O_3$	Methyl-*p*-anisyl furazan peroxide	4.72	B	25	0	1931
	Methyl-*p*-anisyl furoxan	5.59	B	25	0	1931
$C_{10}H_{10}N_2O_6$	2-Hydroxy-4-isopropyl-3,(?)-dinitro-2,4,6-cycloheptatrien-1-one	4.63	B	20	0	1019
	2-Hydroxy-4-isopropyl-5,7-dinitro-2,4,6-cycloheptatrien-1-one	4.79	B	20	-	1013
$C_{10}H_{10}Ni$	*Bis*(cyclopentadienyl)nickel	0 0	B D	25 20	15 -	520 1894
$C_{10}H_{10}O$	1-Phenyl-1-buten-3-one	3.34 3.31 3.37	B B D	n.s. 25 25	0 5 5	675 468 468
$C_{10}H_{10}O_2$	*p*-Diacetylbenzene	2.7_4	B	25	0	2077
	Methyl cinnamate	1.95	D	25	0	588
$C_{10}H_{10}O_4$	*o*-Diacetoxybenzene	2.3_2	CCl_4	25	5	71
	m-Diacetoxybenzene	2.10±0.10 2.1_4	B CCl_4	25 25	0 5	1633 71
	p-Diacetoxybenzene	2.1_0	CCl_4	25	5	71
	2,4-Diacetylresorcinol	3.14 3.54	B D	25 25	15 15	1220 1220
	4,6-Diacetylresorcinol	4.56 4.93	B D	25 25	15 15	1220 1220
	Dimethyl *o*-phthalate	2.8	B	25	-	2088

FORMULA	COMPOUND NAME	μ, D	State, or Solv.	t, °C	Method or P_a, % of P_e	Reference
	Dimethyl *p*-phthalate	2.2	B	25	-	2088,2112
	Hydroquinone diacetate	2.2	B	25	-	2088, 2111-2, 2114
	Methyl acetylsalicylate	2.68	B	25	-	946
	Cyclic phenylene dioxane ether	1.40	n.s.	n.s.	-	701
$C_{10}H_{10}O_4S$	*cis*-*p*-Tolylsulfonylacrylic acid	5.10±0.05	D	25	0?	1973
	trans-*p*-Tolylsulfonylacrylic acid	4.39±0.05	D	25	0?	1973
$C_{10}H_{10}Os$	*Bis*(cyclopentadienyl)osmium	0	B	25	15	520
$C_{10}H_{10}Pb$	*Bis*(cyclopentadienyl)lead	1.63±0.06 2.16 1.29±0.04 1.49±0.06	B D cHx CS_2	25 20 25 25	10 15 15 10	2076 1894 520 385
$C_{10}H_{10}Ru$	*Bis*(cyclopentadienyl)ruthenium	0	B	25	15	520
$C_{10}H_{10}Sn$	*Bis*(cyclopentadienyl)tin	1.02±0.06 1.2 0.96±0.10 0.9 0.96±0.06	B D cHx Hp CS_2	25 20 25 20 25	10 - 15 - 10	2076 1894 520 1894 385
$C_{10}H_{10}V$	*Bis*(cyclopentadienyl)vanadium	0 0.6	B D	25 20	15 15	520 1894
$C_{10}H_{11}Br$	7-Bromotetralin	2.25	B	25	-	1773
$C_{10}H_{11}BrN_2O_2$	1-Acetyl-1-methyl-2-(*p*-bromobenzoyl)-hydrazine	4.72	B	25	-	1042

FORMULA	COMPOUND NAME	μ, D	State, or Solv.	t, °C	Method or P_a, % of P_e	Reference
$C_{10}H_{11}BrO_2$	Methyl 3-(*p*-bromophenyl)propionate	2.41	B	20	0	129
	3-Bromo-2-hydroxy-4-isopropyl-2,4,6-cycloheptatrien-1-one	4.40 4.41±0.10	B B	25 25	0 -	1020 1013
	7-Bromo-2-hydroxy-4-isopropyl-2,4,6-cycloheptatrien-1-one	4.32	B	33	-	991, 1018-9
$C_{10}H_{11}ClO_2$	3-Chloro-2-hydroxy-4-isopropyl-2,4,6-cycloheptatrien-1-one	4.44±0.02 4.74	B B	25 25	- 0	1013 1020
	5-Chloro-2-hydroxy-4-isopropyl-2,4,6-cycloheptatrien-1-one	2.70 2.79±0.02	B B	25 25	0 -	1020 1013
	7-Chloro-2-hydroxy-4-isopropyl-2,4,6-cycloheptatrien-1-one	4.38 4.76±0.04	B B	25 25	0 -	1020 1013
$C_{10}H_{11}CrNO_5$	Chromium pentacarbonyl - piperidine complex	5.8	B	20	-	1891
$C_{10}H_{11}F_3O$	2-Isopropoxytrifluoromethylbenzene	3.44	B	25	0	1378
	4-Isopropoxytrifluoromethylbenzene	3.53	B	25	0	1378
$C_{10}H_{11}MoNO_5$	Molybdenum pentacarbonyl - piperidine complex	5.9	B	20	-	1891
$C_{10}H_{11}N$	1,2-Dimethylindole	2.55	B	25	-	817
	1,3-Dimethylindole	2.06	B	25	-	817
	2,3-Dimethylindole	2.41 3.05±0.05	B D	25 25	- 0	817 1200
	p-Isopropylbenzonitrile	4.71 2.97 4.60	D $CHCl_3$ cHx	24 24 24	0 0 0	228 228 228

FORMULA	COMPOUND NAME	μ, D	State, or Solv.	t, °C	Method or P_c, % of P_c	Reference
	2,4,6-Trimethylbenzonitrile	4.13	B	25	-	1856
$C_{10}H_{11}NO$	*p*-Aminobenzalacetone	4.94	B	25	0	412
		5.59	D	25	0	412
	2,4,6-Trimethylbenzonitrile *N*-oxide	4.38	B	25	-	1856
		4.41	B	37	-	1856
		4.39	D	25	-	1856
$C_{10}H_{11}NO_2$	Diacetanilide	3.1_5	B	25	0	1072
	β-*p*-Dimethyl-β-nitrostyrene	4.55	B	25	15	2033
	1-Nitrotetralin	4.01	B	25	0	2026
	2-Nitrotetralin	4.85	B	25	0	2026
	2-Nitro-1-(*p*-tolyl)-1-propene	4.59	B	25	-	2031
$C_{10}H_{11}NO_3$	β-Methyl-*p*-methoxy-β-nitrostyrene	5.09	B	25	15	2033
	2-Nitro-1-(*p*-methoxyphenyl)-1-propene	5.13	B	25	-	2031
$C_{10}H_{11}NO_4$	2-Hydroxy-4-isopropyl-(?)-nitro-2,4,6-cycloheptatrien-1-one	6.19	B	25	0	991,1019
	2-Hydroxy-4-isopropyl-7-nitro-2,4,6-cycloheptatrien-1-one	6.22±0.04	B	25	-	1013
	Methyl *N*-(2-methoxycarbonylphenyl)-carbamate	3.32	B	25	0	1217
		3.63	D	25	0	1217
	Methyl *N*-(3-methoxycarbonylphenyl)-carbamate	3.91	B	25	0	1217
		4.18	D	25	0	1217

FORMULA	COMPOUND NAME	μ, D	State, or Solv.	t, °C	Method or P_a, % of P_e	Reference
$C_{10}H_{11}NO_4$ (contd.)	Methyl *N*-(4-methoxycarbonylphenyl)-carbamate	4.84	B	25	0	1217
		5.15	D	25	0	1217
$C_{10}H_{11}NO_5W$	Tungsten pentacarbonyl - piperidine complex	6.6	B	20	-	1891
$C_{10}H_{11}N_3O_4$	6-Amino-5,7-dinitro-1,2,3,4-tetrahydro-naphthalene	3.29	B	25	0	1599
		3.02	D	25	0	1599
$C_{10}H_{11}Re$	*Bis*(cyclopentadienyl)rhenium hydride	1.17±0.24	B	25	15	520
$C_{10}H_{12}$	1-(3-Buten-1-ynyl)cyclohexane	0.87	B	25	0	1750
	o,α-Dimethylstyrene	0.8	B	26	-	165
	p-Ethylstyrene	0.61	B	25	-	1462
	Tetralin	0.59	liq	20-60	MA	1566
		0.60	liq	20-30	5-15	22
		0.49	B	10-40	15	1590
		0.52	B	25	-	1773
		1.67±0.04	B	25	-	1538
$C_{10}H_{12}BrNO_2$	4-Bromo-2,3,5,6-tetramethylnitrobenzene	2.38	B	25	-	181
$C_{10}H_{12}Cl_2$	*o*-Dichlorotetramethylbenzene	2.96	B	25	0	1814
	p-Dichlorotetramethylbenzene	~0.25	B	25	0	1814
$C_{10}H_{12}N_2$	Pyrrole - 4-methylpyridine complex	3.75±0.1	CCl_4	n.s.	-	596
$C_{10}H_{12}N_2O_2$	1-Acetyl-1-methyl-2-benzoylhydrazine	5.56	B	25	-	1042
	p-Dimethylamino-β-nitrostyrene	7.66	B	25	-	2031-2
		7.67	B	25	0	412

FORMULA	COMPOUND NAME	μ, D	State, or Solv.	t, °C	Method or P_a, % of P_e	Reference
$C_{10}H_{12}N_2O_3$	1-Acetyl-1-methyl-2-(*o*-hydroxybenzoyl)-hydrazine	6.52	B	25	-	1042
	5,5-Diallylbarbituric acid	1.14±0.01	D	35	0	1847
$C_{10}H_{12}N_2O_4$	1-*t*-Butyl-2,4-dinitrobenzene	4.53	B	25	0	1162
	o-Dinitrotetramethylbenzene	6.91	B	25	0	1814
	p-Dinitrotetramethylbenzene	0.60	B	25	-	181
$C_{10}H_{12}N_2S_2$	*N*-(Dimethylaminomethyl)benzothiazoline-2-thione	4.39	B	25	-	637
$C_{10}H_{12}O$	Butyrophenone	2.83±0.03	B	(25)	-	301
	2,4-Dimethylacetophenone	2.95±0.02	B	25	0	127
	2,5-Dimethylacetophenone	2.85±0.02	B	25	0	127
	3,4-Dimethylacetophenone	3.36±0.02	B	25	0	127
	β- or ω-Ethoxystyrene	1.69	B	18	-	129
	o-Methoxy-α-methylstyrene	1.49	B	27	-	165
	m-Methoxy-α-methylstyrene	1.66	B	24	-	165
	p-Methoxy-α-methylstyrene	1.40	B	26	-	165
	2-Methylpropiophenone	2.89±0.01	B	(25)	-	301
	1,2,3,4-Tetrahydro-1-naphthol	1.74±0.05 1.88±0.02	B B	60 20	- -	771 771
	5,6,7,8-Tetrahydro-1-naphthol	1.69±0.06	B	20	-	771

FORMULA	COMPOUND NAME	μ, D	State, or Solv.	t, °C	Method or P_a, % of P_e	Reference
$C_{10}H_{12}O$ (contd.)	5,6,7,8-Tetrahydro-2-naphthol	1.76	B	40	0	773
		1.86±0.01	B	20	-	771
		1.93	B	20	0	773
		2.01±0.05	B	60	-	771
		1.71±0.03	CCl_4	20,40	0	773
		1.67±0.03	cHx	20,40	0	773
$C_{10}H_{12}O_2$	Ethyl phenylacetate	1.82	n.s.	30	0	1501
	Eugenol	2.46	B	25	0	1152
		2.51	D	25	0	1152
	2-Hydroxy-3-isopropyl-2,4,6-cyclohepta-trien-1-one	3.37±0.03	B	25	0	1021
	2-Hydroxy-4-isopropyl-2,4,6-cyclohepta-trien-1-one	3.93±0.14	B	33	-	1013
		4.04	B	33	0	991, 1018-9
$C_{10}H_{12}O_2S$	*trans*-Methylcinnamylsulfone	4.95	B	30	0	88
$C_{10}H_{12}O_3$	Ethyl anisate	2.48±0.10	B	25	0	1633
	Isopropyl salicylate	2.78	B	n.s.	5	334
$C_{10}H_{12}O_5$	1,2-Dihydroxypropyl salicylate	3.56	B	25	-	946
$C_{10}H_{13}AlBr_3NO_2$	Aluminum bromide - *p*-isobutylnitrobenzene complex	9.12	CS_2	20	0	560
$C_{10}H_{13}Br$	1-Bromo-4-*t*-butylbenzene	1.92±0.10	B	25	0	1633
	2-Bromo-*p*-cymene	1.78	B	25	0	1080
	3-Bromo-*p*-cymene	1.73	B	25	0	1080

FORMULA	COMPOUND NAME	μ, D	State, or Solv.	t, °C	Method or P_a, % of P_e	Reference
	1-Bromo-1,2,5,6-tetramethylbenzene	1.56	B	25	-	181
$C_{10}H_{13}Cl$	2-Chloro-*p*-cymene	1.65 2.1^x	B CCl_4	25 25	0 -	1030 1542
	3-Chloro-*p*-cymene	1.67	B	25	0	1030
$C_{10}H_{13}N$	2-Cyanobicyclo[4.3.0]-5-nonene	3.70	n.s.	n.s.	-	696
	4-Dimethylaminostyrene	2.17	B	25	0	475
$C_{10}H_{13}NO$	*o*-Dimethylaminoacetophenone	3.56	B	25	0	1213
	m-Dimethylaminoacetophenone	3.61	B	25	0	1213
	p-Dimethylaminoacetophenone	5.09	B	25	0	1213
	4-Oxa-9-cyanotricyclo[$5.3.0.0^{3,5}$]decane	3.55	n.s.	n.s.	-	696
$C_{10}H_{13}NO_2$	*o*-Benzbetaine	13.2	B,y	25	-	438
	m-Benzbetaine	12.8	B,y	25	-	438
	o-*t*-Butylnitrobenzene	3.53±0.10	B	25	0	1633
	p-*t*-Butylnitrobenzene	4.65 4.61	B cHx	25 24	0 0	1080 228
	p-Isobutylnitrobenzene	4.26	CS_2	20	0	560
	3,6-Methano-1,2-cyclohexanedicarboximide	2.24±0.03	D	30	0	1059
	Methyl *p*-dimethylaminobenzoate	3.90	B	25	-	2028
	Methyl *N*,*N*-dimethylanthranilate	2.07	B	25	-	438

x) μ^2 value. y) Ethanol.

FORMULA	COMPOUND NAME	μ, D	State, or Solv.	t, °C	Method or P_a, % of P_e	Reference
$C_{10}H_{13}NO_2$	Phenacetin	5.56	liq	165	-	992
		5.75	liq	135	-	992
		5.32	D	25,35	-	992
		5.64	D	150	-	992
	1,2,4,5-Tetramethyl-3-nitrobenzene	3.42	B	25	-	181
		3.62±0.02	B	25	0	930,932
$C_{10}H_{13}NO_3$	2,3,5,6-Tetramethyl-4-nitrophenol	4.11	B	25	-	805a
$C_{10}H_{13}NO_4$	3,4-Dicarbethoxypyrrole	4.08	B	25	5	818
$C_{10}H_{14}$	*n*-Butylbenzene	0.36	liq	20-30	5-15	22
		0.37	B	25	5	359
	Isobutylbenzene	0.31	liq	20-30	5-15	22
	sec-Butylbenzene	0.37	liq	20-30	5-15	22
		0.39	B	25	5	359
	t-Butylbenzene	0.70	gas	185-200	5	84
		0.36	liq	20-30	5-15	22
		0.40	B	25	5	359
		0.41±0.10	B	25	0	1633
		0.52±0.01	B	20	0	1186,1668
		0.53	B	25	0	1080
	1,2-Diethylbenzene	0.59	liq	20-30	5-15	22
	1,3-Diethylbenzene	0.36	liq	20-30	5-15	22
	1,4-Diethylbenzene	0	liq	20-30	5-15	22
		0.2_4	B	25	0	2080
	6,6-Diethylfulvene	1.44±0.03	B	23.7	0	2104

FORMULA	COMPOUND NAME	μ, D	State, or Solv.	t, °C	Method or P_a, % of P_e	Reference
	1,3-Dimethyl-5-ethylbenzene	0	liq	20-30	5-15	22
	1-Methyl-4-isopropylbenzene	0	liq	20-30	5-15	22
		0	B, CCl_4	0-45	0	1080
$C_{10}H_{14}BNO_2$	*B*-Phenyl-*diptych*-boroxazolidine	8.5±0.19	D	25	-	553
$C_{10}H_{14}BeO_4$	*Bis*(acetylacetono)beryllium	0	gas	184-254	-	331
		0.79	B	20	25	1440
		1.17	B	25	-	511
		0.80	D	20	25	1440
		0	CCl_4	20-45	0	1798
		0.71	CCl_4	20	25	1440
		1.15	CCl_4	25	-	511
		1.15^z	CCl_4	25	0	63
		1.14	Hx	25	-	511
		1.13	a	25	-	511
		0.4	$CHCl_3$	25	-	511
		1.13	CS_2	25	-	511
$C_{10}H_{14}BrN$	1-Amino-4-bromotetramethylbenzene	2.78	B	25	-	181
$C_{10}H_{14}Br_2$	Dihydrodicyclopentadiene-*cis*-1,2-dibromide	3.17	B	n.s.	5	414
$C_{10}H_{14}Cl_2O$	α,α'-Dichlorocamphor	4.22±0.04	B	30	0	1010
		4.38±0.06	D	30	0	1010
		4.39±0.01	Hx	30	0	1010
$C_{10}H_{14}CrO_4$	*Bis*(acetylacetono)chromium	1.44	B	25	-	511
$C_{10}H_{14}CuO_4$	*Bis*(acetylacetono)copper	1.17^z	D	25	0	63
		0.31	$CHCl_3$	25	-	511
$C_{10}H_{14}FeO_4$	*Bis*(acetylacetono)iron	0	gas	229	-	331
$C_{10}H_{14}NO_5PS$	*O,O*-Diethyl-*O*-*p*-nitrophenylphosphorothioate	4.98	B	25	-	906

z) "Apparent" μ value caused by atomic polarization. a) Decalin.

FORMULA	COMPOUND NAME	μ, D	State, or Solv.	t, °C	Method or P_a, % of P_e	Reference
$C_{10}H_{14}NO_6P$	Diethyl(*p*-nitrophenyl)phosphate	5.30	B	25	-	906
$C_{10}H_{14}N_2$	3-(2-Piperidyl)-pyridine	2.26	B	20	-	610
		2.46	B	40	-	610
		2.71	B	60	-	610
		2.43	b	40	-	610
	α-Pyridine-β-*N*-methylpyrrolidine	2.64	B	20	10	1741
$C_{10}H_{14}N_2O$	*p*-Nitroso-*N*,*N*-diethylaniline	7.20	B	n.s.	-	704
		7.23	B	25	0	1100
		6.92	CCl_4	25	0	1100
		6.95	CCl_4	45	0	1100
$C_{10}H_{14}N_2O_2$	1-Amino-4-nitrotetramethylbenzene	5.02	B	25	-	181
$C_{10}H_{14}N_2O_3$	5-Allyl-5-isopropylbarbituric acid	1.13±0.01	D	35	0	1847
$C_{10}H_{14}O$	*p*-*t*-Butylphenol	1.62	B	25	0	598
		1.68±0.02	B	20	0	1186,1668
		1.98	D	25	-	451
		1.48	cHx	25	-	451
	6,8(9)-*p*-Menthadien-2-one	2.8	B	15	-	2144
	2-Methyl-5-isopropylphenol	1.55	B	25	0	1080
	3-Methyl-6-isopropylphenol	1.51±0.01	B	20	-	771
		1.55	B	25	0	1080
		1.57±0.01	B	60	-	771
		1.61	B	18	-	962
	2,3,5,6-Tetramethylphenol	1.69	B	25	-	805a
	2,4,6-Trimethylanisole	1.26	B	20	RT	922
		1.27±0.02	B	20	0	1193

b) 2,2,4-Trimethylpentane.

FORMULA	COMPOUND NAME	μ, D	State, or Solv.	t, °C	Method or P_a, % of P_e	Reference
$C_{10}H_{14}O_2$	Camphorquinone	4.4_6	B	25	0	1085
		4.81±0.02	B	30	0	1009
	d-Camphorquinone	4.78	B	35	5	1396
	(-)-Camphorquinone	4.49	B	25	5	432
	o-Diethoxybenzene	1.3_8	B	25	0	2080
		1.7	B	25	-	2088
	m-Diethoxybenzene	1.7_2	B	25	0	2080
	p-Diethoxybenzene	1.7	B	25	-	2111-2, 2114
		1.74	B	20	0	2097
		1.74	B	18	-	671,674
		1.77	B	40	-	671,674
		1.78	B	25	0	2080
		1.79	B	18	0	673
		1.79	B	10	0	1336
		1.80	B	60	-	671,674
		1.82	B	30	0	1336
		1.84	B	50	0	1336
		1.94	B	40	0	2097
		2.02	B	60	0	2097
		1.7_4	CCl_4	25	0	2080
		1.7_6	cHx	25	0	2080
		1.89-1.93	Hp	10-60	0	1336
		1.7_6	CS_2	25	0	2080
$C_{10}H_{14}O_2S$	Methyl 2,4,6-trimethylphenyl sulfone	4.79±0.03	B	20	0	1201
$C_{10}H_{14}O_2Si$	*p*-(Trimethylsilyl)benzcic acid	1.22	B	25	0	1845
$C_{10}H_{14}O_3$	Phenol - dioxane complex	1.82	B	25	5	913

FORMULA	COMPOUND NAME	μ, D	State, or Solv.	t, °C	Method or P_a, % of P_e	Reference
$C_{10}H_{14}O_4$	Dimethyl *cis*-4-cyclohexene-1,2-dicarboxylate	2.38	B	25	0	1751
	Dimethyl *trans*-4-cyclohexene-1,2-dicarboxylate	2.10	B	25	0	1751
$C_{10}H_{14}O_4Th$	*Bis*(acetylacetono)thorium	0 1.18 1.61 2.00	gas B B B	238 20 30 40	- - - -	331 1525 1525 1525
$C_{10}H_{14}O_4Zn$	*Bis*(acetylacetono)zinc	1.35	B	25	-	511
$C_{10}H_{14}O_4Zr$	*Bis*(acetylacetono)zirconium	1.85 2.10 2.16	B B B	20 30 40	- - -	1525 1525 1525
$C_{10}H_{14}O_6U$	*Bis*(acetylacetono)uranium dioxide	5.12	B	n.s.	-	618
$C_{10}H_{14}S$	Methyl 2,4,6-trimethylphenyl sulfide	1.52±0.02	B	20	0	1193
$C_{10}H_{14}Se$	Methyl 2,4,6-trimethylphenyl selenide	1.40±0.02	B	20	0	310
$C_{10}H_{15}Br$	Bromoadamantane	2.27	CCl_4	20	RT	2036
$C_{10}H_{15}BrO$	α-Bromocamphor	3.7 4.06±0.02 4.17±0.01	B,CCl_4 B,Hx D	n.s. 30 30	- 0 0	594 1010 1010
	endo-3-Bromo*endo*camphor	2.49	B	25	0	972
	exo-3-Bromo*exo*camphor	2.68	B	25	0	972
$C_{10}H_{15}BrO_2$	3-Bromocamphorenic acid	1.58	B	35	5	1400
$C_{10}H_{15}BrSi$	*p*-(Trimethylsilylmethylene)bromobenzene	2.14	B	25	10	885

FORMULA	COMPOUND NAME	μ, D	State, or Solv.	t, °C	Method or P_a, % of P_e	Reference
$C_{10}H_{15}Cl$	Chloroadamantane	2.18	CCl_4	20	RT	2036
$C_{10}H_{15}ClO$	α-Chlorocamphor	2.4	"pure"	n.s.	-	594
		3.6	B, CCl_4	n.s.	-	594
		4.11±0.03	B,D,Hx	30	0	1010
	α'-Chlorocamphor	4.14±0.02	B	30	0	1010
		4.10±0.01	D	30	0	1010
		4.11±0.01	Hx	30	0	1010
$C_{10}H_{15}MnO_4P$	Triethylphosphine tetracarbonyl manganese	2.10±0.13	B	25	20	712
$C_{10}H_{15}N$	*N*,*N*-Diethylaniline	1.40	liq	20	5	337
		1.66	B	18	0	962
		1.80±0.02	B	20	0	1183
		1.81	B	25	-	101
		1.81	B	20	5	337
		1.84	B	20	0	336
		1.82	Hx	20	5	337
		1.85	Hx	20	0	336
	N,*N*-Dimethyl-2,4-xylidine	0.71±0.01	B	25	0	524-5
	N,*N*-Dimethyl-2,6-xylidene	0.94±0.01	B	25	0	524-5
	2-*n*-Pentylpyridine	1.90	B	25	0	360
	4-*n*-Pentylpyridine	2.72_1	B	25	0	360
	4-(3-Pentyl)pyridine	2.74_8	B	25	0	360
	2,3,4,6-Tetramethylaniline	1.22	B	25	0	1795
		1.26	D	25	0	1795

FORMULA	COMPOUND NAME	μ, D	State, or Solv.	t, °C	Method or P_a, % of P_e	Reference
$C_{10}H_{15}N$ (contd.)	2,3,5,6-Tetramethylaniline	1.40	B	25	-	181
		1.45	B	25	0	1795
		1.57	D	25	0	1795
$C_{10}H_{15}NO_2$	Aniline - dioxane complex	1.88	B	n.s.	-	509
	2,5-Diethoxyaniline	0.74	B	32	0	30
		1.21[c]	CCl_4	32	0	30
	3-Oxime camphorquinone	3.40	B	35	5	1396
$C_{10}H_{15}NO_2Si$	Trimethyl(*o*-nitrobenzyl)silane	3.62	B	25	15	887
	Trimethyl(*p*-nitrobenzyl)silane	4.86	B	25	15	887
		4.94	B	25	0	1845
$C_{10}H_{15}O_4P$	Diethylphenylphosphate	3.34	B	25	-	906
$C_{10}H_{15}P$	Diethylphenylphosphine	1.35	cHx	25	-	920
$C_{10}H_{16}$	2-*t*-Butyl-1-hexen-3-yne	0.64	n.s.	n.s.	0	1490
	9-Decen-7-yne	0.67	B	20	0	1492
	Limonene	<0.5	B	15	-	2144
	d-Limonene	1.57	B	25	0	1912
	dl- or *d*-Limonene	0.6_3	B	25	0	1084
	4-Methylene-2,6-dimethylhepta-2,5-diene	1.08±0.03	B	30	5	143
	d-Pinene	2.69	B	30	0	1912
	dl-Pinene	0.36	B	30	5	390
		0.6_0	B	25	0	1084

c) Compound formation observed.

FORMULA	COMPOUND NAME	μ, D	State, or Solv.	t, °C	Method or P_a, % of P_e	Reference
$C_{10}H_{16}Cl_2$	2,4-Dichlorocamphene	2.5±0.13	CCl_4	n.s.	-	1030
	2,6-Dichlorocamphene	4.0±0.2	CCl_4	n.s.	-	1030
$C_{10}H_{16}N_2$	Decanedinitrile	4.50	B	25	0	1943
		4.51-4.99	B	25	-	1978-9
		4.63-5.04	B	75	-	1978-9
	N,N,N',N'-Tetramethyl-*p*-phenylenediamine	1.24	B	25	0	2079
		1.29±0.1	B	25	0	267
	N,N,N',N'-Tetramethylphenylenediamine[d]	1.12±0.02	B	25	0	475
$C_{10}H_{16}N_2O_2S$	2,3,5,6-Tetramethyl sulfanilamide	5.3±0.8	D	35	0?	2092
$C_{10}H_{16}N_2O_3$	5-Butyl-5-ethylbarbituric acid	1.11±0.01	D	35	0	1847
$C_{10}H_{16}N_4$	*N,N'*-*Bis*cyanoethylpiperazine	4.23±0.02	B	20	-	568
$C_{10}H_{16}O$	Camphor[e]	2.2	"pure"	n.s.	-	594
		2.26-2.82[f]	solid	50-25	-	826[g]
		2.55-2.68[f]	B	25-50	-	826
		2.97±0.03	B	22	15	2137
		2.98±0.03	B	22	15	416
		3.0	B,CCl_4	n.s.	-	594
		3.06	B	25	5	432
	d-Camphor	2.90±0.03	B	20	-	1672
		3.08	B,Hx	20	0	716
		3.10	B	25	0	1084
	l-Camphor	2.905±0.03	B	20	-	1672
	dl-Camphor	2.91±0.03	B	20	-	1672
		3.10	B	25	0	1084

d) Position of substituents not given. e) Optical activity not specified.
f) Calculated by Jatkar's equation. g) From data of Yager and Morgan, J. Am. Chem. Soc. 57, 2071 (1935).

FORMULA	COMPOUND NAME	μ, D	State, or Solv.	t, °C	Method or P_a, % of P_e	Reference
$C_{10}H_{16}O$ (contd.)	Fenchone	2.92±0.03	B	22	15	2137
		2.94±0.03	n.s.	22	15	416
	2-Methyl-6-isopropylidenecyclohexanone	3.02±0.03	B	25	0	127
	(–)-Piperitone	3.73	B	25	0	1084
	(±)-Piperitone	3.7_8	B	25	0	1084
$C_{10}H_{16}OSi$	Dimethylphenylethoxysilane	1.30	B	30	0	545
$C_{10}H_{16}O_2$	Ascaridole	2.85±0.05	B	25	0	1639
	3-Ethoxy-5,5-dimethyl-2-cyclohexen-1-one	3.79	B	25	-	786
	2-Hydroxy-*p*-menth-1-en-3-one	3.0_4	B	25	0	1085
	2,2,5,5-Tetramethyl-1,3-cyclohexanedione	3.23	B	22	-	1994
$C_{10}H_{16}O_4$	Camphoric acid	2.70	D	35	5	1400
	1,4-Diacetoxycyclohexane	1.47	B	18	0	681a
	cis-1,4-Diacetoxy-2,3-dimethyl-2-butene	2.58	B	25	0	1919
	trans-1,4-Diacetoxy-2,3-dimethyl-2-butene	1.90	B	25	0	1919
	Diethyl cyclobutane-1,1-dicarboxylate	2.24	B	25	0	495
		2.24	B	n.s.	-	1499
	Dimethyl *cis*-1,2-cyclohexanedicarboxylate	2.41	B	25	0	1751
	Dimethyl *trans*-1,2-cyclohexanedicarboxylate	2.14	B	25	0	1751
	Dimethyl *cis*-1,3-cyclohexanedicarboxylate	2.32	n.s.	20	-	1782

FORMULA	COMPOUND NAME	μ, D	State, or Solv.	t, °C	Method or P_a, % of P_e	Reference
	Dimethyl *trans*-1,3-cyclohexanedicarboxylate	2.11	n.s.	20	-	1782
$C_{10}H_{16}Si$	Trimethylbenzylsilane	0.55	B	25	15	887
	Trimethyl-*o*-tolylsilane	0.57 0.61	B B	25 30	5 0	1845 544
	Trimethyl-*m*-tolylsilane	0.46 0.50	B B	25 30	5 0	1845 544
	Trimethyl-*p*-tolylsilane	0.46 0.48	B B	25 30	5 0	1845 544
$C_{10}H_{17}BrO$	*cis*-2-Bromo-4-*t*-butylcyclohexanone	4.272	B	25	0	15
	trans-2-Bromo-4-*t*-butylcyclohexanone	3.196	B	25	0	15
$C_{10}H_{17}Cl$	2-Chlorocamphane	1.4 2.0 2.1±0.11	"pure" B, CCl_4 CCl_4	n.s. n.s. n.s.	- - -	594 594 1030
	2-Chloroisocamphane	2.1±0.11	CCl_4	n.s.	-	1030
	1-Chloro-3,5-dimethyl-2,6-octadiene	2.18	B	20	0	1491
	8-Chloro-4-methyl-2,6-nonadiene	2.20	B	20	0	1491
$C_{10}H_{17}ClO$	*cis*-2-Chloro-4-*t*-butylcyclohexanone	4.29±0.02	B	25	0	13
	trans-2-Chloro-4-*t*-butylcyclohexanone	3.17±0.02	B	25	0	13
$C_{10}H_{17}Cl_3O_2Sn$	*Bis*(ethanol) - phenyltrichlorotin complex	6.47	B	n.s.	-	592
$C_{10}H_{17}NO$	*d*-Camphor oxime	1.16	B	35	5	1396
	o-Cresol - trimethylamine complex	3.02±0.02	B	25	-	789

FORMULA	COMPOUND NAME	μ, D	State, or Solv.	t, °C	Method or P_a, % of P_e	Reference
$C_{10}H_{17}NO$ (contd.)	*p*-Cresol - trimethylamine complex	3.13±0.06	B	25	-	789
	2-Methyl-4-oxodecahydroquinoline (m.p. = 38-39° C)	2.63	B	25	-	1752
	(m.p. = 41-42° C)	2.65	B	25	-	1752
	(m.p. = 62-63° C)	2.66	B	25	-	1752
$C_{10}H_{17}NSi$	Trimethyl-*o*-aminobenzylsilane	1.73	B	25	0	1845
	Trimethyl-*p*-aminobenzylsilane	1.33	B	25	0	1845
$C_{10}H_{18}$	*cis*-Cyclodecene	0.44±0.03	Hp	25	0	10
	trans-Cyclodecene	0	Hp	25	0	10
	Decahydronaphthalene	0 0 2.10	liq liq NB	20-175 25-140 30	- - 0	1052 1907 754
	cis-Decahydronaphthalene	0 ~0	B, CCl_4 B	25 20-100	- -	1538 1727
	trans-Decahydronaphthalene	0 ~0	B B	25 20-100	- -	1538 1727
	5-Decyne	0	Hp	25	11	2094
$C_{10}H_{18}N_2O_4$	3-Methyl-3-nitrosobutan-2-one	3.29	B	20	-	109
	3-Methyl-4-nitrosobutan-2-one	3.32 3.38	B D	25 25	15 15	857 857
$C_{10}H_{18}NiO_2S_4$	Nickel *O,O*-*bis*(isobutylxanthate)	2.3_6	B	22	0	1243

FORMULA	COMPOUND NAME	μ, D	State, or Solv.	t, °C	Method or P_a, % of P_e	Reference
$C_{10}H_{18}O$	Borneol[h]	1.2	"pure"	n.s.	-	594
		1.2	B, CCl_4	n.s.	-	594
		1.7	Ether	20	-	717,723
	d-Borneol	1.6±0.01	B	25	-	1672
	l-Borneol	1.57±0.02	B	22	15	417
		1.59±0.01	B	25	-	1672
		1.66±0.03	B	7	15	665
	dl-Borneol	1.59	B	25	-	1672
		1.66	B, Hx	18	0	716
		1.85	B	35	5	1400
	Cyclodecanone	2.75±0.03	B	20	0	630
	Menthone	2.79	B	15	-	2136
		2.83±0.03	B	22	15	416
		2.83	B	n.s.	-	677
		2.85±0.03	B	22	15	2137
	3-Methyl-1-nonyn-3-ol	1.88±0.01	B	20	0	1184
	1,3,3-Trimethyl-2-oxabicyclo[2.2.2]octane	1.58	gas	200-300	-	539
		1.57	B	25	-	539
$C_{10}H_{18}O_2$	ε-Ethyl-β,δ-dimethylhexanoic acid, ε-lactone	4.36	B	25	-	1259
	Ethyl 3-methylcyclohexanecarboxylate	1.22	n.s.	n.s.	-	1361
	n-Propyl cyclohexanecarboxylate	1.18	n.s.	n.s.	-	1361
$C_{10}H_{18}O_2S_4Zn$	Zinc *O,O*-*bis*-*n*-butylxanthate	2.5_8	B	30	0	1243

h) Optical activity not given.

FORMULA	COMPOUND NAME	μ, D	State, or Solv.	t, °C	Method or P_a, % of P_c	Reference
$C_{10}H_{18}O_4$	Diethyl adipate	2.40±0.03	liq	20-70	0	907
		2.25	B, CCl_4	25-45	-	831
		2.42	B	25	0	1832
		2.42	B	25,50	-	1679
		2.44	B	50	0	1832
		2.44	B	25	-	1763
		2.48	B	20-75	0	907
		2.50	D	20-75	0	907
		2.09	cHx	30	RT	1844
	Diethyl isopropyl malonate	2.42	B	n.s.	-	1499
	Sebacic acid	2.40±0.02	D	25	0	120
		2.49	D	35	-	1581
$C_{10}H_{18}O_6$	*dl*-Diethyl 1,2-dimethoxysuccinate	3.77	B	25	-	2139
	meso-Diethyl 1,2-dimethoxysuccinate	3.37	B	25	-	2139
$C_{10}H_{19}NO$	(±)-1-Epilupinine	2.30	B	25	-	1582
	(±)-3-Epilupinine	2.42	B	25	-	1582
	(±)-1-Lupinine	3.07	B	25	-	1582
	(±)-3-Lupinine	2.83	B	25	-	1582
	trans-2-Oxodecamethylenimine	3.78	B	25	0	783
	2,2,3-Trimethyl-1-oxa-4-azaspiro[4.4]nonane	1.51±0.03	B	30	0	151
$C_{10}H_{20}$	*t*-Butylcyclohexane	0	gas	140-185	-	84
	cis-5-Decene	0.24	liq	20	NRA	23
	trans-5-Decene	0	liq	25	NRA	23

FORMULA	COMPOUND NAME	μ, D	State, or Solv.	t, °C	Method or P_a, % of P_c	Reference
	5-Methyl-4-nonene	0.48	liq	20	0	1180,1189
	2,4,6-Trimethyl-3-heptene	0.70	liq	20	0	1180,1189
$C_{10}H_{20}Br_2$	1,10-Dibromodecane	2.71-2.77	gas	-75 to -45	0	1813
		2.54±0.02	liq	45-95	0	907
		2.55	B	25	0	1943
		2.568±0.004	B	25	0	692
		2.57	B	25,50	0	1835
		2.59	B	20	0	907
		2.609±0.004	B	64	0	692
		2.66	D	20	0	907
		2.60	CCl_4	20	0	907
		2.46	Hx	20	0	907
$C_{10}H_{20}Br_4O_4Zr$	*Bis*(propyl acetate) - zirconium bromide complex	4.00	B	20	10	923
$C_{10}H_{20}Cl_2$	1,10-Dichlorodecane	2.60	B	25	0	1943
$C_{10}H_{20}Cl_4O_4Zr$	Zirconium tetrachloride - di(ethyl propionate) complex	5.23	B	20	15	1442
	Zirconium tetrachloride - di(propyl acetate) complex	6.73	B	20	15	1442
$C_{10}H_{20}Cl_8O_4Zr_2$	Zirconium tetrachloride - ethyl propionate complex dimer	3.56	B	20	0?	924
	Zirconium tetrachloride - propyl acetate complex dimer	3.88	B	20	0?	924
$C_{10}H_{20}CrN_4O_2S_4$	*Bis*(diethyldithiocarbamate)dinitrosyl-chromium	5.2_3	B	23	15	1244

FORMULA	COMPOUND NAME	μ, D	State, or Solv.	t, °C	Method or P_a, % of P_e	Reference
$C_{10}H_{20}I_4O_4Zr$	Zirconium tetraiodide - di(ethyl propionate) complex	3.07	B	(0)	-	1446
	Zirconium tetraiodide - di(propyl acetate) complex	3.96	B	20	-	1446,1444
$C_{10}H_{20}N_2PbS_4$	Lead *bis*diethyldithiocarbamate	2.88	B	25	0	639
$C_{10}H_{20}N_2NiS_4$	Nickel *bis*diethyldithiocarbamate	1.94	B	25	0	639
$C_{10}H_{20}N_2S_4Zn$	Zinc *bis*diethyldithiocarbamate	1.89	B	25	0	639
$C_{10}H_{20}N_4O_4$	3,7-Diethyl-3,7-dinitro-1,5-diazacyclo-öctane	5.4	B	n.s.	-	943
$C_{10}H_{20}O$	*d*-Menthol	1.55±0.02	B	20	0	1672
	l-Menthol	1.55±0.02	B	20	0	1672
		1.58±0.03	B	22	15	417
		1.63±0.03	B	7	15	665
		1.75	B	25	0	1084
	dl-Menthol	1.56±0.02	B	20	-	1672
		1.74	B	25	0	1084
	1,2,4,5-Tetramethylcyclohexanol	1.96	B	18	0	681a
	Octyl vinyl ether	1.2	B	25	-	1763
$C_{10}H_{20}O_2$	2-Acetoxy-*n*-octane	1.9_5	B	25	0	1366
	Isoamyl valerate	1.80[i]	B	28	RT	1401
	Ethyl caprylate	1.68	liq	25	-	1495
		1.75	liq	60	-	1495

i) 3.26 cm frequency.

FORMULA	COMPOUND NAME	μ, D	State, or Solv.	t, °C	Method or P_a, % of P_e	Reference
	5-Ethyl-5-hydroxy-4-methyl-3-heptanone	3.30±0.05	B	20	0	1184
	2-Methyl-2-amyl-1,3-dioxane	1.92	B	25	0	1459
	Pentyl valerate	1.50	B, CCl_4	25-45	-	831
$C_{10}H_{20}O_2S$	Propyl 3-(*n*-butylthio)propionate	2.28	B	25	-	356
$C_{10}H_{20}O_3$	Percapric acid	2.27±0.04	B	30	-	1616
$C_{10}H_{20}O_4S$	Propyl 3-(*n*-butylsulfonyl)propionate	4.27	B	25	-	356
$C_{10}H_{21}Br$	1-Bromodecane	1.90	liq	25	-	707
		2.08	liq	20	-	657
		1.92	B	n.s.	-	699
$C_{10}H_{21}NO$	2-(1-Ethylpentyl)oxazolidine	1.53±0.03	B	30	0	151
$C_{10}H_{22}$	*n*-Decane	~0	liq	-30 to 170	-	418
$C_{10}H_{22}N_2$	1,2,2,4,5,5-Hexamethylpiperazine	0.63	B	25	0	1253
$C_{10}H_{22}O$	*n*-Amyl ether	1.04	liq	30-40	-	1497
	Isoamyl ether	0.98	liq	-17 to 20	-	463
		1.24	B	25	0	1128
	1-Decanol	1.62	B	25	0	1239
		1.71±0.02	B	20	0	1184
		1.7_9	CCl_4	25	5	1108
		1.85	Hx	25	-	1150
	2-Methyl-4-propyl-4-hexanol	1.68±0.02	B	20	0	1184
	4-Propylheptan-4-ol	1.66	B	18	0	681a

FORMULA	COMPOUND NAME	μ, D	State, or Solv.	t, °C	Method or P_a, % of P_e	Reference
$C_{10}H_{22}O$ (contd.)	2,4,6-Trimethyl-4-heptanol	1.71±0.02	B	20	0	1184
$C_{10}H_{22}O_2$	1,10-Decanediol	2.38±0.03	D	15	-	2060
		2.54	D	25,50	0	1833
		2.55	D	20	0	907
$C_{10}H_{22}O_2S$	Diisopentylsulfone	4.57	B	20	0	636
$C_{10}H_{22}O_3$	Tripropyl orthoformate	1.94	B	25	5	60
$C_{10}H_{22}O_6$	Pentaethylene glycol	6.05[j]	liq	20	-	933
		3.42	D	25	10	1999
$C_{10}H_{22}S$	*n*-Amyl sulfide	1.59	B	25,50	0	2059
$C_{10}H_{23}N$	Decylamine	1.28	B	30	-	1579
		1.41	B	30	5	1578
$C_{10}H_{23}O_2P$	Octyl ethylphosphinate	3.30	B	25	-	61
$C_{10}H_{24}Cl_4O_2Sn$	Tin tetrachloride - di(isopentanol) complex	6.10	B	20	0	1439
$C_{10}H_{24}O_6P_2$	Tetraethyl ethylidene-1,2-diphosphonate	4.20	liq	30	-	952
$C_{10}H_{25}NO$	Triethylamine - *n*-butanol complex	2.34	B	25	0	195
	Triethylamine - *t*-butanol complex	2.25	B	25	0	195
$C_{10}H_{25}O_5Sb$	Ethyl antimonate	2.31	B	25	5	60
$C_{10}H_{25}O_5U$	Uranium pentethoxide	1.88	?	0	-	1937
$C_{10}H_{26}O_4Si_3$	1,1,1,2,3,3,3-Heptamethyl-2-acetoxymethyltrisiloxane	2.71	liq	20	RT	37
$C_{10}H_{28}P_2Pt$	*cis*-*Bis*(trimethylphosphine)diethylplatinum	5.8	B	25	15	294

j) $g^{\frac{1}{2}}\mu$ value.

FORMULA	COMPOUND NAME	μ, D	State, or Solv.	t, °C	Method or P_a, % of P_e	Reference
$C_{10}Mn_2O_{10}$	Dimanganese decacarbonyl	0	B, cHx	25	20	1420
$C_{10}O_{10}Re_2$	Dirhenium decacarbonyl	0	B	25	30	1420
$C_{11}H_3Co_3O_9$	Nonacarbonylhydridotricobalt - acetylene complex	0^{k} 1.60^{kk}	B B	27 27	MR 20	1257 1257
$C_{11}H_5MnO_5$	Phenylmanganese pentacarbonyl	0.86±0.07	B	25	0	116
$C_{11}H_5O_5Re$	Phenylrhenium pentacarbonyl	1.25±0.06	B	25	0	116
$C_{11}H_6ClN$	8-Chloro-1-naphthonitrile	5.74	B	19	0	153
$C_{11}H_7Br_2N_3$	3,5-Dibromo-2-phenylazopyridine	1.8	B	n.s.	-	266
$C_{11}H_7CrNO_5$	Chromium pentacarbonyl - aniline complex	4.7	B	20	-	1891
	Chromium pentacarbonyl - 2-methylpyridine complex	6.3	B	20	-	1891
$C_{11}H_7N$	2-Cyanoazulene	5.68	B	25	-	1016
	α-Naphthylcyanide	4.16	B	25	-	2028
	β-Naphthylcyanide	4.36	B	25	-	2028
$C_{11}H_7NO$	1-Naphthylisocyanate	2.32 2.32	B B	20 20	0 MA	344 870
	2-Naphthylisocyanate	2.36	B	20	0	344
$C_{11}H_7NO_5W$	Tungsten pentacarbonyl - aniline complex	5.3	B	20	-	1891
$C_{11}H_7N_3$	*cis*-α-Naphthalenediazocyanide	3.2	B	27	-	1087
	trans-α-Naphthalenediazocyanide	5.6	B	27	-	1087

k) P_E from MR_D. kk) P_E estimated from other carbonyls.

FORMULA	COMPOUND NAME	μ, D	State, or Solv.	t, °C	Method or P_a, % of P_e	Reference
$C_{11}H_7N_3$ (contd.)	*cis*-β-Naphthalenediazocyanide	4.0	B	25	-	1087
	trans-β-Naphthalenediazocyanide	6.9	B	25	-	1087
$C_{11}H_8BrN_3$	5-Bromo-2-phenylazopyridine	2.6	B	n.s.	-	266
$C_{11}H_8ClN_3$	2-(*o*-Chlorophenyl)azopyridine	2.1	B	n.s.	-	266
	3-(*o*-Chlorophenol)azopyridine	2.4	B	n.s.	-	266
$C_{11}H_8CrO_5$	Methyl benzoate - chromium tricarbonyl complex	4.47±0.03	B	25	15	520
$C_{11}H_8N_2$	Methylphenylfumarodinitrile	1.2	B	30	-	117
	Methylphenylmaleodinitrile	6.2	B	30	-	117
	(1,2)-Naphthimidazole	4.15	D	25	0	1451
$C_{11}H_8N_2O_2$	2-Methyl-(?)-5*H*-pyrazolo[a][3,1]benzoxazin-5-one, α-lactone	1.60	B	n.s.	0	2038
	2-Methyl-(?)-5*H*-pyrazolo[a][3,1]benzoxazin-5-one, β-lactone	2.66	B	n.s.	0	2038
$C_{11}H_8N_4O_7$	Pyridine - picric acid complex	8.70 10.43	D D	35 15	0 -	1558 1987
$C_{11}H_8O$	2,3-Benzo-2,4,6-cycloheptatrien-1-one	3.61±0.06	B	25	0	563
	4,5-Benzo-2,4,6-cycloheptatrien-1-one	4.70±0.06	B	25	0	563
$C_{11}H_8OS$	2-Phenyl-4*H*-pyran-4-thione	4.79	B	20	-	1661
$C_{11}H_8O_2$	Cyclopentadiene-*p*-benzoquinone	1.40	B, CCl_4	25,45	0	1065
	1-Formyl-2-naphthol	3.37 3.40	B D	25 25	5 5	1214 1214

FORMULA	COMPOUND NAME	μ, D	State, or Solv.	t, °C	Method or P_a, % of P_e	Reference
	1-Naphthoic acid	1.87	D	35	5	1398
	2-Naphthoic acid	1.95	D	35	5	1398
	2-Phenyl-4*H*-pyran-4-one	4.28	B	20	-	1661
$C_{11}H_9BrS$	4-Bromo-1-naphthyl methyl sulfide	1.68	B	20	0	1197
$C_{11}H_9N$	2-Phenylpyridine	1.77	B	25	0	1842-3
		1.88	B	25	0	1729
		1.92	CCl_4	25	0	1729
	3-Phenylpyridine	2.45	B	25	0	1842-3
	4-Phenylpyridine	2.50	B	25	0	1842-3
		2.55	B	25	0	1729-30
		2.64	CCl_4	25	0	1729-30
$C_{11}H_9NO$	1-Nitroso-2-methoxynaphthalene	4.11	B	(20)	5	1215
		4.19	D	(20)	5	1215
	1-Nitroso-4-methoxynaphthalene	3.56	B	(20)	5	1215
		3.72	D	(20)	5	1215
	2-Nitroso-1-methoxynaphthalene	3.57	B	(20)	5	1215
		3.65	D	(20)	5	1215
	4-Phenoxypyridine	2.56	B	25	-	1763
	4-Phenylpyridine 1-oxide	4.61±0.01	B	25	0	1730
$C_{11}H_9NO_3$	3-Methyl-5-phenyl-4-isoxazolecarboxylic acid	3.26±0.03	D	25	0	1857
	5-Methyl-3-phenyl-4-isoxazolecarboxylic acid	3.35±0.05	D	25	0	1857

FORMULA	COMPOUND NAME	μ, D	State, or Solv.	t, °C	Method or P_a, % of P_e	Reference
$C_{11}H_9NO_3$ (contd.)	1-Nitro-2-methoxynaphthalene	4.66	B	(20)	5	1215
		4.86	D	(20)	5	1215
	1-Nitro-4-methoxynaphthalene	5.19	B	(20)	5	1215
		5.27	D	(20)	5	1215
	2-Nitro-1-methoxynaphthalene	5.04	B	(20)	5	1215
		5.17	D	(20)	5	1215
$C_{11}H_9NO_4$	Ethyl *p*-nitrophenylpropiolate	3.57	B	22	-	129
$C_{11}H_9N_3$	2-Phenylazopyridine	2.2	B	n.s.	-	266
	3-Phenylazopyridine	2.3	B	n.s.	-	266
	4-Phenylazopyridine	2.7	B	n.s.	-	266
$C_{11}H_{10}$	1-Methylnaphthalene	0.51	liq	20-60	MA	1566
		0.23	B	20	-	1673
		0.28	n.s.	n.s.	-	1210
	2-Methylnaphthalene	0.42	liq	20-60	MA	1566
		0.44	n.s.	n.s.	-	1210
$C_{11}H_{10}CrO_3$	*o*-Xylene - chromium tricarbonyl complex	5.41±0.05	B	25	15	520
	m-Xylene - chromium tricarbonyl complex	5.37±0.02	B	25	15	520
	p-Xylene - chromium tricarbonyl complex	5.39±0.05	B	25	15	520
		5.52±0.05	B	n.s.	-	1568
$C_{11}H_{10}F_6O$	4-Propoxy-1,3-*bis*(α,α,α-trifluorotoluene)	4.09	B	25	0	1378
	4-Isopropoxy-1,3-*bis*(α,α,α-trifluorotoluene)	3.96	B	25	0	1378

FORMULA	COMPOUND NAME	μ, D	State, or Solv.	t, °C	Method or P_a, % of P_e	Reference
$C_{11}H_{10}N_2OS$	2-(Benzylthio)-4(3*H*)-pyrimidone	2.86	D	25	0	1716
$C_{11}H_{10}N_2O_2$	Ethyl 4-cinnolinecarboxylate	3.62±0.1	B	25	0	1640
$C_{11}H_{10}N_2O_2S$	*N*-(2-Pyridyl)benzenesulfonamide	4.95±0.02	D	25	0	1550
$C_{11}H_{10}N_2O_4$	*N*-*p*-Nitrobenzylsuccinimide	5.01±0.05	D	25	0	62
$C_{11}H_{10}N_2S$	5,6-Dihydro-3-phenyl-imidazo[2,1-b]-thiazole	4.2±0.2	B	n.s.	-	2129
$C_{11}H_{10}N_4O_5$	1,5-Dimethyl-4-nitro-2-(*p*-nitrophenyl)-3-pyrazolone	4.6	B	30	0	227
$C_{11}H_{10}O$	1-Methoxynaphthalene	0.97	liq	20-60	RT	1618
		1.22	B	20	RT	922
		1.26	B	25	5	1214
		1.28±0.03	B	25	0	477
		1.40	D	25	5	1214
	2-Methoxynaphthalene	0.95	liq	80	RT	1618
		1.11	B	20	RT	922
		1.29	B	25	5	1214
		1.45	D	25	5	1214
	Methyl-2-naphthyl ether	1.12[m]	liq	75-100	-	824
		1.12±0.02[m]	B	25-45	-	824
	1-Methylolnaphthalene	1.70	B	10	-	771
		1.74	B	40	-	771
		1.75	B	20	-	771
		1.78±0.02	B	60	-	771
		1.82±0.04	D	20,60	-	771
$C_{11}H_{10}O_2$	Ethyl phenylpropiolate	2.21	B	21	-	129

m) Calculated from Jatkar's equation.

FORMULA	COMPOUND NAME	μ, D	State, or Solv.	t, °C	Method or P_a, % of P_e	Reference
$C_{11}H_{10}S$	Methyl 1-naphthyl sulfide	1.27	B	20	0	1197
$C_{11}H_{10}Zn$	Cyclopentadienylphenylzinc	1.00 2.9 0.4	B D Hp	20 20 20	15 15 15	1894 1894 1894
$C_{11}H_{11}BrN_2O$	4-Bromo-1,5-dimethyl-3-pyrazolone	6.0_1	B	25	0	227
$C_{11}H_{11}BrO_2$	Ethyl α-bromocinnamate (m.p. = 126° C)	2.83	B	19	-	129
	(m.p. = 131° C)	2.27	B	22	-	129
$C_{11}H_{11}Cr$	Benzene cyclopentadienylchromium	0	B	25	15	520
$C_{11}H_{11}CrNO_3$	Phenyldimethylamine - chromium tricarbonyl complex	6.30±0.05	B	n.s.	-	1568
$C_{11}H_{11}N$	2-Methyl-1-phenylpyrrole	1.94	B	25	-	642
	4'-Methyl-1-phenylpyrrole	1.79±0.04	B	25	0	931
$C_{11}H_{11}NO_2$	*N*-Benzylsuccinimide	1.75±0.02	D	25	0	62
	N-*o*-Tolylusccinimide	1.93±0.03	D	25	0	62
$C_{11}H_{11}NO_3$	*N*-*p*-Methoxyphenylsuccinimide	2.00±0.02	D	25	0	62
$C_{11}H_{11}NO_4$	Ethyl *p*-nitrocinnamate	3.53	B	21	-	129
$C_{11}H_{11}NO_6$	α,α-Diacetoxy-*p*-nitrotoluene	4.1_6	B	25	0	1072
$C_{11}H_{11}N_3O_2S$	2-Sulfanilamidopyridine	6.8±0.2 7.2	D D	25 25	- -	860 1550
$C_{11}H_{11}N_3S$	6-Amino-5,6-dihydro-3-phenylimidazo[2,1-b]-thiazole	2.1±0.2	B	n.s.	-	2129

FORMULA	COMPOUND NAME	μ, D	State, or Solv.	t, °C	Method or P_a, % of P_e	Reference
$C_{11}H_{12}$	*p*-Isopropylphenylacetylene	1.13	B	n.s.	-	1461
$C_{11}H_{12}I_2N_2O$	Antipyrine - iodine complex	7.45[n]	B	25	-	895
$C_{11}H_{12}N_2$	*cis*-*p*-Dimethylaminocinnamonitrile	5.68	B	25	0	412
	trans-*p*-Dimethylaminocinnamonitrile	7.03	B	25	0	412
	3-(*p*-Dimethylaminophenyl) acrylonitrile	6.90	B	20	FL	375
$C_{11}H_{12}N_2O$	3-Methoxy-5-methyl-1-phenylpyrazole	1.82	n.s.	27	15	2037
	1,5-Dimethyl-2-phenyl-3-pyrazolone	5.52 5.5	B B	25 30	15 0	859 227
	2,5-Dimethyl-1-phenylpyrazol-3-one	5.73	n.s.	24	15	2037
$C_{11}H_{12}N_2OS_2$	2-(Morpholinothio) benzothiazole	1.73	B	25	-	637
$C_{11}H_{12}N_2S$	1,5-Dimethyl-2-phenyl-3-pyrazolthione	7.38	B	25	15	859
$C_{11}H_{12}N_4O_2S$	2-Sulfanilamido-4-methylpyrimidine	7.2±0.2	D	25	-	860
$C_{11}H_{12}O_2$	Ethyl *cis*-cinnamate	1.79	B	15	-	129
	Ethyl *trans*-cinnamate	2.10 1.86	liq B	35-40 20	- -	1497 129
	Ethyl cinnamate[p]	1.73[q]	B	28	RT	1401
$C_{11}H_{12}O_4$	α,α-Diacetoxytoluene	2.9_8	B	25	-	1072
	3,5-Diacetyl-2,6-dimethyl-γ-pyrone	4.0_9	B	25	-	1070
$C_{11}H_{12}O_5$	Ethyl *trans*-cinnamate ozonide	2.02	B	23	0	217

n) Lower limit. p) Configuration not given. q) 3.26 cm frequency.

FORMULA	COMPOUND NAME	μ, D	State, or Solv.	t, °C	Method or P_a, % of P_e	Reference
$C_{11}H_{13}BrN_2O_2$	4'-Bromo-2 -nitro-1-phenylpiperidine	4.33	B	25	0	1068
$C_{11}H_{13}N$	*p-t*-Butylbenzonitrile	4.74	D	24	0	228
		4.64	cHx	24	0	228
		2.97	$CHCl_3$	24	0	228
	2-Methyl-3-ethylindole	2.43	B	25	-	817
	3-Methyl-2-ethylindole	2.38	B	25	-	817
$C_{11}H_{13}NO$	*p*-Dimethylaminocinnamaldehyde	5.4	B	21	-	2090
		6.43	B	25	0	412
	2,3,5,6-Tetramethylbenzonitrile *N*-oxide	4.33	B	25	-	1856
		4.34	D	23	-	1856
$C_{11}H_{13}NO_2$	Ethyl *p*-aminocinnamate	3.70	B	25	0	412
		4.23	D	25	0	412
$C_{11}H_{13}NO_4$	Ethyl β-(*p*-nitrophenyl)propionate	4.62	B	20	-	129
$C_{11}H_{13}N_3O$	1-(*p*-Methoxyphenyl)-3,3-dimethyltriazene	2.3_0	B	25	0	1083
$C_{11}H_{14}$	2,4,5-Trimethylstyrene	0.45±0.1	B	25	-	476
	2,4,6-Trimethylstyrene	0.42±0.1	B	25	-	476
$C_{11}H_{14}N_2O_2$	1-Acetyl-1-methyl-2-phenylacetylhydrazine	4.08	B	25	-	1042
	1,2-Diacetyl-1-methyl-2-phenylhydrazine	3.75	B	25	-	1042
	p-Dimethylamino-β-methyl-β-nitrostyrene	7.23	B	25	15	2033
	2-Nitro-1-*p*-dimethylaminophenyl-1-propene	7.22	B	25	-	2031
	1-*p*-Nitrophenylpiperidine	6.8_0	B	25	5	69

FORMULA	COMPOUND NAME	μ, D	State, or Solv.	t, °C	Method or P_a, % of P_e	Reference
$C_{11}H_{14}N_4O_7$	Piperidine - picric acid complex	8.59	D	35	0	1558
$C_{11}H_{14}O$	4(3-Buten-1-ynyl)-3,6-dihydro-2,2-dimethyl-2*H*-pyran	1.30	B	25	-	1750
	2,2-Dimethylpropiophenone	2.70±0.01	B	(25)	-	301
	3-Methyl-3-phenyl-2-butanone	2.63	B	30	0	20
	2,4,5-Trimethylacetophenone	3.22±0.02	B	25	0	127
	2,4,6-Trimethylacetophenone	2.7_9 2.81±0.02	B B	25 25	0 0	1071 127
$C_{11}H_{14}O_2$	Ethyl α-phenylpropionate	1.818	B	30	0	1501
	Ethyl β-phenylpropionate	1.77	B	21	-	129
	2-Methoxy-4-isopropyl-2,4,6-cyclohepta-trien-1-one	4.11±0.02	B	25	0	1022
$C_{11}H_{15}Cl$	Pentamethylchlorobenzene	1.87	B	25	0	1814
$C_{11}H_{15}N$	1-Phenylpiperidine	1.7_4	B	25	5	69
$C_{11}H_{15}NO$	*N*-Benzyl ether of butyraldoxime	2.962	B	23	-	1413
	4-(Cyclohexyloxy)pyridine	3.49	B	25	0	1763
	9-(Dimethylamino)-2,4,6,8-nonatetraenal	8.50±0.04	B	25	0	798
$C_{11}H_{15}NO_4$	5-Methyl-2,3-dicarbethoxypyrrole	2.84	B	25	5	818
	5-Methyl-3,4-dicarbethoxypyrrole	4.31	B	25	5	818
$C_{11}H_{16}$	3-*t*-Butyltoluene	0.32	liq	20-30	~10	22

FORMULA	COMPOUND NAME	μ, D	State, or Solv.	t, °C	Method or P_a, % of P_e	Reference
$C_{11}H_{16}$ (contd.)	4-*t*-Butyltoluene	0.39 0 0.35	gas liq B	205 20-30 25	- 10 0	84 22 1080
	3,5-Diethyltoluene	0	liq	20-30	~10	22
	Methylisobutylfulvene	1.43±0.03	B	30	5	143
$C_{11}H_{16}BNO_2$	*N*-Methyl-*B*-phenyl-*diptych*-boroxazolidine	9.1±0.21	D	25	-	553
$C_{11}H_{16}N_2$	Indole - trimethylamine complex	2.99±0.18	B	25	-	789
$C_{11}H_{16}N_2O$	*N*,*N*-Diethyl-*N*′-phenylurea	3.2±0.1	D	17	0	166
$C_{11}H_{16}N_2OS_2$	5-(*N*-Piperidylmethylene)-3-ethylrhodanine	6.28	B	25	-	1766
$C_{11}H_{16}N_2O_3$	5-Allyl-5-isobutylbarbituric acid	1.19±0.01	D	35	0	1847
$C_{11}H_{16}O$	2-*t*-Butyl-4-methylphenol	1.31±0.02	B	20	0	1186,1668
$C_{11}H_{16}O_3$	Benzyl alcohol - dioxane complex	1.76	B	25	-	913
$C_{11}H_{16}O_6$	2,3,4,6-Tetramethyl-α-methyl-D-glucose	2.29	$CHCl_3$	35	0	1560
$C_{11}H_{17}N$	*N*,*N*-Dimethylmesidine	1.04	B	25	-	805a
	N-Ethyl-*N*-*n*-propylaniline	1.43 1.86	liq B	20 20	5 0	337 336
		1.84	Hx	20	0	336
	Pentamethylaniline	1.11	B	25	-	181
	Phenylethyl-*n*-propyl amine	1.80 1.79	B Hx	20 20	5 5	337 337

FORMULA	COMPOUND NAME	μ, D	State, or Solv.	t, °C	Method or P_a, % of P_e	Reference
$C_{11}H_{17}NO_2Si$	Trimethyl(*o*-nitrophenethyl)silane	3.63	B	25	15	887
	Trimethyl (*p*-nitrophenethyl)silane	4.53	B	25	15	887
$C_{11}H_{17}O_3P$	Diethyl *p*-tolylphosphonate	3.53	liq	30	-	950
$C_{11}H_{18}O_2$	*l*-Bornyl formate	2.06±0.01	n.s.	22	15	417
$C_{11}H_{18}O_4$	Diethyl cyclopentane-1,1-dicarboxylate	2.16 2.16	B B	25 n.s.	0 -	495 1499
	Dimethyl *cis*-4-methylcyclohexane-1,2-dicarboxylate (b.p. 5 mm = 105°C)	2.39	B	25	0	1751
	(b.p. 5 mm = 110°C)	2.39	B	25	0	1751
	Dimethyl *trans*-4-methylcyclohexane-1,2-dicarboxylate (b.p. 4 mm = 110°C)	2.23	B	25	0	1751
	(b.p. 5 mm = 111°C)	2.26	B	25	0	1751
$C_{11}H_{18}Si$	3,3-Dimethyl-1-(1'-cyclohexene)-3-silica-1-butyne	0.89	B	20	-	1493
	Trimethyl(2-phenethyl)silane	0	B	25	15	887
$C_{11}H_{19}NO$	*N,N*-Dimethyl-*p*-(trimethylsilyl)aniline	1.83	B	25	0	1845
	11-Methyl-11-azabicyclo[5.3.1]hendecan-4-one	4.87±0.1	B	25	0	1113
$C_{11}H_{20}BrN$	11-Bromoundecanenitrile	3.95 3.98	B B	25 75	- -	1979 1979
$C_{11}H_{20}N_2O_3$	*N*-Cyclohexyltetrahydro-5-methyl-5-nitro-1,3-oxazine	4.45±0.1	B	20	-	631

FORMULA	COMPOUND NAME	μ, D	State, or Solv.	t, °C	Method or P_a, % of P_e	Reference
$C_{11}H_{20}O_2$	Menthyl formate	2.08±0.03	n.s.	22	15	417
	n-Propyl ester of 3-methylcyclohexane carboxylic acid	1.11	n.s.	n.s.	-	1361
$C_{11}H_{20}O_4$	*dl-erythro*-3,4-Diacetoxyheptane	2.70	liq	25	-	2130
	dl-threo-3,4-Diacetoxyheptane	2.20	liq	25	-	2130
	Diethyl dimethylmalonate	2.12	B	n.s.	-	1499
	Erythritol diisopropylidene ether	2.28±0.07	B	22	15	1437
$C_{11}H_{21}N_2S_2$	1-Piperidine carbodithioic acid, piperidine salt	2.50	B	25	0	639
$C_{11}H_{22}N_2O_2$	Methyl-*n*-octylglyoxime	1.49	D	20	0	1319
$C_{11}H_{22}O$	2-Undecanone	2.71	B	15	10	2141
	6-Undecanone	2.68	CCl_4	25	-	67
$C_{11}H_{22}O_2$	2,2-Diisobutyl-1,3-dioxolane	1.27±0.02	B	30	0	151
$C_{11}H_{22}O_6$	α-1,2,3,4,6-Pentamethyl-D-glucose	2.08 2.11 1.94 1.99	B B D D	20 40 20 40	0 0 0 0	1561-2 1561-2 1561-2 1561-2
	β-1,2,3,4,6-Pentamethyl-D-glucose	2.05 2.06 1.99 2.04	B B D D	20 40 20 40	0 0 0 0	1561-2 1561-2 1561-2 1561-2
$C_{11}H_{23}NO$	2-(1-Isobutyl-3-methylbutylideneamino)-ethanol	2.80±0.05	B	30	0	151

FORMULA	COMPOUND NAME	μ, D	State, or Solv.	t, °C	Method or P_a, % of P_e	Reference
	2-Isobutyl-2,4,5,5-tetramethyloxazolidine	1.63±0.03	B	30	0	151
$C_{11}H_{23}NS_2$	*S*-Ethyl dibutyldithiocarbamate	3.28	B	25	0	639
$C_{11}H_{24}$	Undecane	~0	liq	-11 to 190	-	418
$C_{11}H_{24}NO_4P$	Dipropyl ester of diethylcarbamoylphosphonic acid	3.05	B	25	5	58
	Diisopropyl ester of diethylcarbamoylphosphonic acid	2.91	B	25	5	58
$C_{11}H_{24}O$	4-Propyl-4-octanol	1.74±0.02	B	20	0	1184
	1-Undecanol	1.67 1.74±0.02	B B	25 20	0 0	1234 1184
$C_{11}H_{24}O_2$	2,2-Diethoxyheptane	0.91	B	25	0	1457
	Bis(2,2-dimethylpropoxy)methane	0.89	B	25	0	1457
$C_{11}H_{24}O_4S_2$	1,3-*Bis*(isobutylsulfonyl)propane	4.39	B	20	0	636
$C_{11}H_{24}O_6S_3$	1,3-*Bis*(ethylsulfonyl)-2-(ethylsulfonylmethyl)-2-methylpropane	4.14	B	20	0	636
$C_{11}H_{25}O_2P$	Nonyl ethylphosphinate	3.37	B	25	5	61
$C_{11}H_{26}O_6P_2$	Tetraethyl propylene-1,3-diphosphonate	4.38	liq	30	-	952
$C_{11}H_{26}Si$	Triethylisoamylsilane	0	CCl_4	27	15	1247
$C_{12}Cl_{10}O_2$	*Bis*(pentachlorophenyl)peroxide	1.31	B	25	30	1745
$C_{12}H_4Cl_6S_2$	*Bis*(2,4,5-trichlorophenyl)disulfide	1.15	B	25	-	2030

FORMULA	COMPOUND NAME	μ, D	State, or Solv.	t, °C	Method or P_a, % of P_e	Reference
$C_{12}H_5CrNO_5$	Chromium pentacarbonyl - benzonitrile complex	6.9	B	20	-	1891
$C_{12}H_5MnO_6$	Benzoylmanganese pentacarbonyl	2.09±0.04	B	25	20	116
$C_{12}H_5O_6Re$	Benzoylrhenium pentacarbonyl	2.54±0.07	B	25	0	116
$C_{12}H_6Br_2N_2O_4$	4,4'-Dibromo-2,2'-dinitrobiphenyl	4.88±0.02 4.96	B B	20 25	0 0	1182,1188 1104
$C_{12}H_6Cl_2N_2O$	3,8-Dichlorobenzo[c]cinnoline 5-oxide	5.0_6	B	25	0	250
$C_{12}H_6Cl_2N_2O_4$	4,4'-Dichloro-2,2'-dinitrobiphenyl	4.87±0.02 4.94	B B	20 25	0 0	1182,1188 1104
$C_{12}H_6Cl_2O_2$	2,7-Dichloro-5,10-dioxaanthracene	0.62	B	25	3	729,733, 2009
$C_{12}H_6Cl_2S_2$	2,7-Dichloro-5,10-dithiaanthracene	1.38	B	25	-	167
$C_{12}H_6F_2N_2O_4$	4,4'-Difluoro-2,2'-dinitrobiphenyl	4.87±0.02	B	20	0	1182,1188
$C_{12}H_6N_4O_8$	2,2',4,4'-Tetranitrobiphenyl	4.41	B	25	0	1104
$C_{12}H_6O_2$	1,2-Acenaphthenedione	6.13	B	25	0	251
$C_{12}H_7BrN_2O$	3-Bromobenzo[c]cinnoline 5-oxide	5.3_2	B	25	0	250
$C_{12}H_7Br_3N_2O_2S$	(Phenylsulfonyl)(2,4,6-tribromophenyl)-diimide	4.5_9	B	30	-	541
$C_{12}H_7ClN_2$	2-Chlorophenazine	2.46	B	25	0	1409
$C_{12}H_7ClN_2O$	2-Chlorophenazine *N*-oxide	2.69	B	25	0	1409
$C_{12}H_7ClN_2O_2$	2-Chlorophenazine *N*,*N*-dioxide	3.16	B	25	0	1409

FORMULA	COMPOUND NAME	μ, D	State, or Solv.	t, °C	Method or P_a, % of P_e	Reference
$C_{12}H_7MnO_5$	Benzylmanganese pentacarbonyl	0.83±0.05	B	25	0	116
	Tolylmanganese pentacarbonyl	0.87±0.03	B	25	0	116
$C_{12}H_7N_3O_2$	2-Nitrophenazine	6.61	B	25	0	1409
$C_{12}H_7N_3O_3$	3-Nitrobenzo[c]cinnoline 5-oxide	5.6_4	B	25	0	250
	2-Nitrophenazine *N*-oxide	5.85	B	25	0	1409
$C_{12}H_7N_3O_6$	2,4,6-Trinitrobiphenyl	1.41	B	25	~10	1162
$C_{12}H_8BrNO_2S$	2-Bromo-4'-nitrodiphenyl sulfide	4.40±0.01	B	20	0	1657
	3-Bromo-4'-nitrodiphenyl sulfide	3.84±0.01	B	20	0	1657
	4-Bromo-4'-nitrodiphenyl sulfide	3.42±0.01	B	20	0	1656
$C_{12}H_8BrNO_3$	2-Bromo-4'-nitrodiphenyl ether	4.72±0.01	B	20	0	1657
	3-Bromo-4'-nitrodiphenyl ether	3.85±0.01	B	20	0	1657
	4-Bromo-4'-nitrodiphenyl ether	3.19±0.01	B	20	0	1656
$C_{12}H_8Br_2$	*trans*-1,2-Dibromoacenaphthene	1.63±0.12	B	25	0	1643
	2,2'-Dibromobiphenyl	1.90	B	25	0	1161
	4,4'-Dibromobiphenyl	0	B	25	0	203
$C_{12}H_8Br_2Hg$	*Bis*(*p*-bromophenyl)mercury	0.93	r	142	0	653
$C_{12}H_8Br_2N_2$	4,4'-Dibromoazobenzene	<1	CS_2	21	-	139

r) Decalin.

FORMULA	COMPOUND NAME	μ, D	State, or Solv.	t, °C	Method or P_a, % of P_e	Reference
$C_{12}H_8Br_2O$	*p*-Bromophenyl ether	0.87±0.05	gas	235	5	332
		0.53	B	25	0	654
		0.59	B	18	0	160
		0.60-0.62	B	25,50	0	1836
		0.65	Hx	25	0	1906
		0.62	Hp	25,50	0	1836
$C_{12}H_8Br_2S$	*p*-Bromophenyl sulfide	0.65-0.67	B	25,50	0	1836
$C_{12}H_8Br_2S_2$	*p*-Bromophenyl disulfide	0.75	B	20	0	636
$C_{12}H_8Br_2Se_2$	*p*-Bromophenyl diselenide	0.71±0.3	B	25	-	1638
$C_{12}H_8ClNO_2S$	2-Chloro-4'-nitrodiphenyl sulfide	4.49±0.01	B	20	0	1657
	3-Chloro-4'-nitrodiphenyl sulfide	3.83±0.01	B	20	0	1657
	4-Chloro-4'-nitrodiphenyl sulfide	3.40±0.01	B	20	0	1656
$C_{12}H_8ClNO_3$	2-Chloro-4'-nitrodiphenyl ether	4.80±0.01	B	20	0	1657
	3-Chloro-4'-nitrodiphenyl ether	3.84±0.01	B	20	0	1657
	4-Chloro-4'-nitrodiphenyl ether	3.17±0.01	B	20	0	1656
$C_{12}H_8Cl_2$	*cis*-1,2-Dichloroacenaphthene	2.97±0.12	B	25	0	1643
	trans-1,2-Dichloroacenaphthene	2.04±0.12	B	25	0	1643
	5,6-Dichloroacenaphthene	3.85	B	25	-	807
	2,2'-Dichlorobiphenyl	1.74±0.02	B	25	-	204
		1.7_9	B	25	-	2085
		1.93	B	25	5	656
		1.97	B	25	0	1161
		1.7_3	CCl_4	25	-	2085

FORMULA	COMPOUND NAME	μ, D	State, or Solv.	t, °C	Method or P_a, % of P_e	Reference
	3,3'-Dichlorobiphenyl	1.6_9	B	25	-	2085
		1.82	B	25	5	656
		1.7_3	CCl_4	25	-	2085
	4,4'-Dichlorobiphenyl	0	B	25	-	2088, 2112, 2125
		0	B	25	0	1067
		0.33	B	25	0	203
		0.83	ClB	25	0	1067
		2.02	NB	25	0	1067
$C_{12}H_8Cl_2Hg$	*Bis*(*p*-chlorophenyl)mercury	1.09	s	134	0	653
		1.16	s	142	0	653
$C_{12}H_8Cl_2OS$	*p*-Chlorophenylsulfoxide	2.59	B	25	0	654
		2.73	B	22	-	162
$C_{12}H_8Cl_2O_4S_4$	Diphenyldisulfide-4,4'-disulfonyl chloride	4.68	B	25	-	634
$C_{12}H_8Cl_2O_5S_2$	4,4'-Oxydibenzene sulfonyl chloride	3.54	D	20	-	1412
$C_{12}H_8Cl_2S$	2,2'-Dichlorodiphenyl sulfide	3.29	B	20	-	1659
	2,3'-Dichlorodiphenyl sulfide	2.58	B	20	-	1659
	2,4'-Dichlorodiphenyl sulfide	2.09	B	20	-	1659
	3,3'-Dichlorodiphenyl sulfide	1.93	B	20	-	1659
	3,4'-Dichlorodiphenyl sulfide	1.42	B	20	-	1659
	4,4'-Dichlorodiphenyl sulfide	0.56	B	20	-	1659
		0.58	B	25	0	654
		0.90	B	22	0	160
$C_{12}H_8Cl_2Se$	4,4'-Dichlorodiphenyl selenide	0.78±0.3	B	25	-	1638

s) Decalin.

FORMULA	COMPOUND NAME	μ, D	State, or Solv.	t, °C	Method or P_a, % of P_e	Reference
$C_{12}H_8FNO_2S$	3-Fluoro-4'-nitrodiphenyl sulfide	3.85±0.01	B	20	0	1657
	4-Fluoro-4'-nitrodiphenyl sulfide	3.54±0.01	B	20	0	1656
$C_{12}H_8FNO_3$	2-Fluoro-4'-nitrodiphenyl ether	4.93±0.01	B	20	0	1657
	4-Fluoro-4'-nitrodiphenyl ether	3.34±0.01	B	20	0	1656
$C_{12}H_8F_2$	2,2'-Difluorobiphenyl	1.88	B	25	0	1161
	4,4'-Difluorobiphenyl	0.35	B	25	0	203
$C_{12}H_8F_2Hg$	*Bis*(*p*-fluorophenyl)mercury	0.88	t	142	0	653
$C_{12}H_8F_2O$	*p*-Fluorophenyl ether	0.51	B	25	5	1114
$C_{12}H_8F_2OS$	*Bis*(*p*-fluorophenyl)sulfoxide	2.66	B	25	5	1114
$C_{12}H_8F_2O_2S$	*Bis*(*p*-fluorophenyl)sulfone	3.31	B	25	5	1114
$C_{12}H_8F_2S$	*p*-Fluorophenyl sulfide	0.48	B	25	5	1114
$C_{12}H_8INO_2S$	4-Iodo-4'-nitrodiphenyl sulfide	3.52±0.01	B	20	0	1656
$C_{12}H_8INO_3$	4-Iodo-4'-nitrodiphenyl ether	3.34±0.01	B	20	0	1656
$C_{12}H_8I_2$	2,2'-Diiodobiphenyl	1.71	B	25	0	1161
$C_{12}H_8I_2O$	[illegible]-Iodophenyl ether	2.74	B	27	0	732
		2.80	B	50	0	732
		2.72	Hx	27	0	732
$C_{12}H_8N_2$	Benzo(3,4)cinnoline	3.93	B	25	0	247
	o-Phenanthroline	4.11	B	25	0	510

t) Decalin.

FORMULA	COMPOUND NAME	μ, D	State, or Solv.	t, °C	Method or P_a, % of P_e	Reference
	Phenazine	0	B	15	0	138
		0	B	25	0	265
$C_{12}H_8N_2O$	Benzo(3,4)cinnoline *N*-oxide	5.22	B	25	0	247
	Phenazine *N*-oxide	1.76	B	25	0	1552
$C_{12}H_8N_2O_2$	*N*-2-Naphthylsydnone	6.9_2	B	25	0	427-8
	Phenazine 5,10-dioxide (decomp. 190°C)	2.20	B	25	0	1552
	(decomp. 202°C)	1.40	B	25	0	1552
$C_{12}H_8N_2O_4$	2,2'-Dinitrobiphenyl	5.10	B	25	0	1161
		5.16±0.05	B	25	-	204
		5.23	B	25	0	1104
		5.2_3	B	25	0	298
	2,4-Dinitrobiphenyl	4.34	B	25	0	1162
	2,4'-Dinitrobiphenyl	6.28±0.04	B	20	0	1182,1188
	3,3'-Dinitrobiphenyl	4.0_8	B	25	0	298
		4.17	B	25	0	1161
	4,4'-Dinitrobiphenyl	0	B	25	-	2088, 2112,2125
		0.74	B	40	-	1301
		1.0_3	B	25	-	1107
		0.83	D	40	-	1301
$C_{12}H_8N_2O_4S$	*p*-Nitrophenyl sulfide	3.33	B	20	-	1682
$C_{12}H_8N_2O_4S_2$	*p*-Nitrophenyl disulfide	3.59	B	20	0	160
		4.34	D	20	0	636

FORMULA	COMPOUND NAME	μ, D	State, or Solv.	t, °C	Method or P_a, % of P_e	Reference
$C_{12}H_8N_2O_5$	*o*-Nitrophenyl ether	6.62	B	50	0	732,2011
		6.69	B	27	0	732,2011
		6.77	D	27	0	732,2011
	p-Nitrophenyl ether	2.63	B	25,50	0	654
		2.83	B	25	0	1836
$C_{12}H_8N_3O_5$	4,4'-Dinitrodiphenyl nitroxide	2.7±0.5	n.s.	n.s.	-	263
$C_{12}H_8N_4O_4$	4,4'-Dinitroazobenzene	0	B	25	10	267
$C_{12}H_8N_4O_8S$	2',3,4'-Trinitrobenzenesulfonanilide	3.55	D	n.s.	-	1923
$C_{12}H_8OS$	Phenoxathiine	0.93	B	25	5	1114
		1.10	B, Hx	25	0	729,734
$C_{12}H_8O_2$	Dibenzo-*p*-dioxide	0	B, cHx, CCl_4	20	-	123
		0.64	B	25	0	729,733
		0.57	Hx	25	0	729,733
$C_{12}H_8O_2S_2$	Thianthrene dioxide (m.p. = 246°C)	4.2	B	19	-	162
	(m.p. = 279°C)	1.7	B	19	-	162
$C_{12}H_8S$	Dibenzothiophene	~0	B	24	0	160
		0.83	B	30	0	282
$C_{12}H_8S_2$	Thianthrene	1.42	B	25,50	0	2059
		1.58	B	25	0	265
		1.7	B	22	0	162
		1.55	CCl_4	25	-	124
		1.48	CS_2	24	-	124
$C_{12}H_8Se_2$	Selenanthrene	1.42	B	25	0	265

FORMULA	COMPOUND NAME	μ, D	State, or Solv.	t, °C	Method or P_a, % of P_e	Reference
$C_{12}H_9AsClN$	Phenarsazine chloride	2.28	B	20	0	1346
$C_{12}H_9Br$	2-Bromobiphenyl	1.50	B	25	0	1160
	3-Bromobiphenyl	1.64	B	25	0	1160
	4-Bromobiphenyl	1.65	B	25	0	1068
		1.65±0.02	B	25	0	475
		1.66	B	25	0	1160
		1.72	CCl_4	25	0	298
$C_{12}H_9BrN_2$	4-Bromoazobenzene	1.43	B	20	-	139
$C_{12}H_9BrN_2O$	α-4-Bromoazoxybenzene	2.62_5	B	22	-	565
	β-4-Bromoazoxybenzene	0.92	B	22	-	565
$C_{12}H_9BrN_2O_2S$	(*p*-Bromophenyl)(phenylsulfonyl) diimide	4.40	B	30	0	541
$C_{12}H_9BrO$	*p*-Bromophenyl phenyl ether	1.92±0.02	gas	235	5	332
		1.58	B	25,50	0	1836
		1.59	B	25	0	654
		1.78	B	19	0	160
		1.77	Hx	25	0	1906
		1.67	Hp	25,50	0	1836
$C_{12}H_9Br_2N_3$	4,4'-Dibromodiazoaminobenzene	1.90±0.04	B	25	0	1102
$C_{12}H_9Cl$	5-Chloroacenaphthene	2.50	B	25	-	807
	2-Chlorobiphenyl	1.3_1	B	25	-	2085
		1.42	B	25	0	1160
		1.45	B	n.s.	0	1384
		1.46	B	25	0	656
		1.4_3	CCl_4	25	-	2085

FORMULA	COMPOUND NAME	μ, D	State, or Solv.	t, °C	Method or P_a, % of P_e	Reference
$C_{12}H_9Cl$ (contd.)	3-Chlorobiphenyl	1.65	B	25	0	656
		1.65	B	25	0	1160
		1.8_1	B	25	-	2085
		1.6_2	CCl_4	25	-	2085
	4-Chlorobiphenyl	1.5_4	B	25	-	2085
		1.57	B	25	0	1068
		1.64	B	n.s.	-	1384
		1.64	B	25	0	656
		1.66	B	25	0	1160
		1.68	CCl_4	25	0	298
$C_{12}H_9ClN_2$	4-Chloroazobenzene	1.56	B	21	-	139
$C_{12}H_9ClN_2O$	2-Chloro-4'-hydroxyazobenzene	2.20	B	25	0	1763
	3-Chloro-4'-hydroxyazobenzene	2.56	B	25	0	1763
	4-Chloro-4'-hydroxyazobenzene	2.77	B	25	0	1763
$C_{12}H_9ClN_2O_2S$	(Chlorophenyl)(phenylsulfonyl) diimide	4.3_9	B	30	0	541
$C_{12}H_9ClO$	2-Chlorophenyl phenyl ether	2.16±0.02	B	20	-	1681
	3-Chlorophenyl phenyl ether	1.68±0.01	B	20	-	1681
	4-Chlorophenyl phenyl ether	1.55±0.02	B	20	-	1681
$C_{12}H_9ClOS$	4-Chlorophenyl phenyl sulfoxide	3.50	B	25	0	654
		3.97	B	23	-	162
$C_{12}H_9ClO_2S$	4-Chlorophenyl phenyl sulfone	4.45	B	23	-	162
$C_{12}H_9ClS$	2-Chlorophenyl phenyl sulfide	2.62	B	20	-	1658
	3-Chlorophenyl phenyl sulfide	1.89	B	20	-	1658

FORMULA	COMPOUND NAME	μ, D	State, or Solv.	t, °C	Method or P_a, % of P_e	Reference
	4-Chlorophenyl phenyl sulfide	1.52-1.55	B	25	0	654
		1.52	B	20	-	1658
		1.72	B	21	0	160
		1.78	B	21	-	141
$C_{12}H_9Cl_2N_3$	4,4'-Dichlorodiazoaminobenzene	1.96±0.04	B	25	0	1102
$C_{12}H_9Cl_2NO_2S$	Dichlorosulfonanilide	4.75	D	25	0	1550
$C_{12}H_9F$	2-Fluorobiphenyl	1.36	B	25	0	1160
	3-Fluorobiphenyl	1.57	B	25	0	1160
	4-Fluorobiphenyl	1.49	B	25	0	1160
		1.51	B	25	0	1068
		1.49	CCl_4	25	0	298
$C_{12}H_9FO$	*p*-Fluorophenyl phenyl ether	1.36	B	25	5	1114
$C_{12}H_9FO_2S$	*p*-Fluorophenyl phenyl sulfone	4.29	B	25	5	1114
$C_{12}H_9FS$	*p*-Fluorophenyl phenyl sulfide	1.38	B	25	5	1114
$C_{12}H_9F_2N$	4,4'-Difluorodiphenylamine	2.11	B	25	5	1114
$C_{12}H_9I$	2-Iodobiphenyl	1.27	B	25	0	1160
	3-Iodobiphenyl	1.46	B	25	0	1160
	4-Iodobiphenyl	1.45	B	25	0	1160
		1.38	CCl_4	25	0	919
		1.52	CCl_4	25	0	298
$C_{12}H_9N$	Carbazole	2.11	B	20	0	345

FORMULA	COMPOUND NAME	μ, D	State, or Solv.	t, °C	Method or P_a, % of P_e	Reference
$C_{12}H_9NO_2$	2-Nitrobiphenyl	3.70	B	25	0	1160
		3.81	B	n.s.	0	1384
		3.82±0.01	B	20	0	1182,1188
		3.83	B	25	0	1068
	3-Nitrobiphenyl	3.93	B	n.s.	0	1384
		4.12	B	25	0	1160
	4-Nitrobiphenyl	4.20	B	25	0	1068
		4.31	B	n.s.	0	1384
		4.36	B	25	0	1160
		4.41±0.02	B	25	0	475
		4.16	D	25	0	300
		4.39_5	CCl_4	25	0	300
		4.4_0	CCl_4	25	0	298
$C_{12}H_9NO_2S$	*p*-Nitrophenyl phenyl sulfide	4.26	B	20	-	1682
$C_{12}H_9NO_3$	2-Nitrophenyl phenyl ether	4.45±0.01	B	20	-	1681
	4-Nitrophenyl phenyl ether	4.47±0.18	gas	225-245	5	332
		4.23	B	19	0	160
		4.29-4.31	B	25	0	654
		4.46	B	20	-	1412
		4.33-4.37	CCl_4	25	0	654
$C_{12}H_9NO_3S$	4-(4'-Nitrophenylthio)phenol	4.96	B	20	-	1682
$C_{12}H_9NS$	Phenthiazine	2.15	B	25	5	1114
$C_{12}H_9N_3$	2-Aminophenazine	6.72	B	25	0	1409
$C_{12}H_9N_3O_2$	*p*-Nitroazobenzene	4.45±0.1	B	25	0	267
$C_{12}H_9N_3O_3$	*N*-Nitroso(*o*-nitrophenyl)phenylamine	5.84	B	25	5	1219
		6.02	D	25	5	1219

FORMULA	COMPOUND NAME	μ, D	State, or Solv.	t, °C	Method or P_a, % of P_e	Reference
	N-Nitroso(*m*-nitrophenyl)phenylamine	4.83	B	25	5	1219
		4.96	D	25	5	1219
	N-Nitroso(*p*-nitrophenyl)phenylamine	4.13	B	25	5	1219
		4.35	D	25	5	1219
$C_{12}H_9N_3O_4S$	(*p*-Nitrophenyl)(phenylsulfonyl)diimide	4.7_6	B	25	0	541
$C_{12}H_{10}$	Acenaphthene	0.81	gas	55	0	299
		0.93	gas	190-200	-	807
		0.85	B	25	-	807
		0.85±0.03	B	30	5	150
		0.97±0.02	B	20	0	1181,1189
		1.50	B	25	0	1178
		1.57	B	20	0	1178
		1.6_0	B	30	0	1178
		0.79	D	25	0	299
		0.80	CCl_4	25	0	299
	Biphenyl	0	liq	75-155	-	1052
		0	B	25	0	203,2088, 2112
		0	B	25	0	1067
		0	B	25	0	948
		0	CCl_4	25	0	298
		0	cHx	25	-	949
		0	Hp	20	7	209
		1.90	NB	25	0	1067
		1.88	NB	30	-	754
		0.97	ClB	25	0	1067
$C_{12}H_{10}AlBr_3O$	Aluminum bromide - phenyl ether complex	6.61	B	n.s.	-	1734
$C_{12}H_{10}AsCl$	Diphenylchloroarsine	2.73	Hx	20	13	1346
$C_{12}H_{10}As_2$	Arsenobenzene	0	B	25	-	1090

FORMULA	COMPOUND NAME	μ, D	State, or Solv.	t, °C	Method or P_a, % of P_e	Reference
$C_{12}H_{10}Be$	Diphenylberyllium	1.6_5	B	20	10	1887
		4.3_7	D	20	10	1887
$C_{12}H_{10}BrN$	4-Bromo-4 -aminodiphenyl	3.33	B	25	0	1068
$C_{12}H_{10}BrN_3$	1-Phenyl-3-(*p*-bromophenyl) triazene	2.02±0.04	B	25	0	1102
$C_{12}H_{10}Br_2O_2$	4,8-Dibromo-1,5-dimethoxynaphthalene	0.93±0.03	B	25	0	477
$C_{12}H_{10}Br_2Se$	Diphenylselenium dibromide	3.43	B	25	15	849
$C_{12}H_{10}Br_2Si$	Diphenyldibromosilane	2.77±0.04	B	25	5	1592
$C_{12}H_{10}Cd$	Diphenylcadmium	0.6_8	B	20	10	1886-7
		1.4_6	D	20	10	1886-7
$C_{12}H_{10}Cl_2O_2$	4,8-Dichloro-1,5-dimethoxynaphthalene	0.73	n.s.	n.s.	-	1948
		0.95±0.05	B	25	0	477
$C_{12}H_{10}Cl_2Se$	Diphenylselenium dichloride	3.24	B	25	15	849
		3.50	B	50	0	1811
$C_{12}H_{10}Cl_2Si$	Diphenyldichlorosilane	2.58	B	25	0	373
$C_{12}H_{10}Cl_2Sn$	Diphenyldichlorotin	3.59	B	n.s.	-	592
		4.34	D	n.s.	-	592
		3.65	Hx	n.s.	-	592
$C_{12}H_{10}FN$	(*p*-Fluorophenyl) phenylamine	1.88	B	25	5	1114
$C_{12}H_{10}F_2Si$	Diphenyldifluorosilane	2.59	B	25	0	373
$C_{12}H_{10}Hg$	Diphenylmercury	0	B	14	0	158
		0.2	B	20	10	1886-7
		0.41	B	25	0	653
		0.44±0.03	B	25	5-15	694

FORMULA	COMPOUND NAME	μ, D	State, or Solv.	t, °C	Method or P_a, % of P_e	Reference
		0.72	B	20	-	1700
		0.92	B	30	-	1700
		1.08	B	40	-	1700
		0.4	D	20	10	1886-7
		0.42	D	25	0	372
		0.80	D	20	-	1777
		0.80	D	40	-	1700
		0.85	D	25	-	1700
		0.89	D	20	-	1700
		0.52[u]	CCl_4	25	-	63
		0.68	CCl_4	40	-	1700
		0.85	CCl_4	25	-	1700
		0.93	CCl_4	20	-	1700
		0.44	v	25	0	653
		0.54	v	142	0	653
$C_{12}H_{10}I_2$	Iodine - biphenyl complex	2.9	cHx	25	-	949
$C_{12}H_{10}Mg$	Diphenylmagnesium	4.98	D	20	10	1886-7
$C_{12}H_{10}NO$	Diphenylnitroxide	2.3±0.2	n.s.	n.s.	-	263
$C_{12}H_{10}N_2$	*cis*-Azobenzene	3.0	B	25	-	667-8
	trans-Azobenzene	0	B	25	-	668
		0	B	24	-	139
		0	D	15	0	164
	Benzylmethylfumarodinitrile	0.8	B	30	-	117
	Benzylmethylmaleodinitrile	3.7	B	30	-	117
	3-Methyl-1,2-naphthimidazole	3.89	D	25	0	1451
$C_{12}H_{10}N_2O$	*cis*-Azoxybenzene	4.71	B	22	-	565

u) "Apparent" value caused by atomic polarization. v) Decalin.

FORMULA	COMPOUND NAME	μ, D	State, or Solv.	t, °C	Method or P_a, % of P_e	Reference
$C_{12}H_{10}N_2O$ (contd.)	*trans*-Azoxybenzene	1.57	B	25	0	247
		1.72	B	22	-	565
	o-Hydroxyazobenzene	1.32	B	15	0	164
	p-Hydroxyazobenzene	1.63	B	15	0	164
		1.66	B	25	-	800
		2.06	D	15	0	164
		2.07	D	25	-	800
	N-Nitrosodiphenylamine	3.36	B	25	5	1219
		3.38	B	25	5	1114
		3.42	B	20	0	339
		3.62	D	25	5	1219
$C_{12}H_{10}N_2O_2$	4-Amino-4'-nitrodiphenyl	6.05	B	20	FL	375
		6.42±0.06	B	20	-	1351
		6.51	B	25	0	1068
	cis-4,4'-Dihydroxyazobenzene	2.71	D	25	-	419
	trans-4,4'-Dihydroxyazobenzene	2.61	D	25	-	419
	(*o*-Nitrophenyl)phenylamine	4.15	B	20	-	854
		4.18	B	25	5	1219
		4.23	D	25	5	1219
	(*m*-Nitrophenyl)phenylamine	4.70	B	25	5	1219
		4.93	D	25	5	1219
	(*p*-Nitrophenyl)phenylamine	5.86	B	20	-	854
		6.05	B	25	5	1219
		6.28	D	25	5	1219
$C_{12}H_{10}N_2O_2S$	4-Aminophenyl-4'-nitrophenyl sulfide	5.8±0.03	B	20	-	1658

FORMULA	COMPOUND NAME	μ, D	State, or Solv.	t, °C	Method or P_a, % of P_e	Reference
	Phenyl(phenylsulfonyl)diimide	4.2_0	B	25	0	541
$C_{12}H_{10}N_2O_4S$	Benzenesulfon-*p*-nitroanilide	4.94	D	n.s.	-	1923
	3-Nitrobenzenesulfonanilide	4.78	D	n.s.	-	1923
	4-Nitrobenzenesulfonanilide	4.05	D	n.s.	-	1923
$C_{12}H_{10}N_4O_2$	1-Phenyl-3-(*p*-nitrophenyl)triazene	4.81±0.06	B	25	0	1102
$C_{12}H_{10}N_4O_7$	Aniline - picric acid complex	6.74	D	35	0	1559
		7.0	D	23	-	1987
	α-Picoline - picric acid complex	9.25	D	35	0	1558
	β-Picoline - picric acid complex	9.55	D	35	0	1558
	γ-Picoline - picric acid complex	9.60	D	35	0	1558
$C_{12}H_{10}Ni_2O_2$	Di-μ-carbonyldicyclopentadienyldinickel(I)	0	B	n.s.	-	515
		0	B	25	15	520
$C_{12}H_{10}O$	6-Methyl-7*H*-cycloheptabenzen-7-one	4.25	n.s.	n.s.	-	2048
	Phenyl ether	1.0	gas	100	MB	464
		1.15±0.02	gas	172-210	5	332
		1.36	gas	210	5	625
		0.99	liq	28-50	-	463
		1.14	liq	30-40	-	1497
		1.03	B	20-50	-	463
		1.04	B	25	RT	1710
		1.12	B	25	5	1114
		1.14	B	18	0	160
		1.16±0.02	B	25	0	597
		1.17	B	25	0	654
		1.19	B(?)	20	-	1412

FORMULA	COMPOUND NAME	μ, D	State, or Solv.	t, °C	Method or P_a, % of P_e	Reference
$C_{12}H_{10}O$ (contd.)	Phenyl ether (contd.)	1.20	B	20	RT	1241b
		1.19	Hx	25	0	1906
		0.96	w	20	RT	1617
$C_{12}H_{10}OS$	2-Methyl-6-phenyl-4*H*-pyran-4-thione	5.22	B	20	-	1661
	Phenyl sulfoxide	4.00	B	25	5	1114
		4.03	B	25	0	654
		4.08	B	20	0	636
		4.11	B	23	0	141
		4.20	B	25	0	404
	p-(Phenylthio)phenol	2.21	B	20	-	1682
$C_{12}H_{10}OSe$	Diphenylselenoxide	4.47	B	25	15	851
$C_{12}H_{10}O_2$	1-Acetyl-4-naphthol	4.59	D	25	5	1214
	2-Acetyl-1-naphthol	3.17	B	25	5	1214
		3.15	D	25	5	1214
	4,4'-Dihydroxydiphenyl	2.82	D	20	0	1719
	1-Formyl-2-methoxynaphthalene	3.67	B	25	5	1214
		3.74	D	25	5	1214
	2-Hydroxyphenyl phenyl ether	1.60	B	20	RT	1241b
	2-Methyl-5-phenyl-4-*H*-pyran-4-one	4.61	B	20	-	1661
$C_{12}H_{10}O_2S$	Phenyl sulfone	4.93±0.05	B	25	0	597
		5.06	B	20	0	636
		5.08	B	25	5	1114
		5.09	B	25	0	404
		5.12	B	30	0	88
		5.18	B	23	-	162

w) Nujol.

FORMULA	COMPOUND NAME	μ, D	State, or Solv.	t, °C	Method or P_a, % of P_e	Reference
$C_{12}H_{10}O_3S$	*o*-Hydroxydiphenylsulfone	4.69	B	30	0	47
		5.24	D	30	0	47
	m-Hydroxydiphenylsulfone	5.43	D	30	0	47
	p-Hydroxydiphenylsulfone	6.11	D	30	0	47
$C_{12}H_{10}O_4$	Quinhydrone	~2	D	35	0	1851
$C_{12}H_{10}O_4S_2$	Phenyl disulfone	3.96	B	25	0	1428
		4.04	B	20	0	636
$C_{12}H_{10}S$	Phenyl sulfide	1.46	B	25	5	1114
		1.48	B	21	0	141
		1.51	B	25	0	654
		1.55±0.02	B	25	0	597
		1.57	B	25	0	404
		1.69	B	20	-	1412
$C_{12}H_{10}S_2$	Phenyl disulfide	1.83	B	24	0	160
		1.92	B	20	0	636
$C_{12}H_{10}Se$	Phenyl selenide	1.39	B	20	0	141
		1.54	B	20	-	1412
$C_{12}H_{10}Se_2$	Phenyl diselenide	1.68±0.1	B	25	0	1638
$C_{12}H_{10}Te$	Phenyl telluride	1.14	B	21	0	141
$C_{12}H_{10}Zn$	Diphenyl zinc	0.8_6	B	20	10	1886-7
		1.01-1.18	B	25	0	858
		2.7_3	D	20	10	1886-7
$C_{12}H_{11}AlBr_3N$	Aluminum bromide - diphenylamine complex	6.73	B	20	-	1734

FORMULA	COMPOUND NAME	μ, D	State, or Solv.	t, °C	Method or P_a, % of P_e	Reference
$C_{12}H_{11}N$	2-Aminobiphenyl	1.43	B	n.s.	0	1384
		1.45	B	20	-	1950
	4-Aminobiphenyl	1.75	B	n.s.	0	1384
		1.76	B	25	0	1004
		1.78	B	25	0	1068
		2.09	D	25	0	1004
	Diphenylamine	1.31	solid	20	-	463
		0.81	B	25	RT	1710
	1.08 P_e = 51.73, P_a = 5.17	0.91	B	20	-	1412
		0.95	B, Hx	20	5	337
		1.0	B	25	-	101
		1.01	B	25	5	1114
		1.01	B	25	5	70
		1.03	B, Hx	20	0	336
		1.082[x]	B	25	0	1794
		1.31	B	25	5	1219
		1.317	D	25	0	1794
		1.36	D	25	5	1219
$C_{12}H_{11}NO$	2-Acetylaminoazulene	3.26	B	30	-	1016
	N-Acetyl-1-naphthylamine	3.67	B	25	0	1797
	N-Acetyl-2-naphthylamine	3.68	B	25	0	1797
	2-Aminophenyl phenyl ether	1.28±0.01	B	20	-	1681
$C_{12}H_{11}NO_2$	8-Carbethoxy-8-cyanoheptafulvene	4.43±0.05	B	25	5	2154
$C_{12}H_{11}NO_2S$	Benzenesulfonanilide	4.81	B	20	0	1703
		4.30	D	25	0	1004
		5.02	D	n.s.	-	1923
		5.07±0.02	D	25	0	1550

x) Original work contains other values in various mixed solvents.

FORMULA	COMPOUND NAME	μ, D	State, or Solv.	t, °C	Method or P_a, % of P_c	Reference
$C_{12}H_{11}NS$	*p*-(Phenylthio)aniline	2.97	B	20	-	1682
$C_{12}H_{11}N_3$	*p*-Aminoazobenzene	2.43	B	25	-	800
		2.48	B	20	0	1386
		2.48±0.1	B	25	0	267
		2.55	B	n.s.	-	1342
		2.57±0.02	B	31	0	982
		2.73	B	15	0	164
		3.04	D	25	-	800
		3.11	D	25	0	1386
		2.28	CCl_4	20	0	1386
	Diazoaminobenzene	0.9_1±0.02	B	25	0	1102
	4-Methyl-2-phenylazopyridine	2.7	B	n.s.	-	266
$C_{12}H_{11}O_2Re$	Dicyclopentadienyl rhenium dicarbonyl hydride	3.85±0.09	B	25	15	520,523
$C_{12}H_{12}$	1,2-Dimethylnaphthalene	0.68	n.s.	n.s.	-	1210
	1,3-Dimethylnaphthalene	0.36	n.s.	n.s.	-	1210
	1,5-Dimethylnaphthalene	0.07	n.s.	n.s.	-	1210
	1,6-Dimethylnaphthalene	0.32	n.s.	n.s.	-	1210
	1,7-Dimethylnaphthalene	0.54	n.s.	n.s.	-	1210
	1,8-Dimethylnaphthalene	0.48	n.s.	n.s.	-	1210
	2,3-Dimethylnaphthalene	0.69	n.s.	n.s.	-	1210
	2,6-Dimethylnaphthalene	0.14	n.s.	n.s.	-	1210
	2,7-Dimethylnaphthalene	0.41	n.s.	n.s.	-	1210

FORMULA	COMPOUND NAME	μ, D	State, or Solv.	t, °C	Method or P_a, % of P_e	Reference
$C_{12}H_{12}BrN$	4'-Bromo-2,5-dimethyl-1-phenylpyrrole	0.54±0.03	B	25	0	931
$C_{12}H_{12}ClN$	4'-Chloro-2,5-dimethyl-1-phenylpyrrole	0.50±0.04	B	25	0	931
$C_{12}H_{12}Cr$	Chromium - di(benzene) complex	0	B	n.s.	-	518
		0	B	25	5-10	2076
$C_{12}H_{12}CrO_2$	Cyclopentadienyl-cyclopentenyl chromium dicarbonyl	3.49±0.06	B	25	-	521
$C_{12}H_{12}CrO_3$	1,3,5-Trimethylbenzene - chromium tricarbonyl complex	5.56±0.06	B	25	15	520
		5.81±0.05	B	n.s.	-	1568
$C_{12}H_{12}CuF_6N_2O_2$	*Bis*(trifluoroacetylacetone)ethylene-diiminocopper(II)	9.60±0.15	B	25	-	1286,1288
$C_{12}H_{12}F_6N_2O_2$	*Bis*(trifluoroacetylacetone)ethylenediimine	5.82±0.15	B	25	-	1286
$C_{12}H_{12}F_6O$	5-Butoxy-1,3-*bis*(α,α,α-trifluorotoluene)	3.53	B	25	0	1378
$C_{12}H_{12}FeO$	Acetylferrocene	3.02	B	30	0	1600
$C_{12}H_{12}MoO_3$	Mesitylene - molybdenum tricarbonyl complex	6.36	B	n.s.	-	1568
$C_{12}H_{12}MoO_4$	Cycloöctadiene(1,5)molybdenum tetracarbonyl	4.88±0.06	cHx	n.s.	-	517
$C_{12}H_{12}N_2$	2,2'-Diaminobiphenyl	2.0	B	25	-	204
	4,4'-Diaminobiphenyl	1.3	B	25	-	2088, 2112,2125
		1.375	B	20	-	1673
		1.44	B	25	0	203
		1.54	B	25	5	70
		1.60±0.02	B	25	0	475

FORMULA	COMPOUND NAME	μ, D	State, or Solv.	t, °C	Method or P_a, % of P_e	Reference
	1,1-Diphenylhydrazine	1.89	B	20	15	76
		1.90±0.08	B	15	0	2008
	1,2-Diphenylhydrazine	1.54	B	18	15	76
		1.67	B	20	0	339
		1.70	B	25	5	70
		1.71±0.05	B	15	0	2008
		1.85	B	25	-	1555
$C_{12}H_{12}N_2OS$	2-(Benzylthio)-1-methyl-4-pyrimidone	6.95	D	25	0	1716
	2-(Benzylthio)-3-methyl-4-pyrimidone	2.83	D	25	0	1716
$C_{12}H_{12}N_2OS_2$	5-(Anilinomethylene)-3-ethylrhodanine	4.30	B	25	-	1766
$C_{12}H_{12}N_2O_2$	2,5-Dimethyl-1-(4-nitrophenyl)pyrrole	2.48±0.02	B	25	0	931
		2.50	B	25	-	642
$C_{12}H_{12}N_2O_2S$	*p*-(*p*-Aminophenyl)benzenesulfonamide	6.76	D	25	0	1004
$C_{12}H_{12}N_2O_3$	Phenobarbital	1.18	D	25	-	992
	5-Phenyl-5-ethylbarbituric acid	0.87±0.01	D	35	0	1847
$C_{12}H_{12}N_2O_4$	*m*-*Bis*(β-methyl-β-nitrovinyl)benzene	4.23	B	25	27	2031,2033
	p-*Bis*(β-methyl-β-nitrovinyl)benzene	1.66	B	25	27	2031,2033
	1,3-*Bis*(1-(2-nitro-1-propenyl))benzene	4.23	B	25	-	2031
	1,4-*Bis*(1-(2-nitro-1-propenyl))benzene	1.66	B	25	-	2031
$C_{12}H_{12}N_2O_4S_2$	4'-Sulfamoylbenzenesulfonanilide	5.55	D	25	0	1550
$C_{12}H_{12}N_2O_4S_4$	Phenyldisulfide-4,4'-disulfonamide	6.42	D	25	0	635
$C_{12}H_{12}O$	6-Phenyl-3,5-hexadien-2-one	3.52±0.03	B	25	0	127

FORMULA	COMPOUND NAME	μ, D	State, or Solv.	t, °C	Method or P_a, % of P_e	Reference
$C_{12}H_{12}O_2$	1,4-Dimethoxynaphthalene	2.09±0.02	B	25	0	477
	1,5-Dimethoxynaphthalene	0.36 0.67±0.1	n.s. B	n.s. 25	- 0	1948 477
$C_{12}H_{12}O_2S_2$	Ethylsulfinyl β-naphthalene sulfide	4.98	B	25	0	619
$C_{12}H_{12}O_2Si$	Diphenylsilanediol	1.74	D	25	-	307
$C_{12}H_{12}O_6$	Phloroglucinol triacetate	2.4	B	25	-	2112,2114
$C_{12}H_{12}S_2$	1,4-*Bis*(methylthio)naphthalene	1.93	B	20	0	1197
	1,5-*Bis*(methylthio)naphthalene	1.29±0.05	B	20	0	1378
	1,8-*Bis*(methylthio)naphthalene	1.80	B	20	0	1197
$C_{12}H_{13}BrO$	2-(4-Bromophenyl)cyclohexanone	4.11	B	30	0	784
	2-Bromo-4-phenylcyclohexanone	2.86±0.04 2.88±0.04	D Hx	30 30	0 0	1010 1010
$C_{12}H_{13}ClN_2O_4S$	2-Chlorocyclohexyl 2,4-dinitrophenyl sulfide	5.28	B	25	-	1031
$C_{12}H_{13}ClO$	2-(4-Chlorophenyl)cyclohexanone	4.25	B	30	0	784
$C_{12}H_{13}N$	2,5-Dimethyl-1-phenylpyrrole	1.99 2.00±0.04	B B	25 25	- 0	642 931
$C_{12}H_{13}NO_2$	Ethyl cyanotropylacetate	3.59	B	25	5	2154
$C_{12}H_{13}N_3$	3,3-Dimethyl-1-(2-naphthyl)triazene	2.3₈	B	25	0	479
$C_{12}H_{13}N_3O_4S_2$	4'-Sulfamoylsulfanilanilide	6.99	D	25	0	1550

FORMULA	COMPOUND NAME	μ, D	State, or Solv.	t, °C	Method or P_a, % of P_e	Reference
$C_{12}H_{13}N_3S$	6-Methylamino-5,6-dihydro-3-phenylimidazo-[2,1-b]thiazole	2.3±0.2	B	n.s.	-	2129
$C_{12}H_{14}ClNO_2S$	2-Chlorocyclohexyl 2-nitrophenyl sulfide	5.09	B	25	-	1031
	2-Chlorocyclohexyl 4-nitrophenyl sulfide	5.32	B	25	-	1031
$C_{12}H_{14}Cl_2N_2Zn$	Dichloro*bis*(α-picoline)zinc	8.86±0.04	D	25	10	1707
	Dichloro*bis*(β-picoline)zinc	9.53±0.04 9.54±0.04	B D	25 25	10 10	1707 1707
	Dichloro*bis*(γ-picoline)zinc	9.75±0.1	D	25	10	1707
$C_{12}H_{14}F_6N_2O_2$	*Bis*(trifluoroacetylacetone)ethylene-diimine	5.82±0.02	B	25	10	1287
$C_{12}H_{14}Mg$	*Bis*(methylcyclopentadienyl)magnesium	0.5 5.0 0.5	B D Hp	20 20 20	- - -	1894 1894 1894
$C_{12}H_{14}N_2OS_2$	*N*-(Morpholinomethylene)benzothiazoline-2-thione	4.72	B	25	-	637
$C_{12}H_{14}N_2OS_3$	1-[2-(3-Ethyl-4-thiazolin-2-ylidenyl)]-2-[5-(3-ethyl-2-thione-4-oxo-thiazolyl-idenyl)]ethane	9.18	B	25	0	1767
$C_{12}H_{14}N_2O_2$	1-Acetyl-1-methyl-2-cinnamoylhydrazine	4.34	B	25	-	1042
$C_{12}H_{14}N_2O_2S_3$	3-Ethyl-5-[(3-ethyl-2,4-dioxo-5-thiazoli-dinylidene)-methyl]-4-hydroxyl-2-methyl(thio)thiazolium betaine	7.11	B	25	0	1767
$C_{12}H_{14}N_2O_4S$	Cyclohexyl 2,4-dinitrophenyl sulfide	5.88	B	25	-	1031

FORMULA	COMPOUND NAME	μ, D	State, or Solv.	t, °C	Method or P_a, % of P_e	Reference
$C_{12}H_{14}O$	2-Phenylcyclohexanone[y]	3.13	B	30	0	20
$C_{12}H_{14}O_2$	2-(4-Hydroxyphenyl)cyclohexanone	3.28	D	30	0	784
$C_{12}H_{14}O_4$	Diethyl *o*-phthalate	2.7	B	25	0?	2116
		2.71	B, Hx	25-90	0	205
		2.8_6	B	25	5	71
		2.8	D	25	0?	2116
		2.72	z	25	0	205
		1.92	a	25	0	205
	Diethyl *m*-phthalate	2.4_6	CCl_4	25	0	71
	Diethyl *p*-phthalate	2.5_1	CCl_4	25	0	71
	Diethyl phthalate[b]	2.90	B[c]	28	RT	1401
		2.80	cHx	30	RT	1844
	1,3-Dimethoxy-2,4-diacetylbenzene	3.33	B	25	15	1220
		3.85	D	25	15	1220
	1,3-Dimethoxy-4,6-diacetylbenzene	7.45	B	25	15	1220
		8.07	D	25	15	1220
$C_{12}H_{14}O_5$	*t*-Butyl ester of monoperoxyphthalic acid	3.74	B	50	0	2049
$C_{12}H_{15}AlO_9$	Aluminum acetoacetate	3.99	B	25	0	2041
$C_{12}H_{15}N$	2-*t*-Butylindole	2.46	B	25	-	817
	1,2-Dimethyl-3-ethylindole	2.43	B	25	-	817
	1,3-Dimethyl-2-ethylindole	2.44	B	25	-	817
$C_{12}H_{15}NO$	*N*,*N*-Dimethyl-*p*-aminobenzalacetone	5.3	B	24	-	2090
		5.64	B	25	0	412

y) Phenyl is ε to CO group. z) Decalin. a) Perchloroethane. b) Configuration not given.
c) 3.26 cm frequency.

FORMULA	COMPOUND NAME	μ, D	State, or Solv.	t, °C	Method or P_a, % of P_e	Reference
$C_{12}H_{15}NO_4$	Pyridinium dicarboxymethylide, diethyl ester	4.50±0.07	B	20	-	670
$C_{12}H_{15}N_3O_6$	5-*t*-Butyltrinitro-*m*-xylene	1.15	CCl_4	25,45	0	1064
	2,4,6-Triethyltrinitrobenzene	0.8	B	25	-	841
$C_{12}H_{16}As_2NiO_2$	*o*-Phenylene*bis*(dimethylarsine)nickel dicarbonyl	5.13±0.01 5.24	B n.s.	n.s. n.s.	- -	291 290
$C_{12}H_{16}Cl_2$	*Bis*(chloromethyl)durene	2.08	B	20	RT	1546
$C_{12}H_{16}N_2$	Pyrrole - 2,4,6-trimethylpyridine complex	2.97±0.1	CCl_4	n.s.	-	596
$C_{12}H_{16}N_2OS_3$	1-[2-(3-Ethylthiazolylidenyl)]-2-[5-(3-ethyl-2-thione-4-oxcthiazolylidenyl)]-ethane	8.6	B	25	0	1767
$C_{12}H_{16}N_2S$	*N*-Ethyl-*N*-phenyl-*N'*-allylthiourea	5.1	B	18	0	962
$C_{12}H_{16}N_2S_2$	*N*-(Diethylaminomethylene)benzothiazoline-2-thione	4.38	B	25	-	637
$C_{12}H_{16}NiO_2P_2$	*o*-Phenylene*bis*(dimethylphosphine)nickel dicarbonyl	5.40±0.01	B	n.s.	-	291
$C_{12}H_{16}O$	Cyclohexyl phenyl ether	1.55	B	20	RT	1548
	Mesityl-2-propionone	2.79	B	30	0	20
$C_{12}H_{16}O_2$	Isoamyl benzoate	2.2	CCl_4	n.s.	-	2014
	Methyl 2,3,5,6-tetramethylbenzoate	2.6±0.4	B	26.5	0	2092
$C_{12}H_{16}O_3$	Isoamyl salicylate	2.69	B	25	-	944

FORMULA	COMPOUND NAME	μ, D	State, or Solv.	t, °C	Method or P_a, % of P_e	Reference
$C_{12}H_{17}NO$	Acetylaminodurene	3.86	B	25	0	1797
$C_{12}H_{17}NO_3$	Ethoxy-2,3,5,6-tetramethylnitrobenzene	3.72	B	25	-	805a
$C_{12}H_{17}N_3O_2$	1-Acetyl-1-methyl-2-(*p*-dimethylamino-benzoyl)hydrazine	7.33	B	25	-	1042
$C_{12}H_{18}$	3,5-Dimethyl-*t*-butylbenzene	0.25	B	25	0	1080
	Hexamethylbenzene	0.1	B	20	-	1949
		0.52	B	20	SPC	1300
		0.56	B	40	SPC	1300
	1,3,5-Triethylbenzene	0	liq	20-30	-	22
		0.1	B	25	-	2114
		<0.2	B	25	-	2112
$C_{12}H_{18}BNO_2$	*N*-Ethyl-*B*-phenyl-*diptych*-boroxazolidine	9.1±0.07	D	25	-	553
$C_{12}H_{18}Be_4O_{13}$	Basic beryllium acetate	0	B	25,45	25?	1798
$C_{12}H_{18}CuN_2O_2$	*Bis*(acetylacetonyl)ethylenediimino-copper(II)	4.53±0.15	B	25.25	-	1286,1288
$C_{12}H_{18}Hg_2$	1,4-*Bis*(methylmercuri)-2,3,5,6-tetra-methylbenzene	1.38	CS_2	(20)	-	1700
$C_{12}H_{18}N_2NiO_2$	*Bis*(acetylacetonyl)ethylenediimino-nickel(II)	4.29±0.15	B	25	-	1286,1288
$C_{12}H_{18}N_2O$	2-Diethylaminoacetanilide	3.88±0.04	B	25	0	526
		3.96	B	25	-	1042
$C_{12}H_{18}N_2O_2$	*Bis*(acetylacetonyl)ethylenediimine	3.16±0.12	B	25.25	-	1286
	N,N-Dimethyl-4-amino-2,3,5,6-tetramethyl-nitrobenzene	4.14	B	25	-	805a

FORMULA	COMPOUND NAME	μ, D	State, or Solv.	t, °C	Method or P_a, % of P_e	Reference
$C_{12}H_{18}N_2O_2Pd$	*Bis*(acetylacetonyl)ethylenediimino-palladium(II)	4.72±0.15	B	25.25	-	1286,1288
$C_{12}H_{18}N_2O_3$	5-Cyclohexyl-5-ethylbarbituric acid	1.52±0.01	D	35	0	1847
$C_{12}H_{18}N_4O_7$	Dipropylaminepicrate	11.5	B	30	0	1268
	Triethylaminepicrate	11.7	B	30	0	1268
$C_{12}H_{18}NiO_4S_2$	*Bis*(ethylthioacetoacetato)nickel	4.54±0.1	B	25	0	420
$C_{12}H_{18}O$	6-*t*-Butyl-3-methylanisole	1.3_9	B	25	0	1071
	2,6-Diisopropylphenol	1.1	d	17.5	RT	1531
$C_{12}H_{18}O_2Si$	Methyl *p*-(trimethylsilylmethylene)benzoate	3.44	B	25	10	885
$C_{12}H_{18}O_6$	Hexamethoxybenzene	2.05	B	25	0	479
$C_{12}H_{19}BrSi$	Triethyl(*p*-bromophenyl)silane	1.77	B	25	15	887
$C_{12}H_{20}$	11-Dodecen-9-yne	0.68	B	20	0	1492
$C_{12}H_{20}AuP$	Phenyl(triethylphosphine)gold	6.2	B	25	-	262
$C_{12}H_{20}N_2$	1,10-Decanedicarbonitrile	4.95 4.99 5.04	B B B	25 25 50	0 - -	1943 1979 1979
$C_{12}H_{20}N_2O_2$	*Bis*(acetylacetoneethylenediimine)	3.16±0.03	B	25.25	10	1287
$C_{12}H_{20}N_2O_3$	*p*-Nitrophenol - triethylamine complex	5.57	D	20	-	1987
$C_{12}H_{20}N_4$	*N,N'*-*Bis*(cyanoethyl)-2,5-⇃↑-dimethyl-piperazine	4.32±0.01	B	20	0	568

d) Decalin.

FORMULA	COMPOUND NAME	μ, D	State, or Solv.	t, °C	Method or P_a, % of P_e	Reference
$C_{12}H_{20}O_2$	*l*-Bornyl acetate	1.89±0.02	n.s.	22	15	417
$C_{12}H_{20}O_3Si$	Phenyltriethoxysilane	1.68	B	20	0	1041
$C_{12}H_{20}O_4$	Diethyl cyclohexane-1,1-dicarboxylate	2.25	B	n.s.	-	495,1499
	cis-Diethyl cyclohexane-1,3-dicarboxylate	2.37	n.s.	20	-	1782
	trans-Diethyl cyclohexane-1,3-dicarboxylate	2.15	n.s.	20	-	1782
$C_{12}H_{20}Pb$	Phenyltriethyllead	0.82±0.06	B	22	0	1250
$C_{12}H_{20}S$	2,5-Di-*t*-butylthiophene	0.55±0.20	B	25	0	1641
$C_{12}H_{20}Si$	Phenyltriethylsilane	0.71±0.04	CCl_4	27	15	1247
	1-Phenyl-3-(trimethylsilyl)propane	0.3	B	25	10	885
$C_{12}H_{20}Sn$	Phenyltriethyltin	0.5	CCl_4	22	-	1249
$C_{12}H_{21}AlO_6$	Aluminum tri-*n*-butyrate	3.27	B	25	0	1455
$C_{12}H_{21}AsO_3S_6$	Arsenic *O*-isopropylxanthate	1.5_8	B	20	-	1242
$C_{12}H_{21}CoO_3S_6$	Cobaltic *O*-isopropylxanthate	2.9_5	B	25	0	1243
$C_{12}H_{21}FeO_3S_6$	Ferric *O*-isopropylxanthate	2.9_1	B	20	0	1243
$C_{12}H_{21}N$	Hexahydrojulolidine	0.76	B	25	0	434,1989
$C_{12}H_{22}$	6-Dodecyne	0	B	25	10	2094
$C_{12}H_{22}N_2$	2-Methylperhydropyrido[3,4,5-i,j]-quinolizine	1.30	B	25	0	434

FORMULA	COMPOUND NAME	μ, D	State, or Solv.	t, °C	Method or P_a, % of P_e	Reference
$C_{12}H_{22}N_2O_3$	*N*-Cyclohexyltetrahydro-5-ethyl-5-nitro-1,3-oxazine	4.42±0.1	B	20	-	631
$C_{12}H_{22}N_4NiO_4$	Nickel α-methyl-*n*-propylglyoxime	1.4	B	25	0	277
	Nickel β-methyl-*n*-propylglyoxime	1.5	B	25	0	277
	Nickel methylisopropylglyoxime	2.9	D	25	-	1321
$C_{12}H_{22}O$	Cyclododecanone	2.75±0.03	B	20	0	630
$C_{12}H_{22}O_2$	*n*-Butyl ester of 3-methylcyclohexane carboxylic acid	1.01	n.s.	n.s.	-	1361
	Menthyl acetate	1.85±0.03	B	22	15	417
$C_{12}H_{22}O_2S_4Zn$	Zinc *O*-isoamylxanthate	2.5_3	B	30	0	1243
$C_{12}H_{22}O_{11}$	Sucrose	2.8	e	20	-	1037
		3.4	f	20	-	1037
		3.4	g	20	-	1037
$C_{12}H_{23}N$	Dicyclohexylamine	1.06	Hp	25	-	920
$C_{12}H_{23}NO$	*trans*-2-Oxododecamethylenimine	3.64	B	25	0	783
$C_{12}H_{24}Br_4O_4Zr$	*Bis*(butyl acetate) - zirconium tetrabromide complex	4.24	B	20	10	923
	Bis(ethyl butyrate) - zirconium tetrabromide complex	3.83	B	20	10	923
$C_{12}H_{24}Cl_4O_4Sn$	*Bis*(ethyl butyrate) - tin tetrachloride complex	6.52	B	20	15	1438

e) Pyridine. f) *n*-Butylamine. g) Diethylamine.

FORMULA	COMPOUND NAME	μ, D	State, or Solv.	t, °C	Method or P_a, % of P_e	Reference
$C_{12}H_{24}Cl_4O_4Zr$	*Bis*(butyl acetate) - zirconium tetrachloride complex	6.74	B	20	15	1442
$C_{12}H_{24}I_4O_4Zr$	*Bis*(isobutyl acetate) - zirconium tetraiodide complex	4.26	B	20	0	1444
	Bis(*t*-butyl acetate) - zirconium tetraiodide complex	4.23	B	(0)	0	1446
	Bis(ethyl butyrate) - zirconium tetraiodide complex	3.10	B	20	0	1444
$C_{12}H_{24}N_2O_2$	Dicyclohexylamine - nitrous acid complex	4.14±0.04	B	25	-	1765
$C_{12}H_{24}N_4O$	Phenol - hexamethylenetetramine complex	2.05	CCl_4	27	-	1987
$C_{12}H_{24}N_4O_4$	3,7-Diisopropyl-3,7-dinitro-1,5-diazacyclooöctane	4.6	B	n.s.	-	943
$C_{12}H_{24}O_2$	2,2-Diisobutyl-4-methyl-1,3-dioxolane	1.46±0.02	B	30	0	151
	Ethyl caprate	1.86	B	24	-	978
	5-Hydroxy-2,2,5,6,6-pentamethyl-3-heptanone	3.27±0.04	B	20	0	1184
	Lauric acid	0.76	B	25	15	1468
$C_{12}H_{24}O_3$	Perlauric acid	2.28±0.04 2.36±0.04	B B	30 50	0 0	1163,1616 1163,1616
$C_{12}H_{24}O_4$	Caproic acid dimer	0.7	h	20	ABS,15	1171
$C_{12}H_{24}O_9Si_3$	Cyclic trimer of methyl(acetoxymethyl)-siloxane	3.34	liq	20	RT	37

h) Xylene.

FORMULA	COMPOUND NAME	μ, D	State, or Solv.	t, °C	Method or P_a, % of P_e	Reference
$C_{12}H_{25}Br$	1-Bromododecane	1.89	liq	25	-	707
		2.01	liq	20	-	657
$C_{12}H_{25}BrHg$	Dodecylmercury bromide	2.79±0.06	B	40	-	872-3
		3.70±0.06	D	40	-	872-3
$C_{12}H_{25}Cl$	1-Chlorododecane	2.11	liq	25	-	707
		1.94	B	20	RT	921
		2.01	Hx	20	RT	921
$C_{12}H_{25}ClHg$	Dodecylmercury chloride	2.79±0.06	B	40	-	872-3
		3.0±0.06	D	40	-	872-3
$C_{12}H_{25}I$	1-Iodododecane	1.8_7	CCl_4	20	0	78
$C_{12}H_{25}Li$	*n*-Dodecyllithium	1.00	Hx	25	0	2034
$C_{12}H_{26}$	Dodecane	~0	liq	-10 to 210	-	418
$C_{12}H_{26}O$	1-Dodecanol	1.52	B	20	-	977
		1.7_3	CCl_4	25	5	1108
	5-Isopropyl-5-nonanol	1.68±0.02	B	25	0	1184
$C_{12}H_{26}O_3$	1,1,1-Tripropoxypropane	1.94	B	25	5	60
$C_{12}H_{26}O_7$	Hexaethylene glycol	6.14[i]	liq	20	-	933
		3.46	D	25	10	1999
$C_{12}H_{27}As$	Tributylarsine	0.93	B	20	0	1028
$C_{12}H_{27}BO_3$	*n*-Butylborate	0.78	B	25	0	1458
		0.80	D	25	0	1458
		0.7_7	CCl_4	25	0	72

i) $g^{\frac{1}{2}}\mu$ value.

FORMULA	COMPOUND NAME	μ, D	State, or Solv.	t, °C	Method or P_a, % of P_e	Reference
$C_{12}H_{27}BO_3$ (contd.)	*sec*-Butylborate	0.86	B	25	0	1122
		0.8_5	CCl_4	25	0	72
	Isobutylborate	0.86	B	25	0	1122
		0.7_7	CCl_4	25	0	72
$C_{12}H_{27}ClSn$	Tri-*n*-butylchlorotin	3.64	B	n.s.	-	592
		4.03	D	n.s.	-	592
		3.58	Hx	n.s.	-	592
$C_{12}H_{27}N$	Tri-*n*-butylamine	0.77	B	25	5	1344
		0.78	B	25	0	68
$C_{12}H_{27}NO_3S$	Tri-*n*-butylamine - sulfur trioxide complex	7.2	B	25	5	1344
		7.2-8.4	CCl_4	25	5	1344
$C_{12}H_{27}O_2P$	Decyl ethylphosphinate	3.45	B	25	5	61
$C_{12}H_{27}O_3P$	Diethyl octylphosphonate	2.83±0.02	liq	32	-	951
	Dibutyl butanephosphonite	2.90	CCl_4	20	-	50
	Tributyl phosphite	1.92	CCl_4	20	-	50
$C_{12}H_{27}O_3PS$	Tributyl thiophosphate	3.02	liq	25	5	473
		2.84	B	25	5	473
		3.07	B	20	0	59
$C_{12}H_{12}O_4P$	Tributyl phosphate	3.35	liq	25	5	473
		3.07	B	25	5	473
		3.05	CCl_4	20	-	50
$C_{12}H_{27}O_4V$	Tri-*n*-butyl vanadate	1.12	B	25	0	271
	Triisobutyl vanadate	1.10	B	25	0	271

FORMULA	COMPOUND NAME	μ, D	State, or Solv.	t, °C	Method or P_a, % of P_c	Reference
	Tri-*sec*-butyl vanadate	1.10	B	25	0	271
	Tri-*t*-butyl vanadate	1.16	B	25	0	271
$C_{12}H_{27}P$	Tributylphosphine	2.22	cHx	25	-	920
$C_{12}H_{28}BrN$	Tri-*n*-butylammonium bromide	7.66	B	25	-	564
$C_{12}H_{28}Br_2PtS_2$	α-*Bis*(di-*n*-propylsulfide)platinum dibromide	2.34	B	20	-	844
$C_{12}H_{28}ClN$	Tri-*n*-butylammonium chloride	7.22	B	25	-	564
$C_{12}H_{28}Cl_2PtS_2$	α-*Bis*(di-*n*-propylsulfide)platinum dichloride	2.37	B	20	-	844
	α-*Bis*(diisopropylsulfide)platinum dichloride	2.41	B	20	-	844
$C_{12}H_{28}IN$	Tri-*n*-butylammonium iodide	8.15 7.6-7.7	B j	25 n.s.	- RT	564 389
$C_{12}H_{28}N_2O_4PtS_2$	α-*Bis*(di-*n*-propylsulfide)platinum dinitrite	2.50	B	20	-	844
	β-*Bis*(di-*n*-propylsulfide)platinum dinitrite	13.2	B	20	-	844
$C_{12}H_{28}N_2O_6PtS_2$	α-*Bis*(di-*n*-propylsulfide)platinum dinitrate	3.96	B	20	-	844
	β-*Bis*(di-*n*-propylsulfide)platinum dinitrate	12.0	B	20	-	844
$C_{12}H_{28}NiO_4P_2S_4$	Nickel di-*n*-propyldithiophosphate	2.0	B	31	-	1246
$C_{12}H_{28}NiP_2S_4$	Nickel *bis*(diisopropyldithiophosphinate)	1.68	B	35.3	-	1246
$C_{12}H_{28}O_4Si$	Tetra-*n*-propoxysilane	1.67 1.85	B B	25 20	5 0	53 1041

j) Benzene/*m*-xylene 1/1.

FORMULA	COMPOUND NAME	μ, D	State, or Solv.	t, °C	Method or P_a, % of P_e	Reference
$C_{12}H_{28}O_4Si$ (contd.)	Tetraisopropoxysilane	1.82	B	20	0	1041
$C_{12}H_{28}O_4Ti$	Tetrapropyl titanate	1.20	Hx	25	-	274
$C_{12}H_{28}O_6P_2$	Tetraethyl tetramethylene-1,4-diphosphonate	4.28	liq	30	-	952
$C_{12}H_{30}As_2Cl_2Pt$	α-*Bis*(triethylarsine) platinum dichloride	~0	B	20	-	845
	β-*Bis*(triethylarsine) platinum dichloride	10.6	B	20	-	845
$C_{12}H_{30}B_3N_3$	Hexaethylborazole	0.37	B	25	5	2065
$C_{12}H_{30}Br_2NiO_2P_2$	*Bis*(triethylphosphine oxide) nickel bromide	9.63	B	n.s.	-	811
$C_{12}H_{30}Br_2NiP_2$	*Bis*(triethylphosphine) nickel dibromide	~0	B	20	0	846
$C_{12}H_{30}Br_2P_2Pt$	α-*Bis*(triethylphosphine) platinum dibromide	~0	B	20	-	845
	β-*Bis*(triethylphosphine) platinum dibromide	11.3	B	20	-	845
$C_{12}H_{30}Br_3NiP_2$	*Bis*(triethylphosphine) nickel tribromide	2.30	B	25	-	862
$C_{12}H_{30}Cl_2CoP_2$	Cobaltous chloride - *bis*(triethylphosphine) complex	8.8	B	20	-	847
$C_{12}H_{30}Cl_2CrO_2P_2$	Dichloro*bis*(triethylphosphine oxide)-chromium	4.9	B	25	15	812
$C_{12}H_{30}Cl_2O_6P_2Pt$	*Bis*(diethyl ethylphosphate) - platinum dichloride complex	9.12	B	n.s.	-	57
$C_{12}H_{30}Cl_2P_2Pt$	α-*Bis*(triethylphosphine) platinum dichloride	~0	B	20	-	845

FORMULA	COMPOUND NAME	μ, D	State, or Solv.	t, °C	Method or P_a, % of P_e	Reference
	β-*Bis*(triethylphosphine) platinum dichloride	10.8	B	20	-	845
$C_{12}H_{30}Cl_2P_2Zn$	Dichloro*bis*(triethylphosphine)zinc	7.57±0.1	B	25	10	1707
$C_{12}H_{30}Cl_2PtSb_2$	β-*Bis*(triethylstibine) platinum dichloride	9.3	B	20	-	845
$C_{12}H_{30}CoI_2O_2P_2$	*Bis*(triethylphosphine oxide) cobalt diiodide	10.69	B	n.s.	-	811
$C_{12}H_{30}CoN_2O_8P_2$	*Bis*(triethylphosphine oxide) cobalt dinitrate	10.91	B	n.s.	-	811
$C_{12}H_{30}I_2P_2Pt$	α-*Bis*(triethylphosphine) platinum diiodide	~0	B	20	-	845
	β-*Bis*(triethylphosphine) platinum diiodide	8.3	B	20	-	845
$C_{12}H_{30}I_2PtSb_2$	α-*Bis*(triethylstibine) platinum diiodide	~0	B	20	-	845
$C_{12}H_{30}N_2NiO_6P_2$	*Bis*(triethylphosphine) nickel dinitrate	8.91	B	20	0	846
$C_{12}H_{30}N_2NiO_8P_2$	*Bis*(triethylphosphine oxide) nickel dinitrate	10.51	B	n.s.	-	811
$C_{12}H_{30}N_2O_4P_2Pt$	α-*Bis*(triethylphosphine) platinum dinitrite	~0	B	20	-	845
$C_{12}H_{30}N_2O_4PtSb_2$	α-*Bis*(triethylstibine) platinum dinitrite	~0	B	20	-	845
$C_{12}H_{30}N_2O_6P_2Pt$	α-*Bis*(triethylphosphine) platinum dinitrate	2.78	B	20	-	845
$C_{12}H_{30}Na_2Zn_2$	Ethylsodium - diethylzinc complex dimer	6.5-7.5	B	25	5-15	694
$C_{12}H_{30}OSi_2$	Hexaethyldisiloxane	0.65 0.63 0.66	B CCl_4 Hx	20 24.2 25	0 - 0	1041 1248 627-8

FORMULA	COMPOUND NAME	μ, D	State, or Solv.	t, °C	Method or P_a, % of P_e	Reference
$C_{12}H_{30}P_2Pt$	Dimethyl[1,2-*bis*(diethylphosphino)ethane]-platinum II)	6.7±0.1	B	25	15	293
$C_{12}H_{31}ClP_2Pt$	Chlorohydridobis(triethylphosphine)platinum	4.2	n.s.	n.s.	-	286
$C_{12}H_{32}P_2Pt$	*cis*-*Bis*(trimethylphosphine)di-*n*-propyl-platinum	5.7	B	25	15	294
$C_{12}H_{33}NO_2Si_3$	1,1,1,2,3,3,3-Heptamethyl-2-(diethylamino-methyl)trisiloxane	1.23	liq	20	RT	595
$C_{13}H_5N_3O_7$	2,4,7-Trinitro-9-fluorenone	1.46±0.10	B	25	0	1633
$C_{13}H_6Br_2O$	2,7-Dibromo-9-fluorenone	4.47	B	25	0	774
$C_{13}H_6Br_2O_2$	2,7-Dibromoxanthone	4.13	B	25	0	1069
$C_{13}H_6N_2O_5$	2,5-Dinitro-9-fluorenone	~6	B	25	0	774
	2,7-Dinitro-9-fluorenone	4.84	B	25	0	774
$C_{13}H_6N_2O_6$	2,4-Dinitroxanthone	3.01	B	25	0	1069
	2,7-Dinitroxanthone	5.76	B	25	0	1069
$C_{13}H_7BrO$	2-Bromofluorenone	3.80±0.04	B	30	5	130
	3-Bromofluorenone	2.52±0.04	B	30	5	130
	10-Bromo*peri*naphthindenone	3.27	D	n.s.	-	2174
$C_{13}H_7ClN_2O_2$	9-Chloro-6-nitroacridine	4.98±0.02	B	25	0	942
	9-Chloro-7-nitroacridine	4.53±0.02	B	25	0	942
$C_{13}H_7ClN_2O_3$	9-Chloro-3-nitroacridine *N*-oxide	6.08	B	25	0	1552

FORMULA	COMPOUND NAME	μ, D	State, or Solv.	t, °C	Method or P_a, % of P_e	Reference
$C_{13}H_7Cl_2N$	3,9-Dichloroacridine	2.01±0.02	B	25	10	1549
$C_{13}H_7MnO_6$	Phenylacetylmanganese pentacarbonyl	2.18±0.04	B	25	0	116
	p-Toluylmanganese pentacarbonyl	2.08±0.04	B	25	20	116
$C_{13}H_7NO_3$	2-Nitro-9-fluorenone	6.09	B	25	0	774
$C_{13}H_8Br_2$	2,7-Dibromofluorene	0 0.22	B B	15 25	0 0	137 774
$C_{13}H_8Br_2O$	*p,p'*-Dibromobenzophenone	1.70 1.93	B B	22 n.s.	0 -	138 1223
	3,7-Dibromo-2-phenyl-2,4,6-cyclohepta-trien-1-one	3.67±0.01	B	25	0	1022
$C_{13}H_8Br_2S$	*p,p'*-Dibromothiobenzophenone	1.71	B	n.s.	-	1223
$C_{13}H_8ClN$	9-Chloroacridine	0	B	25	10	1549
$C_{13}H_8ClNO$	9-Chloroacridine *N*-oxide	2.62	B	25	0	1552
$C_{13}H_8Cl_2$	9,9-Dichlorofluorene	1.87	B	20	0	137
$C_{13}H_8Cl_2N_2$	*Bis*(*p*-chlorophenyl) carbodiimide	0.0	D	35	0	1713
	Di-(*p*-chlorophenyl) diazomethane	0.62	B	25	-	1774
$C_{13}H_8Cl_2O$	*p,p'*-Dichlorobenzophenone	1.58±0.03 1.65 1.79 1.72	B B B Hx	22 13 n.s. 25	0-15 0 - 0	555 138 1223 1906
$C_{13}H_8Cl_2S$	*p,p'*-Dichlorothiobenzophenone	1.58	B	n.s.	-	1223

FORMULA	COMPOUND NAME	μ, D	State, or Solv.	t, °C	Method or P_a, % of P_e	Reference
$C_{13}H_8Cl_4$	*Bis*(*p*-chlorophenyl)dichloromethane	0.48	B	17	0	142
$C_{13}H_8Cl_5OSb$	Antimony pentachloride - *peri*naphthindenone complex	8.49	D	n.s.	-	2174
$C_{13}H_8CrO_3$	Naphthalene - chromium tricarbonyl complex	6.33±0.03	B	25	15	520
$C_{13}H_8F_2O$	*p,p'*-Difluorobenzophenone	1.76	B	25	5	1114
$C_{13}H_8N_2O_2S$	4-Cyano-4 -nitrodiphenyl sulfide	3.36±0.01	B	20	0	1656
$C_{13}H_8N_2O_3$	4-Cyano-4 -nitrodiphenyl ether	2.71±0.01	B	20	0	1656
$C_{13}H_8N_2O_4$	2,5-Dinitrofluorene	7.11	B	25	0	774
	2,7-Dinitrofluorene	~1.7 ~2.4	B n.s.	25 25	0 0	774 1922
$C_{13}H_8O$	9-Fluorenone	3.32 3.36±0.02 3.38	B B B	18 20-60 25	0 MA 0	137 1517 774
	*peri*Naphthindenone	3.89 3.99 3.98	B D k	n.s. n.s. n.s.	- - -	2174 2174 2174
$C_{13}H_8OS$	Thioxanthone	5.4	D	17	-	2091
	Xanthione	5.4	B	28	-	2091
$C_{13}H_8O_2$	3-Hydroxy*peri*naphthindenone	4.28	D	n.s.	-	2174
	Xanthone	2.94-2.98 2.95±0.02 3.10 3.14	B B B B	10-40 15 14 25	- 15 0 0	1584 2023 164 1069

k) Anisole.

FORMULA	COMPOUND NAME	μ, D	State, or Solv.	t, °C	Method or P_a, % of P_e	Reference
$C_{13}H_8S_2$	Thioxanthione	5.2	D	31	-	2091
$C_{13}H_9Br$	2-Bromofluorene	1.68±0.02	B	30?	5	131
$C_{13}H_9BrO$	*p*-Bromobenzophenone	2.78	B	20	0	138
	4-Bromo-2-phenyl-2,4,6-cycloheptatrien-1-one	2.73±0.01	B	25	0	1022
$C_{13}H_9Br_2N$	Acridine - bromine complex	3.79[kk]	B	25	-	895
$C_{13}H_9Cl$	9-Chlorofluorene	1.78	B	14	0	137
$C_{13}H_9ClN_2$	9-Amino-3-chloroacridine	4.84±0.02	B	25	10	1549
$C_{13}H_9ClN_2O_2$	4-Chloro-4'-nitro-*N*-benzylideneaniline	3.86	B	25	-	498
	4'-Chloro-4-nitro-*N*-benzylideneaniline	2.79	B	25	-	498
$C_{13}H_9ClO$	*p*-Chlorobenzophenone	2.73 2.74 2.66	B Hx n.s.	13 25 n.s.	0 0 -	138 1906 701
$C_{13}H_9Cl_2N$	*syn*-*p*-Chlorobenzophenonechloroimine	2.70±0.05	B	25	-	1941
	anti-*p*-Chlorobenzophenonechloroimine	2.49±0.06	B	25	-	1941
	p-Chlorobenzylidene-*p*-chloroaniline	1.57	B	25	0	398
	3,3'-Dichloro-*N*-benzylideneaniline	2.36	B	25	-	498
	3,4'-Dichloro-*N*-benzylideneaniline	1.89	B	25	-	498
	3',4-Dichloro-*N*-benzylideneaniline	2.29	B	25	-	498
$C_{13}H_9Cl_2NO$	*p*,*p*'-Dichlorobenzanilide	3.26	B	30	0	2148

kk) Lower limit.

FORMULA	COMPOUND NAME	μ, D	State, or Solv.	t, °C	Method or P_a, % of P_e	Reference
$C_{13}H_9FO$	*p*-Fluorobenzophenone	2.65	B	25	5	1114
$C_{13}H_9I_2N$	Acridine - iodine complex	3.51^m	B	25	-	895
$C_{13}H_9N$	Acridine	1.94±0.02	B	25	10	1549
		1.97	B	14	0	138
		2.13±0.03	n.s.	n.s.	-	2
	4-Cyanobiphenyl	4.33±0.02	B	25	0	475
	Phenanthridine	1.5_1	B	25	0	399
	o-Phenylbenzonitrile	3.84	B	18	-	129
$C_{13}H_9NO$	Acridine *N*-oxide	3.90±0.02	n.s.	n.s.	-	2
		4.08	B	25	0	1552
	2-Biphenylisocyanate	2.19	B	20	MA	870
	4-Cyanodiphenyl ether	4.23±0.01	B	20	0	1656
$C_{13}H_9NO_2$	2-Nitrofluorene	4.5	D	25	0	1922
		5.48	D	11	0	137
$C_{13}H_9NS$	4-Cyanodiphenyl sulfide	4.14±0.01	B	20	0	1656
$C_{13}H_9N_3$	*cis*-4-Diphenyldiazocyanide	5.53	B	25	0	538
	trans-4-Diphenyldiazocyanide	4.52	B	25	0	538
$C_{13}H_9N_3O_4$	*p*-Nitrobenzylidene-*p*-nitroaniline	3.59	B	25	15	856
$C_{13}H_{10}$	Fluorene	0.28	B	14	0	137
		0.28	B	20	-	1673
		0.53	B	25	-	1921

m) Lower limit.

FORMULA	COMPOUND NAME	μ, D	State, or Solv.	t, °C	Method or P_a, % of P_e	Reference
		0.58	B	20	-	1181,1189
		0.83	B	25	0	774
		0.65	D	25	-	1921
$C_{13}H_{10}AlBr_3O$	Aluminum bromide - benzophenone complex	8.47	B	20	-	1734
$C_{13}H_{10}AlCl_3O$	Aluminum chloride - benzophenone complex	8.38±0.14	B	20	-	1415
		8.74	B	20	-	2006-7
$C_{13}H_{10}AsN$	Diphenylcyanoarsine	4.22	Hx	20	15	1346
$C_{13}H_{10}BrN$	Benzophenonebromoimine	2.86±0.03	B	25	-	1941
	5-Bromosalicylideneaniline	1.20	B	25	0	398
$C_{13}H_{10}Br_2$	*Bis*(*p*-bromophenyl)methane	1.81	B	17	0	142
		1.88	B	25	0	654
$C_{13}H_{10}Br_2O_2S_2$	*p*-Bromophenylsulfinyl *p*-bromobenzyl sulfide	3.38	B	25,40	0	619
$C_{13}H_{10}ClN$	Benzophenonechloroimine	2.99±0.04	B	25	-	1941
	2-Chloro-*N*-benzylideneaniline	2.48	B	25	-	498
	2'-Chloro-*N*-benzylideneaniline	1.11	B	25	-	498
	3-Chloro-*N*-benzylideneaniline	2.40	B	25	-	498
	3'-Chloro-*N*-benzylideneaniline	2.01	B	25	-	498
	4-Chloro-*N*-benzylideneaniline	1.79	B	25	0	398
		2.58	B	25	-	498
$C_{13}H_{10}ClNO$	α-*o*-Chlorobenzophenone oxime	1.62	D	n.s.	-	1470
	β-*o*-Chlorobenzophenone oxime	1.62	D	n.s.	-	1470

FORMULA	COMPOUND NAME	μ, D	State, or Solv.	t, °C	Method or P_a, % of P_e	Reference
$C_{13}H_{10}ClNO$ (contd.)	α-*m*-Chlorobenzophenone oxime	1.51	B, CCl_4	n.s.	-	1470
	β-*m*-Chlorobenzophenone oxime	1.62	B, CCl_4	n.s.	-	1470
	α-*p*-Chlorobenzophenone oxime	2.34	D	n.s.	-	1470
	β-*p*-Chlorobenzophenone oxime	2.40	D	n.s.	-	1470
	N-*p*-Chlorophenylbenzamide	4.35	B	30	0	2148
	Diphenyl carbamoyl chloride	3.75	B	20	-	1412
	N-Phenyl-*p*-chlorobenzamide	3.17	B	30	0	2148
	Salicylidene-*p*-chloroaniline	2.27 2.49	B D	25 25	5 5	368 368
$C_{13}H_{10}Cl_2$	4-Chloro-4'-(chloromethyl)biphenyl	1.9_1	B	25	-	2085
	Diphenyldichloromethane	2.41	B	17	0	142
$C_{13}H_{10}Cl_2O_2S_2$	*p*-Chlorophenylsulfinyl *p*-chlorobenzyl sulfide	3.35 3.39	B B	25 40	0 0	619 619
$C_{13}H_{10}N_2$	9-Aminoacridine	4.13±0.02	B	25	10	942,1549
	Carbodianil	1.91	B	18	0	156
	Diphenyldiazomethane	1.43	CCl_4	0	-	1774
$C_{13}H_{10}N_2O$	Pyocyanine	7.0	D	25	-	861
$C_{13}H_{10}N_2O_2$	2-Amino-7-nitrofluorene	5.75	B	20	FL	375
	Benzal-*p*-nitroaniline	5.04±0.05	B	n.s.	15	705

FORMULA	COMPOUND NAME	μ, D	State, or Solv.	t, °C	Method or P_a, % of P_e	Reference
	2,7-Nitroaminofluorene	6.8	D	25	0	1922
	p-Nitrobenzalaniline	4.18±0.05	B	n.s.	-	705
$C_{13}H_{10}N_2O_4$	Di-*p*-nitrophenylmethane	4.32	B	24	0	142
	2-Methyl-5,5'-dinitrobiphenyl	4.55	B	25	-	1697
$C_{13}H_{10}N_2S_2$	Phenylbenzothiazolesulfenamide	2.38	B	25	-	637
$C_{13}H_{10}N_4$	1,5-Diphenyltetrazole	5.95	B	25	0	892
$C_{13}H_{10}O$	Benzophenone	2.5	gas	100	-	464
		3.09	liq	50-60	-	1497
		2.5	B	20-50	-	463
		2.5	B	30	Relax	1375,1377
		2.86	B	28	-	979
		2.87	B	30	ABS	1580
		2.90	n	n.s.	-	258
		2.95	B	20	-	314
		2.96	B	25	5	1114
		2.98±0.03	B	25	15?	416,2137
		2.98	B	13	0	138
		2.98	B,Hx[p]	20	-	609
		2.99	B	20	-	187
		3.03±0.02	B	22	15	555
		3.16	B	18	0	672
		2.97	pXy	30.6	5	593
		3.2	n.s.	n.s.	-	2048
	Fluorenol	2.10±0.13	B	60	-	771
		1.66±0.05	D	20	-	771
		1.78±0.05	D	60	-	771
	2-Phenyl-2,4,6-cycloheptatrien-1-one	3.82±0.05	B	25	0	1022

n) Benzene saturated with water.   p) Also toluene and CS_2.

FORMULA	COMPOUND NAME	μ, D	State, or Solv.	t, °C	Method or P_a, % of P_e	Reference
$C_{13}H_{10}O$ (contd.)	Xanthene	1.29	B	28	0	164
$C_{13}H_{10}O_2$	4-Hydroxybenzophenone	3.96	D	32	0	1271
	Phenyl benzoate	1.83_1 ±0.01	B	22	15	413
		1.94	B	30	-	2062
$C_{13}H_{10}O_3$	4,4'-Dihydroxybenzophenone	4.49	D	32	0	1271
	o-Phenoxybenzoic acid	2.73	B	25	-	945
	Phenyl salicylate	2.27	B	n.s.	5	334
		3.01	B	25	-	944
		3.18	B	40	-	766,876
$C_{13}H_{10}O_4$	9-Fluorenylidenedihydroperoxide	2.4	B	25	-	731
	4-Methyl-ε-acetylumbelliferone	3.33[q]	B	25-45	-	827
$C_{13}H_{10}S$	Thiobenzophenone	3.40	B	20	0	792
$C_{13}H_{11}BrO$	*p*-Bromophenyl *p*-tolyl ether	2.41±0.02	gas	230-245	5	332
		2.00	Hx	25	0	1906
	1-(2-Bromo-1-propanoyl)naphthalene	2.88±0.02	B	25	-	260
	2-(2-Bromo-1-propanoyl)naphthalene	3.19±0.02	B	25	-	260
$C_{13}H_{11}Br_2N_3$	4,4'-Dibromo-*N*-methyldiazoaminobenzene	2.54±0.04	B	25	0	1102
$C_{13}H_{11}Cl$	Chlorodiphenylmethane	1.85	B	25	5	319
		1.91	B	25	-	2085
		1.93	D	20	0	1969
	3-α-Naphthyl-1-chloro-1-propene	1.48±0.02	liq,B	32	5	1035
		1.28±0.02	solid	32	5	1035

q) Calculated from Jatkar's equation.

FORMULA	COMPOUND NAME	μ, D	State, or Solv.	t, °C	Method or P_a, % of P_e	Reference
$C_{13}H_{11}N$	*N*-Benzylideneaniline	1.56±0.1	B	n.s.	15	705
		1.57	B	25	-	498
		1.58	B	25	0	398
$C_{13}H_{11}NO$	Benzanilide	3.38	B	30	0	2148
		3.83	D	25,40	0	959
	Salicylideneaniline	2.40	B	25	5	368
		2.47	B	25	0	398
		2.57	D	25	5	368
$C_{13}H_{11}NO_2$	Salicylanilide	4.09	B	25	-	1042
$C_{13}H_{11}NO_2S$	2-Methyl-4'-nitrodiphenyl sulfide	4.25±0.01	B	20	0	1657
	3-Methyl-4'-nitrodiphenyl sulfide	4.50±0.01	B	20	0	1657
	4-Methyl-4'-nitrodiphenyl sulfide	4.67±0.01	B	20	0	1656
$C_{13}H_{11}NO_3$	2-Methyl-4'-nitrodiphenyl ether	4.40±0.01	B	20	0	1657
	3-Methyl-4'-nitrodiphenyl ether	4.57±0.01	B	20	0	1657
	4-Methyl-4'-nitrodiphenyl ether	4.70±0.01	B	20	0	1656
	p-(2-Furyl)-β-methyl-β-nitrostyrene	4.59	B	25	23	2033
$C_{13}H_{11}NO_3S$	*p*-(*p*-Nitrophenylthio)anisole	4.98	B	20	-	1682
$C_{13}H_{11}NO_4$	Ethyl 5-nitro-2-naphthoate	4.05	B	25	5	1533
	Ethyl 8-nitro-2-naphthoate	3.99	B	25	5	1533
$C_{13}H_{11}NS_2$	Diphenyldithiocarbamic acid	4.48	B	25	0	639
$C_{13}H_{11}N_3$	9-Hydrazinoacridine	5.99	B	25	0	1409

FORMULA	COMPOUND NAME	μ, D	State, or Solv.	t, °C	Method or P_a, % of P_e	Reference
$C_{13}H_{11}N_3O$	9-Hydrazinoacridine *N*-oxide	6.32	B	25	0	1409
$C_{13}H_{11}N_3O_3$	4-Methoxy-4'-nitroazobenzene	6.5	B	17	-	2091
$C_{13}H_{12}$	Diphenylmethane	<0.4	gas	100	MB	464
		0.26	liq	30-60	RT	1617
		~0	B	12	0	142
		0.22-0.26	B	25	0	654
		0.33_5	B	25	0	1601
		0.37	B	20,50	-	463
		0.77	B	20	-	1412
	6-Methyl-6-phenylfulvene	1.20±0.02	B	20	-	967
	α-Naphthylmethylethylene	~0	B	27	-	165
	5-Phenyl-1,5-heptadien-3-yne	0.70	B	25	-	1750
$C_{13}H_{12}BrNO$	5-Bromo-1-phenethyl-2-pyridone	3.38	B	14?	-	1952
$C_{13}H_{12}ClNO$	4-Chloro-1-phenethyl-2-pyridone	5.36	B	14?	-	1952
	5-Chloro-1-phenethyl-2-pyridone	3.24	B	14?	-	1952
$C_{13}H_{12}INO$	5-Iodo-1-phenethyl-2-pyridone	3.59	B	14?	-	1952
$C_{13}H_{12}N_2$	Benzaldehyde phenylhydrazone	1.91	B	20	0	339
		1.99	B	19	15	76
		2.08	B	15	0	2008
	Benzophenonehydrazone	2.04	B	16	0	164
	2,9-Diaminofluorene	1.98	D	18	0	137
	N,N'-Diphenylformamidine	2.22	D	25	0	1451

FORMULA	COMPOUND NAME	μ, D	State, or Solv.	t, °C	Method or P_a, % of P_e	Reference
	1-Ethyl-(1,2)-naphthimidazole	4.07	D	25	0	1451
$C_{13}H_{12}N_2O$	1,1-Diphenylurea	2.7±0.1	D	17	0	166
		4.11	D	20	-	562
		4.20±0.04	D	25	0	120
		4.74	r	25	-	562
		3.84	s	25	-	562
		3.88	t	25	-	562
		4.18	u	25	-	562
	1,3-Diphenylurea	3.94±0.04	D	25	0	120
		4.41	D	25	-	562
		4.6	D	20	0	792
		6.94	r	25	-	562
		5.92	s	25	-	562
		6.02	t	25	-	562
		5.77	u	25	-	562
	2-Hydroxy-5-methylazobenzene	1.25	B	25	-	800
		1.23	D	25	-	800
	4-Hydroxy-2-methylazobenzene	1.69	B	25	-	800
		1.88	D	25	-	800
	4-Hydroxy-3-methylazobenzene	1.55	B	25	-	800
		2.11	D	25	-	800
	p-Methoxyazobenzene	1.30	B	23	0	164
		1.58	B	25	-	800
		1.55	D	25	-	800
	Phenyl-*p*-tolylnitrosamine	3.57	B	25	5	1114
$C_{13}H_{12}N_2O_2$	3-Ethyl-2-methyl-5*H*-pyrazolo-[a][3,1]-benzoxazin-5-one,α-lactone	1.33	B	n.s.	-	2038

r) Methylcellosolve. s) Acetone. t) Ethanol. u) Methanol.

FORMULA	COMPOUND NAME	μ, D	State, or Solv.	t, °C	Method or P_a, % of P_e	Reference
$C_{13}H_{12}N_2O_2$ (contd.)	3-Ethyl-2-methyl-5*H*-pyrazolo-[a][3,1]-benzoxazin-5-one, β-lactone	2.53	B	n.s.	-	2038
	N-Methyl(*o*-nitrophenyl)phenylamine	3.67 3.84	B D	25 25	5 5	1219 1219
	N-Methyl(*p*-nitrophenyl)phenylamine	6.69 6.83	B D	25 25	5 5	1219 1219
	1'-Phenylbenzeneazomethane dioxide	3.36	B	20	-	567
$C_{13}H_{12}N_2O_3$	5-Nitro-1-phenethyl-2-pyridone	3.54	B	14?	-	1952
$C_{13}H_{12}N_2O_3S$	Phenylazoxy-*p*-tolylsulfone	6.0_1	B	30	0	540
$C_{13}H_{12}N_2S$	1,3-Diphenylthiourea	4.89 4.9 $5.05_1 \pm 0.002$	D D D	20 26 25	0 0 5	792 166 275-6
$C_{13}H_{12}N_4O$	*sym*-Diphenylcarbazone	3.84	D	35	0	1569
$C_{13}H_{12}N_4S$	*sym*-Diphenylthiocarbazone	2.54	D	35	0	1569
	N-Phenylamino-*N'*-anilinothiourea	3.53	B	20	10	424
$C_{13}H_{12}O$	Benzohydrol	1.60 1.62	B B	25 25	5 -	319 194
	Benzyl phenyl ether	1.13	liq	40-60	RT	1617
	2,7-Dimethyl-4,5-benzo-2,4,6-cyclohepta-trien-1-one	3.66-3.84 3.21 3.54	B CCl_4 cHx	25 25 25	0 0 0	563 563 563

FORMULA	COMPOUND NAME	μ, D	State, or Solv.	t, °C	Method or P_a, % of P_e	Reference
	6,8-Dimethyl-7*H*-benzocyclohepten-7-one	3.66±0.03	B	25	0	127
		3.67	B	25	0	1303
		3.7	n.s.	n.s.	-	2048
	6-(4-Methoxyphenyl)fulvene	2.17	B	20	-	967
	2-Methylphenyl phenyl ether	0.94±0.01	B	20	-	1681
		1.03	B	20	RT	1241b
	3-Methylphenyl phenyl ether	1.21±0.01	B	20	-	1681
	4-Methylphenyl phenyl ether	1.26±0.01	B	20	-	1681
		1.32	B	20	0	160
$C_{13}H_{12}OS$	*p*-(Phenylthio)anisole	2.17	B	20	-	1682
$C_{13}H_{12}O_2$	1-Acetyl-4-methoxynaphthalene	4.02	B	25	5	1214
		4.14	D	25	5	1214
	2-Acetyl-1-methoxynaphthalene	3.35	B	25	5	1214
		3.39	D	25	5	1214
$C_{13}H_{12}O_2S_2$	Phenylsulfinyl benzyl sulfide	5.31	B	25	0	619
		5.49	D	25	0	619
$C_{13}H_{12}O_3S$	*o*-Methoxyphenyl phenyl sulfone	6.01	B	30	0	47
$C_{13}H_{12}S$	2-Tolyl phenyl sulfide	1.34	B	20	-	1658
	3-Tolyl phenyl sulfide	1.62	B	20	-	1658
	4-Tolyl phenyl sulfide	1.76	B	20	-	1658
		1.77	B	25	0	308
$C_{13}H_{13}BrN_2O_4S$	*endo*,*exo*-2-Bromo-3-(2,4-dinitrophenylthio)norbornane	5.40	B	25	-	1031

FORMULA	COMPOUND NAME	μ, D	State, or Solv.	t, °C	Method or P_a, % of P_e	Reference
$C_{13}H_{13}BrO$	2-(4-Bromobenzylidene)cyclohexanone	2.89	B	30	0	784
$C_{13}H_{13}ClN_2O_4S$	*endo,exo*-2-Chloro-3-(2,4-dinitrophenylthio)norbornane	5.41	B	25	-	1031
$C_{13}H_{13}ClO$	2-(4-Chlorobenzylidene)cyclohexanone	2.75	B	30	0	784
$C_{13}H_{13}ClSi$	Methyldiphenylchlorosilane	2.08	Hx	-40 to 20	-	1964
$C_{13}H_{13}IO$	2-(4-Iodobenzylidene)cyclohexanone	2.76	B	30	0	784
$C_{13}H_{13}N$	*N*-Methyldiphenylamine	1.56 1.62	B D	25 25	5 5	1219 1219
	Phenyl-*p*-toluidine	1.02	B	25	5	1114
$C_{13}H_{13}NO$	*p*-Homosalicylideneaniline (red)	2.94	B	25	0	399
	p-Homosalicylideneaniline (yellow)	2.98	B	25	0	399
	1-Phenethyl-2-pyridone	3.91 4.2	B B	27 14?	0 -	1934 1952
$C_{13}H_{13}NO_2S$	*p*-Benzenesulfonotoluidide	5.30	B	20	0	1703
	N-Methylbenzenesulfonanilide	5.08	B	20	0	1703
	p-Toluenesulfonanilide	5.25	B	20	0	1703
$C_{13}H_{13}NO_3$	Ethyl 3-methyl-5-phenyl-4-isoxazolecarboxylate	3.11±0.04	D	25	0	1857
	Ethyl 3-phenyl-5-methyl-4-isoxazolecarboxylate	3.21±0.04	D	25	0	1857
$C_{13}H_{13}N_3$	*N*-Methyldiazoaminobenzene	1.50±0.03	B	25	0	1102

FORMULA	COMPOUND NAME	μ, D	State, or Solv.	t, °C	Method or P_a, % of P_e	Reference
$C_{13}H_{14}ClNO_2S$	*endo*,*exo*-2-Chloro-3-(2-nitrophenylthio)-norbornane	4.59	B	25	-	1031
	endo,*exo*-2-Chloro-3-(4-nitrophenylthio)-norbornane	5.04	B	25	-	1031
$C_{13}H_{14}CrO_3$	1,2,4,5-Tetramethylbenzene - chromium tricarbonyl complex	6.04	B	n.s.	-	1568
$C_{13}H_{14}CuF_6N_2O_2$	*Bis*trifluoroacetylacetonepropylenediimino-copper(II)	9.73±0.03	B	25	10	762
$C_{13}H_{14}F_6N_2NiO_2$	*Bis*trifluoroacetylacetonepropylenediimino-nickel(II)	9.30±0.03	B	25	10	762
$C_{13}H_{14}N_2OS$	1-Ethyl-2-(benzylthio)-4(1*H*)-pyrimidone	6.91	D	25	0	1716
$C_{13}H_{14}N_2OS_2$	5-(*N*-Methylanilinomethylene)-3-ethyl-rhodanine	5.33	B	25	-	1766
$C_{13}H_{14}N_4O$	*sym*-Diphenylcarbohydrazide	3.80	D	35	0	1569
$C_{13}H_{14}O$	2-Benzylidenecyclohexanone	3.14±0.03	B	25	0	127
$C_{13}H_{15}BrO_2$	2-(4-Bromo-α-hydroxybenzyl)cyclohexanone	3.88	B	n.s.	-	785
$C_{13}H_{15}ClN_2$	7-Chloro-4-butylaminoquinoline	5.70±0.05	D	n.s.	-	1380
$C_{13}H_{15}ClO$	2-(4-Chlorobenzyl)cyclohexanone	3.31	B	30	0	784
$C_{13}H_{15}ClO_2$	2-(4-Chloro-α-hydroxybenzyl)cyclohexanone	3.64	B	n.s.	-	785
$C_{13}H_{15}IO_2$	2-(4-Iodo-α-hydroxybenzyl)cyclohexanone	3.73	B	n.s.	-	785
$C_{13}H_{15}N$	2,5,4'-Trimethyl-1-phenylpyrrole	2.34±0.02	B	25	0	931

FORMULA	COMPOUND NAME	μ, D	State, or Solv.	t, °C	Method or P_a, % of P_e	Reference
$C_{13}H_{15}NO_2$	2-Ethyl-2-phenylglutarimide	2.83±0.03	D	30	0	1059
$C_{13}H_{16}Cl_2N_2Pt$	Propylenedichloroplatinum(II) - dipyridine complex	6.05	n.s.	n.s.	-	4
$C_{13}H_{16}F_6N_2O_2$	*Bis*(trifluoroacetylacetone)propylenediimine	5.70±0.02	B	25	5	761
$C_{13}H_{16}N_2O_2$	2-Ethyl-2-(4-aminophenyl)glutarimide	3.64±0.03	D	30	0	1059
$C_{13}H_{16}N_2S_2$	*N*-Cyclohexylbenzothiazolesulfenamide	2.87	B	25	-	637
$C_{13}H_{16}O$	2-Benzylcyclohexanone	2.90±0.01	B?	n.s.	-	302
	2,2-Dimethyl-4-pentenophenone	2.68±0.01	B	(25)	-	301
	2-Ethyl-4-pentenophenone	2.80±0.01	B	(25)	-	301
	2-(2-Tolyl)cyclohexanone	3.31	B	30	0	784
	2-(3-Tolyl)cyclohexanone	2.99	B	30	0	784
	2-(4-Tolyl)cyclohexanone	2.91	B	30	0	784
$C_{13}H_{16}O_4$	Diethyl phenylmalonate	2.543	B	30	0	1501
$C_{13}H_{17}N$	1-Ethyl-3,3-dimethyl-2-methyleneindoline	1.13	B	25	0	1767
	1-Methyl-3-*t*-butylindole	2.05	B	25	-	817
$C_{13}H_{17}NO$	1-Naphthol - trimethylamine complex	3.41±0.05	B	25	-	789
	2-Naphthol - trimethylamine complex	3.80±0.08	B	25	-	789
$C_{13}H_{17}NO_2$	Ethyl *p*-dimethylaminocinnamate	4.18 4.6	B B	25 19	0 -	412 2090

FORMULA	COMPOUND NAME	μ, D	State, or Solv.	t, °C	Method or P_a, % of P_e	Reference
$C_{13}H_{17}N_3O$	Pyramidone	5.2	B	25	0	227
$C_{13}H_{17}N_3O_3$	Dihydroxypyramidone[v]	4.42	B	25	-	1042
$C_{13}H_{18}N_2OS_2$	1-[2-(1-Ethylpyrrolidinyl)]-2-[5-(3-ethyl-2-thione-4-oxothiazolylidenyl)]ethane	8.56	B	25	0	1767
	5-(2'-*N*-Piperidylvinylmethylene)-3-ethylrhodanine	8.22	B	25	-	1766
$C_{13}H_{19}NO_2$	2,3,5,6-Tetramethylphenyl urethan	3.1±0.08 3.2±0.04	B D	35 35	- -	2092 2092
$C_{13}H_{20}BNO_2$	*N*-Isopropyl-*B*-phenyl-*diptych*-boroxazolidine	9.6±0.08	D	25	-	553
$C_{13}H_{20}CuN_2O_2$	*Bis*(acetylacetone)propylenediiminocopper(II)	4.63	B	25.25	-	1286,1288
	Bis(acetylacetone)trimethylenediimino-copper(II)	4.23	B	25.25	-	1286,1288
$C_{13}H_{20}N_2NiO_2$	*Bis*(acetylacetone)propylenediiminonickel(II)	4.43	B	25.25	-	1286,1288
$C_{13}H_{20}N_2O$	α-Diethylamino-*N*-methylacetanilide	3.42±0.05	B	25	0	526
$C_{13}H_{20}N_2O_2$	β-Diethylaminoethyl *p*-aminobenzoate	3.7	n.s.	n.s.	-	1299
$C_{13}H_{20}N_4O$	1-Acetyl-1-methyl-2-isonicotinoylhydrazine	4.24	B	25	-	1042
$C_{13}H_{20}O$	β-Ionone	3.50	B	25	-	896
$C_{13}H_{20}O_2Si$	Methyl 4-(2-(trimethylsilyl)ethyl)benzoate	3.23	B	25	10	885
$C_{13}H_{20}O_8$	Pentaerythritol tetraacetate	1.9 2.20±0.03 2.6 ~2.8 2.3_7	B B B B CCl_4	25 22 n.s. n.s. 25	- 15 - - 0	2112 1437 431 429 1081

v) Placement of OH groups not given.

FORMULA	COMPOUND NAME	μ, D	State, or Solv.	t, °C	Method or P_a, % of P_e	Reference
$C_{13}H_{20}O_8$ (contd.)	Tetraethyl methanetetracarboxylate	>0	B	n.s.	-	429
$C_{13}H_{21}As$	Butylethyl-*p*-tolylarsine	1.30	B	20	0	1028
$C_{13}H_{21}BrSi$	Triethyl-*p*-bromobenzylsilane	2.08	B	25	15	887
$C_{13}H_{22}N_2O_2$	*Bis*(acetylacetone)propylenediimine	2.98	B	25	10	1286-7
$C_{13}H_{22}O_2$	*l*-Bornyl propionate	1.86±0.05	B	22	15	417
$C_{13}H_{24}N_2O_3$	*N*-Cyclohexyltetrahydro-5-*n*-propyl-5-nitro-1,3-oxazine	4.41±0.1	B	20	-	631
	N-Cyclohexyltetrahydro-5-isopropyl-5-nitro-1,3-oxazine	4.56±0.1	B	20	-	631
$C_{13}H_{24}O$	Cyclotridecanone	2.76±0.03	B	20	0	630
$C_{13}H_{24}O_2$	*l*-Menthyl propionate	1.79_5±0.015	B	22	15	413
$C_{13}H_{24}O_4$	Brassylic acid	2.68	D	35	-	1581
	Diethyl azelate	2.36	liq	30-40	-	1497
	Diethyl di-*n*-propylmalonate	2.17	B	n.s.	-	1499
$C_{13}H_{25}NO$	*trans*-2-Oxotridecamethylenimine	3.68	B	25	0	783
$C_{13}H_{26}O$	4,8-Dimethyl-6-undecanone	2.59	CCl_4	25	-	67
	7-Tridecanone	2.61	CCl_4	25	-	67
$C_{13}H_{26}O_2$	Ethyl *n*-undecylate	1.91	Hp	25	0	1122
$C_{13}H_{26}O_2Si_3$	1,1,1,2,3,3,3-Heptamethyl-2-phenyltrisiloxane	1.00	liq	20	RT	595

FORMULA	COMPOUND NAME	μ, D	State, or Solv.	t, °C	Method or P_a, % of P_e	Reference
$C_{13}H_{28}NO_4P$	Dibutyl ester of diethylcarbamoylphosphonic acid	3.12	B	25	5	58
$C_{13}H_{28}O_2$	2,2-Di-*n*-propoxy-*n*-heptane	0.93	B	25	0	1457
$C_{13}H_{28}O_4$	Pentaerythritol tetraethyl ether	1.1	B	n.s.	-	431
$C_{13}H_{28}O_4S_2$	1,3-*Bis*(isopentylsulfonyl)propane	4.56	B	20	0	636
$C_{13}H_{29}NO_2$	Dodecylamine formate	1.78	B	30	5	1757
$C_{13}H_{30}O_6P_2$	Tetraethyl pentamethylene diphosphonate	4.28	liq	30	-	952
$C_{13}H_{32}Si_2$	1,1,1,3,3,3-Hexaethyl-1,3-disilicapropane	0.57	Hx	25	0	627
$C_{13}H_{33}BrP_2Pt$	*trans*-Methylbromo*bis*(triethylphosphine)-platinum(II)	3.7±0.1	B	25	15	293
$C_{13}H_{33}ClP_2Pt$	*cis*-Methylchloro*bis*(triethylphosphine)-platinum(II)	8.4±0.1	B	25	15	293
	trans-Methylchloro*bis*(triethylphosphine)-platinum(II)	3.4	B	25	15	293
$C_{13}H_{33}IP_2Pt$	*trans*-Methyliodo*bis*(triethylphosphine)-platinum(II)	4.1±0.1	B	25	15	293
$C_{13}H_{33}NO_3P_2Pt$	*trans*-Methylnitrato*bis*(triethylphosphine)-platinum(II)	6.05±0.1	B	25	15	293
$C_{14}H_4Cl_4O_2$	2,4,7,9-Tetrachlorobenzofuro[2,3-b]-benzofuran	1.06±0.24	B	25	0	1608
$C_{14}H_6Cl_2O_2$	1,8-Dichloroanthraquinone	2.85	B	22	0	514
	2,3-Dichloroanthraquinone	2.54	B	22	0	514

FORMULA	COMPOUND NAME	μ, D	State, or Solv.	t, °C	Method or P_a, % of P_e	Reference
$C_{14}H_6Cl_4O_2$	2,4,7,9-Tetrachloro-5a,10b-dihydrobenzo-furo[2,3-b]benzofuran	3.31±0.19	B	25	0	1608
$C_{14}H_6N_4O_{12}$	2,2',4,4'-Tetranitrodiphenic acid	5.86±0.04	D	20	0	1182,1188
$C_{14}H_7ClO_2$	1-Chloroanthraquinone	1.54	B	22	0	514
		1.8	B	25	0	169
		1.56	D	22	0	514
		1.9	D	30	0	168
		1.9	D	25	0	169
	2-Chloroanthraquinone	1.72	B	22	0	514
$C_{14}H_7Cl_3$	1,8,10-Trichloroanthracene	1.62±0.12	B	25	0	1643
$C_{14}H_7CrNO_5$	Chromium pentacarbonyl - isoquinoline complex	6.9	B	20	-	1891
$C_{14}H_8ClNO_2$	3-Chloro-*N*-phenylphthalimide	2.45±0.03	D	25	0	1192
	N-*o*-Chlorophenylphthalimide	2.26±0.02	D	25	0	62
	N-*m*-Chlorophenylphthalimide	3.47±0.03	D	20	0	62
	N-*p*-Chlorophenylphthalimide	3.99±0.04	D	20	0	62
$C_{14}H_8Cl_2$	1,8-Dichloroanthracene	3.2	D	23	0	168
	2,2'-Dichlorodiphenylacetylene	1.92	B	25	5	320
	3,3'-Dichlorodiphenylacetylene	1.91	B	25	5	320
$C_{14}H_8Cl_2I_2N_2Pd$	*p*-Chlorophenylisonitrile - palladium iodide complex	1.10	B	25	-	38
$C_{14}H_8Cl_2N_2O_2$	*C*,*N*-*Bis*(*p*-chlorophenyl) sydnone	4.61±0.02	B	25?	0	737

FORMULA	COMPOUND NAME	μ, D	State, or Solv.	t, °C	Method or P_a, % of P_e	Reference
$C_{14}H_8Cl_2O_4$	*Bis*-*p*-chlorobenzoyl peroxide	1.33 1.36	B B	30 20	0 0	1164 1164
$C_{14}H_8Cl_4$	*cis*-1,5-Dichloroanthracene 9,10-dichloride	3.7	w	25	0	168
	trans-1,8-Dichloroanthracene 9,10-dichloride	2.4	D	45	0	168
	1,1-Dichloro-2-(2-chlorophenyl)-2-(4-chlorophenyl)ethylene	2.25	Hp	25	-	808
	1,1-Dichloro-2,2-*bis*(4-chlorophenyl)-ethylene	0 <0.3 0	B B Hp	27 20 25	0 5 -	993 1205 808
$C_{14}H_8Cl_6$	1,1,1,2-Tetrachloro-2,2-*bis*-(4-chlorophenyl)ethane	0.50	Hp	25	-	808
$C_{14}H_8F_6O$	4-Phenoxy-1,3-*bis*(trifluoromethyl)benzene	3.55	B	25	0	1378
$C_{14}H_8N_2$	4,4'-Dicyanobiphenyl	1.3_0	B	25	-	1107
$C_{14}H_8N_2O$	Benzoylenebenzimidazole	1.99	B	25	0	1922
$C_{14}H_8N_2O_4$	*N*-*p*-Nitrophenylphthalimide	6.45±0.1	D	25	0	62
	4-Nitro-*N*-phenylphthalimide	2.61±0.03	D	25	0	1192
$C_{14}H_8N_2S_3$	2,2'-Thio*bis*benzothiazole	3.10	B	25	-	2029
$C_{14}H_8N_2S_4$	2,2'-Dithio*bis*benzothiazole	3.55	B	25	-	2029
$C_{14}H_8O_2$	Anthraquinone	0 0	B B	22 25	0 0	514 948

w) 1-Methylnaphthalene.

FORMULA	COMPOUND NAME	μ, D	State, or Solv.	t, °C	Method or P_a, % of P_e	Reference
$C_{14}H_8O_2$ (contd.)	Phenanthrenequinone	5.33±0.03	B	20-60	MA	1517
		5.6_1	B	25	0	251
		5.7_0	D	25	0	251
		5.5_4	CS_2	25	0	251
		4.7_5	$CHCl_3$	25	0	251
$C_{14}H_8O_2S_2$	*cis*-Di(thiosalicylide)	6.39	n.s.	n.s.	-	85
$C_{14}H_8O_3$	2,2'-Biphenyldicarboxylic anhydride	5.33	B	25	0	1104
$C_{14}H_8O_4$	*cis*-Disalicylide	6.26±0.02	B	25?	-	435,437
$C_{14}H_9Br$	9-Bromoanthracene	1.5_1	B	25	0	1366
$C_{14}H_9Br_2Cl_3$	2,2-*Bis*(4-bromophenyl)-1,1,1-trichloroethane	1.20	B	20	0	2107
$C_{14}H_9ClN_2O$	α-Oximinophenylacetonitrile *o*-chlorophenyl *N*-ether	6.9_8	B	25	5	104
	β-Oximinophenylacetonitrile *o*-chlorophenyl *N*-ether	1.2_3	B	25	5	104
	α-Oximinophenylacetonitrile *m*-chlorophenyl *N*-ether	6.2_7	B	25	5	104
	β-Oximinophenylacetonitrile *m*-chlorophenyl *N*-ether	1.8_0	B	25	5	104
	α-Oximinophenylacetonitrile *p*-chlorophenyl *N*-ether	5.6_3	B	25	5	104
	β-Oximinophenylacetonitrile *p*-chlorophenyl *N*-ether	1.5_4	B	25	5	104

FORMULA	COMPOUND NAME	μ, D	State, or Solv.	t, °C	Method or P_a, % of P_e	Reference
$C_{14}H_9ClN_2O_2$	*C*-(*p*-Chlorophenyl)-*N*-phenylsydnone	6.06±0.02	B	25?	0	737
	N-(*p*-Chlorophenyl)-*C*-phenylsydnone	5.16±0.02 5.17±0.02	B B	25? 25	0 -	737 736
$C_{14}H_9ClN_2O_3$	9-Chloro-2-methoxy-6-nitroacridine	6.26±0.02	B	25	0	942
	9-Chloro-2-methoxy-7-nitroacridine	5.10±0.02	B	25	0	942
$C_{14}H_9Cl_2NO$	6,9-Dichloro-2-methoxyacridine	2.91±0.02	B	25	10	1549
$C_{14}H_9Cl_2NO_2$	6,9-Dichloro-2-methoxyacridine *N*-oxide	5.51	B	25	0	1552
$C_{14}H_9Cl_5$	1,1,1-Trichloro-2-*o*-chlorophenyl-2-*p*-chlorophenylethane	2.26 2.07 2.12	B CCl_4 Hp	20 25 25	0 - -	2107 808 808
	1,1,1-Trichloro-2-*m*-chlorophenyl-2-*p*-chlorophenylethane	1.79	B	20	0	2107
	1,1,1-Trichloro-2,2-*bis*(*p*-chlorophenyl)-ethane	1.03±0.02 0.91 1.07 1.13 1.19 1.12 0.93	liq B B B D CCl_4 Hp	104-145 20 20 20 25 25 25	0 - 0 0 0 0 -	993 753 993 2107 993 993 808
$C_{14}H_9NO_2$	9-Nitroanthracene	3.43 3.71	B CCl_4	25 25	0 0	300 300
	N-Phenylphthalimide	2.34±0.02	D	20	0	62
$C_{14}H_{10}$	Anthracene	0	B,D	25-30	0	265,1178

FORMULA	COMPOUND NAME	μ, D	State, or Solv.	t, °C	Method or P_a, % of P_e	Reference
$C_{14}H_{10}$ (contd.)	Diphenylacetylene	0	B	18	-	129,161
		0.30	B	25	-	2085
		1.13	B	10-70	0	1808
	Phenanthrene	0	B, Hp	20	0	209,1181, 1189
$C_{14}H_{10}BrCl$	1-Bromo-2-*p*-chlorophenyl-2-phenylethylene (low melting)	1.29	B	18	0	138
	(high melting)	2.29	B	17	0	138
$C_{14}H_{10}BrNO_2$	*cis*-4-Bromo-4'-nitrostilbene	4.52±0.02	B	25	0	475
	trans-4-Bromo-4'-nitrostilbene	3.11±0.02	B	25	0	475
$C_{14}H_{10}Br_2$	1-Bromo-2-*p*-bromophenyl-2-phenylethylene (low melting)	1.23	B	20	0	138
	(high melting)	2.45	B	21	0	138
	1,1-Dibromo-2,2-diphenylethylene	1.63	B	15	0	138
		1.67±0.03	B	20	5	1205
		2.55	B	20	-	129
	1,1-*Bis*(*p*-bromophenyl)ethylene	1.44	n.s.	n.s.	5	321
$C_{14}H_{10}ClNO$	9-Chloro-2-methoxyacridine	1.70±0.02	B	25	10	1549
$C_{14}H_{10}ClNO_2$	9-Chloro-2-methoxyacridine *N*-oxide	3.38	B	25	0	1552
$C_{14}H_{10}Cl_2$	1,1-*Bis*(*p*-chlorophenyl)ethylene	1.40	B	13	0	138
		1.44	B	25	5	321
	1,1-Dichloro-2,2-diphenylethylene	1.77±0.03	B	20	5	1205
		1.78_5	B	13	0	138

FORMULA	COMPOUND NAME	μ, D	State, or Solv.	t, °C	Method or P_a, % of P_c	Reference
	1,1-Dichloro-2,2-diphenylethylene (m.p. = 60°C)	2.71	B	20	-	129
	(m.p. = 144°C)	0	B	15	-	129
$C_{14}H_{10}Cl_2N_2Pt$	*Bis*(benzonitrile)platinum dichloride	12.5±1	B	20	-	848
$C_{14}H_{10}Cl_2N_4$	1,4-*Bis*(*p*-chlorophenyl)-1,4-dihydro-1,2,4,5-tetrazine	0.65±0.05	B	25	0	436
$C_{14}H_{10}Cl_2O_2$	*cis*-1,5-Dichloro-9,10-dihydroxy-9,10-dihydroanthracene	2.98	D	25	0?	168
$C_{14}H_{10}Cl_4$	1,1-Dichloro-2-*o*-chlorophenyl-2-*p*-chlorophenylethane	2.38	CCl_4	25	-	808
	1,1-Dichloro-2,2-*bis*(*p*-chlorophenyl)ethane	1.05	B	25	-	808
		1.13	CCl_4	25	-	808
	1,1,1-Trichloro-2-*p*-chlorophenyl-2-phenylethane	1.80	Hp	25	-	808
$C_{14}H_{10}Cl_4N_2Sn$	Stannic chloride - di(benzonitrile) complex	6.60-7.11	B	15	0	2003
$C_{14}H_{10}F_2$	1,1-*Bis*(*p*-fluorophenyl)ethylene	1.53	B	25	5	321
$C_{14}H_{10}I_2N_2Pd$	Phenylisonitrile - palladium iodide complex	1.53	B	25	-	38
$C_{14}H_{10}MoN_2O_4$	Molybdenum tetracarbonyl - dipyridine complex	9.3	B	20	-	1891
$C_{14}H_{10}N_2O$	Diphenylazoxime	1.573	D	20	0	1320

FORMULA	COMPOUND NAME	μ, D	State, or Solv.	t, °C	Method or P_a, % of P_e	Reference
$C_{14}H_{10}N_2O$ (contd.)	2,5-Diphenyl-1,3,4-oxadiazole	3.48	B	25	15	549
		3.89	D	20	0	1320
	3,4-Diphenyl-1,2,5-oxadiazole	4.20	B	25	-	1856
		4.779	D	20	0	1320
	α-Oximinophenylacetonitrile *N*-phenyl ether	6.3±0.1	B	25	-	103
		6.3_5	B	25	5	104
	β-Oximinophenylacetonitrile *N*-phenyl ether	1.0_2	B	25	5	104
		1.0_7	B	25	-	103
	Phenylbenzoyldiazomethane	2.8_2	B	25	0	36
$C_{14}H_{10}N_2O_2$	Azodibenzoyl	2.87	B	25	15	549
	3,4-Diphenyl-1,2,5-oxadiazole peroxide	5.23	B	25	-	1931
	Diphenyldioxadiazine	0.91	B	25	-	1856
	C,*N*-Diphenylsydnone	6.55±0.02	B	25?	0	737
		6.6	B	25	0	428
		6.6_0	B	25	-	427
		6.61±0.02	B	25	-	736
$C_{14}H_{10}N_2O_4$	α,β-Dinitrostilbene	0	B	16	-	129
	p,*p*′-Dinitrostilbene	~0	x	16	-	129
	1,1-Diphenyl-2,2-dinitroethylene	5.53	B	15	0	138
	cis-2-Nitro-2-phenyl-1-(*p*-nitrophenyl)-ethylene	1.65±0.03	B	20	5-10	421
$C_{14}H_{10}N_2O_4W$	Tungsten tetracarbonyl - dipyridine complex	10.8	B	20	-	1891

x) 1-Methylnaphthalene.

FORMULA	COMPOUND NAME	μ, D	State, or Solv.	t, °C	Method or P_a, % of P_e	Reference
$C_{14}H_{10}N_2O_5$	*o,p'*-Dinitrostilbene oxide	5.00	B	15	0	157
	p,p'-Dinitrostilbene oxide (low melting)	5.79	D	17	0	157
	(high melting)	~2.1	y	16	0	157
$C_{14}H_{10}N_2S_2$	2,3-Diphenyl-2,5-*endo*thio-1,3,4-thiadiazoline	8.8	B	25	15	859
		9.1	D	25	15	859
$C_{14}H_{10}O$	1-Anthrol	1.45	B	20	0	1045
	2-Anthrol	1.54	B	20	0	1045
	9-Anthrol	1.72	B	6.3	0	1425
		1.99	B	15	0	1425
		1.82	D	15	0	1425
		1.88	Tol	15	0	1425
	Anthrone	3.46	B	25	0	43
		3.62±0.02	B	20-60	MA	1517
		3.66	B	25	-	2028
		3.69	B	20	0	329
	Diphenylketene	1.7_6	B	25	0	41
		1.9_2	B	25	0	788
	2-Methyl-9-fluorenone	3.32±0.03	B	30	5	145
	3-Methyl-9-fluorenone	3.58±0.03	B	30	5	145
$C_{14}H_{10}O_2$	Dibenzoyl	0.35	liq	60-75	RT	1617
		3.2	B	25	-	1696
		3.61	B	25,50	10	727
		3.74	B	18	0	673
		3.7_9	B	25	0	251
		3.81	D	25	0	251

y) 1-Methylnaphthalene.

FORMULA	COMPOUND NAME	μ, D	State, or Solv.	t, °C	Method or P_a, % of P_e	Reference
$C_{14}H_{10}O_2$ (contd.)	Dibenzoyl (contd.)	3.45	CCl_4	50	10	727
		3.49	CCl_4	25	10	727
		3.63	CCl_4	25	0	251
		3.44	Hx	25	10	727
		3.5_2	Hx	25	0	251
		3.3_6	z	25	0	251
		3.4_9	CS_2	25	0	251
		3.24	$CHCl_3$	25	0	251
$C_{14}H_{10}O_2S_2$	Benzoyl disulfide	1.1	B	25	0	1428
		1.1	B	25,40	-	2030
		1.4	B	45	0	1428
$C_{14}H_{10}O_3$	Benzoic anhydride	4.18	B	25	0	1428
	Benzoyl ether	1.11	liq	20-60	RT	1617
	endo-1,2,3,4-Tetrahydro-1,4-ethanonaphthalene-2,3-dicarboxylic anhydride	4.05	B	25	-	1924
	exo-1,2,3,4-Tetrahydro-1,4-ethanonaphthalene-2,3-dicarboxylic anhydride	4.32	B	25	-	1924
$C_{14}H_{10}O_4$	Benzoyl peroxide	1.55	B	30	0	1163
		1.59	B	45	0	1428
		1.60	B	30	0	1164
	endo-1,2,3,4-Tetrahydro-9-oxo-1,4-ethanonaphthalene-2,3-dicarboxylic anhydride	6.54	B	25	-	1924
	exo-1,2,3,4-Tetrahydro-9-oxo-1,4-ethanonaphthalene-2,3-dicarboxylic anhydride	2.86	B	25	-	1924
$C_{14}H_{11}Br$	1-Bromo-1,2-diphenylethylene (m.p. < room temperature)	1.31	B	23	-	129
	(m.p. = 31°C)	1.39	B	21	-	129

z) Decalin.

FORMULA	COMPOUND NAME	μ, D	State, or Solv.	t, °C	Method or P_a, % of P_e	Reference
	1-Bromo-2,2-diphenylethylene	1.44±0.03	B	20	5	1205
		1.52	B	22	0	138
	1-(4-Bromophenyl)-1-phenylethylene	1.51	B	25	5	321
	1-(4-Bromophenyl)-2-phenylethylene	1.44	n.s.	n.s.	-	1904
		1.67±0.02	B	25	0	475
$C_{14}H_{11}Cl$	1-Chloro-2,2-diphenylethylene	1.49±0.03	B	20	5	1205
	1-(4-Chlorophenyl)-1-phenylethylene	1.50	B	25	5	321
	1-(?-Chlorophenyl)-2-phenylethylene	2.04	D	20	0	1969
$C_{14}H_{11}ClN_2O$	9-Amino-6-chloro-2-methoxyacridine	6.73±0.02	B	25	0-10	942,1549
$C_{14}H_{11}ClN_2O_2$	1-Benzoyl-2-(*p*-chlorobenzoyl)hydrazide	3.85	D	25	15	549
$C_{14}H_{11}Cl_3$	1,1,1-Trichloro-2,2-diphenylethane	1.96	B	20	0	2107
		1.95±0.04	B	20	5	1205
		1.77	Hp	25	-	808
$C_{14}H_{11}Cl_4NO_2$	Chloroanil - dimethylaniline complex	2.6_4	CCl_4	20	-	211
$C_{14}H_{11}N$	2-Phenylindole	2.03	B	25	-	817
	3-Phenylindole	2.23	B	25	-	817
$C_{14}H_{11}NO$	*N*-Methylacridone	3.5	B	30	-	2091
$C_{14}H_{11}NO_2$	4-Nitrostilbene	4.41	n.s.	n.s.	-	1904
		4.56±0.02	B	25	0	475
	N-Piperonylideneaniline	1.8_5	B	25	0	1089
$C_{14}H_{11}NO_3$	4-Nitrostilbene oxide	4.16	B	17	0	157

FORMULA	COMPOUND NAME	μ, D	State, or Solv.	t, °C	Method or P_a, % of P_e	Reference
$C_{14}H_{11}NO_4$	4-(4-Nitrophenoxy)acetophenone	3.16	B	20	-	1412
$C_{14}H_{11}NS$	*N*-Methylthioacridine	5.2	a	17	-	2091
$C_{14}H_{11}N_3O_2$	1,4-Diphenyl-3,5-dioxo-1,2,4-triazolidine	2.00 2.74	B D	25 25	15 15	859 859
$C_{14}H_{11}N_3O_4$	α-Benzoyl-β-(4-nitrobenzoyl)hydrazide	5.61	B	25	15	549
$C_{14}H_{12}$	9,10-Dihydroanthracene	~0.4	B	25	0	265
	1,1-Diphenylethylene	0-0.38 0.5	B B	25 10-70	5 0	321 1808
	1,2-Diphenylethylene	0	B	25	0	475
	cis-1,2-Diphenylethylene	0	B	18	-	129
	trans-1,2-Diphenylethylene	0	B	10-70	0	1808
$C_{14}H_{12}BrN$	4-Amino-4'-bromostilbene	3.56±0.02	B	25	0	475
$C_{14}H_{12}Br_2$	?-Dibromo-2,2'-dimethylbiphenyl	5.81±0.03	B	20	0	1188
	dl-1,2-Dibromo-1,2-diphenylethane	2.8_4	B	25	-	2077
	meso-1,2-Dibromo-1,2-diphenylethane	0.4-0.9	B	25	-	2077
	2,2'-*Bis*(bromomethyl)biphenyl	2.81	B	20	0	1182
$C_{14}H_{12}ClN$	3-Chloro-3'-methyl-*N*-benzylideneaniline	2.72	B	25	-	498
	3'-Chloro-4-methyl-*N*-benzylideneaniline	2.19	B	25	-	498
	4-Chloro-4'-methyl-*N*-benzylideneaniline	3.01	B	25	-	498

a) 1-Methylnaphthalene.

FORMULA	COMPOUND NAME	μ, D	State, or Solv.	t, °C	Method or P_a, % of P_e	Reference
	4'-Chloro-4-methyl-*N*-benzylideneaniline	2.08	B	25	0	398
$C_{14}H_{12}ClNO$	*o*-Methoxybenzylidene-*p*-chloroaniline	3.85	B	25	5	368
$C_{14}H_{12}ClN_3O$	6-Chloro-9-hydrazino-2-methoxyacridine	7.45	B	25	0	1409
$C_{14}H_{12}ClN_3O_2$	6-Chloro-9-hydrazino-2-methoxyacridine *N*-oxide	8.23	B	25	0	1409
$C_{14}H_{12}Cl_2$	4,4'-Dichloro-2,2'-dimethylbiphenyl	0.76	B	25	0	1104
	5,5'-Dichloro-2,2'-dimethylbiphenyl	2.37	B	25	-	1697
	α-Dichlorostilbene	1.2_8	B	25	0	2081
		1.46	B	25	5	727
		1.33	CCl_4	25	5	727
	β-Dichlorostilbene	2.7_8	B	25	0	2081
$C_{14}H_{12}Cl_4O_2Sn$	Tin chloride - di(benzaldehyde) complex	7.5-8.1	B	18	0	2003
$C_{14}H_{12}Cl_4O_4Sn$	Tin chloride - di(benzoic acid) complex	5.73	B	n.s.	-	1449
$C_{14}H_{12}N_2$	Di(benzylidene)hydrazine	0.89	B	n.s.	-	677
		1.01	B	18	15	76
		1.05	B	15	0	2008
$C_{14}H_{12}N_2O$	9-Amino-2-methoxyacridine	4.65±0.02	B	25	10	1549
	cis-Benzoylformaldehyde phenylhydrazone	1.72	B	20	-	854
	trans-Benzoylformaldehyde phenylhydrazone	2.74	B	20	-	854
	5-Cyano-1-phenethyl-2-pyridone	3.66	B	14?	-	1952
$C_{14}H_{12}N_2O_2$	*p*-Acetoxyazobenzene	1.63	B	25	-	800

FORMULA	COMPOUND NAME	μ, D	State, or Solv.	t, °C	Method or P_a, % of P_e	Reference
$C_{14}H_{12}N_2O_2$ (contd.)	4-Amino-4'-nitrostilbene	6.50 6.83±0.07	B B	20 20	FL -	375 1351
	α,β-Dibenzoylhydrazide	2.65	D	25	15	549
	α-Diphenylglyoxime	1.502	D	20	-	1319
	β-Diphenylglyoxime	2.14	D	20	-	1319
	γ-Diphenylglyoxime	1.56	D	20	-	1319
$C_{14}H_{12}N_2O_3$	*N*-Acetyl(*o*-nitrophenyl)phenylamine	5.80 5.96	B D	25 25	5 5	1219 1219
	N-Acetyl(*m*-nitrophenyl)phenylamine	5.71 5.83	B D	25 25	5 5	1219 1219
	N-Acetyl(*p*-nitrophenyl)phenylamine	6.21 6.35	B D	25 25	5 5	1219 1219
	α-Methyl *N*-ether of *p*-nitrobenzophenone-oxime	6.65±0.1	B	25	0	1908
	β-Methyl *N*-ether of *p*-nitrobenzophenone-oxime	1.10±0.1	B	25	0	1908
	α-Methyl *O*-ether of *p*-nitrobenzophenone-oxime	3.78	B	25	0	1938
	β-Methyl *O*-ether of *p*-nitrobenzophenone-oxime	4.29	B	25	0	1938
$C_{14}H_{12}N_2O_4$	2,2'-Dimethyl-4,4'-dinitrobiphenyl	1.31	B	25	0	1104
	2,2'-Dimethyl-5,5'-dinitrobiphenyl	5.53	B	25	-	1697

FORMULA	COMPOUND NAME	μ, D	State, or Solv.	t, °C	Method or P_a, % of P_e	Reference
$C_{14}H_{12}N_2O_4S_2$	*p*-Nitrobenzyl disulfide	4.97	B	25	-	2030
$C_{14}H_{12}N_2S_2$	*N*-Phenyl-*N*-methylbenzothiazolesulfenamide	2.44	B	25	-	637
$C_{14}H_{12}N_4$	3-Anilino-1-phenyl-1,2,4-1*H*-triazole	3.54±0.02	B	25	0	436
	1,4-Dihydro-1,4-diphenyl-1,2,4,5-tetrazine	0.8±0.05	B	25	0	436
$C_{14}H_{12}O$	Stilbene oxide	1.75	B	14	0	157
$C_{14}H_{12}O_2$	Benzoin	3.49 3.57-3.61	B B	18 18-60	0 -	673 671,674
	2,6-Dimethyldi-*o*-phenylene dioxide -1,2-dioxa-4,6,8-nonatriene	0.61	B	20	0	729,733
	2-Methyl-2-phenyl-*o*-phenylene methylene dioxide	1.06	n.s.	n.s.	-	186
	2-*p*-Phenoxymethyl-2,4,6-cycloheptatrien-1-one	3.63±0.05	B	25	0	1022
$C_{14}H_{12}O_2S$	Phenylstyrylsulfone	5.93	B	30	0	88
$C_{14}H_{12}S_2$	Dimethylthianthrene	1.62	B	25	-	527
$C_{14}H_{13}I$	4-Iodo-2,2'-dimethylbiphenyl	1.54	CCl_4	25	0	919
	4-Iodo-3,3'-dimethylbiphenyl	1.54	CCl_4	25	0	919
$C_{14}H_{13}IO_2$	4-Iodo-3,3'-dimethoxybiphenyl	2.80	CCl_4	25	0	919
$C_{14}H_{13}N$	2-Aminostilbene	1.10 1.49±0.02	n.s. B	n.s. 25	- 0	1904 475
	3-Aminostilbene	1.33 1.51±0.02	n.s. B	n.s. 25	- 0	1904 475

FORMULA	COMPOUND NAME	μ, D	State, or Solv.	t, °C	Method or P_a, % of P_e	Reference
$C_{14}H_{13}N$ (contd.)	4-Aminostilbene	1.82 2.07±0.02	n.s. B	n.s. 25	- 0	1904 475
	3-Methyl-*N*-benzylideneaniline	1.51	B	25	-	498
	4-Methyl-*N*-benzylideneaniline	1.53	B	25	-	498
	4'-Methyl-*N*-benzylideneaniline	1.80	B	25	-	498
$C_{14}H_{13}NO$	4-Acetamidobiphenyl	3.86	B	25	0	1068
	N-Acetyldiphenylamine	3.84 3.94	B D	25 25	5 5	1219 1219
	Benzophenone methyloxime	0.389±0.005	B	20	5	1593
	N-Benzylbenzamide	3.90 3.93	D D	40 25	0 0	959 959
	o-Methoxybenzylideneaniline	3.05	B	25	0	398
	Phenylacetanilide	3.36 3.77	B D	25 25,40	- 5	1042 959
	Salicylidene-*m*-toluidine	2.61	B	25	0	398
	Salicylidene-*p*-toluidine	2.63 2.74	B D	25 25	5 5	368 368
$C_{14}H_{13}NO_3$	1-(3,4-Methylenedioxyphenethyl)-2(1*H*)-pyridone	4.17	B	27	0	1934
$C_{14}H_{13}N_2O_7PS$	*O*-Ethyl-*O*,*O*-di(*p*-nitrophenyl)phosphorothioate	4.97	D	25	-	906
$C_{14}H_{13}N_2O_8P$	Ethyldi(*p*-nitrophenyl)phosphate	6.0[b]	B	25	-	906

b) Original author states "unreliable because of low solubility."

FORMULA	COMPOUND NAME	μ, D	State, or Solv.	t, °C	Method or P_a, % of P_e	Reference
$C_{14}H_{13}N_3O$	*p*-Acetamidoazobenzene	3.99	B	25	-	800
$C_{14}H_{14}$	2,2'-Dimethylbiphenyl	0.66 0.47	B CCl_4	25 25	0 0	1104 919
	3,3'-Dimethylbiphenyl	0.50	CCl_4	25	0	919
	1,2-Diphenylethane	0 0.45_5	liq B	58-178 25	- 0	1052 1601
	6-Methyl-6-*p*-tolylfulvene	1.34	B	20	-	967
$C_{14}H_{14}Br_2S$	Bromobenzyl sulfide	~5.4	B	25	0	849
$C_{14}H_{14}Br_2Te$	Di-*p*-tolyltellurium dibromide	3.24	B	25	15	849
$C_{14}H_{14}ClN_3$	*p*-Chloro-*p'*-(dimethylamino)azobenzene	4.89	B	25	0	269
$C_{14}H_{14}Cl_2Te$	Di-*p*-tolyltellurium dichloride	3.01	B	25	15	849
$C_{14}H_{14}Cr$	Ditolylchromium	0	B	25	15	520
$C_{14}H_{14}FeO_2$	Diacetylferrocene	4.23	B	30	0	1600
$C_{14}H_{14}Hg$	Di-*p*-tolylmercury	0.75	c	142	-	653
$C_{14}H_{14}IN_3$	*p*-Iodo-*p'*-(dimethylamino)azobenzene	4.76	B	25	0	267
$C_{14}H_{14}I_2S$	Iodobenzyl sulfide	4.4	B	10	15	849
$C_{14}H_{14}NO_3$	*p,p'*-Dianisyl nitroxide	3.3±0.3	n.s.	n.s.	-	263
$C_{14}H_{14}N_2$	*trans*-*p*-Azotoluene	~0	B	25	-	668
	9,10-Dimethyl-9,10-dihydrophenazine	~0.4	B	25	0	265

c) Decalin.

FORMULA	COMPOUND NAME	μ, D	State, or Solv.	t, °C	Method or P_a, % of P_e	Reference
$C_{14}H_{14}N_2O$	*cis-o,o'*-Azoxytoluene	4.39_5	B	22	-	565
	trans-o,o'-Azoxytoluene	1.75	B	22	-	565
	cis-p,p'-Azoxytoluene	5.10	B	22	-	565
	trans-p,p'-Azoxytoluene	1.75_5	B	22	-	565
	α-Benzoyl-β-*p*-tolylhydrazide	3.41	B	25	15	549
	Di-*p*-tolylnitrosamine	3.82	B	25	5	1114
$C_{14}H_{14}N_2OS$	5-(2'-Anilinovinylmethylene)-3-ethyl-rhodanine	5.89	B	25	0	1766
$C_{14}H_{14}N_2O_2$	Dibenzyl hyponitrite	0.4	B	20	-	793
	4-(Dimethylamino)-4'-nitrodiphenyl	6.60 6.93±0.07	B B	20 20	FL -	375 1351
	1,1'-Diphenylazomethane dioxide	3.05	B	20	-	567
$C_{14}H_{14}N_2O_2S$	*p*-Nitrophenyl-*p'*-(dimethylaminophenyl) sulfide	6.21	B	20	-	1682
$C_{14}H_{14}N_2O_2S_2$	5-Anilino-(*N*-acetylmethylene)-3-ethyl-rhodanine	4.42	B	25	0	1766
$C_{14}H_{14}N_2O_3$	*cis-o,o'*-Azoxyanisole	6.22	B	22	-	565
	trans-o,o'-Azoxyanisole	2.43_5	B	22	-	565
	p,p'-Azoxyanisole	2.3 2.22 2.36 2.48	solid, liq liq B B	110-202 70-140 20 20	- - - -	456 1241e 1241c 1991

FORMULA	COMPOUND NAME	μ, D	State, or Solv.	t, °C	Method or P_a, % of P_e	Reference
$C_{14}H_{14}N_4O_2$	4-Nitro-4'-(dimethylamino)azobenzene	8.16 8.1	B D	25 18	0 -	269 2091
$C_{14}H_{14}N_4O_7$	Dimethylaniline picrate	6.06	D	35	0	1559
$C_{14}H_{14}N_4S$	*cis-cis-N*-Phenylimino-*N'*-anilino-*S*-methylisothiourea	4.84	B	20	10	424
	trans-trans-N-Phenylimino-*N'*-anilino-*S*-methylisothiourea	2.26	B	20	10	424
$C_{14}H_{14}O$	Benzyl ether	1.39	B	21	0	141
	3,4-Dimethylphenyl phenyl ether	1.54	B	30	0	732,2010
	6-Methyl-6-*p*-(methoxyphenyl)fulvene	1.90	B	20	-	967
	o-Methylphenyl ether	0.74 0.73	B Hx	27,50 31	5 5	732,2011 2011
	m-Methylphenyl ether	1.41	B	30	0	732,2010
	m-Methylphenyl *p*-methylphenyl ether	1.43	B	30	0	732,2010
	p-Tolyl ether	1.45±0.03 1.42-1.43 1.47	gas B B	230 25 30	5 0 0	332 654 732,2010
$C_{14}H_{14}OS$	Benzyl sulfoxide	3.91	B	23	0	141
	p-Tolyl sulfoxide	4.41 4.43	B B	25 25	5 0	1114 654
$C_{14}H_{14}OTe$	Di-*p*-tolyltellurium oxide	3.96	D	40	15	851
$C_{14}H_{14}O_2$	2,2'-Dimethoxybiphenyl	1.53±0.002	B	25	-	204

FORMULA	COMPOUND NAME	μ, D	State, or Solv.	t, °C	Method or P_a, % of P_e	Reference
$C_{14}H_{14}O_2$ (contd.)	3,3'-Dimethoxybiphenyl	1.73	B	20	RT	922
		1.74	CCl_4	25	0	919
	4,4'-Dimethoxybiphenyl	1.53	B	25	0	203
		1.68	B	20	RT	922
		1.84	CCl_4	25	0	919
	meso-cis-1,2-Diphenyl-1,2-ethanediol	2.08	B	25	0	2082
		2.33	B	60	0	679
		2.35	B	25	0	679
		2.50	B	25	0	441
	dl-trans-1,2-Diphenyl-1,2-ethanediol	2.41	B	18	-	671
		2.67	B	60	0	679
		2.69	B	25	0	2082
		2.73	B	18	0	679
		2.75	B	25	0	441
	l-trans-1,2-Diphenyl-1,2-ethanediol	2.73	B	25	0	441
	4-Methoxybenzhydrol	2.03	B	25	5	319
$C_{14}H_{14}O_2S$	Dibenzyl sulfone	5.00	B	30	0	88
$C_{14}H_{14}O_2S_2$	*p*-Methoxyphenyl disulfide	3.11	B	25	-	2030
	Tolylsulfinyl tolyl sulfide	5.92	B	25,40	0	619
		6.07	D	25	0	619
$C_{14}H_{14}O_3$	*p*-Methoxyphenyl ether	2.18	B	10-40	0	955
$C_{14}H_{14}S$	Benzyl sulfide	1.39	B	21	0	141
	o-Tolyl sulfide	1.09±0.01	B	25	0	1683
	m-Tolyl sulfide	1.65±0.03	B	25	0	1683

FORMULA	COMPOUND NAME	μ, D	State, or Solv.	t, °C	Method or P_a, % of P_e	Reference
	p-Tolyl sulfide	1.95	B	25	0	654
		1.97±0.02	B	25	-	1658
	o-Tolyl *m*-tolyl sulfide	1.45±0.02	B	25	0	1683
	o-Tolyl *p*-tolyl sulfide	1.40±0.01	B	25	0	1683
	m-Tolyl *p*-tolyl sulfide	1.84±0.01	B	25	0	1683
$C_{14}H_{14}S_2$	Benzyl disulfide	1.90	B	25	-	2030
	p-Tolyl disulfide	2.52	B	20	0	636
$C_{14}H_{14}Se$	*p*-Tolyl selenide	1.83±0.1	B	25	-	1638
$C_{14}H_{14}Se_2$	Benzyl diselenide	1.55±0.1	B	25	-	1638
	p-Tolyl diselenide	2.31±0.1	B	25	-	1638
$C_{14}H_{15}IO_4$	Diphenyliodylhydroxide - acetic acid complex	3.5-5	B	25	0	1072
$C_{14}H_{15}N$	Dibenzylamine	0.97	liq	20	5	337
		1.02	B	20	5	337
		1.00	Hx	20	5	337
	6-(*p*-Dimethylaminophenyl)fulvene	3.65	B	20	-	967
	N,N-Dimethyl-4-biphenylamine	2.04±0.02	B	25	5	475
	Di-*p*-tolylamine	0.94	B	25	5	1114
$C_{14}H_{15}NO_2$	Quinone - dimethylamine complex	1.5_3	CCl_4	20	-	211
$C_{14}H_{15}NO_2S$	*N*-Methyl-*p*-toluenesulfonanilide	5.44	B	20,35	0	1703

FORMULA	COMPOUND NAME	μ, D	State, or Solv.	t, °C	Method or P_a, % of P_c	Reference
$C_{14}H_{15}NO_2S$ (contd.)	*p*-Toluenesulfono-*p*-toluidide	5.32	B	20	0	1703
$C_{14}H_{15}NS$	*N*,*N*-Dimethyl(*p*-phenylthio)aniline	3.35	B	20	-	1682
$C_{14}H_{15}N_3$	*p*-Dimethylaminoazobenzene	3.14	B	n.s.	-	1342
		3.17±0.02	B	31	0	982
		3.18	B	25	-	800
		3.22	B	25	0	269
		3.30	B	20	0	1385
		3.71	B	27	0	164
		3.45	D	20	0	1386
		3.03	CCl_4	15	0	1386
		2.94	Hp	15	0	1386
	3,3-Dimethyl-1-(2'-biphenyl)triazene	2.3_2	B	25	0	1083
	3,3-Dimethyl-1-(4'-biphenyl)triazene	2.6_1	B	25	0	1083
	4,4'-Dimethyldiazoaminobenzene	0.9_1±0.04	B	25	0	1102
$C_{14}H_{15}P$	Diphenylethylphosphine	1.35	cHx	25	-	920
$C_{14}H_{16}$	1-Butylnaphthalene	0.68	n.s.	n.s.	-	1210
	2-Butylnaphthalene	0.75	n.s.	n.s.	-	1210
$C_{14}H_{16}BNO$	*B*,*B*-Diphenylboroxazolidine	8.7±0.01	D	25	-	553
$C_{14}H_{16}Cr$	Chromium - ditoluene complex	0	B	25	5-10	2076
$C_{14}H_{16}N_2$	6,6'-Diamino-2,2'-dimethylbiphenyl	1.67	B	20	0	136
$C_{14}H_{16}N_2OS_2$	1-[2-(1-Ethyl-2(1*H*)-pyridylidenyl)]-2-[5-(3-ethyl-2-thione-4-oxothiazolylidenyl)]ethane	9.21	B	25	-	1767

FORMULA	COMPOUND NAME	μ, D	State, or Solv.	t, °C	Method or P_a, % of P_e	Reference
$C_{14}H_{16}N_2O_2$	*Bis*(*p*-methoxyphenyl)hydrazine	2.27	B	25	-	1555
$C_{14}H_{16}Si$	Dimethyldiphenylsilane	0.34	B	30	0	544
$C_{14}H_{17}ClO$	5-(4-Chlorophenyl)cycloöctanone	3.39±0.02	B	25	0	19
$C_{14}H_{17}N$	2,2',5,5'-Tetramethyl-1-phenylpyrrole	2.07±0.02	B	25	0	931
$C_{14}H_{17}NO$	*p*-(Dimethylamino)cinnamylideneacetone (m.p. = 120°C)	6.7	B	25	-	2090
	(m.p. = 215°C)	2.4	D	20	-	2090
$C_{14}H_{18}Cl_4O_2$	Tetrachlorodibutoxybenzene	2.14±0.06	CCl_4	20-110	RT	2158
$C_{14}H_{18}N_2O_5$	4-*t*-Butyl-2,6-dimethyl-3,5-dinitroacetophenone	2.7_7	B	25	0	1071
$C_{14}H_{18}O$	2-Ethyl-2-methyl-4-pentenophenone	2.77±0.01	B	(25)	-	301
$C_{14}H_{18}O_4$	Diethyl phenylethylmalonate	2.52	B	30	-	1501
	Diethyl tropylmalonate	2.44±0.05 2.46	B B	25 25	5 5	334a 2154
$C_{14}H_{19}ClO$	Methylchlorofluorenone	3.14±0.04	B	30	5	130
$C_{14}H_{20}AuP$	Styryl(triethylphosphine)gold	6.55	B	25	-	262
$C_{14}H_{20}Br_2$	2,5-Dibromo-1,4-di-*t*-butylbenzene	<0.4[d]	B	25	0	929
$C_{14}H_{20}O_2$	2,5-Di-*t*-butyl-1,4-benzoquinone	0.81±0.03	B	25	0	928
$C_{14}H_{20}O_6$	Ethylcarbethoxyketene dimer	3.8_0	B	25	0	41

d) Probably between 0 and 0.28.

FORMULA	COMPOUND NAME	μ, D	State, or Solv.	t, °C	Method or P_a, % of P_e	Reference
$C_{14}H_{21}Cl_3O_2Sn$	*Bis*(tetrahydrofuran) - phenyltrichlorotin complex	6.29	B	n.s.	-	592
$C_{14}H_{21}NO$	*N*-Benzyl ether of heptanaloxime	3.348	B	23	-	1413
	O-Benzyl ether of heptanaloxime	1.603	B	23	-	1413
$C_{14}H_{21}NO_2$	2-Nitro-1,4-di-*t*-butylbenzene	3.70±0.01	B	25	0	932
$C_{14}H_{22}$	1,4-Di-*t*-butylbenzene	0.58 0.64	B B	20 40	SPC SPC	1300 1300
$C_{14}H_{22}BNO_2$	*N*-*n*-Butyl-*B*-phenyl-*diptych*-boroxazolidine	11.2±0.16	D	25	-	553
$C_{14}H_{22}N_2O$	α-Diethylaminoacet-2,6-dimethylanilide	4.16±0.03 4.0	B n.s.	25 n.s.	0 -	526 1299
$C_{14}H_{22}N_4O_7$	Dibutylamine picrate	11.5 12.1	B D	30 30	0 0	1268 1268
	Octylamine picrate	12.1	D	30	0	1268
$C_{14}H_{22}O$	2,4-Di-*t*-butylphenol	1.39±0.02	B	20	0	1186,1668
	4-(2,6,6-Trimethyl-1-cyclohexene-1-yl)-2-methyl-2-butenal	3.59	B	25	0	896
$C_{14}H_{22}O_2$	2,5-Di-*t*-butylhydroquinone	1.68±0.20	B	25	0	1429
	Hydroquinone di-*n*-butyl ether	1.8_1	B	25	0	2080
$C_{14}H_{23}BrSi$	Triethyl(*p*-bromophenethyl)silane	1.89	B	25	15	887
$C_{14}H_{23}P$	Dibutylphenylphosphine	1.33	cHx	25	-	920
$C_{14}H_{24}NP$	Dibutyl-*p*-aminophenylphosphine	2.58	cHx	25	-	920

FORMULA	COMPOUND NAME	μ, D	State, or Solv.	t, °C	Method or P_a, % of P_e	Reference
	Diethyl-*p*-diethylaminophenylphosphine	2.83	cHx	25	-	920
$C_{14}H_{24}N_2O_2$	*Bis*(acetylacetone)tetramethyldiimine	3.93±0.1	B	25	10	761
$C_{14}H_{24}O_2$	*l*-Bornyl *n*-butyrate	1.80_4±0.015	B	22	15	413
	Pentylketene dimer	3.58	B	25	0	1621
$C_{14}H_{26}N_2O_3$	*N*-Cyclohexyltetrahydro-5-*n*-butyl-5-nitro-1,3-oxazine	4.46±0.1	B	20	-	631
$C_{14}H_{26}N_4NiO_4$	α-Nickel methyl-*n*-butylglyoxime acid	1.3	B	25	0	277
	β-Nickel methyl-*n*-butylglyoxime acid	1.3	B	25	0	277
$C_{14}H_{26}O$	Cyclotetradecanone	2.74±0.03	B	20	5	630
$C_{14}H_{26}O_2$	*l*-Menthyl *n*-butyrate	1.67_4±0.015	B	22	15	413
$C_{14}H_{26}O_4$	Diethyl sebacate	2.38	liq	20	0	907
		2.42	liq	30-40	-	1497
		2.42	liq	45	0	907
		2.45	liq	70	0	907
		2.52	B	25,50	0	1832
		2.53	B	20-70	0	907
		2.56	D	20-70	0	907
$C_{14}H_{28}ClNO_2Si_3$	1,1,1,2,3,3,3-Heptamethyl-2-*o*-chlorophenylaminomethyltrisiloxane	2.33	liq	20	RT	595
$C_{14}H_{28}N_2NiS_4$	Nickel(II) *N*,*N*-isohexyldithiocarbamate	2.0_4	B	23	0	1243
$C_{14}H_{28}N_2S_4Zn$	Zinc(II) *N*,*N*-isohexyldithiocarbamate	1.9_3	B	23	0	1243
	Zinc(II) *N*,*N*-di-*n*-propyldithiocarbamate	1.4_9	B	22	0	1243

FORMULA	COMPOUND NAME	μ, D	State, or Solv.	t, °C	Method or P_a, % of P_e	Reference
$C_{14}H_{28}O_2$	Ethyl laurate	1.3	liq	20-145	-	1052
	Myristic acid	0.76 0.77	B B	25 25	15 15	1793 1468
$C_{14}H_{28}O_3$	Permyristic acid	2.27±0.04	B	30	0	1616
$C_{14}H_{29}Br$	1-Bromotetradecane	1.83 1.92	liq liq	25 20	- -	707 657
$C_{14}H_{29}NO_2Si_3$	1,1,1,2,3,3,3-Heptamethyl-2-phenylamino-methyltrisiloxane	2.09	liq	20	RT	595
$C_{14}H_{30}O$	Tetradecanol	1.6_9	CCl_4	25	5	1108
$C_{14}H_{30}O_6S_3$	2-Methyl-1,3-*bis*(propylsulfonyl)-2-(propylsulfonylmethyl)propane	4.18	B	20	0	636
$C_{14}H_{30}O_8$	Heptaethylene glycol	6.35 3.65	liq D	20 25	- 10	933 1999
$C_{14}H_{31}NO_2$	Dodecylamine acetate	2.24	B	30	5	1757
$C_{14}H_{31}O_3P$	Diethyldecylphosphonate	2.78±0.02	liq	32	-	951
$C_{14}H_{32}Si_2$	*trans*-1,2-*Bis*(triethylsilyl)ethylene	0	B	25	10	885
$C_{14}H_{33}NP_2PtS$	*trans*-Methylthiocyanato*bis*(triethyl-phosphine)platinum(II)	6.65±0.1	B	25	15	293
$C_{14}H_{34}O_7Si_4$	1,1,1,2,3,4,4,4-Octamethyl-2,3-*bis*-(acetoxymethyl)tetrasiloxane	3.38	liq	20	RT	37
$C_{14}H_{34}Si_2$	1,1,1,4,4,4-Hexaethyl-1,4-disilicabutane	0.59	Hx	25	0	627

FORMULA	COMPOUND NAME	μ, D	State, or Solv.	t, °C	Method or P_a, % of P_e	Reference
$C_{14}H_{35}ClP_2Pt$	*trans-Bis*(triethylphosphine)chloro(ethyl)-platinum	3.7	B	25	15	294
$C_{14}H_{35}IP_2Pt$	*trans-Bis*(triethylphosphine)iodo(ethyl)-platinum	4.15	B	25	15	294
$C_{14}H_{36}Cl_2P_2Pt$	Dimethyldichloro*bis*(triethylphosphine)-platinum(II)	5.3±0.1	B	25	15	293
$C_{14}H_{36}I_2P_2Pt$	Dimethyldiiodo*bis*(triethylphosphine)-platinum(II)	5.85±0.1	B	25	15	293
$C_{14}H_{36}P_2Pt$	*cis*-Dimethyl*bis*(triethylphosphine)-platinum(II)	5.55±0.1	B	25	15	293
	cis-Bis(trimethylphosphine)di-*n*-butyl-platinum	5.65	B	25	15	294
$C_{14}H_{42}O_5Si_6$	Tetradecamethylhexasiloxane	1.58	liq	20	RT	37
$C_{15}H_9Br_2N$	2,4-Dibromo-α-cyanostilbene	3.42±0.17	B	25-55	-	2022
$C_{15}H_{10}BrF_3$	*cis*-1-(*p*-Trifluoromethylphenyl)-1-phenyl-2-bromoethylene	3.46	n.s.	n.s.	-	1667
	trans-1-(*p*-Trifluoromethylphenyl)-1-phenyl-2-bromoethylene	2.26	n.s.	n.s.	-	1667
$C_{15}H_{10}BrN$	4-Bromo-α-cyanostilbene	3.34±0.04	B	25-55	-	2022
$C_{15}H_{10}Br_2O$	α,β-Dibromobenzylideneacetophenone	3.20	B	22	-	129
	p,p′-Dibromobenzylideneacetophenone	2.05	D	17	-	129
$C_{15}H_{10}CrO_3$	Biphenyl - chromium tricarbonyl complex	5.35±0.05	B	n.s.	-	1568

FORMULA	COMPOUND NAME	μ, D	State, or Solv.	t, °C	Method or P_a, % of P_e	Reference
$C_{15}H_{10}N_4O_7$	Isoquinoline - picric acid complex	7.92	D	35	0	1558
	Quinoline - picric acid complex	8.30	D	35	0	1558
$C_{15}H_{10}O$	2,3,6,7-Dibenzo-2,4,6-cycloheptatrien-1-one	3.20±0.03	B	30	5	146
	Diphenylcyclopropenone	5.08	B	25	-	1748
$C_{15}H_{10}O_2$	3-Phenyl-1,2-benzopyrone	4.3₃	B	25	0	1070
$C_{15}H_{11}BrO$	Benzylidene-*p*-bromoacetophenone	2.96	B	18	-	129
	α-Bromobenzylideneacetophenone	3.90	B	20	-	129
	β-Bromobenzylideneacetophenone	3.62	B	21	-	129
	p-Bromobenzylideneacetophenone	2.49	B	22	-	129
$C_{15}H_{11}ClN_2O_3$	9-Chloro-2-ethoxy-6-nitroacridine	6.01±0.02	B	25	0	942
$C_{15}H_{11}ClO_5$	2-Phenylbenzopyrylium perchlorate	~8	e	25	-	1066
$C_{15}H_{11}N$	α-Cyanostilbene	3.37±0.08	B	25-55	-	2022
	3-Cyanostilbene	4.12±0.02	B	25	0	475
	4-Cyanostilbene	4.42±0.02 4.30	B n.s.	25 n.s.	0 -	475 1904
	2-Phenylquinoline	1.76±0.03	B	25	0	1843
	3-Phenylquinoline	2.22±0.03	B	25	0	1843
	4-Phenylquinoline	2.33±0.03	B	25	0	1843

e) Dimethylaniline.

FORMULA	COMPOUND NAME	μ, D	State, or Solv.	t, °C	Method or P_a, % of P_e	Reference
$C_{15}H_{11}NO$	Diphenylisoxazole	4.98	B	25	-	1856
	3,4-Diphenylisoxazole	3.36	B	n.s.	0	1932
	3,5-Diphenylisoxazole	2.99	B	n.s.	-	1515
$C_{15}H_{11}NO_2$	2-(Anilinomethylene)-1,3-indandione	2.42	B	25	-	1766
	N-*o*-Tolylphthalimide	2.47±0.03	D	25	0	62
$C_{15}H_{12}Cl_2$	1,1-*Bis*(*p*-chlorophenyl)cyclopropane	2.05	B	23.7	5	593
		2.09	B	25	0	557
		2.01	pXy	30.6	5	593
		1.99	f	23.7	5	593
	trans-1,2-*Bis*(*p*-chlorophenyl)cyclopropane	1.46	B	25	0	557
$C_{15}H_{12}F_{16}O_4$	*Bis*(2,2,3,3,4,4,5,5-octafluoropentyl)ester of glutaric acid	3.25	B	20	0	1666
$C_{15}H_{12}N_2$	4,5-Diphenylimidazole	4.37	D	25	0	1451
$C_{15}H_{12}N_2O$	*cis*-*o*-Tolyl *N*-ether of oximinophenylacetonitrile	6.4_7	B	25	5	104
	trans-*o*-Tolyl *N*-ether of oximinophenylacetonitrile	0.9_7	B	25	5	104
	cis-*m*-Tolyl *N*-ether of oximinophenylacetonitrile	6.8_8	B	25	5	104
	trans-*m*-Tolyl *N*-ether of oximinophenylacetonitrile	1.0_7	B	25	5	104
	cis-*p*-Tolyl *N*-ether of oximinophenylacetonitrile	6.8_9	B	25	5	104

f) Tetrachloroethylene.

FORMULA	COMPOUND NAME	μ, D	State, or Solv.	t, °C	Method or P_a, % of P_e	Reference
$C_{15}H_{12}N_2O$ (contd.)	*trans*-*p*-Tolyl *N*-ether of oximinophenylacetonitrile	1.0_6	B	25	5	104
$C_{15}H_{12}N_2O_2$	*C*-Phenyl-*N*-*p*-tolylsydnone	6.96±0.02	B	25?	0	737
	N-Phenyl-*C*-*p*-tolylsydnone	6.72±0.02	B	25?	0	737
$C_{15}H_{12}N_2O_3$	Hydrofuramide	2.68±0.01	B	35	0	1849
$C_{15}H_{12}O$	Benzylideneacetophenone (m.p. not given)	2.95 3.03	B B	20 20	- 0	129 442
	(m.p. = 49°C)	3.05	B	20	0	442
	(m.p. = 57°C)	3.01	B	20	0	442
	2,3,6,7-Dibenzo-2,6-cycloheptadien-1-one	3.28±0.03	B	30	5	146
	2,3-Diphenylcyclopropanone	5.08	n.s.	n.s.	-	2047
	9-Methoxyanthracene	1.46	B	25	0	43
$C_{15}H_{12}O_2$	Benzylideneacetophenone oxide	3.89	B	15	0	157
$C_{15}H_{12}O_4$	Phenyl acetylsalicylate	2.54	B	25	-	946
$C_{15}H_{13}Cl$	3-Chloro-1,1-diphenylpropene	1.95	D	20	0	1969
$C_{15}H_{13}IN_2S_2$	5-Methylmercapto-2,3-diphenyl-1,3,4-thiadiazolinium iodide	13.1	$CHCl_3$	25	15	859
$C_{15}H_{13}NO$	Cinnamanilide	3.83	B	25	-	1042
$C_{15}H_{13}NO_2$	4-Methyl-4'-nitrostilbene	5.27±0.05	B	20	-	1351
	cis-2-Nitro-2-phenyl-1-*p*-tolylethylene	4.79±0.02	B	20	5-10	421

FORMULA	COMPOUND NAME	μ, D	State, or Solv.	t, °C	Method or P_a, % of P_e	Reference
$C_{15}H_{13}NO_3$	4-Methoxy-4'-nitrostilbene	5.51±0.02	B	25	0	475
		5.61±0.06	B	20	-	1351
		7.8	B	17	-	2091
$C_{15}H_{13}N_3OS$	1,4-Diphenyl-5-methylmercapto-3,5-endoxy-1,2,4-triazoline	7.7	D	25	15	859
$C_{15}H_{13}N_3S$	1,4-Diphenyl-5-methyl-3,5-*endo*thio-1,2,4-triazoline	8.4	$CHCl_3$	25	15	859
$C_{15}H_{14}$	1,1-Diphenylcyclopropane	0.54	B	25	0	557
		0.2-0.5	pXy	23.7	5	593
	1,2-Diphenylcyclopropane	0.52	B	25	0	557
$C_{15}H_{14}N_2$	*p,p'*-Dimethylcarbodianil	1.98	B	17	0	156
	Di-*p*-tolyldiazomethane	1.98	CCl_4	0	-	1774
$C_{15}H_{14}N_4S$	*p*-Thiocyano-*p'*-(dimethylamino)azobenzene	6.65	B	25	0	269
$C_{15}H_{14}N_4Se$	*p*-Selenocyano-*p'*-(dimethylamino)azobenzene	6.78	B	25	0	269
$C_{15}H_{14}O$	*p,p'*-Dimethylbenzophenone	3.45	B	n.s.	-	1223
	1,3-Diphenyl-2-propanone	2.67	B	18	0?	673
		2.95	B	30	0	20
$C_{15}H_{14}O_2$	Ethyl *p*-phenylbenzoate	2.03	B	15	-	129
$C_{15}H_{14}O_2S$	*p,p'*-Dimethoxythiobenzophenone	4.47±0.02	B	25	15	416,2137
	trans-Phenylcinnamylsulfone	5.69	B	30	0	88
$C_{15}H_{14}O_3$	2,4-Dimethoxybenzophenone	4.16	B	25	5	319

FORMULA	COMPOUND NAME	μ, D	State, or Solv.	t, °C	Method or P_a, % of P_e	Reference
$C_{15}H_{14}O_3$ (contd.)	4,4'-Dimethoxybenzophenone	3.93±0.02	B	25	15	416,2137
$C_{15}H_{14}O_4$	8-Acetyl-6-ethyl-4-methylumbelliferone	3.33-3.58[g]	B	25-45	-	827
$C_{15}H_{14}S$	Dimethylsulfonium-9-fluorenylidide	6.2±0.2	B	25	5	1502
	p,p'-Dimethylthiobenzophenone	3.45	B	n.s.	-	1223
$C_{15}H_{15}Bi$	Bismuth *tris*(cyclopentadienyl)	1.17	B	25	15	520
$C_{15}H_{15}ClN_2$	4-Chloro-4'-dimethylamino-*N*-benzylidene-aniline	5.23	B	25	-	498
	4'-Chloro-4-dimethylamino-*N*-benzylidene-aniline	3.70	B	25	-	498
$C_{15}H_{15}Cl_3N_3Tl$	Thallium trichloride - *tris*(pyridine) complex	2.10	B	20	10	1739
$C_{15}H_{15}N$	4,4'-Dimethyl-*N*-benzylideneaniline	1.58	B	25	-	498
$C_{15}H_{15}NO$	*N*-Benzyl-2-phenylacetamide	3.93	D	25	0	959
	Phenylpropionanilide	3.65	D	25	0	959
$C_{15}H_{15}NO_2$	2-(*N*-Piperidylmethylene)-1,3-indandione	3.49	B	25	-	1766
$C_{15}H_{15}NS_2$	*S*-Ethyl diphenyldithiocarbamate	3.15	B	25	0	639
$C_{15}H_{15}N_2O_3$	*N*-Nitrosophenylglycine, benzyl ester	4.06	B	25	0	428
$C_{15}H_{15}N_3O_2$	4-Dimethylamino-4'-nitro-*N*-benzylidene-aniline	6.7 8.6±0.2	B B	25 n.s.	- 15	498 705
	4'-Dimethylamino-4-nitro-*N*-benzylidene-aniline	6.92±0.05	B	n.s.	15	705

g) Calculated by Jatkar's equation.

FORMULA	COMPOUND NAME	μ, D	State, or Solv.	t, °C	Method or P_a, % of P_e	Reference
$C_{15}H_{16}$	1,3-Diphenylpropane	0.55	B	25	0	1601
$C_{15}H_{16}N_2$	Benzylidene-*p*-dimethylaminoaniline	2.67±0.1	B	n.s.	15	705
	4-Dimethylaminobenzylideneaniline	3.6±0.05	B	n.s.	15	705
$C_{15}H_{16}N_2O$	*sym*-Dimethyldiphenylurea	3.32±0.04	B	25	0	120
		3.20±0.04	D	25	0	120
		3.27	D	25	-	562
		3.6±0.1	D	24	0	166
		3.38±0.04	CCl_4	25	0	120
		3.14±0.04	Hp	25	0	120
		3.24	h	25	-	562
	1,1-Di-*o*-tolylurea	4.13±0.04	D	25	0	120
$C_{15}H_{16}N_4S$	*cis-cis-N*-(2-Methylphenylimino)-*N'*-anilino-*S*-methylisothiourea	3.07	B	20	10	424
$C_{15}H_{16}O$	6,8-Diethyl-7*H*-benzocyclohepten-7-one	3.47	B	25	0	1303
$C_{15}H_{16}O_2$	2-Methoxy-4'-methylbenzohydrol	1.73	B	25	5	319
	Bis(*p*-methoxyphenyl)methane	1.62±0.07	B	25	15	416,2137
		1.87±0.02	B	20	0	1187
$C_{15}H_{16}O_3$	2,4-Dimethoxybenzohydrol	2.69	B	25	5	319
	3,4-Dimethoxybenzohydrol	2.11	B	25	5	319
$C_{15}H_{16}O_{10}$	2,3,4,6-Tetraacetyl-α-methyl-D-glucose	2.37	$CHCl_3$	35	0	1560
		2.45	$CHCl_3$	25	0	1560
	2,3,4,6-Tetraacetyl-β-methyl-D-glucose	3.24	$CHCl_3$	25	0	1560
		3.37	$CHCl_3$	35	0	1560

h) Ethanol.

FORMULA	COMPOUND NAME	μ, D	State, or Solv.	t, °C	Method or P_a, % of P_e	Reference
$C_{15}H_{17}NO$	Ethyl-1-phenethyl-2(1*H*)-pyridone	3.22	B	27	0	1934
$C_{15}H_{17}N_3$	2-Methyl-4'-dimethylaminoazobenzene	3.11 3.18	B B	n.s. 31	- 0	1342 982
	4-Methyl-4'-dimethylaminoazobenzene	2.75	B	25	0	269
$C_{15}H_{17}N_3O$	4-Methoxy-4'-dimethylaminoazobenzene	2.77	B	25	0	269
$C_{15}H_{17}N_3O_3S$	*N*,*N*-Dimethyl-*p*-(*p*-tolylsulfonylazoxy)-aniline	8.4_6	B	30	0	540
$C_{15}H_{18}BNO$	*B*,*B*-Diphenyltetrahydroboroxazine	9.0±0.22	D	25	-	553
	N-Methyl-*B*,*B*-diphenylboroxazolidine	8.8±0.34	D	25	-	553
$C_{15}H_{18}CrO_3$	Hexamethylbenzene - chromium tricarbonyl complex	6.22±0.02 6.48±0.05	B B	25 n.s.	15 -	520 1568
$C_{15}H_{18}N_2O$	1-Benzyldihydronicotinamide	3.89	D	25	0	312
$C_{15}H_{18}O_3$	Santonin	3.7	B	21,42	-	992
$C_{15}H_{19}N$	2,5,2',4',5'-Pentamethyl-1-phenylpyrrole	2.06±0.04	B	25	0	931
$C_{15}H_{20}Br_2$	*cis*-Dihydro-α-tricyclopentadien-1,2-dibromide	3.23	B	n.s.	15	414
	cis-Dihydro-β-tricyclopentadiene-1,2-dibromide	3.21	B	n.s.	15	414
	trans-Dihydro-β-tricyclopentadiene-1,2-dibromide	1.94	B	n.s.	15	414
$C_{15}H_{20}ClN_3$	7-Chloro-4-(2-diethylaminoethylamino)-quinoline	5.90±0.05	D	n.s.	-	1380

FORMULA	COMPOUND NAME	μ, D	State, or Solv.	t, °C	Method or P_a, % of P_e	Reference
$C_{15}H_{20}N_2OS_2$	5-(4'-*N*-Piperidylbutadienylmethylene)-3-ethylrhodanine	9.06	B	25	-	1766
$C_{15}H_{20}O$	2,5-Dicyclopentylidenecyclopentanone	2.67±0.03	B	25	0	127
	2,2-Diethyl-4-pentenophenone	2.66±0.01	B	(25)	-	301
	2-(Isopropylphenyl)cyclohexanone	2.93	B	30	0	784
$C_{15}H_{20}O_4$	Diethyl β-phenylglutarate	2.505	B	30	0	1501
$C_{15}H_{21}AlO_6$	Aluminum(III) acetylacetonate	0	gas	229-245	-	331
		1.08	B	20	20	1440
		1.43	B	n.s.	-	1149
		1.4_3 [i]	B	25	0	63
		1.11	D	20	20	1440
		1.3_2 [i]	D	25	0	63
		1.03	CCl_4	20	20	1440
$C_{15}H_{21}CoO_6$	Cobalt(III) acetylacetonate	1.35	B	n.s.	-	1149
$C_{15}H_{21}CrO_6$	Chromium(III) acetylacetonate	0	gas	235-245	-	331
		1.6	B	n.s.	-	1149
$C_{15}H_{21}FeO_6$	Iron(III) acetylacetonate	1.05	B	40	-	1301
		1.25	B	30	-	1301
		1.35	B	25	-	1301
		1.47	B	n.s.	-	1149
		1.48	B	20	-	1301
$C_{15}H_{22}$	Pentacyclo[$9.2.1.1^{3,9}.0^{2,10}.0^{4,8}$]pentadecane	<0.52	B	n.s.	0	414
$C_{15}H_{22}Cl_3NPt$	*trans*-Dichloro(2,2,5,5-tetramethylhex-3-yne)(4-chloropyridine)platinum(II)	1.5±0.2	B	25	15	287

i) "Apparent" values caused by atomic polarization.

FORMULA	COMPOUND NAME	μ, D	State, or Solv.	t, °C	Method or P_a, % of P_e	Reference
$C_{15}H_{22}O$	Aromadendrone	2.11	B	23	-	1402
	cis-Vetivone	3.87	n.s.	23	-	1403
	trans-Vetivone	3.73	n.s.	23	-	1403
$C_{15}H_{22}O_3$	Bithone	3.74	j	22	-	1994
$C_{15}H_{22}O_{10}$	2,3,4,6-Tetraacetyl-α-methyl-D-glucose	2.25	B	30	-	1561
		2.41	B	40	-	1561
		2.48	D	30	-	1562
		2.64	D	40	-	1562
	2,3,4,6-Tetraacetyl-β-methyl-D-glucose	3.63	B	30	-	1561
		3.69	B	40	-	1561
		3.58	D	30	-	1562
		3.64	D	40	-	1562
	α-Tetraacetylmethylglucoside	2.38	n.s.	n.s.	-	677
	β-Tetraacetylmethylglucoside	3.08	n.s.	n.s.	-	677
$C_{15}H_{23}Cl_2NPt$	*trans*-Dichloro(2,2,5,5-tetramethylhex-3-yne)(pyridine)platinum(II)	2.75±0.2	B	25	15	287
$C_{15}H_{23}N_3O_2$	1-(Diethylaminoacetyl)-1-phenyl-2-acetyl-2-methylhydrazine	3.8_2	B	25	-	1042
$C_{15}H_{24}$	Aromadendrene	0.93	B	23	-	1402
	Cedrene	0.38	liq	15-35	-	879
	Sesquiterpene from *cis*-vitivone	1.12	n.s.	23	-	1403
	Sesquiterpene from *tran*-vitivone	0.97	n.s.	23	-	1403

j) Ethanol.

FORMULA	COMPOUND NAME	μ, D	State, or Solv.	t, °C	Method or P_a, % of P_e	Reference
$C_{15}H_{24}O$	2,6-Di-*t*-butyl-4-methylphenol	1.68±0.02	B	20	0	1186,1668
	3-Methyl-5-(2,6,6-trimethyl-1-cyclohexen-1-yl)-2,4-pentadien-1-ol	2.11	B	25	0	896
$C_{15}H_{26}$	Dihydroaromadendrene	0.79	B	23	-	1402
$C_{15}H_{26}Br_2$	(-)Cadinene dihydrobromide	4.20	B	25	0	835
$C_{15}H_{26}Cl_2$	(-)Cadinene dihydrochloride	4.08	B	25	0	835
$C_{15}H_{26}N_2$	Allomatridine	1.27	B	25	-	434,1989
	Matridine	0.63	B	25	-	434,1989
	Sparteine	0.80	B	25	-	434
	α-Isosparteine	1.95	B	25	-	434
$C_{15}H_{26}O_3Si$	Phenyltri-*n*-propoxysilane	1.58	B	20	0	1041
	Phenyltriisopropoxysilane	1.56	B	20	0	1041
$C_{15}H_{27}AlO_6$	Aluminum tri(*n*-valerate)	3.47	B	25	0	1455
$C_{15}H_{27}N_5O_6$	3,7,10-Triethyl-3,7,10-trinitro-1,5-diazabicyclo[3.3.3]undecane	5.7	B	n.s.	-	943
$C_{15}H_{28}$	*cis*-Vetivane	0.93	n.s.	23	-	1403
	trans-Vetivane	0.73	n.s.	23	-	1403
$C_{15}H_{28}O_4$	Tridecane-1,13-dicarboxylic acid	2.35	D	35	-	1581
$C_{15}H_{29}NO$	*trans*-2-Oxopentadecamethylenimine	3.69	B	25	0	783

FORMULA	COMPOUND NAME	μ, D	State, or Solv.	t, °C	Method or P_a, % of P_e	Reference
$C_{15}H_{30}AsN_3S_6$	Arsenic *tris*(*N*,*N*-diethyldithiocarbamate)	4.50	B, CS_2	20	15	1242
$C_{15}H_{30}BiN_3S_6$	Bismuth *tris*(*N*,*N*-diethyldithiocarbamate)	4.08	B	25	0	639
$C_{15}H_{30}O$	8-Pentadecanone	2.73	CCl_4	25	-	67
$C_{15}H_{30}O_2$	Methyl myristate	1.62	B	25	15	1793
$C_{15}H_{32}NO_4P$	Diisopentyl ester of diethylcarbamoyl-phosphoric acid	2.95	B	25	5	58
$C_{15}H_{32}O_3$	1,1,1-Tributoxypropane	1.86	B	25	5	60
$C_{15}H_{33}BO_3$	Tri-*n*-pentyl borate	0.80	B	25	0	1458
		0.77	D	25	0	1458
		0.84	CCl_4	25	0	72
	Triisopentyl borate	0.82	B	20	-	343
		0.81	CCl_4	25	0	72
$C_{15}H_{33}NO_2$	Dodecylamine *n*-propionate	1.79	B	30	5	1757
$C_{15}H_{33}O_4V$	Tri-*t*-pentyl vanadate	1.11	B	25	0	271
$C_{16}H_{10}$	Acepleiadylene	0.49	B	30	0	1516
	Fluoranthene	0	B	n.s.	0	149
	Pyrene	0	B	30	0	1516
$C_{16}H_{10}N_2$	*cis*-Diphenylmaleodinitrile	5.6_8	CCl_4	25	0	2109
	trans-Diphenylmaleodinitrile	1.0_7	CCl_4	25	0	2109
$C_{16}H_{10}N_2O_2$	2-Phenyl-5*H*-pyrazolo[a][3,1]-benzoxazin-5-one, α-lactone	2.80	B	20?	0	2038

FORMULA	COMPOUND NAME	μ, D	State, or Solv.	t, °C	Method or P_a, % of P_e	Reference
	2-Phenyl-5*H*-pyrazolo[a][3,1]-benzoxazin-5-one, β-lactone	1.89	B	20?	0	2038
$C_{16}H_{10}O_2$	*cis*-1,4-Diphenyl-2-butyne-1,4-dione	3.39	n.s.	n.s.	-	1706[k]
	trans-1,4-Diphenyl-2-butyne-1,4-dione	2.60	n.s.	n.s.	-	1706[k]
$C_{16}H_{11}BrN_2$	*o*-Bromophenylazoazulene	3.57	B	25	0	571
	p-Bromophenylazoazulene	3.96	B	25	0	571
$C_{16}H_{11}ClN_2$	*o*-Chlorophenylazoazulene	3.55	B	25	0	571
	p-Chlorophenylazoazulene	4.01	B	25	0	571
$C_{16}H_{11}Cl_7O_2$	1,1,1-Trichloro-2,2-*bis*(2,4-dichloro-5-methoxyphenyl)ethane	3.10	B	25	0	1607
	1,1,1-Trichloro-2,2-*bis*(3,5-dichloro-2-methoxyphenyl)ethane	2.50	B	25	0	1607
$C_{16}H_{11}IN_2$	*o*-Iodophenylazoazulene	3.56	B	25	0	571
	p-Iodophenylazoazulene	3.87	B	25	0	571
$C_{16}H_{11}NO_3$	3,5-Diphenyl-4-isoxazolecarboxylic acid	3.42±0.04	D	25	0	1857
$C_{16}H_{11}N_3O_2$	*o*-Nitrophenylazoazulene	4.97	B	25	0	571
	p-Nitrophenylazoazulene	7.20	B	25	0	571
$C_{16}H_{11}N_3O_6$	1,3,5-Trinitrobenzene - naphthalene complex	0.69±0.06	CCl_4	20	-	212
$C_{16}H_{11}N_3O_7$	Naphthalene - picric acid complex	2.18	D	35	0	1558

k) Value from J. prakt. chem. [2], 151, 212 (1938).

FORMULA	COMPOUND NAME	μ, D	State, or Solv.	t, °C	Method or P_a, % of P_e	Reference
$C_{16}H_{12}$	Benzylideneindene	0.32±0.02 0.96±0.03	B B	20 30	0 0	1189 144
	1-Methylene-2,3,6,7-dibenzo-2,4,6-cycloheptatriene	0.5±0.1	B	30	5	146
	Phenylbenzofulvene	1.30±0.04	B	20	0	1199
$C_{16}H_{12}BrN_3$	4-(4'-Bromophenylazo)-1-naphthylamine	3.83 4.52	B D	25 25	0 0	801 801
$C_{16}H_{12}Br_2O_4$	Dimethyl 4,4'-dibromodiphenate	2.19	B	25	0	1104
$C_{16}H_{12}ClN_3$	4-(4'-Chlorophenylazo)-1-naphthylamine	3.65 4.51	B D	25 25	0 0	801 801
$C_{16}H_{12}Cl_2$	*cis*-11,12-Dichloro-9,10-dihydro-9,10-ethanoanthracene	3.16±0.12	B	25	0	1643
	trans-11,12-Dichloro-9,10-dihydro-9,10-ethanoanthracene	2.17±0.12	B	25	0	1643
$C_{16}H_{12}Cl_6O_2$	1,1-Dichloro-2,2-*bis*(2,4-dichloro-5-methoxyphenyl)ethane	1.67	B	25	0	1607
	1,1-Dichloro-2,2-*bis*(3,5-dichloro-2-methoxyphenyl)ethane	1.96	B	25	0	1607
$C_{16}H_{12}N_2$	Phenylazoazulene	2.63	B	25	0	571
$C_{16}H_{12}N_2O$	1-Benzeneazo-2-naphthol	1.61 1.4	B CS_2	25 16	- 0	668 164
	2-Benzeneazo-1-naphthol	1.8	B	28	0	164
	4-Benzeneazo-1-naphthol	2.1	D	28	0	164

FORMULA	COMPOUND NAME	μ, D	State, or Solv.	t, °C	Method or P_a, % of P_e	Reference
	o-Hydroxyphenylazoazulene	3.38	B	25	0	571
		3.49	D	25	0	571
	p-Hydroxyphenylazoazulene	2.72	D	25	0	571
$C_{16}H_{12}N_2O_8$	Dimethyl 4,4'-dinitrodiphenate	2.05	B	25	0	1104
$C_{16}H_{12}N_4O_7$	1-Naphthylamine picrate	5.08	D	35	0	1559
	2-Naphthylamine picrate	6.96	D	35	0	1559
$C_{16}H_{12}O_4$	Di-*o*-cresotide	6.34±0.02	B	25?	0	437
	Di-*m*-cresotide	6.74±0.02	B	25?	0	437
	Di-*p*-cresotide	6.67±0.02	B	25?	0	437
$C_{16}H_{13}Br_2Cl_3$	1,1,1-Trichloro-2,2-*bis*(4-bromo-2-methylphenyl)ethane	2.38	B	25	-	1612
	1,1,1-Trichloro-2,2-*bis*(4-bromo-3-methylphenyl)ethane	1.85	B	25	-	1612
	1,1,1-Trichloro-2,2-*bis*(5-bromo-2-methylphenyl)ethane	3.22	B	25	-	1612
$C_{16}H_{13}ClO_5$	2-Phenyl-3-methylbenzopyrylium perchlorate	~7	m	25	-	1066
$C_{16}H_{13}Cl_2NO_3$	2-*p*-Chlorobenzamidoethyl *p*-chlorobenzoate	3.28	B	30	0	2148
$C_{16}H_{13}Cl_5$	1,1,1-Trichloro-2,2-*bis*(2-methyl-4-chlorophenyl)ethane	2.46	B	25	0	1607
	1,1,1-Trichloro-2,2-*bis*(2-methyl-5-chlorophenyl)ethane	3.13	B	25	0	1607

m) Dimethylaniline.

FORMULA	COMPOUND NAME	μ, D	State, or Solv.	t, °C	Method or P_a, % of P_e	Reference
$C_{16}H_{13}Cl_5O_2$	1-Chloro-2,2-*bis*(2,4-dichloro-5-methoxy-phenyl)ethane	2.86	B	25	0	1607
	1-Chloro-2,2-*bis*(3,5-dichloro-2-methoxy-phenyl)ethane	2.08	B	25	0	1607
$C_{16}H_{13}N$	Phenyl-2-naphthylamine	1.56±0.2	B	35	0	1850
$C_{16}H_{13}NO$	2-Methyl-4,5-diphenyloxazole	1.7	B	25	0	1430
$C_{16}H_{13}NO_2$	1-(4-Nitrophenyl)-4-phenyl-1,3-butadiene	4.53 4.75±0.02	n.s. B	n.s. 25	- 0	1904 475
$C_{16}H_{13}NO_2S$	2-Naphthalenesulfonanilide	4.70	B	20	0	1703
$C_{16}H_{13}NO_4$	Methyl *p*-nitro-α-phenylcinnamate	3.81	B	17	-	129
$C_{16}H_{13}N_3$	1-Benzeneazo-2-naphthylamine	2.16	B	16	0	164
	4-Benzeneazo-1-naphthylamine	2.38 2.58 2.97	B B D	25 16 25	0 0 0	801 164 801
$C_{16}H_{14}Br_2$	*cis*-4,4'-Dibromo-α,α'-dimethylstilbene	3.00	B	25	-	1756
	trans-4,4'-Dibromo-α,α'-dimethylstilbene	0.84	B	25	-	1756
$C_{16}H_{14}Br_2N_2O_2Pd$	*p*-Methoxyphenylisonitrile - palladium bromide complex	2.79	B	25	-	38
$C_{16}H_{14}Cl_2$	*cis*-4,4'-Dichloro-α,α'-dimethylstilbene	2.80	B	25	-	1756
	trans-4,4'-Dichloro-α,α'-dimethylstilbene	0.94	B	25	-	1756
$C_{16}H_{14}I_2$	*cis*-4,4'-Diiodo-α,α'-dimethylstilbene	2.88	B	25	-	1756

FORMULA	COMPOUND NAME	μ, D	State, or Solv.	t, °C	Method or P_a, % of P_e	Reference
	trans-4,4'-Diiodo-α,α'-dimethylstilbene	0.64	B	25	-	1756
$C_{16}H_{14}I_2N_2O_2Pd$	*p*-Methoxyphenylisonitrile - palladium iodide complex	2.53	B	25	-	38
$C_{16}H_{14}I_2N_2Pd$	*p*-Tolylisonitrile - palladium iodide complex	1.93	B	25	-	38
$C_{16}H_{14}F_{16}O_4$	*Bis*(2,2,3,3,4,4,5,5-octafluoropentyl)ester of 3-methylglutaric acid	3.09	B	20	0	1666
$C_{16}H_{14}N_2O_2$	*C,N*-Di-*p*-tolylsydnone	7.15±0.02	B	25?	0	737
$C_{16}H_{14}N_2O_4$	*cis*-4,4'-Dinitro-α,α'-dimethylstilbene	6.88	B	25	-	1756
	trans-4,4'-Dinitro-α,α'-dimethylstilbene	0.68	B	25	-	1756
$C_{16}H_{14}O$	4-Acetylstilbene	3.03 3.20±0.02	n.s. B	n.s. 25	- 0	1904 475
$C_{16}H_{14}O_2$	4,4'-Diacetylbiphenyl	1.9	B	25	-	2112
	1,5-Dimethoxyanthracene	0.63	B	25	0	479
	9,10-Dimethoxyanthracene	1.73	B	25	0	479
	α-1,4-Diphenyl-2-butyne-1,4-diol	2.40±0.02	D	20	0	1185
	Methyl α-phenylcinnamate	1.94	B	17	-	129
$C_{16}H_{14}O_2S$	*cis*-β,*p*-Tolylmercaptocinnamic acid	3.27	D	25	0	1972
	trans-β,*p*-Tolylmercaptocinnamic acid	3.20	D	25	0	1972
$C_{16}H_{14}O_4$	2,2'-Dihydroxybiphenyldiacetate	2.2_0	B	25	-	204

FORMULA	COMPOUND NAME	μ, D	State, or Solv.	t, °C	Method or P_a, % of P_e	Reference
$C_{16}H_{14}O_4$ (contd.)	4,4'-Dihydroxybiphenyl diacetate	1.9	B	25	-	2088
	Dimethyl 2,2'-biphenyldicarboxylate	2.3	B	25	-	2088
		2.38	B	25	-	204
		2.44	B	25	0	1104
	Dimethyl 4,4'-biphenyldicarboxylate	2.2	B	25	-	2088
$C_{16}H_{14}O_4S$	*cis*-*p*-Tolylsulfonylcinnamic acid	4.35	D	25	0	1972
	trans-*p*-Tolylsulfonylcinnamic acid	3.85	D	25	0	294
$C_{16}H_{14}O_4S_2$	*p*-Methoxybenzoyl disulfide	2.50	B	25	-	2030
$C_{16}H_{15}Cl_3$	1,1,1-Trichloro-2,2-di-*p*-tolylethane	2.21±0.1	B	20	0	2107
$C_{16}H_{15}Cl_3O_2$	1,1,1-Trichloro-2,2-di-*p*-anisylethane	2.90±0.1	B	20	0	2107
$C_{16}H_{15}N$	1,2-Dimethyl-3-phenylindole	2.63	B	25	-	817
	1,3-Dimethyl-2-phenylindole	2.20	B	25	-	817
	2-Methyl-3-benzylindole	2.51	B	25	-	817
	3-Methyl-2-benzylindole	2.13	B	25	-	817
$C_{16}H_{15}NO_3$	Ethyl salicylidene-*p*-aminobenzoate (red or yellow)	2.71	B	25	0	399
		2.75	B	25	15	856
$C_{16}H_{15}N_3O$	*anti*-*N*-*p*-Dimethylaminophenyl ether of oximinophenylacetonitrile	2.97	B	25	5	104
$C_{16}H_{16}$	1,2,5,6-Dibenzocycloöctadiene	0.5-0.6	B	25	-	1567
	1,1-Diphenyl-2,2-dimethylethylene	0.51	B	30	0	599
		0.65	CCl_4	30	0	599

FORMULA	COMPOUND NAME	μ, D	State, or Solv.	t, °C	Method or P_a, % of P_e	Reference
$C_{16}H_{16}ClN$	2-Chloro-*N*,*N*-dimethyl-4'-stilbenamine	2.11	B	25	0	480
		2.65±0.02	B	25	0	475,481
	4-Chloro-*N*,*N*-dimethyl-4'-stilbenamine	4.02±0.02	B	25	0	475
$C_{16}H_{16}Cl_4O_2Sn$	Stannic chloride - di(acetophenone) complex	8.7	B	18	0	2003
$C_{16}H_{16}CuN_2O_2$	*Bis*(*N*-methylsalicylaldimine)copper(II)	1.57	B	25	10	1671
		1.45	D	25	10	1671
$C_{16}H_{16}N_2NiO_2$	*Bis*(*N*-methylsalicylaldimine)nickel(II)	0.86	B	25	10	1671
		1.29	D	25	10	1671
$C_{16}H_{16}N_2OS_2$	5-(4'-Anilinobutadienylmethylene)-3-ethylrhodanine	6.33	B	25	-	1766
$C_{16}H_{16}N_2O_2$	Azine of anisaldehyde	1.84	B+ $CHCl_3$	20	-	1991
	1,2-Diacetyl-1,2-diphenylhydrazine	4.51	B	25	-	1042
	4-Dimethylamino-4'-nitrostilbene	7.15	B	20	FL	375
		7.42±0.02	B	25	0	475
		7.61±0.08	B	20	-	1351
		8.3	n	17	-	2091
		7.25	D	20	FL	375
	p-Dimethylamino-ω-nitrostilbene	7.65	B	20	FL	375
$C_{16}H_{16}N_2O_2S_2$	5-(2'-*N*-Acetylanilinovinylmethylene)-3-ethylrhodanine	5.10	B	25	-	1766
	1-[2-(3-Ethylbenzoxazolinylidenyl)]-2-[5-(3-ethyl-2-thione-4-oxothiazolylidenyl)]ethane	8.37	B	25	0	1767

n) 1-Methylnaphthalene.

FORMULA	COMPOUND NAME	μ, D	State, or Solv.	t, °C	Method or P_a, % of P_e	Reference
$C_{16}H_{16}N_2O_3$	1-Acetyl-1-(*o*-hydroxybenzoyl)-2-methyl-2-phenylhydrazine	6.52	B	25	-	1042
$C_{16}H_{16}N_2S_4$	*Bis*(methylphenylthiocarbamoyl)disulfide	2.57	B	25	0	639
$C_{16}H_{16}N_3S_2I$	5-Methylmercapto-5-methyl-1,4-diphenyl-1,3,4-triazolinium iodide	11.1	$CHCl_3$	25	15	859
$C_{16}H_{16}N_4$	1,4-Dihydro-1,4-*p*-tolyl-1,2,4,5-tetrazine	0.75±0.05	B	25	12	436
$C_{16}H_{16}N_4O_4$	1,4-*Bis*(*p*-nitrophenyl)piperazine	~6	B	25	5	69
$C_{16}H_{16}O$	1,1-Diphenyl-2-butanone	2.92	B	30	0	20
	1,3-Diphenyl-2-butanone	2.71	B	30	0	20
	3,3-Diphenyl-2-butanone	2.70	B	30	0	20
	2,7-Pentamethylene-4,5-benzo-2,4,6-cycloheptatrien-1-one	3.09±0.06	B	25	0	563
	2,4,6-Trimethylbenzophenone	2.65	B	25	0	41
$C_{16}H_{16}O_2$	Ethyl diphenylacetate	1.76	B	30	0	1501
$C_{16}H_{16}O_4$	1,2-Diethoxycarbonylazulene	1.72	B	30	-	1016
$C_{16}H_{16}O_{11}$	2,3,4,6-Tetraacetyl-β-acetyl-D-galactose	2.94	$CHCl_3$	35	0	1560
	2,3,4,6-Tetraacetyl-α-acetyl-D-glucose	3.05	$CHCl_3$	25,35	0	1560
	2,3,4,6-Tetraacetyl-β-acetyl-D-glucose	2.50 2.56	$CHCl_3$ $CHCl_3$	25 35	0 0	1560 1560
$C_{16}H_{17}N$	2-Dimethylaminostilbene	0.64 1.07±0.02	n.s. B	n.s. 25	- 0	1904 475

FORMULA	COMPOUND NAME	μ, D	State, or Solv.	t, °C	Method or P_a, % of P_e	Reference
	3-Dimethylaminostilbene	1.47 1.67±0.02	n.s. B	n.s. 25	- 0	1904 475
	4-Dimethylaminostilbene	2.08 2.41±0.02	n.s. B	n.s. 25	- 0	1904 475
$C_{16}H_{17}NO$	*N*-Benzoyl-2-phenylisopropylamine	3.74	B	25	5	432
	N-Benzyl-3-phenylpropionamide	3.73 3.77	D D	25 40	0 0	959 959
	N-Phenethyl-2-phenylacetamide	3.99 4.06	D D	25 40	0 0	959 959
$C_{16}H_{17}NO_3$	3-Ethyl-1-(3,4-methylenedioxyphenethyl)-2(1*H*)-pyridone	3.68	B	27	0	1934
$C_{16}H_{17}N_3O_6$	1,3,5-Trinitrobenzene - 1,2,4,5-tetramethylbenzene complex	0.55±0.05	CCl_4	20	-	212
$C_{16}H_{18}$	1,4-Diphenylbutane	0.52	B	25	0	1601
$C_{16}H_{18}N_2$	*N*,*N*′-Diphenylpiperazine	0.85	B	20	-	568
	1,4-Diphenylpiperazine	0.59	B	25	5	69
$C_{16}H_{18}N_2O_3$	4,4'-Azoxydiphenetole	2.55	B	20	-	1991
	4,4'-Diethoxyazoxybenzene	2.42	B	20	-	1241c
$C_{16}H_{18}O_2$	3,3'-Diethoxybiphenyl	1.82	CCl_4	25	0	919
	4,4'-Diethoxybiphenyl	1.9	B	25	-	2088, 2112,2125

FORMULA	COMPOUND NAME	μ, D	State, or Solv.	t, °C	Method or P_a, % of P_e	Reference
$C_{16}H_{18}O_4$	3,3',5,5'-Tetramethoxybiphenyl	2.26	CCl_4	25	0	919
$C_{16}H_{18}O_5$	*Bis*(tropolone methyl ether)·H_2O	6.80±0.1	B	20-40	0	1022
$C_{16}H_{19}N_3$	4-Diethylaminoazobenzene	3.14±0.02 3.15	B B	31 n.s.	0 -	982 1342
$C_{16}H_{20}BNO$	*B,B*-Diphenyl-*N*-ethylboroxazolidine	10.1±0.07	D	25	-	553
	B,B-Diphenyltetrahydroboroxazepine	10.6±0.12	D	25	-	553
$C_{16}H_{20}Cr$	Di-*o*-xylyl chromium	0	B	25	15	520
$C_{16}H_{20}N_2$	2,2'-*Bis*(dimethylamino)biphenyl	1.55 1.57±0.01	B B	20 20	0 0	1182 1188
	4,4'-*Bis*(dimethylamino)biphenyl	1.21±0.02 1.26	B B	25 25	0 0	475 2079
$C_{16}H_{20}N_2O_2$	*Bis*(acetylacetone)-*m*-phenylenediimine	3.93±0.1	B	25	10	761
	4,4'-Diethoxy-2,2'-biphenylamine	3.11±0.02	B	20	0	1182,1188
$C_{16}H_{20}N_4$	4,4'-Azo*bis*(*N,N*-dimethylaniline)	1.95	B	25	10	267
$C_{16}H_{20}O$	2-Allyl-2-ethyl-4-pentenophenone	2.66±0.01	B	(25)	-	301
$C_{16}H_{22}O_4$	Dibutyl *m*-phthalate	2.75	B	20	-	2133
$C_{16}H_{22}O_6$	Di-*t*-butyl peroxyphthalate	3.20	B	30	0	2049
$C_{16}H_{22}O_{10}$	Quercitolpenta(aeeee)acetate	2.32	B	25	0	40
	Quercitolpenta(eeeee)acetate	2.28	B	25	0	40

FORMULA	COMPOUND NAME	μ, D	State, or Solv.	t, °C	Method or P_a, % of P_e	Reference
$C_{16}H_{22}O_{11}$	2,3,4,6-Tetraacetyl-β-acetyl-D-galactose	2.75	B	30,40	0	1561-2
		2.63	D	30,40	0	1561-2
	2,3,4,6-Tetraacetyl-α-acetyl-D-glucose	3.30	B	30	0	1561-2
		3.47	B	40	0	1561-2
		3.77	D	30,40	0	1561-2
	2,3,4,6-Tetraacetyl-β-acetyl-D-glucose	2.56	B	30,40	0	1561-2
		3.10	D	30	0	1561-2
		3.18	D	40	0	1561-2
	Pentaacetyl-D-glucose	2.45	B, $CHCl_3$	20	0	1676
	α-Pentaacetyl-D-glucose	3.47	n.s.	n.s.	-	677
	β-Pentaacetyl-D-glucose	2.425	B	20	0	1063
		2.489	B	35	0	1063
		2.538	B	50	0	1063
		2.276	$CHCl_3$	20	0	1063
		2.314	$CHCl_3$	35	0	1063
		2.420	$CHCl_3$	50	0	1063
$C_{16}H_{24}$	*syn*-Octamethyltricycloöctadiene	0.63±0.2	n.s.	n.s.	-	354
	anti-Octamethyltricycloöctadiene	0	n.s.	n.s.	-	354
$C_{16}H_{24}NiO_2P_2$	[*o*-Phenylene*bis*(diethylphosphine)]nickel dicarbonyl	5.48±0.01	B	n.s.	-	291
		5.42	n.s.	n.s.	-	290
$C_{16}H_{25}Cl_2NPt$	*trans*-Dichloro(2,2,5,5-tetramethylhex-3-yne)(4-methylpyridine)platinum(II)	3.3±0.2	B	25	15	287
$C_{16}H_{26}O_2$	2,5-Di-*t*-butyl-1,4-dimethoxybenzene	1.23±0.01	B	20	0	1186
		1.47±0.13	B	25	0	1429

FORMULA	COMPOUND NAME	μ, D	State, or Solv.	t, °C	Method or P_a, % of P_e	Reference
$C_{16}H_{30}O$	Cyclohexadecanone	2.77±0.03 2.75±0.03	B D	20 20	0 0	630 630
$C_{16}H_{30}O_2$	2,7-Dimethyl-3,6-diisopropyl-4-octyne-3,6-diol	2.31±0.02	D	20	0	1185
	4,7-Dipropyl-5-decyne-4,7-diol	2.46±0.03	D	20	0	1185
$C_{16}H_{31}NO$	*trans*-2-Oxohexadecamethylenimine	3.67	B	25	0	783
$C_{16}H_{32}Br_4O_8Si$	Silicon tetrabromide - tetradioxane complex	0.55±0.09	D	25	-	899
$C_{16}H_{32}Cl_4O_4Sn$	Tin tetrachloride - di(butyl butyrate) complex	6.44	B	20	15	1438
$C_{16}H_{32}O_2$	Palmitic acid	0.73 0.77 0.78 1.76 1.77 0.82	B B B D D cHx	25 23 25 25 23 23	15 26 15 0 - -	1793 1869 1468 178 1869 1869
$C_{16}H_{32}O_3$	Perpalmitic acid	2.30±0.04	B	30	0	1616
$C_{16}H_{32}O_5$	8,9,15-Trihydroxypalmitic acid	4.31	D	40	0	178
$C_{16}H_{32}P_2Pt$	*trans-Bis*(triethylphosphine)ethynylplatinum	0.7	B	25	15	294
$C_{16}H_{33}Br$	1-Bromohexadecane	1.87 1.98	liq liq	25 20	- -	707 657
$C_{16}H_{33}BrHg$	Hexadecylmercury bromide	2.33±0.06	B	40?	-	872-3
$C_{16}H_{33}I$	1-Iodohexadecane	1.83	CCl_4	20	-	78

FORMULA	COMPOUND NAME	μ, D	State, or Solv.	t, °C	Method or P_a, % of P_e	Reference
$C_{16}H_{33}NO_2Si_3$	1,1,1,2,3,3,3-Heptamethyl-2-ethylphenyl-aminomethyltrisiloxane	2.14	liq	20	RT	595
$C_{16}H_{34}O$	Hexadecanol	1.67	B	25	0	1366
		1.73	B	20-60	-	719
		1.68	CCl_4	25	5	1108
		1.32-1.60	Hx	20-60	-	719
		1.83	Ether	0-20	-	719,723
$C_{16}H_{34}O_2S$	*Bis*(1-methylheptyl)sulfone	4.56	B	20	0	636
	Di-*n*-octylsulfone	4.54	B	20	0	636
$C_{16}H_{34}S_2$	*n*-Octyl disulfide	1.99	B	25	-	2030
$C_{16}H_{35}NO_2$	Dodecylamine *n*-butyrate	1.32	B	30	5	1757
$C_{16}H_{35}O_3P$	Diethyldodecylphosphonate	2.73±0.02	liq	32	-	951
$C_{16}H_{36}BrN$	Tetra-*n*-butylammonium bromide	5.1-5.4	B	20	-	389
		11.7	B	25	-	564
		11.9	p	25	-	582
$C_{16}H_{36}Br_2PtS_2$	α-*Bis*(di-*n*-butylsulfide)platinum dibromide	2.21	B	20	-	844
$C_{16}H_{36}ClNO_4$	Tetra-*n*-butylammonium perchlorate	14.2	B	25	-	564
$C_{16}H_{36}Cl_2PtS_2$	α-*Bis*(di-*n*-butylsulfide)platinum dichloride	2.37	B	20	-	844
	β-*Bis*(di-*n*-butylsulfide)platinum dichloride	9.3	B	20	-	844
	α-*Bis*(diisobutylsulfide)platinum dichloride	2.43	B	20	-	844
	β-*Bis*(diisobutylsulfide)platinum dichloride	9.1	B	20	-	844

p) Bromobenzene.

FORMULA	COMPOUND NAME	μ, D	State, or Solv.	t, °C	Method or P_a, % of P_e	Reference
$C_{16}H_{36}Cl_2PtS_2$ (contd.)	α-*Bis*(di-*sec*-butylsulfide)platinum dichloride	2.41	B	20	-	844
$C_{16}H_{36}I_2PtS_2$	α-*Bis*(di-*n*-butylsulfide)platinum diiodide	2.20	B	20	-	844
$C_{16}H_{36}N_2O_4PtS_2$	α-*Bis*(diisobutylsulfide)platinum dinitrite	2.53	B	20	-	844
	β-*Bis*(diisobutylsulfide)platinum dinitrite	13.8	B	20	-	844
$C_{16}H_{36}O_4Si$	Tetra-*n*-butoxysilane	1.66 1.80	B B	20 20	5 0	53 1041
$C_{16}H_{36}O_4Ti$	Tetra-*n*-butoxytitanate	1.70 1.15	B Hx	20 25	5 0	53,51 274
$C_{16}H_{38}Cl_6O_4Sn_2$	Butoxytrichlorotin - *n*-butanol dimer	3.22	B	20	0	1439
	Butoxytrichlorotin - isobutanol dimer	3.24	B	20	0	1439
$C_{16}H_{38}N_4O_2$	Capric acid - hexamethylenetetramine complex	1.51	CCl_4	27	-	1987
$C_{16}H_{40}P_2Pt$	*cis*-*Bis*(triethylphosphine)diethylplatinum	5.5	B	25	15	294
$C_{17}H_9BrO$	7-Bromobenzanthrone	2.18	D	n.s.	-	2174
$C_{17}H_{10}O$	Benzanthrone	3.31 3.52 3.37	B B D	n.s. 25 n.s.	- 0 -	2174 1761 2174
$C_{17}H_{12}OS$	2,6-Diphenyl-4*H*-pyran-4-thione	5.31	B	20	-	1661
	2,6-Diphenyl-4*H*-1-thiapyran-4-one	4.41 4.42 4.44	B B n.s.	20 20 n.s.	- 0 -	1661 64 1471

FORMULA	COMPOUND NAME	μ, D	State, or Solv.	t, °C	Method or P_a, % of P_e	Reference
$C_{17}H_{12}O_2$	Dibenzospiropyran	1.20±0.02	B	30	0	170
	2,6-Diphenyl-4*H*-pyran-4-one	3.85	B	20	0	792
		4.74	B	20	-	1661
$C_{17}H_{12}O_3$	2-Naphthyl salicylate	3.08	B	25	-	944
$C_{17}H_{12}O_3S$	2,6-Diphenyl-4*H*-1-thiapyran-4-one 1-dioxide	0.94	B	20	0	64
$C_{17}H_{12}S_2$	2,6-Diphenyl-4*H*-1-thiapyran-4-thione	4.95	B	20	-	1661
$C_{17}H_{13}BrO$	9-(2-Bromo-1-propanoyl)anthracene	3.03±0.02	B	25	-	260
	9-(2-Bromo-1-propanoyl)phenanthrene	2.97±0.02	B	25	-	260
$C_{17}H_{13}Br_3N_2$	2,4,6-Tribromo-4'-dimethylamino-α-cyano-stilbene	4.10±0.03	B	25-55	-	2022
$C_{17}H_{13}N$	2,6-Diphenylpyridine	1.37	B	n.s.	0	1842-3
		1.55	B	25	0	1729
		1.53	CCl_4	25	0	1729
$C_{17}H_{13}NO_2$	2-(*N*-Methylanilinomethylene)-1,3-indandione	2.64	B	25	-	1766
$C_{17}H_{14}Br_2N_2$	2,6-Dibromo-4'-dimethylamino-α-cyano-stilbene	4.00±0.14	B	25-55	-	2022
$C_{17}H_{14}N_2$	2,2-Ditropylpropanedinitrile	4.35±0.05	B	25	5	2154
	N-Ethyl(9,10)phenanthrenoimidazole	4.14	D	25	0	1451
	o-Tolylazoazulene	2.54	B	25	0	571
	p-Tolylazoazulene	2.29	B	25	0	571

FORMULA	COMPOUND NAME	μ, D	State, or Solv.	t, °C	Method or P_a, % of P_e	Reference
$C_{17}H_{14}N_2O$	4-Benzeneazo-1-methoxynaphthalene	0.94	B	25	0	164
	o-Methoxyphenylazoazulene	1.93	B	25	0	571
	p-Methoxyphenylazoazulene	2.27	B	25	0	571
	1-*p*-Tolueneazo-2-naphthol	1.67	B	25	-	668
$C_{17}H_{14}N_2O_2$	*N*-(10-Acetyl-9-acridanylidene)acetamide	2.53±0.02	B	25	10	1549
$C_{17}H_{14}O$	1,5-Diphenyl-1,4-pentadien-3-one	3.31 3.3	B n.s.	n.s. n.s.	0 -	675 2048
$C_{17}H_{15}BrN_2$	4-Bromo-4 -dimethylamino-α-cyanostilbene	5.86±0.23	B	25-55	-	2022
$C_{17}H_{15}N$	*N*-Benzylpyridine-*p*-cyclopentadienilide	10.3	B	25	-	1748
$C_{17}H_{15}NO_2S$	*N*-(1-Naphthyl)-*p*-toluenesulfonamide	5.17	B	20	0	1703
	N-(2-Naphthyl)-*p*-toluenesulfonamide	5.05	B	20	0	1703
$C_{17}H_{15}N_3$	4-Tolylazo-1-naphthylamine	1.53 2.29	B D	25 25	0 0	801 801
$C_{17}H_{15}N_3O$	4-(*p*-Methoxyphenylazo)-1-naphthylamine	1.93 2.93	B D	25 25	0 0	801 801
$C_{17}H_{15}Ni_3O_2$	Tricyclopentadienyl trinickel dicarbonyl	0	B	25	15	520
$C_{17}H_{16}N_2$	4-Dimethylamino-4'-cyanostilbene	6.95 7.05	B D	20 20	FL FL	375 375
	N-Ethyl-4,5-diphenylimidazole	4.14	D	25	0	1451
$C_{17}H_{16}N_2S$	3-Ethyl-2-(phenyliminoethylidene)benzo-thiazoline	2.37±0.6	B	25	0	222

FORMULA	COMPOUND NAME	μ, D	State, or Solv.	t, °C	Method or P_a, % of P_e	Reference
$C_{17}H_{16}O$	Mesityl phenyl ketene	1.7_4 1.8_3	B B	30 25	0 0	41 788
$C_{17}H_{16}OS$	*cis*-2,6-Diphenylthio-γ-pyran-4-one	1.65	B	20	0	64
	trans-2,6-Diphenylthio-γ-pyran-4-one	1.63	B	20	0	64
$C_{17}H_{16}O_2$	2-Ethoxy-1,3-diphenyl-2-propen-1-one[q]	3.30-3.33[q]	B	20	0	442
	Ethyl β-phenylcinnamate	2.00	B	19	-	129
$C_{17}H_{17}BrN_2O_2$	4-Bromo-3'-nitro-4'-piperidylbiphenyl	4.33	B	25	0	1068
$C_{17}H_{17}NO_2$	2-(2'-*N*-Piperidylvinylmethylene)-1,3-indandione	6.33	B	25	-	1766
$C_{17}H_{18}$	2,4,6-Trimethylstilbene	<0.4	B	25	0	475
$C_{17}H_{18}F_{16}O_2$	2,2,3,3,4,4,5,5,6,6,7,7,8,8,9,9-Hexadecafluorononyl ester of 2-ethylhexanoic acid	2.80	B	20	0	1666
$C_{17}H_{18}N_2$	"Troger's base"	1.0_1	B	25	-	66
$C_{17}H_{18}N_2OS_2$	α-Ethoxy-5-methylmercapto-2,3-diphenyl-1,3,4-thiadiazoline	1.85	B	25	15	859
$C_{17}H_{18}O_2$	Phenyl thymol ketone	3.34 3.59	B D	32 32	0 0	1271 1271
$C_{17}H_{19}N$	4-Dimethylamino-2'-methylstilbene	2.15 2.51	n.s. B	n.s. 25	- 0	1904 475,480
$C_{17}H_{19}N_3OS_2$	1-[2-(1-Ethyl-3-methylbenzimidazolinylidenyl)]-2-[5-(3-ethyl-2-thione-4-oxothiazolylidenyl)]ethane	10	B	25	0	1767

q) Four isomers melting from 63°C to 81°C are included in the range given.

FORMULA	COMPOUND NAME	μ, D	State, or Solv.	t, °C	Method or P_a, % of P_e	Reference
$C_{17}H_{20}$	1,5-Diphenylpentane	0.48	B	25	0	1601
$C_{17}H_{20}N_2O$	4,4'-*Bis*(dimethylamino)benzophenone	4.99 5.16 5.18 5.23±0.10	B B B B	20 n.s. 25 25	RT - 0 0	1547 1223 1176 1633
$C_{17}H_{20}N_2S$	4,4'-*Bis*(dimethylamino)thiobenzophenone	5.93±0.10 6.12	B B	25 n.s.	0 -	1633 1223
$C_{17}H_{20}N_4O_7$	Diethylbenzylamine picrate	11.8	B	30	0	1268
$C_{17}H_{20}O$	6,8-Diisopropyl-7*H*-benzocyclohepten-7-one	3.16 3.14	B cHx	25 25	0 0	1303 1303
	6,8-Dipropyl-7*H*-benzocyclohepten-7-one	3.50	B	25	0	1303
$C_{17}H_{21}N_3O_3S$	*N,N*-Diethyl-4-(4-tolylsulfonylazoxy)-aniline	8.8_9	B	30	0	540
$C_{17}H_{22}BNO$	*N*-Isopropyl-*B,B*-diphenylboroxazolidine	10.3±0.02	D	25	-	553
$C_{17}H_{23}BrO$	2-Isobornyl-4-bromo-6-methylphenol	2.80	B	25	0	941
	4-Isobornyl-2-bromo-6-methylphenol	1.84	B	25	0	941
$C_{17}H_{24}O$	*o*-Isobornylcresol	1.44	B	25	0	941
	p-Isobornylcresol	1.46	B	25	0	941
$C_{17}H_{24}O_2$	5-Methyl-7-(2,6,6-trimethyl-1-cyclohexene-1-yl)-2,4,6-heptatrienoic acid	2.88	B	25	0	896
$C_{17}H_{24}O_{11}$	Pinitol pentaacetate	4.13	B	25	0	40
	Quebrachitol pentaacetate	3.35	B	25	0	40

FORMULA	COMPOUND NAME	μ, D	State, or Solv.	t, °C	Method or P_a, % of P_e	Reference
$C_{17}H_{25}N_3O$	6-Methoxy-8-(3-diethylamino)propylamino-quinoline	1.61	B	25	-	1953
$C_{17}H_{34}O$	9-Heptadecanone	2.40	liq	51	-	1363
$C_{17}H_{34}O_4$	Monomyristin	3.02	B	30	0	590
$C_{17}H_{34}O_5$	Methyl 8,9,15-trihydroxypalmitate	4.30	D	20	0	178
$C_{17}H_{36}NO_4P$	Dihexyl ester of diethylcarbamoylphosphonic acid	2.97	B	25	5	58
$C_{17}H_{36}N_2O$	Tetrabutyl urea	3.90	"pure"	25	-	562
		3.64±0.04	B	25	0	120
		3.43	D	25	-	562
		3.46±0.04	D	25	0	120
		3.53±0.04	CCl_4	25	0	120
		3.31±0.04	Hp	25	0	120
$C_{17}H_{36}N_2S$	Tetra-*n*-butylammonium thiocyanate	6.3-6.9	B	19	-	389
$C_{17}H_{36}O_4$	Tetrabutoxymethane	1.01	B	25	5	56
$C_{17}H_{36}O_6S_3$	1,3-*Bis*(butylsulfonyl)-2-(butylsulfonyl-methyl)-2-methylpropane	4.23	B	20	0	636
$C_{18}H_{10}Cl_2$	5,11-Dichloronaphthacene	2.5	B	20	0	1178
$C_{18}H_{10}F_{16}O_4$	*Bis*(2,2,3,3,4,4,5,5-octafluoropentyl)ester of phthalic acid	3.80	B	20	0	1666
$C_{18}H_{10}O_2$	Naphthacene-9,10-quinone	2.3	B	25	0	1178
$C_{18}H_{11}NO_2$	Quinoline yellow	3.67	D	14	0	164
$C_{18}H_{12}$	Benzo[c]phenanthrene	0.70±0.07	B	30	5	150

FORMULA	COMPOUND NAME	μ, D	State, or Solv.	t, °C	Method or P_a, % of P_e	Reference
$C_{18}H_{12}$ (contd.)	Naphthacene	0	B	27	0	1178
	Triphenylene	0	B	30	5	150
$C_{18}H_{12}Br_2O_2$	*p-Bis*(*p*-bromophenoxy) benzene	0.89	B	40	0	957
		0.90	B	25	0	957
		0.93	B	10	0	957
$C_{18}H_{12}Br_3N$	*Tris*(*p*-bromophenyl) amine	0.69	B	30	0	1536
		0.57	CCl_4	30	0	1536
$C_{18}H_{12}Cl_2$	6,6-*Bis*(*p*-chlorophenyl) fulvene	0.68±0.1	B	23.7	0	2104
$C_{18}H_{12}Cl_3N$	*Tris*(*p*-chlorophenyl) amine	0.64	B	30	0	1536
		0.50	CCl_4	30	0	1536
$C_{18}H_{12}Cl_3OP$	*Tris*(*p*-chlorophenyl) phosphine oxide	2.95±0.03	B	25	5	1502
$C_{18}H_{12}Cl_3P$	*Tris*(*p*-chlorophenyl) phosphine	0.65±0.07	B	25	5	1502
$C_{18}H_{12}F_3N$	*Tris*(*p*-fluorophenyl) amine	0.82	B	30	0	1536
		0.64	CCl_4	30	0	1536
$C_{18}H_{12}I_3N$	*Tris*(*p*-iodophenyl) amine	0.54	B	30	0	1536
		0.75	CCl_4	30	0	1536
$C_{18}H_{12}N_2O_6$	*p-Bis*(*m*-nitrophenoxy) benzene	4.85±0.05	B	25	0	597
$C_{18}H_{12}N_2O_7S$	*p*-(*m*-Nitrophenylsulfonyl)-*m*-nitrophenoxy-benzene	6.05±0.08	B	25	0	597
$C_{18}H_{12}N_3O_9PS$	*O,O,O*-Tri-(*p*-nitrophenyl) phosphorothioate	3.33	D	25	-	906
$C_{18}H_{12}N_5O_6$	α,α-Diphenyl-β-picrylhydrazyl	4.92	B	25	0	1995
$C_{18}H_{13}ClOS$	*p*-(*p*-Chlorophenylthio)-phenoxybenzene	2.04±0.02	B	25	0	597

FORMULA	COMPOUND NAME	μ, D	State, or Solv.	t, °C	Method or P_a, % of P_e	Reference
$C_{18}H_{13}ClO_3S$	*p*-(*p*-Chlorophenylsulfonyl)-phenoxybenzene	4.38±0.04	B	25	0	597
$C_{18}H_{13}N$	Pyridinium fluoren-9-ylide	4.13±0.08	B	20	0	670
$C_{18}H_{13}NO_2S_2$	*p*-(*m*-Nitrophenylthio)-phenylthiobenzene	4.83±0.05	B	25	0	597
	p-(*p*-Nitrophenylthio)-phenylthiobenzene	4.25±0.02	B	25	0	597
$C_{18}H_{13}NO_3S$	*p*-(*m*-Nitrophenylthio)-phenoxybenzene	5.15±0.05	B	25	0	597
	p-(*p*-Nitrophenylthio)-phenoxybenzene	4.39±0.04	B	25	0	597
	p-(*p*-Nitrophenoxy)-phenylthiobenzene	3.82±0.03	B	25	0	597
$C_{18}H_{13}NO_4$	*p*-(*p*-Nitrophenoxy)-phenoxybenzene	4.20±0.04	B	25	0	597
$C_{18}H_{13}NO_5S$	*p*-(*m*-Nitrobenzenesulfonyl)-phenoxybenzene	5.9±0.1	B	25	0	597
	p-(*p*-Nitrobenzenesulfonyl)-phenoxybenzene	4.63±0.03	B	25	0	597
	p-(*m*-Nitrophenoxy)-benzenesulfonylbenzene	5.48+0.08	B	25	0	597
	p-(*p*-Nitrophenoxy)-benzenesulfonylbenzene	4.70±0.02	B	25	0	597
$C_{18}H_{13}N_2O_7PS$	*O*-Phenyl-*O*,*O*-*bis*(*p*-nitrophenyl)phosphorothioate	4.78	D	25	-	906
$C_{18}H_{13}N_3O$	5,10-Dihydro-5-isonicotinylphenazine	6.18	B	25	-	1551
$C_{18}H_{13}N_3O_6S$	*N*,*N*-Diphenyl-3,5-dinitrobenzenesulfonamide	4.24	D	n.s.	-	1923
$C_{18}H_{13}N_3O_7$	Acenaphthane - picric acid complex	2.25	D	35	0	1558
$C_{18}H_{13}N_5O_6$	α,α-Diphenyl-β-picrylhydrazine	3.59	B	25	0	1995

FORMULA	COMPOUND NAME	μ, D	State, or Solv.	t, °C	Method or P_a, % of P_e	Reference
$C_{18}H_{14}$	1-Cinnamylideneindene	1.38±0.03	B	30	0	144
	5,10-Dimethylindeno[2,1-a]indene	0.6	B	n.s.	0	149
	1,4-Diphenylbenzene	0.60 0.73	B B	20 40	SPC SPC	1300 1300
	6,6-Diphenylfulvene	1.34±0.04 1.34±0.04 1.34±0.03	B B B	23.7 20 30	0 0 0	2104 1189 144
$C_{18}H_{14}Cl_2O_4$	*cis*-1,5-Dichloro-9,10-dihydro-9,10-anthradiol diacetate	1.48±0.12	B	25	0	1643
	trans-1,5-Dichloro-9,10-dihydro-9,10-anthradiol diacetate	1.20±0.12	B	25	0	1643
$C_{18}H_{14}Cl_3N_2Tl$	Thallium trichloride - di(quinone) complex	2.17	D	20	10	1739
$C_{18}H_{14}FN$	(*p*-Fluorophenyl)diphenylamine	1.41	B	25	5	1114
$C_{18}H_{14}Fe$	Di(indenyl) iron	0	B	25	15	520
$C_{18}H_{14}NO_5PS$	*O*,*O*-Diphenyl-*O*-*p*-nitrophenylphosphorothioate	4.52	D	25	-	906
$C_{18}H_{14}N_4O_{12}$	Diethyl 2,2',4,4'-tetranitrodiphenate	5.10±0.03	B	20	0	1188
	Diethyl 4,4',6,6'-tetranitrodiphenate	5.10	B	20	0	1182
$C_{18}H_{14}OS$	*p*-(Phenylthio)-phenoxybenzene	1.82±0.02	B	25	0	597
$C_{18}H_{14}O_2$	1,4-Diphenoxybenzene	1.20±0.02 1.43 1.44	B B B	25 10-40 10-40	0 0 0	597 957 955

FORMULA	COMPOUND NAME	μ, D	State, or Solv.	t, °C	Method or P_a, % of P_e	Reference
$C_{18}H_{14}O_3S$	4-(Phenylsulfonyl)phenyl phenyl ether	5.15±0.03	B	25	0	597
$C_{18}H_{14}O_4$	Butyne-1,4-diol dibenzoate	3.25±0.03	B	(20)	0	1968
$C_{18}H_{14}Ru$	Diindenylruthenium	0	B	25	5-10	2076
$C_{18}H_{14}S_2$	1,4-Di(phenylthio)benzene	1.96±0.02	B	25	0	597
$C_{18}H_{15}Al$	Triphenylaluminum	1.1_8 3.7_5	B D	20 20	15 15	1888 1888
$C_{18}H_{15}As$	Triphenylarsine	1.08	B	18	0	158
$C_{18}H_{15}AsCl_2$	Triphenylarsine dichloride	0	B	25	45	850
$C_{18}H_{15}AsO$	Triphenylarsine oxide	5.54	B	25	15	851
$C_{18}H_{15}AuClO_3P$	Diphenyl phenylphosphate - gold chloride complex	6.27	B	n.s.	-	57
$C_{18}H_{15}BCl_3P$	Boron trichloride - triphenylphosphine complex	7.06±0.06	B	25	5	1502
$C_{18}H_{15}Bi$	Triphenylbismuth	0	B	16	0	158
$C_{18}H_{15}BiCl_2$	Triphenylbismuth dichloride	0 1.17	B B	25 25	30 0	850 1427
$C_{18}H_{15}BiN_2O_6$	Triphenylbismuth dinitrate	3.29±0.15	B	25	30	850
$C_{18}H_{15}BrGe$	Triphenylgermanium bromide	2.35±0.11	B	25	0	1427
$C_{18}H_{15}BrPb$	Triphenyllead bromide	4.24±0.10	B	25	-	1118
$C_{18}H_{15}BrSi$	Triphenylbromosilane	2.26±0.04	B	25	5	1592

FORMULA	COMPOUND NAME	μ, D	State, or Solv.	t, °C	Method or P_a, % of P_e	Reference
$C_{18}H_{15}BrSn$	Triphenyltin bromide	3.18	B	22	-	1249
$C_{18}H_{15}ClPb$	Triphenyllead chloride	4.24±0.10	B	25	-	1118
		4.35	B	22	-	1250
$C_{18}H_{15}ClSi$	Triphenylchlorosilane	2.16	B	25	0	373
$C_{18}H_{15}ClSn$	Triphenyltin chloride	3.28±0.03	B	25	0	1427
		3.31	B	n.s.	-	592
		3.94	D	n.s.	-	592
$C_{18}H_{15}Cl_2Sb$	Triphenylstibine dichloride	0	B	25	30	850
		1.19±0.05	B	25	0	1427
$C_{18}H_{15}FSi$	Triphenylfluorosilane	1.85	B	25	0	373
$C_{18}H_{15}Ga$	Triphenylgallium	0.4	B	20	15	1888
		2.1_6	D	20	15	1888
$C_{18}H_{15}IPb$	Triphenyllead iodide	3.76±0.10	B	25	-	1118
$C_{18}H_{15}In$	Triphenylindium	0.80	B	20	15	1888
		0.8_8	D	20	15	1888
$C_{18}H_{15}N$	Triphenylamine	0.26	B	15	0	158
		0.47	B	20	5	337
		0.52	B	25	5	70
		0.55	B	25	5	1114
		0.65	B	25	-	920
		0.72	B	30	0	1536
		0.64	CCl_4	30	0	1536
$C_{18}H_{15}NO_2S$	*N*,*N*-Diphenylbenzenesulfonamide	4.72	B	20	0	1703
	N-Phenyldibenzenesulfonamide	3.15	B	20	0	1703

FORMULA	COMPOUND NAME	μ, D	State, or Solv.	t, °C	Method or P_a, % of P_e	Reference
$C_{18}H_{15}NO_3$	Ethyl 3,5-diphenyl-4-isoxazolecarboxylate	3.22±0.03	D	25	0	1857
$C_{18}H_{15}N_3O$	4-(*p*-Acetylphenylazo)-1-naphthylamine	5.34	B	25	0	801
		5.96	D	25	0	801
$C_{18}H_{15}N_3O_2$	4-(*p*-Methoxycarbonylphenylazo)-1-naphthyl-amine	3.61	B	25	0	801
		4.69	D	25	0	801
$C_{18}H_{15}OP$	Triphenylphosphine oxide	4.31±0.02	B	25	5	1502
		4.34	B	25	15	851
$C_{18}H_{15}O_3P$	Triphenylmetaphosphate	1.60	B	n.s.	-	57
	Triphenylphosphite	2.04	B	25	0	1122
		1.59	CCl_4	20	-	50
$C_{18}H_{15}O_3PS$	*O,O,O*-Triphenyl phosphorothioate	2.58	B	25	-	906
		2.60	B	25	0	1122
$C_{18}H_{15}O_4P$	Triphenylphosphate	2.79	B	25	-	906
		2.82	B	20	-	343
		2.84	B	25	0	1122
		2.89	CCl_4	20	-	50
$C_{18}H_{15}P$	Triphenylphosphine	1.40±0.04	B	25	5	1502
		1.46	B	17	0	158
		1.44	cHx	25	-	920
$C_{18}H_{15}PS$	Triphenylphosphine sulfide	4.78	B	25	15	851
$C_{18}H_{15}SSb$	Triphenylstibine sulfide	5.44	B	25	15	851
$C_{18}H_{15}Sb$	Triphenylstibine	0.5_7	B	14	0	158
		0.83±0.03	B	25	5-15	694
$C_{18}H_{15}Tl$	Triphenylthallium	0.7	B,D	20	15	1888

FORMULA	COMPOUND NAME	μ, D	State, or Solv.	t, °C	Method or P_a, % of P_e	Reference
$C_{18}H_{16}AsClO$	Triphenylarsine hydroxychloride	9.2	D	40	15	851
$C_{18}H_{16}ClOSb$	Triphenylstibine hydroxychloride	2.99±0.05	D	40	15	851
$C_{18}H_{16}N_2$	*N,N′*-Diphenyl-*p*-phenylenediamine	1.61±0.2	B	35	0	1850
$C_{18}H_{16}N_2O_3S$	1-(3,4-Methylenedioxyphenethyl)-5-(2-methyl-5-thiazolyl)-2(1*H*)-pyridone	4.24	B	27	0	1934
$C_{18}H_{16}N_2O_4$	Azine diacetate of *p*-hydroxybenzaldehyde	2.31	B + $CHCl_3$	20	-	1991
$C_{18}H_{16}N_2O_4S_2$	*m*-Benzenedisulfonanilide	4.43	B	20	0	1703
$C_{18}H_{16}OPb$	Triphenyllead hydroxide	~2.4	B	22	-	1250
$C_{18}H_{16}OSi$	Triphenylsilanol	1.46 1.51 1.46	B D CCl_4	25 25 22.6	- - -	307 307 1248
$C_{18}H_{16}OSn$	Triphenyltin hydroxide	2.00±0.08	B	22	-	1249
$C_{18}H_{16}O_4$	*cis*-Dimethyl 9,10-dihydroanthracene-9,10-dicarboxylate	2.6	B	24	0?	168
	trans-Dimethyl 9,10-dihydroanthracene-9,10-dicarboxylate	1.7	B	24	0?	168
$C_{18}H_{16}Si$	Triphenylsilane	0.66±0.04	CCl_4	25	15	1247
$C_{18}H_{17}NO$	Pyridine - diphenylmethanol complex	3.71	B	25	0	318
$C_{18}H_{17}O_2P$	Triphenylphosphine oxide hydrate	4.60	B	25	15	851
$C_{18}H_{17}O_2Sb$	Triphenylstibine dihydroxide	0	B	25	15	850

FORMULA	COMPOUND NAME	μ, D	State, or Solv.	t, °C	Method or P_a, % of P_e	Reference
$C_{18}H_{18}Cl_4O_2$	Chloranil - hexamethylbenzene complex	1.00	CCl_4	20	-	211
$C_{18}H_{18}N_2OS_2$	1-[2-(1-Ethyl-2(1*H*)-quinolylidenyl)]-2-[5-(3-ethylthione-4-oxothiazolylidenyl)]-ethane	7.85	B	25	-	1767
$C_{18}H_{18}N_2O_2S_2$	5-(4'-*N*-Acetylanilinobutadienylmethylene)-3-ethylrhodanine	5.48	B	25	-	1766
$C_{18}H_{18}N_2O_4$	Antipyrine - salicylic acid complex	7.26	B	25	-	945
$C_{18}H_{18}N_2O_5$	Diethyl azoxydibenzoate	3.11	B	20	-	1991
$C_{18}H_{18}O_2$	α-2,5-Diphenyl-3-hexyne-2,5-diol	2.12±0.02	D	20	0	1185
	β-2,5-Diphenyl-3-hexyne-2,5-diol	2.23±0.02	D	20	0	1185
$C_{18}H_{18}O_4$	4,4'-Diethoxybenzil	5.4	B	25	-	1696
$C_{18}H_{19}Cl_3$	2,2-*Bis*(2,4-dimethylphenyl)-1,1,1-trichloroethane	2.52±0.1	B	20	0	2107
	2,2-*Bis*(2,5-dimethylphenyl)-1,1,1-trichloroethane	1.96±0.1	B	20	0	2107
	2,2-*Bis*(3,4-dimethylphenyl)-1,1,1-trichloroethane	2.51±0.1	B	20	0	2107
	2,2-*Bis*(*p*-ethylphenyl)-1,1,1-trichloroethane	2.21±0.1	B	20	0	2107
$C_{18}H_{19}Cl_3O_2$	2,2-*Bis*(*p*-ethoxyphenyl)-1,1,1-trichloroethane	3.00±0.1	B	20	0	2107
$C_{18}H_{19}N$	1-(*p*-Dimethylaminophenyl)-4-phenyl-1,3-butadiene	2.58	B	25	0	475

FORMULA	COMPOUND NAME	μ, D	State, or Solv.	t, °C	Method or P_a, % of P_e	Reference
$C_{18}H_{20}Cl_4O_4Zr$	Zirconium tetrachloride - di(benzylacetate) complex	6.54	B	20	15	1442
$C_{18}H_{20}CuN_2O_2$	*Bis*(*N*-ethylsalicylaldimine) copper(II)	2.06	B	25	10	1671
		1.66	D	25	10	1671
$C_{18}H_{20}N_2$	*p*-*Bis*(2,5-dimethyl-1-pyrryl) benzene	0.77±0.03	B	25	0	931
$C_{18}H_{20}N_2NiO_2$	*Bis*(*N*-ethylsalicylaldimine) nickel(II)	0.67	B	25	10	1671
		0.74	D	25	10	1671
$C_{18}H_{20}N_2OS_3$	1-[2-(3-Butylbenzothiazolinylidenyl)]-2-[5-(3-ethyl-2-thione-4-oxothiazolylidenyl)]ethane	7.73	B	25	0	1767
$C_{18}H_{20}N_2O_2Pd$	*trans*-*Bis*(*N*-ethylsalicylaldimine)-palladium(II)	0.5	B	25	10	1671
		0	D	25	10	1671
$C_{18}H_{20}N_2O_2S_2$	1-[2-(3-Isobutylbenzoxazolinylidenyl)]-2-[5-(3-ethyl-2-thione-4-oxothiazolylidenyl)]ethane	8.28	B	25	0	1767
$C_{18}H_{20}O_2$	*cis*-4,4-Dimethoxy-α,α'-dimethylstilbene	1.77	B	25	-	1756
	trans-4,4-Dimethoxy-α,α'-dimethylstilbene	1.80	B	25	-	1756
$C_{18}H_{21}La$	*Tris*(methylcyclopentadienyl) lanthanum	1.54	B	20	15	1894
		3.32	D	20	15	1894
$C_{18}H_{21}N$	4-Diethylaminostilbene	2.69	B	25	0	475
	4-Dimethylamino-2,2'-dimethylstilbene	2.25	B	25	0	475,480
	4-Dimethylamino-2,5'-dimethylstilbene	2.00	n.s.	n.s.	-	1904
		2.31	B	25	0	475,480

FORMULA	COMPOUND NAME	μ, D	State, or Solv.	t, °C	Method or P_a, % of P_e	Reference
	4-Dimethylamino-2'-ethylstilbene	2.21 2.45	n.s. B	n.s. 25	- 0	1904 475,480
	6-Methyl-6-(*p*-piperidylphenyl)fulvene	2.75	B	20	-	967
$C_{18}H_{21}N_3O_6$	1,3,5-Trinitrobenzene - hexamethylbenzene complex	0.87±0.03 0.90	CCl_4 CCl_4	20 n.s.	- -	212 210
$C_{18}H_{22}$	1,6-Diphenylhexane	0.52	B	25	0	1601
$C_{18}H_{22}Cl_2O_6P_2Pt$	*Bis*(diisopropyl isopropylphosphate) - platinum dichloride complex	9.05	B	n.s.	-	57
$C_{18}H_{22}N_2$	1,4-Di-*o*-tolylpiperazine	0.56	B	25	5	69
$C_{18}H_{22}N_2O_3$	4,4'-Di-*n*-propoxyazoxybenzene	2.41	B	20	-	1241c
$C_{18}H_{22}O_2$	Estra-3,17-dione	2.90	B	25	-	17
$C_{18}H_{23}N_3$	*p*-*t*-Butyl-*p*'-(dimethylamino)azobenzene	2.62	B	25	0	269
$C_{18}H_{24}$	1-Octylnaphthalene	0.72	n.s.	n.s.	-	1210
$C_{18}H_{24}O_2$	1,3,3,5,5,7-Hexamethyl-1,4-etheno-1,2,3,4,5,6,11,12-octahydronaphthalene-2,6-dione	1.71	B	24	0	229
$C_{18}H_{24}O_{12}$	Inositolhexa(aeeeea)acetate	3.03	B	25	0	40
	Inositolhexa(eaeaee)acetate	4.12	B	25	0	40
	Inositolhexa(eaeeee)acetate	2.79	B	25	0	40
	Inositolhexa(eeeeee)acetate	2.3	D	25	0	40

FORMULA	COMPOUND NAME	μ, D	State, or Solv.	t, °C	Method or P_a, % of P_e	Reference
$C_{18}H_{26}O_2$	Methyl 5-methyl-7-(2,6,6-trimethyl-1-cyclohexene-1-yl)-2,4,6-heptatrienoate	2.95	B	25	-	896
$C_{18}H_{26}O_4$	Di(isopentyl)phthalate	2.61[r]	B	28	RT	1401
$C_{18}H_{27}N_3$	8-(1-Methyl-4-diethylamino)butylamino-quinoline	0.70	B	25	-	1953
$C_{18}H_{29}NO_2$	2,3,5,6-Tetraisopropylnitrobenzene	3.57	B	25	-	1393
$C_{18}H_{30}N_3O_7$	Tri-*n*-butylamine picrate	11.9	B	30	0	1268
		13.2	B	25	-	564
		12.2	D	30	0	1268
		11	pXy	20	RT	389
$C_{18}H_{30}O$	2,4,6-Tri-*t*-butylphenol	1.41	Solid	n.s.	-	1294
		1.63	B	25	0	598
		1.57	D	25	-	451
		1.55	cHx	25	-	451
$C_{18}H_{30}O_2$	Linolenic acid	1.56	B	18	-	1872
		1.73	D	18	-	1872
$C_{18}H_{32}O_2$	Linoleic acid	1.218	B	30	0	1465
		1.51	B	18	-	1872
		1.522	B	20,75	0	2044-5
		1.23	D	19,23	-	1869, 1871,2046
		1.71	D	18	-	1872
		1.40	Hx	18	-	1868
		1.36	cHx	18	-	1868, 1871,1869
		1.214	s	18	-	1868
		1.350	Tol	n.s.	0	2044-5
$C_{18}H_{32}O_3Si$	Phenyltri-*n*-butoxysilane	1.53	B	20	0	1041

r) 3.26 cm frequency. s) Methylcyclohexane.

FORMULA	COMPOUND NAME	μ, D	State, or Solv.	t, °C	Method or P_a, % of P_e	Reference
$C_{18}H_{33}N$	Tricyclohexylamine	0.42	B	25	-	920
$C_{18}H_{34}O$	Cycloöctadecanone	2.70±0.03	B	20	0	630
$C_{18}H_{34}O_2$	Elaidic acid	1.55	liq	44	-	1495
		1.61	liq	60	-	1495
		1.74	D	25,40	-	1496
	Oleic acid	0.77 t	liq	(20)	-	1419
		1.09 t	liq	(20)	-	1419
		1.44	liq	25	-	1495
		1.52	liq	60	-	1495
		1.019	B	30	0	1465
		1.42	B	18	-	1872
		1.452	B	2	-	2044-5
		1.69	D	18	-	1869, 1871-2 2046
		1.76	D	25,40	-	1496
		1.17	Hx	40	-	1868
		1.23	Hx	18	-	1868
		1.13	cHx	25	-	1870
		1.14	cHx	18	-	1868-9, 1871
		1.35	cHx	50	-	1870
		1.52	cHx	70	-	1870
		1.61	cHx	79	-	1870
		1.10	u	18	-	1868
$C_{18}H_{34}O_3$	Ricinelaidic acid	2.08	liq	53	-	1495
		1.99	D	25,40	-	1496
	Ricinoleic acid	2.04	liq	25	-	1495
		2.07	liq	60	-	1495
		1.94	D	25,60	-	1496

t) After 8 hours electric discharge in H_2.   u) Methylcyclohexane.

FORMULA	COMPOUND NAME	μ, D	State, or Solv.	t, °C	Method or P_a, % of P_e	Reference
$C_{18}H_{34}O_4$	Hexadecane-1,16-dicarboxylic acid	2.75	D	35	-	1581
$C_{18}H_{35}AlO_4$	Aluminum monoöleate	2.20	B[v]	20	5	1513
$C_{18}H_{35}As_2BrPt$	*trans*-*Bis*(triethylarsine)bromo(phenyl)-platinum	2.9	B	25	15	294
$C_{18}H_{35}ClP_2Pt$	*cis*-*Bis*(triethylphosphine)chloro(phenyl)-platinum	9.05	B	25	15	294
	trans-*Bis*(triethylphosphine)chloro-(phenyl)platinum	2.6	B	25	15	294
$C_{18}H_{35}NO$	*trans*-2-Oxoöctadecamethylenimine	3.67	B	25	0	783
$C_{18}H_{36}CdN_2S_4$	Cadmium *N*,*N*-diisobutyldithiocarbamate	1.6_9	B	21	0	1243
$C_{18}H_{36}Cl_4O_4Sn$	Tin tetrachloride - di(pentylbutyrate) complex	6.38	B	20	15	1438
$C_{18}H_{36}FeN_3OS_4$	Iron(III) nitrosyl diisobutyldithio-carbamate	4.2_9	B	20	0	1244
$C_{18}H_{36}N_2NiS_4$	Nickel(II) *N*,*N*-diisobutyldithiocarbamate	1.8_0	B	30	0	1243
$C_{18}H_{36}N_2S_4$	*Bis*(dibutylthiocarbamoyl)disulfide	2.42	B	25	0	639
$C_{18}H_{36}O$	*cis*-9-Octadecen-1-ol	1.74	B	20	-	2108
	trans-9-Octadecen-1-ol	1.72	B	20	-	2108
$C_{18}H_{36}O_2$	Ethyl palmitate	1.2 1.89	liq Hp	30-182 25	- 0	1052 1122

v) Plus water.

FORMULA	COMPOUND NAME	μ, D	State, or Solv.	t, °C	Method or P_a, % of P_e	Reference
	Stearic acid	0.87	B	20?	-	1663
		1.04	B	18	-	1872
		1.80	B	30	-	1276
		1.516	D	2	-	2045
		1.66	D	23	-	1869, 1871-2
		1.76	D	25	0	2126
		0.67	cHx	25	-	1870
		0.79	cHx	50	-	1870
		0.90	cHx	70	-	1870
		1.05	cHx	23	-	1869,1871
		1.06	cHx	79	-	1870
		0.79	CS_2	20?	-	1663
$C_{18}H_{36}O_5$	Ethyl aleurate	4.34	D	25	0	178
$C_{18}H_{37}BrHg$	Octadecylmercury bromide	2.0±0.06	B	40	-	872-3
$C_{18}H_{37}Cl$	1-Chloroöctadecane	1.94	B	20	RT	921
		2.02	Hx	20	RT	921
$C_{18}H_{38}O$	Octadecanol	1.7_4	CCl_4	25	5	1108
		1.72	W	55	-	1865
$C_{18}H_{39}BO$	Tri-*n*-hexyl borate	0.8_6	CCl_4	25	0	72
$C_{18}H_{39}NO_2$	Tetra-*n*-butylammonium acetate	11.3	B	25	-	564
$C_{18}H_{39}O_3P$	Diethyltetradecylphosphonate	2.62±0.02	liq	32	-	951
$C_{18}H_{39}O_3PS$	Trihexylthiophosphate	3.21	B	20	0	59
$C_{18}H_{42}Cl_2NiP_2$	*Bis*(tri-*n*-propylphosphine)nickel dichloride	~0	B	20	-	846
$C_{18}H_{42}Cl_2P_2Pt$	α-*Bis*(tri-*n*-propylphosphine)platinum dichloride	~0	B	20	-	845

w) *p*-Dichlorobenzene.

FORMULA	COMPOUND NAME	μ, D	State, or Solv.	t, °C	Method or P_a, % of P_e	Reference
$C_{18}H_{42}Cl_2P_2Pt$ (contd.)	β-*Bis*(tri-*n*-propylphosphine)platinum dichloride	11.6	B	20	-	845
$C_{18}H_{42}OSi_2$	Hexapropyldisiloxane	0.67	Hx	25	0	627-8
$C_{18}H_{42}O_{10}Si_5$	1,1,1,2,3,4,5,5,5-Nonamethyl-2,3,4-*tris*-(acetoxymethyl)pentasiloxane	3.59	liq	20	RT	37
$C_{18}H_{48}N_2O_3Si_4$	1,1,1,2,3,4,4,4-Octamethyl-2,3-*bis*-(diethylaminomethyl)tetrasiloxane	1.40	liq	20	RT	595
$C_{19}H_{13}N$	9-Phenylacridine	2.49±0.03	n.s.	n.s.	-	2
$C_{19}H_{13}NO$	9-Phenoxyacridine	2.40	B	25	0	1409
	9-Phenylacridine *N*-oxide	4.13±0.01	n.s.	n.s.	-	2
$C_{19}H_{13}NO_2$	9-Phenoxyacridine *N*-oxide	4.67	B	25	0	1409
$C_{19}H_{13}NO_3$	3-Nitro-4'-phenylbenzophenone	4.39±0.04	B	20	0	1182,1188
$C_{19}H_{13}N_3O_3$	5,10-Dihydro-5-*p*-nitrobenzoylphenazine	4.90	B	25	-	1551
$C_{19}H_{13}N_3O_6$	Tri-*p*-nitrophenylmethane	3.26	D	16	0	142
$C_{19}H_{14}ClN$	Diphenylmethylene-*p*-chloroaniline	2.94	B	12	0	138
$C_{19}H_{14}Cl_2$	*p*-Chlorophenyldiphenylchloromethane	1.94_5	B	17	0	142
$C_{19}H_{14}N_2$	9-Fluorenonephenylhydrazone	2.14	B	16	0	164
$C_{19}H_{14}N_2O$	5-Benzoyl-5,10-dihydrophenazine	3.34	B	25	-	1551
$C_{19}H_{14}N_2O_2$	*p*-Benzoyloxyazobenzene	1.95	B	30	-	2062

FORMULA	COMPOUND NAME	μ, D	State, or Solv.	t, °C	Method or P_a, % of P_e	Reference
$C_{19}H_{14}O$	Fuchsone	5.80	B	32	0	1271
		5.83	D	32	0	1271
	4-Phenylbenzophenone	3.18±0.03	B	20	0	1182,1188
$C_{19}H_{14}O_2$	Benzaurin	6.85	D	32	0	1271
$C_{19}H_{14}O_3$	Aurin	7.96	D	32	0	1271
	Phenyl *o*-phenoxybenzoate	2.76	B	25	-	946
$C_{19}H_{15}Br$	Bromotriphenylmethane	2.08	B	25	0	1405
		2.1	B	25	0	493
$C_{19}H_{15}Br_5Sn$	Tin tetrabromide - triphenylbromomethane complex	2.7	B	25	0	493
$C_{19}H_{15}Cl$	Chlorotriphenylmethane	1.97	B	10-70	0	1807
		2.00	B	25	0	1405
		2.11	B	17	0	142
		1.96	D	20	0	1969
$C_{19}H_{15}F$	Fluorotriphenylmethane	1.89	B	25	0	1405
$C_{19}H_{15}I$	Iodotriphenylmethane	1.81	B	25	0	1405
$C_{19}H_{15}N$	Diphenylmethyleneaniline	1.99	B	13	0	138
		2.0_5	B	25	0	399
$C_{19}H_{15}NOS$	Triphenylmethylthionitrite	1.93	B	20	-	969
		1.86	CCl_4	20	-	969
$C_{19}H_{15}N_3$	*N*-Benzylidene-*p*-phenylazoaniline	2.03	B	25	10	267
$C_{19}H_{15}Na$	Triphenylmethylsodium	7.16	D	20	0	1405

FORMULA	COMPOUND NAME	μ, D	State, or Solv.	t, °C	Method or P_a, % of P_e	Reference
$C_{19}H_{15}NaO$	Sodium derivative of triphenylmethanol	1.63±0.02	D	25	0	1405
$C_{19}H_{16}$	Triphenylmethane	0	liq	94-175	-	1052
		0.21	B	(25)	-	1921
		0.46	D	(25)	-	1921
		0.3	CCl_4	25	5	70
		0.62	CS_2	15	0	142
$C_{19}H_{16}N_2$	Benzophenonephenylhydrazone	2.24	B	15	0	164
$C_{19}H_{16}O$	2,5-Dibenzylidenecyclopentanone	3.36±0.03	B	25	0	127
	6-Phenyl-6-(*p*-methoxyphenyl)fulvene	2.08	B	20	-	967
	Triphenylmethanol	1.47	B	25	-	789
		1.81	B	25	0	1405
		2.13	B	10-70	0	1807
$C_{19}H_{16}O_2$	4-Phenoxybenzhydrol	1.91	B	25	5	319
$C_{19}H_{17}NO_2S$	*N*,*N*-Diphenyl-*p*-toluenesulfonamide	5.14	B	20	0	1703
$C_{19}H_{18}AuP$	Methyl(triphenylphosphine)gold	5.6	B	25	-	262
$C_{19}H_{18}IO_3P$	Methyl iodide - triphenylphosphite complex	6.39	B	n.s.	-	52
$C_{19}H_{18}N_2O$	Sempervirine hydrate	7.5	B	22	-	852
		8.5	D	22	-	852
$C_{19}H_{19}NO$	*p*-Dimethylaminocinnamylideneacetophenone	5.4	D	26	-	2090
$C_{19}H_{19}NO_2$	Ethyl cyanoditropylacetate	3.38	B	25	5	2154
	2-(4'-*N*-Piperidylbutadienylmethylene)-1,3-indandione	7.66	B	25	-	1766

FORMULA	COMPOUND NAME	μ, D	State, or Solv.	t, °C	Method or P_a, % of P_e	Reference
$C_{19}H_{20}N_2S$	3-Ethyl-2-(phenyliminobutenylidene)-benzothiazoline	4.17±0.12	B	25	0	222
$C_{19}H_{20}O_2$	2,6-Diphenyl-3,5-dimethyltetrahydro-γ-pyrone	1.82	n.s.	25	0	1070
$C_{19}H_{20}O_4$	*O,O′,O′′,O′′′*-Dibenzylidenepentaerythritol	2.4_3	CCl_4	25	0	1081
	Diethyl diphenylmalonate	4.433	B	30	0	1501
$C_{19}H_{22}N_2OS_2$	1-[2-(1-Ethyl-3,3-dimethylindolinyl-idenyl)]-2-[5-(3-ethyl-2-thione-4-oxothiazolylidenyl)]ethane	6.32	B	25	0	1767
$C_{19}H_{22}N_2O_3$	1-{3,4-Methylenedioxyphenethyl)-5-(1-methyl-2-pyrrolidinyl)-2(1*H*)-pyridone	4.39	B	27	0	1934
$C_{19}H_{23}N$	4-Dimethylamino-2',4',6'-trimethylstilbene	2.11±0.02	B	25	0	475,480
$C_{19}H_{24}O$	6,8-Di-*t*-butyl-7*H*-benzocyclohepten-7-one	2.32±0.1	B	25	0	1303
		2.08±0.1	cHx	25	0	1303
$C_{19}H_{26}N_2S_2$	*N,N*-Dicyclohexylbenzothiazolesulfenamide	3.01	B	25	-	637
$C_{19}H_{26}O_2$	Δ^4-Androstene-3,17-dione	3.35	D	25	-	1002
$C_{19}H_{28}O_2$	Androstane-3,17-dione	2.94	B	25	-	17
		3.1	B	25	0	1381
		3.28	D	25	-	997
	Δ^5-Androstene-3-*cis*-ol-17-one	2.48	D	25	-	1002
	Etiocholane-3,17-dione	3.5	B	25	-	1381
	Testane-11,17-dione	4.1	B	25	-	1382

FORMULA	COMPOUND NAME	μ, D	State, or Solv.	t, °C	Method or P_a, % of P_e	Reference
$C_{19}H_{28}O_2$ (contd.)	*cis*-Testerone	5.21	D	25	-	1002
	Testerone	4.35	D	25	-	1002
$C_{19}H_{30}O$	Androstane-17-one	2.98	B	25	-	17
		3.0	B	25	-	1382
	Testane-17-one	3.0	B	25	-	1382
$C_{19}H_{30}O_2$	Δ^5-Androstene-3-*cis*-17-*cis*-diol	2.71	D	25	-	1002
	Δ^5-Androstene-3-*cis*-17-*trans*-diol	2.92	D	25	-	1002
	Androsterone	3.73	D	25	-	1002
	β-Androsterone	2.98	D	25	-	1002
$C_{19}H_{32}O_2$	Androstane-3-*trans*-17-*trans*-diol	2.31	D	25	-	997
	Androstane-3-*cis*-17-*trans*-diol	3.02	D	25	-	997
$C_{19}H_{34}O$	Tricyclohexylmethanol	1.62[x]	Solid	n.s.	-	1294
$C_{19}H_{37}As_2ClPt$	*trans*-*Bis*(triethylarsine)chloro(benzyl)-platinum	2.8	B	25	15	294
$C_{19}H_{37}ClNiP_2$	*trans*-*Bis*(triethylphosphine)chloro-(*o*-tolyl)nickel	2.35	B	25	-	295
$C_{19}H_{37}ClP_2Pt$	*cis*-*Bis*(triethylphosphine)chloro-(*o*-tolyl)platinum	9.15	B	25	15	294
	cis-*Bis*(triethylphosphine)chloro-(*p*-tolyl)platinum	8.95	B	25	15	294

x) Special method.

FORMULA	COMPOUND NAME	μ, D	State, or Solv.	t, °C	Method or P_a, % of P_e	Reference
	trans-*Bis*(triethylphosphine)chloro(benzyl)-platinum	2.8	B	25	15	294
$C_{19}H_{44}Si_2$	1,1,1,3,3,3-Hexapropyl-1,3-disilicapropane	0.56	Hx	25	0	627
$C_{19}H_{45}IP_2Pt$	*trans*-Methyliodo*bis*(tri-*n*-propylphosphine)-platinum(II)	3.9±0.1	B	25	15	293
$C_{20}H_{12}$	Perylene	0.45±0.15	B	30	5	150
		1.3	B	25	0	1178
		1.9	B	27	0	1178
		2.1	B	30	0	1178
$C_{20}H_{12}FeO_4$	Diphenylcyclopentadienonetricarbonyliron	3.1±0.2	B	25	0	1116
$C_{20}H_{12}N_2O_4$	*cis*-*p*-Nitrobenzylidene-2-nitrofluorene	7.85±0.01	B	30	5	143
	trans-*p*-Nitrobenzylidene-2-nitrofluorene	4.45±0.01	B	30	5	143
$C_{20}H_{12}S$	Dinaphthalenethiophene	0.8	pXy	30	0	282
$C_{20}H_{13}Br$	*p*-Bromobenzylidenefluorene	1.30±0.03	B	30	5	143
$C_{20}H_{13}ClO$	*p*-(Chlorobenzal)fluorene oxide	1.90	B	14.6	0	157
	9-*p*-Chlorobenzylidenexanthene	1.48±0.03	B	30	5	132
$C_{20}H_{13}ClS$	10-*p*-Chlorobenzylidenethiaxanthene	1.48±0.03	B	30	5	148
$C_{20}H_{13}NO_3$	*p*-(Nitrobenzal)fluorene oxide	4.03	B	12.7	0	157
$C_{20}H_{13}N_3O_7$	Phenanthrene - picric acid complex	2.97	D	35	-	1558
$C_{20}H_{14}$	Benzylidenefluorene	0.73±0.03	B	30	0	144

FORMULA	COMPOUND NAME	μ, D	State, or Solv.	t, °C	Method or P_a, % of P_e	Reference
$C_{20}H_{14}ClNO_3$	6-Chloro-2-methoxy-9-phenoxyacridine *N*-oxide	6.60	B	25	0	1409
$C_{20}H_{14}F_{24}O_4$	*Bis*(2,2,3,3,4,4,5,5,6,6,7,7-dodecafluoroheptyl) ester of 3-methylglutaric acid	3.28	B	20	0	1666
$C_{20}H_{14}N_2$	1,1'-Azonaphthalene	0.59	B	25	-	1763
	trans-1,1'-Azonaphthalene	0	Tol	0	-	536
	2,2'-Azonaphthalene	1.08	B	25	-	1763
	cis-2,2'-Azonaphthalene	3.0±0.05	Tol	0	-	536
	trans-2,2'-Azonaphthalene	0	Tol	0	-	536
	1-Naphthylazoazulene	2.82	B	25	0	571
	2-Naphthylazoazulene	2.63	B	25	0	571
$C_{20}H_{14}O$	9-Benzylidenexanthene	0.51±0.05	B	30	5	132
$C_{20}H_{14}S$	10-Benzylidenethiaxanthene	1.22±0.03	B	30	5	148
$C_{20}H_{15}Br$	1-Bromo-1,2,2-triphenylethylene	1.66±0.05	B	20	5	1205
$C_{20}H_{15}Br_2Cl$	1,2-Dibromo-1-*p*-chlorophenyl-1,2-diphenylethane (m.p. = 113°C)	1.58	B	16	-	129
	(m.p. = 160°C)	2.63	B	17	-	129
$C_{20}H_{15}Cl$	1-(*p*-Chlorophenyl)-2,2-diphenylethylene	1.57±0.02	B	30	5	146
	1-Chloro-1,2,2-triphenylethylene	1.68±0.05	B	20	5	1205
$C_{20}H_{15}ClO$	2-Chloro-2,2-diphenylacetophenone	2.91±0.03	B	20	5	1205

FORMULA	COMPOUND NAME	μ, D	State, or Solv.	t, °C	Method or P_a, % of P_e	Reference
$C_{20}H_{15}NO_2$	2-(4'-Anilinobutadienylmethylene)-1,3-indandione	6.36	B	25	-	1766
$C_{20}H_{15}NO_3$	2-Methoxy-9-phenoxyacridine *N*-oxide	5.33	B	25	0	1409
$C_{20}H_{15}N_3O_6$	1,3,5-Trinitrobenzene - stilbene complex	0.82±0.09	CCl_4	20	-	212
$C_{20}H_{16}$	Triphenylethylene	<0.3 <0.3 0.6	B B B	25 20 10-70	0 5 0	475 1205 1808
$C_{20}H_{16}Br_2O$	2,6-*Bis*(*p*-bromobenzylidene)cyclohexanone	2.29	B	30	0	784
$C_{20}H_{16}Cl_2$	*dl*-1,4-*Bis*(α-chlorobenzyl)benzene	2.5_0	B	25	-	2078,2085
	meso-1,4-*Bis*(α-chlorobenzyl)benzene	2.3_0	B	25	-	2078,2085
	1,2-Dichloro-1,1,2-triphenylethane	1.54	B	18	-	129
$C_{20}H_{16}Cl_2O$	2,6-*Bis*(*p*-chlorobenzylidene)cyclohexanone	2.23	B	30	0	784
$C_{20}H_{16}I_2O$	2,6-*Bis*(*p*-iodobenzylidene)cyclohexanone	2.31	B	30	0	784
$C_{20}H_{16}N_4$	4,5-Dihydro-1,4-diphenyl-3,5-phenylimino-1,2,4-triazole	7.2	B	30	-	2063
$C_{20}H_{16}O_2$	7,12-Dimethoxybenz[a]anthracene	1.65	B	25	-	479
$C_{20}H_{17}N_3$	*N*-*p*-Methylbenzylidene-*p*-phenylazoaniline	2.47	B	25	10	267
$C_{20}H_{17}N_3O$	*N*-Anisylidene-*p*-phenylazoaniline	2.58	B	20	-	1991
$C_{20}H_{18}$	1,1,1-Triphenylethane	0.4	B	10-70	0	1808
$C_{20}H_{18}NiO_4$	Nickel *bis*(benzoylacetonate)	1.6[y]	D	25	0	63

y) "Apparent" value caused by atomic polarization.

FORMULA	COMPOUND NAME	μ, D	State, or Solv.	t, °C	Method or P_a, % of P_e	Reference
$C_{20}H_{18}O$	2,6-Dibenzylidenecyclohexanone	3.09±0.03	B	25	0	127
$C_{20}H_{18}O_2$	1,4-Di-*p*-toloxybenzene	1.82	B	10-40	0	957
$C_{20}H_{18}O_6U$	Uranyl *bis*(benzoylacetonate)	3.12 3.09	B CCl_4	n.s. n.s.	- -	618 618
$C_{20}H_{19}N$	1-(4-Dimethylaminophenyl)-2-(1-naphthyl)-ethylene	1.98	n.s.	n.s.	-	1904
$C_{20}H_{20}Cl_2O$	2,6-*Bis*(*p*-chlorobenzyl)cyclohexanone	3.03	B	30	0	784
$C_{20}H_{20}Co_2O_4$	Cycloöctatriene - cobalt carbonyl complex dimer	3.02	B	n.s.	-	519
$C_{20}H_{20}Cr_4O_4$	Cyclopentadienyl chromium oxide tetramer	0	B	25	-	522
$C_{20}H_{20}N_2O_4S_2$	*N*,*N'*-Dimethyl-*m*-benzenedisulfonanilide	4.81	B	20	0	1703
$C_{20}H_{20}OSi$	Triphenylethoxysilane	1.25	B	30	0	545
$C_{20}H_{20}Pb$	Ethyltriphenyllead	0.87±0.06	B	22	0	1250
$C_{20}H_{20}Sn$	Ethyltriphenyltin	0.74±0.03	B	22	0	1249
$C_{20}H_{22}N_4NiO_4$	Nickel α-methylbenzylglyoxime, acid	1.3	B	25	0	277
	Nickel β-methylbenzylglyoxime, acid	1.6	B	25	0	277
$C_{20}H_{22}O$	2,2-Dibenzylcyclohexanone	2.71±0.02	B?	n.s.	-	302
	cis-2,6-Dibenzylcyclohexanone	2.73±0.03	B?	n.s.	-	302
	trans-2,6-Dibenzylcyclohexanone	2.78±0.05	B?	n.s.	-	302

FORMULA	COMPOUND NAME	μ, D	State, or Solv.	t, °C	Method or P_a, % of P_e	Reference
$C_{20}H_{22}O_4$	Dehydrodiisoeugenol	2.7_9	B	25	0	1152
		2.88	D	25	0	1152
$C_{20}H_{24}CuN_2O_2$	*Bis*(*N*-propylsalicylaldimine)copper(II)	1.92	B	25	10	1671
		1.82	D	25	10	1671
$C_{20}H_{24}N_2NiO_2$	*Bis*(*N*-propylsalicylaldimine)nickel(II)	0.75	B	25	10	1671
		0.97	D	25	10	1671
$C_{20}H_{24}N_2O_2Pd$	*trans*-*Bis*(*N*-propylsalicylaldimine)-palladium(II)	0.77	B	25	10	1671
		0	D	25	10	1671
$C_{20}H_{26}$	1,8-Diphenyloctane	0.5±0.02	B	25	0	1601
$C_{20}H_{26}N_2O_3$	4,4'-*Bis*(butoxy)azoxybenzene	2.38	B	20	-	1241c
$C_{20}H_{26}O_2$	4,4-Dimethylestra-3,17-dione	2.68	B	25	-	17
$C_{20}H_{28}N_2$	2,2'-*Bis*(diethylamino)biphenyl	1.75±0.02	B	20	0	1188
		1.82	B	20	0	1182
$C_{20}H_{28}O_8Th$	Thorium acetylacetonate	1.60	B	20	5	1440
		1.83	B	25	-	511
		1.57	D	20	5	1440
		1.58	CCl_4	20	5	1440
$C_{20}H_{28}O_8Zr$	Zirconium acetylacetonate	1.39	B	2	10	1440
		1.42	B	10	10	1440
		1.46	B	20	10	1440
		1.58	B	40	10	1440
		1.70	B	60	10	1440
		1.75	B	25	-	511
		1.52	D	20	10	1440
		1.45	CCl_4	20	10	1440

FORMULA	COMPOUND NAME	μ, D	State, or Solv.	t, °C	Method or P_a, % of P_e	Reference
$C_{20}H_{30}Cl_2NiP_2$	*Bis*(diethylphenylphosphine)nickel dichloride	~0	B	20	0	846
$C_{20}H_{30}Cl_2P_2Pt$	α-*Bis*(diethylphenylphosphine)platinum dichloride	~0	B	20	-	845
$C_{20}H_{30}O_2$	17-Methyltestosterone	4.20	D	25	-	997
$C_{20}H_{30}O_4$	Dihexylphthalate	2.70±0.06	CCl_4	-30 to 120	RT	2158
$C_{20}H_{32}O_2$	17-Methyl-Δ^5-androstene-3-*cis*-17-*trans*-diol	2.81	D	25	-	997
$C_{20}H_{38}O_2$	4,7-Diisobutyl-2,9-dimethyl-5-decyne-4,7-diol	2.50±0.03	D	20	0	1185
	5,8-Dibutyl-6-dodecyne-5,8-diol	2.35±0.05	D	20	0	1185
	Ethyl oleate	1.36 1.83 1.92	liq liq liq	28-150 20 60	- - -	1052 1495 1495
$C_{20}H_{38}O_3$	Ethyl ricinoleate	2.25 2.29	liq liq	25 60	- -	1495 1495
	Ethyl ricinelaidate	2.49 2.51	liq liq	25 60	- -	1495 1495
$C_{20}H_{40}O_2$	Ethyl stearate	1.2 1.65 1.90	liq liq Hp	48-167 40-50 25	- - 0	1052 1497 1122
$C_{20}H_{42}Cl_2N_2P_2Pt_2S_2$	α-*Bis*(tripropylphosphinethiocyanato)-platinum chloride dimer	1.57	n.s.	n.s.	-	288

FORMULA	COMPOUND NAME	μ, D	State, or Solv.	t, °C	Method or P_a, % of P_e	Reference
	β-*Bis*(tripropylphosphinethiocyanato)-platinum chloride dimer	1.36	n.s.	n.s.	-	288
$C_{20}H_{42}O_6S_3$	1,3-*Bis*(isoamylsulfonyl)-2-(isoamylsulfonylmethyl)-2-methylpropane	4.44	B	20	0	636
$C_{20}H_{43}O_3P$	Diethylhexadecylphosphonate	2.55±0.02	liq	32	-	951
$C_{20}H_{44}BrN$	Tetraisoamylammonium bromide	14.8 >14	B B	25 25	- -	963 759
$C_{20}H_{46}Cl_6O_4Sn_2$	Pentoxytrichlorotin - isopentanol dimer	3.10	B	20	0	1439
$C_{20}H_{46}Si_2$	1,1,1,4,4,4-Hexapropyl-1,4-disilicabutane	0.55	Hx	25	0	627
$C_{20}H_{47}Cl_3P_2Pd_2S$	Unnamed complex, *cis*-form	8.3±0.1	B	25	20	289
$C_{20}H_{47}Cl_3P_2Pt_2S$	Unnamed complex, *cis*-form	13.0±0.1	B	25	20	289
$C_{20}H_{48}I_2P_2Pt$	Dimethyldiiodo*bis*(tri-*n*-propylphosphine)-platinum(II)	5.8±0.1	B	25	15	293
$C_{21}H_{12}O_3S_3$	Tri(thiosalicylide)	1.54±0.02	n.s.	n.s.	-	85
$C_{21}H_{12}O_6$	Trisalicylide	2.42 2.95±0.03	B B	n.s. 25?	- 0	435 437
$C_{21}H_{14}F_{24}O_6$	*Tris*(2,2,3,3,4,4,5,5-octafluoropentyl)ester of tricarballylic acid	3.73	B	20	0	1666
$C_{21}H_{14}O$	2,3-Diphenylindone	3.31	B	20	-	154
$C_{21}H_{14}O_2$	Spiro[1,2*H*-benzopyran-3'-[3*H*]-naphtho-[2,1-*b*]pyran]	1.4_2	B	25	-	787
$C_{21}H_{15}AsNiO_3$	Triphenylarsinenickel tricarbonyl	3.59±0.01	B	n.s.	-	291

FORMULA	COMPOUND NAME	μ, D	State, or Solv.	t, °C	Method or P_a, % of P_e	Reference
$C_{21}H_{15}Br_3S_3$	α(aee)-2,4,6-*Tris*(*p*-bromophenyl)-*s*-trithiane	2.17	B	20	-	689
	β(eee)-2,4,6-*Tris*(*p*-bromophenyl)-*s*-trithiane	3.70	CS_2	20	-	689
$C_{21}H_{15}Cl_3S_3$	α(aee)-2,4,6-*Tris*(*p*-chlorophenyl)-*s*-trithiane	2.21	B	20	-	689
	β(eee)-2,4,6-*Tris*(*p*-chlorophenyl)-*s*-trithiane	3.67	CS_2	20	-	689
$C_{21}H_{15}NiO_3P$	Triphenylphosphinetricarbonylnickel(II)	3.83±0.01	B	n.s.	-	290-1
$C_{21}H_{16}N_2O$	5-Cinnomyl-5,10-dihydrophenazine	2.88	B	25	-	1551
$C_{21}H_{18}Cl_4O_6Sn$	Tin tetrachloride - *tris*(benzoic acid) complex	6.16	B	n.s.	15	1449
$C_{21}H_{18}N_2O_2$	1-Acetyl-2-benzoyl-1,2-diphenylhydrazine	4.85	B	25	-	1042
	1,2-Dibenzoyl-1-methyl-2-phenylhydrazine	5.85	B	25	-	1042
$C_{21}H_{18}O_2$	*o*-Cresolbenzein	6.70	D	32	0	1271
$C_{21}H_{18}S_3$	2a4e6e-Triphenyl-*s*-trithiane	2.09	CCl_4	25	0	688
	2e4e6e-Triphenyl-*s*-trithiane	2.08	CCl_4	25	0	688
$C_{21}H_{20}NS_2$	3-Ethyl-2-(6-phenylimino-2,4-hexadienylidene)benzothiazoline	5.32±0.10	B	25	0	222
$C_{21}H_{20}N_4$	*N*,*N*-Dimethyl-α-(*p*-phenylazophenylimino)-*p*-toluidine	4.51±0.1	B	25	0	267
$C_{21}H_{21}As$	Tri(*p*-tolyl)arsine	1.76	B	20	0	1028

FORMULA	COMPOUND NAME	μ, D	State, or Solv.	t, °C	Method or P_a, % of P_e	Reference
$C_{21}H_{21}N$	Tribenzylamine	0.65	B	20	5	337
$C_{21}H_{21}N_3$	Tri(methyleneaniline)	1.180±0.012	B	30	0	528-9
		1.174±0.013	CCl_4	30	0	528-9
		1.160±0.020	z	30	0	528-9
$C_{21}H_{21}OSb$	Tri-*p*-tolylstibine oxide	2.0±0.1	B	25	15	851
		2.3±0.1	D	40	15	851
$C_{21}H_{21}O_4P$	Tricresyl phosphate	2.84±0.06	CCl_4	-30 to 120	-	2158
$C_{21}H_{21}P$	Tri-*p*-tolylphosphine	1.92	cHx	25	-	920
$C_{21}H_{24}B_3N_3$	*B*-Trimethyl-*N*-triphenylborazole	0.18	B	25	5	2065
$C_{21}H_{24}N_2OS_2$	3-Ethyl-5[4-(1-ethyl-3,3-dimethyl-2(3*H*)-indolinylidene)-2-butenylidene]rhodanine	6.91±0.10	B	30	0	1027
$C_{21}H_{24}O_4$	Diethyl β,β-diphenylglutarate	2.43	B	30	-	1501
$C_{21}H_{24}Si_2$	1,1,2-Tri-*p*-tolyldisilane	0.80±0.02	B	30	0	1488
$C_{21}H_{28}N_2$	1,3-Diphenyl-2-hexylimidazolidine	1.16±0.04	B	30	0	151
		1.11±0.03	pXy	30	5	512
		1.17±0.04	pXy	80	5	512
$C_{21}H_{30}O$	$\Delta^{4,6}$-3,17-Dimethylandrostadiene-17-*trans*-ol	1.83	D	25	-	997
$C_{21}H_{32}O$	Cardol	2.3	n.s.	n.s.	-	980
$C_{21}H_{32}O_2$	4,4-Dimethylandrostane-3,17-dione	2.28	B	25	-	17
$C_{21}H_{34}O$	3,17-Dimethylandrostane-17-*trans*-ol	2.08	D	25	-	997

z) Ligroin - a hydrocarbon mixture.

FORMULA	COMPOUND NAME	μ, D	State, or Solv.	t, °C	Method or P_a, % of P_e	Reference
$C_{21}H_{34}O_2$	3,17-Dimethylandrostane-3-*cis*-17-*trans*-diol	2.41	D	25	-	997
	3,17-Dimethylandrostane-3-*trans*-17-*trans*-diol	2.16	D	25	-	997
$C_{21}H_{36}N_4O_7$	Triisoamylammonium picrate	11.9	B	30	0	1268
		12.99	B	25	-	963
		13.4	B	25	-	564
		14.0	B	25	-	759
$C_{21}H_{41}BrNiP_2$	*trans*-*Bis*(triethylphosphine)bromo(mesityl)-nickel	2.9	B	25	-	295
$C_{21}H_{41}BrP_2Pt$	*cis*-*Bis*(triethylphosphine)bromo(mesityl)-platinum	9.15	B	25	15	294
$C_{21}H_{41}ClNiP_2$	*trans*-*Bis*(triethylphosphine)chloro-(mesityl)nickel	2.4	B	25	-	295
$C_{21}H_{42}AsN_3S_6$	Arsenic *N*,*N*-di-*n*-propyldithiocarbamate	4.9_2	B	20	15	1242
$C_{21}H_{42}CoN_3S_6$	Cobaltic *N*,*N*-di-*n*-propyldithiocarbamate	1.6_2	B	20	0	1243
$C_{21}H_{42}CuN_3S_6$	Cupric *N*,*N*-di-*n*-propyldithiocarbamate	1.6_8	B	30	0	1243
$C_{21}H_{42}FeN_3S_6$	Ferric *N*,*N*-di-*n*-propyldithiocarbamate	1.6_2	B	20	0	1243
$C_{21}H_{42}O_3$	2-Methoxyethanol stearate	2.07	liq	50	-	1497
$C_{21}H_{42}O_4$	Glycerol 1-monostearate	3.07	B	30	0	590
$C_{21}H_{44}N_2S$	Tetraisoamylammonium thiocyanate	15.5	B	25	-	564
$C_{21}H_{44}O_3$	1,1,1-Trihexoxypropane	1.92	B	25	5	60

FORMULA	COMPOUND NAME	μ, D	State, or Solv.	t, °C	Method or P_a, % of P_e	Reference
$C_{21}H_{44}O_8S_4$	1,3-*Bis*(butylsulfonyl)-2,2-*bis*(butylsulfonylmethyl)propane	3.23	B	20	0	636
$C_{22}H_{15}Br$	1-(*p*-Bromobenzylidene)-2,3,6,7-dibenzo-2,4,6-cycloheptatriene	1.60±0.02	B	30	5	146
$C_{22}H_{15}Cl$	1-(*p*-Chlorobenzylidene)-2,3,6,7-dibenzo-2,4,6-cycloheptatriene	1.36±0.04	B	30	5	146
	5-(*p*-Chlorobenzylidene)-2,3,6,7-dibenzo-2,4,6-cycloheptatriene	1.37±0.04	n.s.	n.s.	-	1545
$C_{22}H_{15}MnO_4P$	Triphenylphosphinemanganese tetracarbonyl	1.14±0.05	B	25	20?	712
$C_{22}H_{16}$	1-Benzylidene-2,3,6,7-dibenzo-2,4,6-cycloheptatriene	0.55±0.06	B	30	5	146
	5-Benzylidene-2,3,6,7-dibenzo-2,4,6-cycloheptatriene	0.2±0.2	n.s.	n.s.	-	1545
	Cinnamylidenefluorene	1.20	B	30	0	144
	Diphenylbenzofulvene	1.30±0.04	B	20	0	1189,1199
$C_{22}H_{16}Cl_2Fe$	*Bis*(*p*-chlorophenylcyclopentadienyl)iron	3.12±0.03	B	24.9	-	1723
$C_{22}H_{16}O_2$	6,12-Diacetylnaphthacene	3.3_3 3.3_8	B B	25 27	0 0	1178 1178
	5,11-Diacetylnaphthacene	3.35±0.3	B	26	0	1178
	3-Methylbenzonaphtho-*bis*spiro-2-pyran	1.4_7	B	25	-	787
	3'-Methylbenzonaphtho-*bis*spiro-2-pyran	1.6	B	25	-	787

FORMULA	COMPOUND NAME	μ, D	State, or Solv.	t, °C	Method or P_a, % of P_e	Reference
$C_{22}H_{17}NO$	*N*-Methylquinolino-β-naphthopyrrylospiran	10.4±0.5 7.9±0.2	B a	30 30	0 0	170 170
$C_{22}H_{18}$	1-Benzylidene-2,3,6,7-dibenzo-2,6-cycloheptadiene	0.97±0.05	B	30	5	146
$C_{22}H_{18}Si$	5,5,5-Triphenyl-5-silica-1-penten-3-yne	0.56	B	20	-	1493
$C_{22}H_{21}Cl$	Tri-(*p*-tolyl)chloromethane	2.40	B	25	0	1405
$C_{22}H_{22}O_2$	Thymolbenzein	6.76	D	32	0	1271
$C_{22}H_{24}N_2O_2$	*Bis*(benzoylacetone)ethylenediimine	3.21±0.05	B	25.25	10	1286-7
$C_{22}H_{24}O_4$	3-Hydroxy-2-*p*-cymene carboxylic acid bimolecular cyclic ester	6.63	n.s.	n.s.	-	86
$C_{22}H_{25}NO$	Triphenylmethanol - trimethylamine complex	1.65±0.06	B	25	-	789
$C_{22}H_{28}CuN_2O_2$	*Bis*(*N*-butylsalicylaldimine)copper(II)	2.04 1.85	B D	25 25	10 10	1671 1671
$C_{22}H_{28}N_2NiO_2$	*Bis*(*N*-butylsalicylaldimine)nickel(II)	0.68 0.84	B D	25 25	10 10	1671 1671
$C_{22}H_{28}N_2O_2Pd$	*trans*-*Bis*(*N*-butylsalicylaldimine)-palladium(II)	0.45 0	B D	25 25	10 10	1671 1671
$C_{22}H_{30}N_2O_3$	4,4'-*Bis*(pentoxy)azoxybenzene	2.35	B	20	-	1241c
$C_{22}H_{30}O_2$	*dd*,*ll*-2,5-Dimethyl-2,5-dimethoxy-3,4-diphenylhexane	1.69 1.95 2.11 1.83-1.99	B B B CCl_4	40 27 20 20-40	- - - -	1699 1699 1699 1699

a) 1-Methylnaphthalene.

FORMULA	COMPOUND NAME	μ, D	State, or Solv.	t, °C	Method or P_a, % of P_e	Reference
	meso-2,5-Dimethyl-2,5-dimethoxy-3,4-diphenylhexane	1.33-1.36	B	20-40	-	1699
		1.40-1.44	CCl_4	20-40	-	1699
$C_{22}H_{32}O_3$	Anarcardic acid	3.3	n.s.	n.s.	-	980
$C_{22}H_{32}O_{14}$(?)	Octaacetate of β-D-galactose	3.68	B	30	0	1561
		3.72	B	40	0	1561
		3.33	D	40	0	1562
		3.47	D	30	0	1562
$C_{22}H_{34}P_2Pt$	1,2-*Bis*(diethylphosphino)ethanediphenylplatinum	8.4	B	25	15	294
$C_{22}H_{38}Cl_2N_2O_3Si_4$	1,1,1,2,3,4,4,4-Octamethyl-2,3-*bis*-(*o*-chlorophenylaminomethyl)tetrasiloxane	2.84	liq	20	RT	595
$C_{22}H_{38}N_4O_7$	Tetra-*n*-butylammonium picrate	15.7	B	25	-	582
		17.9	B	25	-	564
		15.0	D	18	RT	389
		14.4	ClB	25	-	582
		14.3	b	25	-	582
		13.8	c	25	-	582
		16.8	d	25	-	582
$C_{22}H_{38}O_2$	Dioctyl ether of hydroquinone	1.64	B	25	0	2080
$C_{22}H_{40}N_2O_3Si_4$	1,1,1,2,3,4,4,4-Octamethyl-2,3-*bis*(phenylaminomethyl)tetrasiloxane	2.81	liq	20	RT	595
$C_{22}H_{42}N_4NiO_4$	Nickel methyloctylglyoxime	2.8	D	25	-	1321
$C_{22}H_{42}O_4$	Diethyl octadecanedioate	2.50	B	25,50	0	1832
$C_{22}H_{44}O_2$	Butyl stearate	1.88	B	24	-	978
		1.89	cHx	30	RT	1844

b) Bromobenzene. c) 25 mole per cent dichlorobenzene plus benzene.
d) 50 mole per cent dichlorobenzene plus benzene.

FORMULA	COMPOUND NAME	μ, D	State, or Solv.	t, °C	Method or P_a, % of P_e	Reference
$C_{22}H_{47}O_3P$	Diethyloctadecylphosphonate	2.52±0.02	liq	32	-	951
$C_{22}H_{50}O_{13}Si_6$	1,1,1,2,3,4,5,6,6,6-Decamethyl-2,3,4,5-*tetrakis*(acetoxymethyl)hexasiloxane	3.77	liq	20	RT	37
$C_{22}H_{52}Cl_2P_2Pd_2S_2$	Unnamed complex	10.7±0.1	B	25	-	289
$C_{22}H_{52}Cl_2P_2Pt_2S_2$	α-*Bis*(tripropylphosphino ethyl sulfide) platirum chloride dimer	10.3±0.1 10.35	B n.s.	25 n.s.	- -	289 288
	β-*Bis*(tripropylphosphino ethyl sulfide) platirum chloride dimer	0 0	B n.s.	25 n.s.	- -	289 288
$C_{23}H_{16}O$	6,8-Diphenyl-7*H*-cycloheptabenzen-7-one	3.69±0.03	B	25	0	127
$C_{23}H_{17}N$	2,4,6-Triphenylpyridine	1.69	B	n.s.	0	1842-3
$C_{23}H_{18}$	5,5,5-Triphenyl-1-penten-3-yne	0.85	B	20	-	1493
$C_{23}H_{18}N_2O$	1,2-Naphthoquinone-2-benzylphenylhydrazone	2.2	B	29	0	164
$C_{23}H_{19}P$	Triphenylphosphoniumcyclopentadienylide	6.99	B	25	0	1565
$C_{23}H_{21}NO$	1,3,3-Trimethylindolino-β-naphthopyrrylospiran	1.38±0.02 1.43±0.02 1.52±0.02	B Tol Tol	30 30 90	0 0 0	170 170 170
$C_{23}H_{22}Br_2CuN_2O_2$	*Bis*(*p*-bromobenzoylacetone)propylenediiminocopper(II)	7.28±0.1	B	25	35	762
$C_{23}H_{22}Br_2N_2NiO_2$	*Bis*(*p*-bromobenzoylacetone)propylenediiminonickel(II)	7.19±0.1	B	25	35	762
$C_{23}H_{22}N_4NiO_6$	*Bis*(*m*-nitrobenzoylacetone)propylenediiminonickel(II)	9.43±0.1	B	25	35	762

FORMULA	COMPOUND NAME	μ, D	State, or Solv.	t, °C	Method or P_a, % of P_e	Reference
$C_{23}H_{24}$	1,3,5-Triphenylpentane	0.99_5	B	25	0	1601
$C_{23}H_{24}Br_2N_2O_2$	*Bis*(*p*-bromobenzoylacetone)propylenediimine	3.99±0.1	B	25	10	761
$C_{23}H_{24}CoN_2O_2$	*Bis*(benzoylacetone)propylenediimino-cobalt(II)	4.03±0.15	B	25.25	-	1286,1288
$C_{23}H_{24}CuN_2O_2$	*Bis*(benzoylacetone)propylenediimino-copper(II)	4.54±0.15	B	25.25	-	1286,1288
	Bis(benzoylacetone)trimethylenediimino-copper(II)	4.24±0.15	B	25.25	-	1286,1288
$C_{23}H_{24}CuN_2O_3$	*Bis*(benzoylacetone)-1,3-diimino-2-propanolo-copper(II)	5.17±0.1	B	25	35	762
$C_{23}H_{24}N_2NiO_2$	*Bis*(benzoylacetone)propylenediimino-nickel(II)	4.35±0.15	B	25.25	-	1286,1288
$C_{23}H_{24}N_4O_6$	*Bis*(*m*-nitrobenzoylacetone)trimethylene-diimine	6.96±0.1	B	25	10	761
$C_{23}H_{26}N_2$	4,4'-*Bis*(dimethylamino)triphenylmethane	1.5_8	B	25	-	400
$C_{23}H_{26}N_2O_2$	*Bis*(benzoylacetone)propylenediimine	2.39±0.07	B	25.25	10	1286-7
	Bis(benzoylacetone)trimethylenediimine	3.71±0.04	B	25.25	10	1286-7
$C_{23}H_{26}N_2O_3$	*Bis*(benzoylacetone)-1,3-diimino-2-propanol	3.52±0.1	B	25.25	10	761
$C_{23}H_{28}N_4O_3$	1,3-Diethyl-5-[4-(1,3-diethyl-2(3*H*)-benzimidazolylidene)-2-butenylidene]-barbituric acid	17.7±1.3	D	30	0	1027
$C_{23}H_{29}ClN_4O_3$	3-Chloro-9-(4-diethylamino-1-methylbutyl-amino)-7-methoxy-2-nitroacridine	6.88±0.02	B	25	0	942

FORMULA	COMPOUND NAME	μ, D	State, or Solv.	t, °C	Method or P_a, % of P_e	Reference
$C_{23}H_{30}ClN_3O$	3-Chloro-9-(4-diethylamino-1-methylbutylamino)-7-methoxyacridine	5.72±0.02	B	25	0	942
	6-Chloro-9-(4-diethylamino-1-methylbutylamino)-2-methoxyacridine	5.90±0.02	B	25	0	942
$C_{23}H_{30}O$	2,7-Dodecamethylene-4,5-benzo-2,4,6-cycloheptatrien-1-one	3.47±0.06 3.23	B cHx	25 25	0 0	563 563
$C_{23}H_{31}ClN_4O$	2-Amino-3-chloro-9-(4-diethylamino-1-methylbutylamino)-7-methoxyacridine	5.16±0.02	B	25	0	942
$C_{23}H_{32}O$	6,8-Dihexyl-7*H*-benzocyclohepten-7-one	3.43	B	25	0	1303
$C_{23}H_{44}O_2$	8,16-Tricosanedione	3.6	n.s.	n.s.	-	430
$C_{23}H_{46}O$	12-*n*-Tricosanone	2.50	liq	69	-	1363
$C_{24}H_{16}N_6O_8$	*Tetrakis*(*p*-nitrophenyl)hydrazine	4.15	B	25	-	1555
$C_{24}H_{17}Br$	2-Bromo-1,3,5-triphenylbenzene	1.53	B[e]	20	0	255
$C_{24}H_{18}N_2O_5$	5-(1,3-Dioxolo[g]quinolin-6-yl)-1-(3,4-methylenedioxyphenethyl)-2(1*H*)pyridone	5.66	B	27	0	1934
$C_{24}H_{18}OSi$	10,10-Diphenylphenoxsilane	0.97±0.03	B	25	0	743
$C_{24}H_{18}O_3$	4,4'-*Bis*(*p*-phenoxyphenyl) ether	1.61 1.63	B B	10,25 40	0 0	955 955
$C_{24}H_{18}O_6$	Tri-*o*-cresotide	4.28±0.04	B	25?	0	437
	Tri-*m*-cresotide	3.08±0.03	B	25?	0	437
	Tri-*p*-cresotide	3.18±0.04	B	25?	0	437

e) Saturated with water.

FORMULA	COMPOUND NAME	μ, D	State, or Solv.	t, °C	Method or P_a, % of P_e	Reference
$C_{24}H_{18}S_2$	*p*-Biphenyl disulfide	2.20	B	25	-	2030
$C_{24}H_{20}$	Triphenylcyclopentadienylmethane	1.1	B	25	10	669
$C_{24}H_{20}AuP$	Phenyl(triphenylphosphine)gold	6.2	B	25	-	262
$C_{24}H_{20}Cl_2PtS_2$	*cis*-*Bis*(diphenylsulfide)platinum dichloride	5.5	B	20	-	848
$C_{24}H_{20}Cl_2PtTe_2$	*cis*-*Bis*(diphenyltelluride)platinum dichloride	6.0±0.5	B	20	-	848
$C_{24}H_{20}Ge$	Tetraphenylgermanium	0	B,D	20	15	1892
$C_{24}H_{20}NP$	Triphenylphosphinephenylimide	4.43±0.03 4.48	B B	25 25	5 15	1502 851
$C_{24}H_{20}N_2$	Tetraphenylhydrazine	1.27	B	25	-	1555
$C_{24}H_{20}O_2$	7,14-Dimethoxydibenz[a,d]anthracene	1.67	B	25	0	479
$C_{24}H_{20}Pb$	Tetraphenyllead	0	B,D	20	15	1892
$C_{24}H_{20}Si$	Tetraphenylsilane	0	B,D	20	15	1892
$C_{24}H_{20}Sn$	Tetraphenyltin	0	B,D	20	15	1892
$C_{24}H_{25}N_3$	*Bis*(*p*-dimethylaminophenyl)phenylmethyl cyanide	1.1_4	B	25	0	400
$C_{24}H_{26}CuN_2O_2$	*Bis*(benzoylacetone)tetramethylenediimino-copper(II)	4.39±0.1	B	25	50	762
$C_{24}H_{27}N_3$	Tri(methylene-*p*-toluidine)	0.896±0.026 0.893±0.018 0.873±0.009	B CCl_4 f	30 30 30	0 0 0	528-9 528-9 528-9

f) Ligroin - a hydrocarbon mixture.

FORMULA	COMPOUND NAME	μ, D	State, or Solv.	t, °C	Method or P_a, % of P_e	Reference
$C_{24}H_{27}N_3O_6S_3$	*N*-Methylene-*p*-toluenesulfonamide trimer	8.04	B	50	0	1703
		8.15	B	35	0	1703
		8.17	B	20	0	1703
$C_{24}H_{28}Cl_2N_4NiO_4$	Nickel α-*p*-chlorophenyl-*n*-butylglyoxime acid	1.8	B	25	0	277
$C_{24}H_{28}N_2O_2$	*Bis*(benzoylacetone)tetramethylenediimine	3.81±0.1	B	25	10	761
$C_{24}H_{29}N_3O_2S$	1,3-Diethyl-5-[4-(1-ethyl-3,3-dimethyl-2(1*H*)-indolinylidene)-2-butenylidene]-2-thiobarbituric acid	9.70±0.15	B	30	0	1027
$C_{24}H_{30}B_3N_3$	*B*-Triethyl-*N*-triphenylborazole	0.47	B	25	5	2065
$C_{24}H_{32}CuN_2O_2$	*Bis*(*N*-amylsalicylaldimine)copper(II)	2.06	B	25	10	1671
		1.93	D	25	10	1671
$C_{24}H_{32}N_2NiO_2$	*Bis*(*N*-amylsalicylaldimine)nickel(II)	0.80	B	25	10	1671
		1.15	D	25	10	1671
$C_{24}H_{32}N_2O_2Pd$	*trans*-*Bis*(*N*-amylsalicylaldimine)-palladium(II)	0	B,D	25	10	1671
$C_{24}H_{34}N_2O_2$	4,4'-Di-*n*-hexoxyazobenzene	1.88	B	n.s.	10	1241d
$C_{24}H_{34}N_2O_3$	4,4'-Di-*n*-hexoxyazoxybenzene	2.35	B	20	-	1241c
$C_{24}H_{34}O_5$	Dehydrocholic acid	5.67	D	25	0	1003
$C_{24}H_{36}O_4$	Dehydrodesoxycholic acid	4.86	D	25	0	1003
$C_{24}H_{36}O_5$	Reductodehydrocholic acid	5.20	D	25	0	1003
$C_{24}H_{38}Cl_2P_2Pt$	*cis*-*Bis*(triethylphosphine)di(*p*-chlorophenyl)platinum	4.25	B	25	15	294

FORMULA	COMPOUND NAME	μ, D	State, or Solv.	t, °C	Method or P_a, % of P_e	Reference
	trans-*Bis*(triethylphosphine)di(*p*-chlorophenyl)platinum	0	B	25	15	294
$C_{24}H_{38}I_2P_2Pt$	*cis*-*Bis*(triethylphosphine)di(*p*-iodophenyl)-platinum	4.95	B	25	15	294
$C_{24}H_{38}O_3$	Dehydrolithocholic acid	3.75	D	25	0	1003
$C_{24}H_{38}O_4$	Apocholic acid	3.01	D	25	0	1003
	Dioctyl phthalate	3.06±0.06	CCl_4	-30 to 120	RT	2158
	3-Hydroxy-12-ketocholic acid	4.29	D	25	0	1003
$C_{24}H_{40}As_2Pt$	*cis*-*Bis*(triethylarsine)diphenylplatinum	7.25	B	25	15	294
	trans-*Bis*(triethylarsine)diphenylplatinum	0	B	25	15	294
$C_{24}H_{40}O_3$	Lithocholic acid	2.52	D	25	0	1003
$C_{24}H_{40}O_4$	Desoxycholic acid	3.25	D	25	0	1003
	Hypodesoxycholic acid	3.15	D	25	0	1003
$C_{24}H_{40}O_5$	Cholic acid	3.87	D	25	0	1003
$C_{24}H_{40}P_2Pt$	*cis*-*Bis*(triethylphosphine)diphenylplatinum	7.2	B	25	15	294
	trans-*Bis*(triethylphosphine)diphenyl-platinum	0	B	25	15	294
$C_{24}H_{42}N_4O_7$	Octadecylamine picrate	12.3	D	30	0	1268

FORMULA	COMPOUND NAME	μ, D	State, or Solv.	t, °C	Method or P_a, % of P_e	Reference
$C_{24}H_{46}Cl_3NO_2P_2Pt_2S$	*cis*-PSP-*cis*-*sym*-μ-Chloro-μ'-*p*-nitrophenylthiodichloro-*bis*-(tri-*n*-propylphosphine)diplatinum	12.4	B	25	10	292
$C_{24}H_{46}FeO_4$	Iron laurate	1.76	B	n.s.	-	90
$C_{24}H_{46}O_4$	Dilauroyl peroxide	1.31	B	20	0	1164
$C_{24}H_{47}Cl_3P_2Pt_2S$	*cis*-PSP-*cis*-*sym*-μ-Chloro-μ'-phenylthiodichloro-*bis*-(tri-*n*-propylphosphine)-diplatinum	13.5	B	25	10	292
$C_{24}H_{48}Ni_2O_4P_4$	Tetracarbonyl-μ,μ'-di-1,2-*bis*(diethylphosphinoethane)dinickel	1.6	B	n.s.	-	291
$C_{24}H_{51}BO_3$	Tri-*n*-octyl borate	0.9_7	CCl_4	25	0	72
$C_{24}H_{51}N$	Tri-*n*-octylamine	0.80	B	25	0	68
$C_{24}H_{51}O_3PS$	Trioctylthiophosphate	3.11	B	20	0	59
$C_{24}H_{52}O_4Si$	Tetrahexylorthosilicate	1.67	B	20	5	53
$C_{24}H_{52}O_4Ti$	Tetrahexylorthotitanate	1.63 1.63	n.s. B	n.s. 20	- 5	51 53
$C_{24}H_{54}As_2Cl_2Pt$	α-*Bis*(tri-*n*-butylarsine)platinum dichloride	~0	B	20	-	845
$C_{24}H_{54}Cl_2P_2Pt$	α-*Bis*(tri-*n*-butylphosphine)platinum dichloride	~0	B	20	-	845
	β-*Bis*(tri-*n*-butylphosphine)platinum dichloride	11.6	B	20	-	845

FORMULA	COMPOUND NAME	μ, D	State, or Solv.	t, °C	Method or P_a, % of P_e	Reference
$C_{24}H_{54}Cl_2PtSb_2$	β-*Bis*(tri-*n*-butylstibine)platinum dichloride	11.0	B	20	-	845
$C_{24}H_{54}OSi_2$	Hexabutyldisiloxane	0.71	Hx	25	0	627-8
$C_{25}H_{16}O_2$	Dinaphtho-*bis*spiro-2-pyran	1.6_4	B	25	-	787
$C_{25}H_{20}$	Tetraphenylmethane	0	B,D	20	15	1892
$C_{25}H_{25}Co_2$	Pentacyclopentadienyl dicobalt	0	B	25	5	2076
$C_{25}H_{30}N_2O_2$	*Bis*(benzoylacetone)pentamethylenediimine	4.23±0.1	B	25	10	761
$C_{25}H_{49}Cl_3OP_2Pt_2S$	*cis*-PSP-*cis*-*sym*-μ-Chloro-μ'-*p*-methoxyphenylthiodichloro-*bis*-(tri-*n*-propylphosphine)diplatinum	12.8	B	25	10	292
$C_{25}H_{52}O_4$	Tetrahexylorthocarbonate	1.01	B	25	5	56
$C_{25}H_{56}Si_2$	1,1,1,3,3,3-Hexabutyl-1,3-disilicapropane	0.60	Hx	25	0	627
$C_{26}H_{14}Br_2$	2,2'-Dibromo-$\Delta^{9,9'}$-bifluorene	2.28±0.07	B	30?	5	131
$C_{26}H_{14}F_2$	2,2'-Difluoro-$\Delta^{9,9'}$-bifluorene (mostly *cis*)	2.53	B	15	0	128
$C_{26}H_{16}$	$\Delta^{9,9'}$-Bifluorene	0 0	B B	30 16	0 0	144,149 128
	Dibenzo[a,c]triphenylene	0	B	30	5	149,150
	Phenanthro[3,4-c]phenanthrene	0.70	B	20	-	1300
$C_{26}H_{16}Cl_2$	9,9'-[*Bis*(*p*-chlorophenyl)methylene]-fluorene	1.27±0.03	B	30	5	147

FORMULA	COMPOUND NAME	μ, D	State, or Solv.	t, °C	Method or P_a, % of P_e	Reference
$C_{26}H_{16}O_2$	$\Delta^{9,9'}$-Bixanthene	0 1.15±0.01	B B	25 30	0 5	948 143
$C_{26}H_{18}$	Benzhydrylidenefluorene	1.03	B	30	0	144
	Diphenyldibenzofulvene	1.20±0.04	B	20	0	1189
$C_{26}H_{18}Cl_2$	*cis*-9,10-Dichloro-9,10-diphenylanthracene	3.0	B	38	0?	168
$C_{26}H_{18}N_2O_4$	*cis*-1,2-*Bis*(*m*-nitrophenyl)-1,2-diphenylethylene	4.9	B	20	0?	193
	trans-1,2-*Bis*(*m*-nitrophenyl)-1,2-diphenylethylene	5.7	B	20	0?	193
	cis-1,2-*Bis*(*p*-nitrophenyl)-1,2-diphenylethylene	7.0	B	20	0?	193
	trans-1,2-*Bis*(*p*-nitrophenyl)-1,2-diphenylethylene	1.9	B	20	0?	193
$C_{26}H_{20}$	Tetraphenylethylene	0 0.33 0.99 0.77	B B B D	10-70 20 40 20	0 SPC SPC SPC	1808 1300 1300 1300
$C_{26}H_{20}N_2$	*N,N'*-Dibenzylidenebenzidine	1.4	B	20	-	1991
	N,N'-Di-2-naphthyl-*p*-phenylenediamine	2.76	D	35	0	1850
$C_{26}H_{20}N_2O_2$	1,2-Dibenzoyl-1,2-diphenylhydrazine	4.53	B	25	-	1042
$C_{26}H_{20}N_2S_4$	*Bis*(diphenylthiocarbamoyl)disulfide	3.4	B	25	0	639
$C_{26}H_{20}N_4$	1,4,5-Triphenyl-3,5-*endo*anilino-1,2,4-triazoline	9.1 8.8	B D	25 25	15 15	859 859

FORMULA	COMPOUND NAME	μ, D	State, or Solv.	t, °C	Method or P_a, % of P_e	Reference
$C_{26}H_{22}N_2$	Benzophenone benzylphenylhydrazone	2.62	B	26	0	164
$C_{26}H_{22}OSi$	2,7-Dimethyl-10,10-diphenylphenoxsilane	1.01±0.03	B	25	0	743
$C_{26}H_{22}O_3$	4,4'-*Bis*(*p*-methylphenoxyphenyl) ether	2.06 2.09	B B	25 40	0 0	955 955
$C_{26}H_{32}CoN_2O_2$	*Bis*(*N*-cyclohexylsalicylaldimine)cobalt(II)	4.86	B	25	10	1671
$C_{26}H_{38}N_2O_2$	4,4'-*Bis*(*n*-heptoxy)azobenzene	1.85	B	n.s.	10	1241d
$C_{26}H_{38}N_2O_3$	4,4'-*Bis*(*n*-heptoxy)azoxybenzene	2.36	B	20	-	1241c
$C_{26}H_{42}O_4$	Dinonyl phthalate	2.65±0.06	CCl_4	-30 to 120	RT	2158
	Gitogenin	2.66	D	25	-	997
$C_{26}H_{44}P_2Pt$	*cis*-*Bis*(triethylphosphine)dibenzyl-platinum	6.7	B	25	15	294
	cis-*Bis*(triethylphosphine)di-*o*-tolyl-platinum	7.5	B	25	15	294
	trans-*Bis*(triethylphosphine)di-*o*-tolyl-platinum	0	B	25	15	294
	cis-*Bis*(triethylphosphine)di-*p*-tolyl-platinum	6.75	B	25	15	294
$C_{26}H_{46}N_4O_7$	Tetraisoamylammonium picrate	18.1 18.4 19.5	B B B	25 25 25	- - -	963 564 759
$C_{26}H_{48}N_2O_3Si_4$	1,1,1,2,3,4,4,4-Octamethyl-2,3-*bis*(ethyl-phenylaminomethyl)tetrasiloxane	2.81	liq	20	RT	595

FORMULA	COMPOUND NAME	μ, D	State, or Solv.	t, °C	Method or P_a, % of P_e	Reference
$C_{26}H_{51}Cl_2NO_2P_2Pt_2S_2$	*cis*-P(S-Ethyl)P-*cis*-*sym*-μ-ethylthio-μ'-*p*-nitrophenylthiodichloro-*bis*(tri-*n*-propylphosphine)-diplatinum	14.8	B	25	10	292
$C_{26}H_{52}Cl_2P_2Pt_2S_2$	*cis*-P(S-Ethyl)P-*cis*-*sym*-μ-ethylthio-μ'-phenylthiodichloro-*bis*(tri-*n*-propylphosphine)diplatinum	11.1	B	25	10	292
	cis-P(S-Phenyl)P-*cis*-*sym*-μ-ethylthio-μ'-phenylthiodichloro-*bis*(tri-*n*-propylphosphine)diplatinum	11.05	B	25	10	292
$C_{26}H_{58}Si_2$	1,1,1,4,4,4-Hexabutyl-1,4-disilicabutane	0.56	Hx	25	-	627
$C_{26}H_{62}P_2Pt_2S_4$	*cis*-*sym*-di-μ-Ethylthio-diethylthio-*bis*-(tri-*n*-propylphosphine)diplatinum	7.61	B	25	10	292
	trans-*sym*-di-μ-Ethylthio-diethylthio-*bis*-(tri-*n*-propylphosphine)diplatinum	0	B	25	10	292
$C_{27}H_{16}O_2$	Xanthylidene anthrone	2.88±0.08 2.92±0.04	Tol Tol	90 30	5 5	143 143
$C_{27}H_{18}Br_2$	1,3-Di-*p*-bromophenyl-1,3-diphenylpropadiene	1.94	B	19	0	152
$C_{27}H_{18}Cl_2$	1,3-Di-*p*-chlorophenyl-1,3-diphenylpropadiene	1.58	B	25	0	152
$C_{27}H_{18}O$	Benzhydrylidene anthrone	3.60±0.04	Tol	30	5	143
$C_{27}H_{18}O_2$	α-Naphthol benzein	5.76 6.07	B D	32 32	0 0	1271 1271
$C_{27}H_{19}Cl$	1-*p*-Chlorophenyl-1,3,3-triphenylpropadiene	1.56	B	22	0	152

FORMULA	COMPOUND NAME	μ, D	State, or Solv.	t, °C	Method or P_a, % of P_e	Reference
$C_{27}H_{20}$	Tetraphenylpropadiene	0	B	19	0	152
$C_{27}H_{20}N_2$	2,3-Diphenylindonephenylhydrazone	1.95	D	14	0	164
$C_{27}H_{24}N_4O_7$	Tribenzylamine picrate	12.0	B	30,40	0	1269
$C_{27}H_{28}O$	2,2,6-Tribenzylcyclohexanone	2.69±0.03	B?	n.s.	-	302
$C_{27}H_{32}$	1,5,9-Triphenylnonane	0.86	B	25	0	1601
$C_{27}H_{44}O_3$	Tigogenin	2.38	D	25	-	997
$C_{27}H_{44}O_4$	Chlorogenin	2.69	D	25	-	997
$C_{27}H_{45}FO$	2-Fluorocholestan-3-one	4.39±0.03	B	25	10	16
$C_{27}H_{46}O$	3-Cholestanone	3.01±0.03	B	25	10	16
	Cholesterol	2.01	D	25	-	997
$C_{27}H_{46}O_2$	Cholestane-3-*cis*-ol-7-one	3.01	D	25	-	1002
	Δ^5-Cholestene-3-*cis*-ol-7-one	3.82	D	25	-	1002
$C_{27}H_{48}O$	Dihydrocholesterol	1.83	D	25	-	997
$C_{27}H_{48}O_2$	Cholestane-3-*cis*-7-*cis*-diol	2.57	D	25	-	1002
	Cholestane-3-*cis*-7-*trans*-diol	2.32	D	25	-	1002
$C_{27}H_{54}CoN_3S_6$	Cobalt(III) *N,N*-diisobutyldithiocarbamate	1.9_4	B	23	0	1243
$C_{27}H_{54}CrN_3S_6$	Chromium(III) *N,N*-di-*n*-butyldithiocarbamate	2.0_0	B	30	0	1243
$C_{27}H_{54}O$	14-Heptacosanone	2.32	CCl_4	25	-	67

FORMULA	COMPOUND NAME	μ, D	State, or Solv.	t, °C	Method or P_a, % of P_e	Reference
$C_{27}H_{56}O_3$	Trioctyl orthopropionate	2.02	B	25	5	60
$C_{27}H_{57}BO_3$	Tri-*n*-nonyl borate	0.9_3	CCl_4	25	0	72
$C_{27}H_{63}Ag_3Br_3O_9P_3$	Diisopropyl isopropylphosphate - silver bromide complex trimer	3.53	B	n.s.	-	57
$C_{27}H_{63}Ag_3I_3O_9P_3$	Diisopropyl isopropylphosphate - silver iodide complex trimer	3.43	B	n.s.	-	57
$C_{27}H_{63}Br_3Cu_3O_9P_3$	Diisopropyl isopropylphosphate - copper bromide complex trimer	2.78	B	n.s.	-	57
$C_{27}H_{63}Cu_3I_3O_9P_3$	Diisopropyl isopropylphosphate - copper iodide complex trimer	2.57	B	n.s.	-	57
$C_{28}H_{15}BrN_2$	9-Bromoanthrazine	1.5_1	B	25	0	1366
$C_{28}H_{16}O_2$	$\Delta^{10,10'}$-Bianthrone	0	B	25	0	948
$C_{28}H_{16}O_8$	Tetrasalicylide	2.07±0.05	B	25	0	437
$C_{28}H_{18}$	5-(9-Fluorenylidene)-dibenzo[a,e]cycloheptatriene	0.83±0.1	B?	30?	5	1544
$C_{28}H_{18}Br_2$	*trans*-9-Bromoanthracene dimer	0.36±0.04 0.36	B B	25 25	0 0	259,261 49
$C_{28}H_{18}Cl_2$	9-Chloroanthracene dimer	0.60	B	25	0	49
$C_{28}H_{20}$	1-Biphenylene-4,4-diphenylbutadiene	1.19±0.03	B	30	0	144
$C_{28}H_{20}N_2O_4$	*cis*-1,4-*Bis*(*m*-nitrophenyl)-1,4-diphenylbutadiene	5.2	B	20	0	193

FORMULA	COMPOUND NAME	μ, D	State, or Solv.	t, °C	Method or P_a, % of P_e	Reference
	trans-1,4-*Bis*(*m*-nitrophenyl)-1,4-diphenyl-butadiene	5.9	B	20	0	193
$C_{28}H_{20}S$	Tetraphenylthiophene	0.60	B	30	0	282
$C_{28}H_{22}O_2$	Tetraphenylbutynediol	2.51±0.05	D	20	0	1185
$C_{28}H_{24}$	1↓,2,3,4↑-Tetraphenylcyclobutane (m.p. = 149°C)	0.69	B	20	SPC	1300
		1.41	B	40	SPC	1300
		1.27	D	20	SPC	1300
		1.92	D	40	SPC	1300
	1↓,2↓,3↑,4↑-Tetraphenylcyclobutane (m.p. = 163°C)	0.44	B	20	SPC	1300
		0.93	B	40	SPC	1300
		0.90	D	20	SPC	1300
		1.46	D	40	SPC	1300
$C_{28}H_{24}As_2NiO_2$	1,2-*Bis*(diphenylarsino)ethanenickel dicarbonyl	5.07±0.01	B	n.s.	-	291
$C_{28}H_{24}NiO_2P_2$	1,2-*Bis*(diphenylphosphino)ethanenickel dicarbonyl	4.81±0.01	B	n.s.	-	291
$C_{28}H_{26}$	1,8,15,20-Tetramethylquaterphenyl	0.67	CCl_4	25	0	919
	2,9,14,21-Tetramethylquaterphenyl	0.71	CCl_4	25	0	919
$C_{28}H_{26}O_4$	Tetramethoxyquaterphenyl	2.31	CCl_4	25	0	919
$C_{28}H_{28}Br_2PtS_2$	β-*Bis*(dibenzylsulfide)platinum dibromide	8.7	B	20	-	844
$C_{28}H_{28}Cl_2OTe_2$	*Bis*(di-*p*-tolyltellurium chloride) oxide	6.1±0.2	D	25	0	849
$C_{28}H_{28}Cl_2PtS_2$	β-*Bis*(dibenzylsulfide)platinum dichloride	7.8	B	20	-	844
$C_{28}H_{28}I_2PtS_2$	α-*Bis*(dibenzylsulfide)platinum diiodide	2.39	B	20	-	844

FORMULA	COMPOUND NAME	μ, D	State, or Solv.	t, °C	Method or P_a, % of P_e	Reference
$C_{28}H_{28}N_2O_4$	*Tetrakis*(*p*-methoxyphenyl)hydrazine	1.30	B	25	-	1555
$C_{28}H_{33}N_3$	*Bis*(4-diethylaminophenyl)phenylmethyl cyanide	1.74	B	25	0	400
$C_{28}H_{38}O_{19}$	α-Cellubiose octaacetate	2.21 2.80 2.88 3.11	$CHCl_3$ $CHCl_3$ $CHCl_3$ $CHCl_3$	20 20 35 50	- 0 0 0	1676 1063 1063 1063
$C_{28}H_{40}As_2Pt$	*trans*-*Bis*(triethylarsine)di(phenylethynyl)-platinum	1.3	B	25	15	294
$C_{28}H_{40}NiP_2$	*trans*-*Bis*(triethylphosphine)di(phenylethynyl)nickel	0.45	B	25	-	295
$C_{28}H_{40}P_2Pt$	*trans*-*Bis*(triethylphosphine)di(phenylethynyl)platinum	0.9	B	25	15	294
$C_{28}H_{46}Br_2NiP_2$	*Bis*-(di-*n*-butylphenylphosphine)nickel dibromide	1.9±0.15	B	25	15	335
$C_{28}H_{46}Cl_2NiP_2$	*Bis*(di-*n*-butylphenylphosphine)nickel dichloride	1.3±0.1	B	25	15	335
$C_{28}H_{46}I_2NiP_2$	*Bis*(di-*n*-butylphenylphosphine)nickel diiodide	3.1±0.1	B	25	15	335
$C_{28}H_{46}N_2NiO_6P_2$	*Bis*(di-*n*-butylphenylphosphine)nickel dinitrate	9.8±0.1	B	25	15	335
$C_{28}H_{58}S_2$	Di-*n*-tetradecyl disulfide	1.97	B	25	-	2030
$C_{28}H_{60}O_4Si$	Tetraheptyl orthosilicate	1.65	B	20	5	53
$C_{28}H_{60}O_4Ti$	Tetraheptyl orthotitanate	1.61	B	20	5	51,53

FORMULA	COMPOUND NAME	μ, D	State, or Solv.	t, °C	Method or P_a, % of P_c	Reference
$C_{28}H_{70}O_5Si_6$	Tetradecaethylhexasiloxane	1.41	Hx	25	0	627
$C_{29}H_{16}Cl_4O$	*Tetrakis*(4-chlorophenyl)cyclopentadienone	2.26±0.1	B	40	0	407
$C_{29}H_{18}Cl_2O$	2,5-*Bis*(4-chlorophenyl)-3,4-diphenylcyclopentadienone	4.61±0.1	B	40	0	407
$C_{29}H_{20}O$	Tetraphenylcyclopentadienone	3.43±0.1 3.61±0.04 3.55	B B B[g]	40 30 20	0 5 0	407 143 255
$C_{29}H_{22}N_3O_2P$	Triphenylphosphonium-(2-*p*-nitrophenylazo)-cyclopentadienylide	9.09	B	25	0	1565
$C_{29}H_{23}N_2P$	Triphenylphosphonium-(2-phenylazo)-cyclopentadienylide	6.52	B	25	0	1565
$C_{29}H_{30}N_2O_3V$	*Bis*(benzoylacetone)propylenediiminovanadium(IV) - benzene complex	6.05±0.1	B	25	35	762
$C_{29}H_{30}N_4O_2S_3$	1,3-Diethyl-5-[1,3-*bis*(3-ethyl-2-benzothiazolinylidene)-2-propylidene]-2-thiobarbituric acid	13.3	D	30	0	1486
$C_{29}H_{50}O$	4,4-Dimethylcholestan-3-one	2.78	B	25	-	17
$C_{29}H_{58}O$	15-Nonacosanone	2.14	B	25	15	1468
$C_{30}H_{18}Cl_2O_2$	*trans* dimer of 9-Anthroyl chloride	0.9±0.3	B	n.s.	-	48
$C_{30}H_{18}CoN_3O_6$	*Tris*(1-nitroso-2-naphthol)cobalt(III)	3.3 3.8	B B,Hx	25 20	0 -	63 716
$C_{30}H_{18}FeN_3O_6$	*Tris*(1-nitroso-2-naphthol)iron(III)	2.7 3.1	B B,Hp	25 20	0 -	63 716

g) Saturated with water.

FORMULA	COMPOUND NAME	μ, D	State, or Solv.	t, °C	Method or P_a, % of P_e	Reference
$C_{30}H_{18}N_2$	*trans*-9-Cyanoanthracene dimer	1.2±0.3	B	25	0	259,261
$C_{30}H_{20}O_2$	*trans* dimer of 9-Anthraldehyde	1.5±0.1	B	n.s.	-	48
$C_{30}H_{22}O_4$	1,4-*Bis*(phenoxyphenoxy)benzene	1.82	B	40	0	955
$C_{30}H_{24}O_5V$	*Bis*(dibenzoylmethano)oxovanadium(IV)	3.61±0.09	B	25	0	760
$C_{30}H_{24}$	2,7,2',7'-Tetramethyl-$\Delta^{9,9'}$-bifluorene	0	B	n.s.	0	149
$C_{30}H_{25}O_5Sb$	Pentaphenyl stibnate	2.31	n.s.	n.s.	-	51
$C_{30}H_{25}O_5W$	Pentaphenyl tungstate	2.42	n.s.	n.s.	-	51
$C_{30}H_{28}N_2$	9-*Bis*(*p*-dimethylaminophenyl)methylene-fluorene	3.90±0.03	B	30	5	147
$C_{30}H_{36}O_4$	*cis*-Diethylhexamethyl-21,22,23,24-tetroxa-pentacyclo[16.2.2.1^{3,6}.1^{8,11}.1^{13,16}]-tetracosa-3,5,8,10,13,15,18,20-octaene	0.77	B	n.s.	0	3
	trans-Diethylhexamethyl-21,22,23,24-tetroxapentacyclo[16.2.2.1^{3,6}.1^{8,11}.1^{13,16}]-tetracosa-3,5,8,10,13,15,18,20-octaene	0	B	n.s.	0	3
$C_{30}H_{37}BrNiP_2$	*trans*-*Bis*(diethylphenylphosphine)bromo-(1-naphthyl)nickel	3.05	B	25	-	295
$C_{30}H_{46}N_2NiS_2P_2$	*Bis*(di-*n*-butylphenylphosphine)nickel thiocyanate	1.8±0.1	B	25	15	335
$C_{30}H_{50}Cl_2N_2O_4P_2Pt_2S_2$	*cis*-Di-μ-*p*-nitrophenylthiodichloro-*bis*(tri-*n*-propylphosphine)-diplatinum	13.0	B	25	10	292

FORMULA	COMPOUND NAME	μ, D	State, or Solv.	t, °C	Method or P_a, % of P_e	Reference
	trans-Di-μ-*p*-nitrophenylthiodichloro-*bis*(tri-*n*-propylphosphine)diplatinum	4.68	B	25	10	292
$C_{30}H_{50}O$	Friedlin	2.83±0.03	B	32	5	1035
	Lupanone	2.99±0.03	B	25	-	1111
$C_{30}H_{50}O_2$	Cerin	2.41±0.05	B	50	5	1035
$C_{30}H_{51}N_3O_4Si_5$	1,1,1,2,3,4,5,5,5-Nonamethyl-2,3,4-*tris*-(phenylaminomethyl)pentasiloxane	3.33	liq	20	RT	595
$C_{30}H_{52}$	Lupane	0±0.02	B	25	-	1111
$C_{30}H_{52}As_2Pt$	*trans*-*Bis*(tri-*n*-propylarsine)diphenylplatinum	0	B	25	15	294
$C_{30}H_{52}Cl_2P_2Pt_2S_2$	*cis*-Di-μ-phenylthiodichloro-*bis*(tri-*n*-propylphosphine)diplatinum	11.2	B	25	10	292
	trans-Di-μ-phenylthiodichloro-*bis*(tri-*n*-propylphosphine)diplatinum	0.75	B	25	10	292
$C_{30}H_{52}O$	Friedlinol (low melting)	1.80±0.07	B	32	5	1035
	(high melting)	1.83±0.05	B	50	5	1035
$C_{30}H_{58}O_4$	Ethylene glycol dimyristate	2.32	B	30	0	590
$C_3H_{63}N$	Tri-*n*-decylamine	0.72	B	25	0	68
$C_{30}H_{64}Cl_2P_2Pt_2S_2$	*cis*-Di-μ-cyclohexylthiodichloro-*bis*-(tri-*n*-propylphosphine)diplatinum	10.6	B	25	10	292
	trans-Di-μ-cyclohexylthiodichloro-*bis*-(tri-*n*-propylphosphine)diplatinum	1.35	B	25	10	292

FORMULA	COMPOUND NAME	μ, D	State, or Solv.	t, °C	Method or P_a, % of P_e	Reference
$C_{31}H_{23}P$	Fluoren-9-ylidenetriphenylphosphorane	7.09	B	25	-	864
$C_{31}H_{32}$	1,3,5,7-Tetraphenylheptane	1.52	B	25	0	1601
$C_{31}H_{34}N_4O_5S_2$	1,3-*Bis*(2-methoxyethyl)-5-[1,3-*bis*(3-ethyl-2-benzothiazolinylidene)-2-propylidene]-barbituric acid	9.8	D	30	0	1486
$C_{31}H_{49}NO$	nitrile of Lupanone carboxylic acid	3.93±0.03	B	25	-	1111
$C_{31}H_{51}N$	nitrile of Lupane carboxylic acid	3.77±0.03	B	25	-	1111
$C_{32}H_{24}$	3,6-*Bis*(diphenylmethylene)-1,4-cyclohexadiene	0	B	n.s.	0	149
	Tetraphenyl-*p*-phenylenedimethylene	0	B	30	5	1543
$C_{32}H_{24}As_2NiO_2$	*o*-Phenylene*bis*(diphenylarsine)nickel dicarbonyl	5.14±0.01	B	n.s.	-	291
$C_{32}H_{24}NiO_2P_2$	*o*-Phenylene*bis*(diphenylphosphine)nickel dicarbonyl	5.39±0.01	B	n.s.	-	291
$C_{32}H_{24}O_8$	Tetra-*o*-cresotide	1.25±0.1	B	25?	0	437
$C_{32}H_{38}Br_2NiP_2$	*Bis*(*n*-butyldiphenylphosphine)nickel dibromide	5.9±0.02	B	25	15	335
$C_{32}H_{38}Cl_2NiP_2$	*Bis*(*n*-butyldiphenylphosphine)nickel dichloride	3.3±0.01	B	25	15	335
$C_{32}H_{38}I_2NiP_2$	*Bis*(*n*-butyldiphenylphosphine)nickel diiodide	6.8±0.02	B	25	15	335
$C_{32}H_{38}N_2NiO_6P_2$	*Bis*(*n*-butyldiphenylphosphine)nickel dinitrate	9.8±0.1	B	25	15	335

FORMULA	COMPOUND NAME	μ, D	State, or Solv.	t, °C	Method or P_a, % of P_e	Reference
$C_{32}H_{56}Cl_2O_2P_2Pt_2S_2$	*cis*-Di-μ-*p*-methoxyphenylthio-dichloro-*bis*(tri-*n*-propylphosphine)-diplatinum	11.3	B	25	10	292
	trans-Di-μ-*p*-methoxyphenylthio-dichloro-*bis*(tri-*n*-propylphosphine)-diplatinum	2.06	B	25	10	292
$C_{32}H_{62}MgO_2$	Magnesium palmitate	3.35	B	25	-	1455
$C_{32}H_{66}S$	*n*-Hexadecyl sulfide	1.47±0.02	B	30	0	2145
$C_{32}H_{66}S_2$	*n*-Hexadecyl disulfide	2.00±0.02	B	30	0	2145
$C_{32}H_{66}S_3$	*n*-Hexadecyl trisulfide	1.63±0.02	B	30	0	2145
$C_{32}H_{66}S_4$	*n*-Hexadecyl tetrasulfide	2.16±0.02	B	30	0	2145
$C_{32}H_{68}O_4Si$	Tetraoctyl orthosilicate	1.61	B	20	5	53
$C_{32}H_{68}O_4Ti$	Tetraoctyl orthotitanate	1.68	B	20	5	51,53
$C_{32}H_{80}O_6Si_7$	Hexadecaethylheptasiloxane	1.55	Hx	25	0	627
$C_{33}H_{36}O_6$	3-Hydroxy-2-*p*-cymenecarboxylic acid (trimolecular cyclic ester)	4.13	n.s.	n.s.	-	86
$C_{33}H_{45}AlO_6$	Hydroxymethylenecamphoroaluminum(III)	2.31	B	n.s.	-	1149
$C_{33}H_{45}CoO_6$	Hydroxymethylenecamphorocobalt(III)	2.35	B	n.s.	-	1149
$C_{33}H_{45}CrO_6$	Hydroxymethylenecamphorochromium(III)	2.3	B	n.s.	-	1149
$C_{33}H_{45}FeO_6$	Hydroxymethylenecamphoroiron(III)	2.4	B	n.s.	-	1149
$C_{33}H_{68}O_4$	Tetraoctyl orthocarbonate	0.94	B	25	5	56

FORMULA	COMPOUND NAME	μ, D	State, or Solv.	t, °C	Method or P_a, % of P_c	Reference
$C_{34}H_{20}Fe_2O_6$	*Bis*(diphenylacetylene)hexacarbonyldiiron	3.3±0.2	B	n.s.	-	768
$C_{34}H_{34}O$	2,2,6,6-Tetrabenzylcyclohexanone	2.33±0.07	B?	n.s.	-	302
$C_{34}H_{35}ClO_{13}S_4$	1-Chloro-2,3,4,6-tetra-*p*-toluenesulfonyl-D-glucoside	6.45	B	20	10	171
$C_{34}H_{38}N_2NiP_2S_2$	*Bis*(*n*-butyldiphenylphosphine)nickel thiocyanate	2.0±0.1	B	25	15	335
$C_{34}H_{44}NiP_2$	*trans*-*Bis*(diethylphenylphosphine)*bis*-(*o*-tolyl)nickel	0	B	25	-	295
$C_{34}H_{52}BNO$	Tetra-*n*-butylammonium hydroxide - triphenylboron complex	19.8	B	25	-	564
$C_{36}H_{24}N_4O_4Zr$	Zirconium 8-hydroxyquinolate	1.97	D	25	-	1440
$C_{36}H_{26}$	1,4-*Bis*(diphenylmethylene)-1,4-dihydronaphthalene	0	B	30	5	145,1543
$C_{36}H_{30}BrCuO_6P_2$	*Bis*(diphenyl phenylphosphate) - copper(I) bromide complex	1.65	B	n.s.	-	57
$C_{36}H_{30}Br_2CoP_2$	*Bis*(triphenylphosphine)cobalt dibromide	7.6±0.1	B	(25)	0	231
$C_{36}H_{30}Br_2NiP_2$	*Bis*(triphenylphosphine)nickel dibromide	5.9±0.3 6.8±0.2	B B	25 (25)	0 0	2040 231
$C_{36}H_{30}ClCuO_6P_2$	*Bis*(diphenyl phenylphosphate) - copper(I) chloride complex	1.70	B	n.s.	-	57
$C_{36}H_{30}Cl_2PtSb_2$	β-*Bis*(triphenylstibine)platinum dichloride	9.3	B	20	-	845
$C_{36}H_{30}CoI_2P_2$	*Bis*(triphenylphosphine)cobalt diiodide	9.6±0.1	B	(25)	0	231

FORMULA	COMPOUND NAME	μ, D	State, or Solv.	t, °C	Method or P_a, % of P_e	Reference
$C_{36}H_{30}CuIO_6P_2$	*Bis*(diphenyl phenylphosphate) - copper(I) iodide complex	1.60	B	n.s.	-	57
$C_{36}H_{30}I_2NiP_2$	*Bis*(triphenylphosphine)nickel diiodide	8.5±0.3	B	25	0	2040
		8.95±0.15	B	(25)	0	231
$C_{36}H_{30}N_2NiO_2P_2$	Dinitrosyl*bis*(triphenylphosphine)nickel(0)	4.26	B	30	-	617
$C_{36}H_{30}O_5V$	*Bis*(dibenzoylmethanooxovanadium(IV)) - benzene complex	3.61±0.1	B	25	35	762
$C_{36}H_{30}O_6W$	Hexaphenyltungstate	2.15	B	25	5	60
		2.55	n.s.	n.s.	-	51
$C_{36}H_{30}Pb_2$	Triphenyllead dimer	0	B	25	0	1118
$C_{36}H_{40}ClNiP_2$	*trans*-*Bis*(diethylphenylphosphine)chloro-(phenylethynyl)nickel	2.0	B	25	-	295
$C_{36}H_{63}N_3O_4Si_5$	1,1,1,2,3,4,5,5,5-Nonamethyl-2,3,4-*tris*-(ethylphenylaminomethyl)pentasiloxane	3.32	liq	20	RT	595
$C_{36}H_{66}CaO_4$	Calcium(II) oleate	4.53[h]	B	30	0	89
$C_{36}H_{66}CdO_4$	Cadmium(II) oleate	4.40	B	25	-	1455
$C_{36}H_{66}CrO_4$	Chromium(II) oleate	4.39	B	n.s.	-	90
$C_{36}H_{66}CuO_4$	Copper(II) oleate	1.21[h]	B	30	0	89
$C_{36}H_{66}FeO_4$	Iron(II) oleate	2.88	B	n.s.	-	90
$C_{36}H_{66}MgO_4$	Magnesium(II) oleate	1.67[h]	B	30	0	89
		2.19	B[i]	20	5	1513
		2.99	B	25	-	1455

h) For criticism see Nature 171, 1075 (1953).   i) Plus water.

FORMULA	COMPOUND NAME	μ, D	State, or Solv.	t, °C	Method or P_a, % of P_e	Reference
$C_{36}H_{66}NiO_4$	Nickel(II) oleate	2.67[hh]	B	30	0	89
$C_{36}H_{66}O_4Pb$	Lead(II) oleate	4.32[hh]	B	30	0	89
$C_{36}H_{66}O_4Zn$	Zinc(II) oleate	0.29[hh]	B	30	0	89
$C_{36}H_{66}N_2NiO_8P_2$	*Bis*(tricyclohexylphosphine oxide)nickel dinitrate	11.21	B	n.s.	-	922
$C_{36}H_{67}AlO_5$(?)	Aluminum(III) dioleate	2.59 3.11	B[ii] B	20 n.s.	5 -	1513 90
$C_{36}H_{69}AlO_6$	Aluminum(III) laurate	4.09	B	25	-	1455
$C_{36}H_{70}CdO_4$	Cadmium(II) stearate	4.84	B	25	-	1455
$C_{36}H_{70}FeO_4$	Iron(II) stearate	3.49	B	n.s.	-	90
$C_{36}H_{70}MgO_4$	Magnesium(II) stearate	3.70	B	25	-	1455
$C_{36}H_{70}NiO_4S_2$	*Bis*(hexadecylthioglycolato)nickel	2.55±0.1	B	25	0	420
$C_{36}H_{74}S_2$	*n*-Octadecyl disulfide	2.07	B	25	-	2030
$C_{36}H_{76}O_4Si$	Tetranonyl orthosilicate	1.77	B	20	5	53
$C_{36}H_{76}O_4Ti$	Tetranonyl orthotitanate	1.70	B	20	5	51,53
$C_{36}H_{90}O_7Si_8$	Octadecaethyloctasiloxane	1.67	Hx	25	0	627
$C_{37}H_{27}$	Tribiphenylmethyl	<0.7	B	8-35	-	883
$C_{37}H_{76}O_4$	Tetranonyl orthocarbonate	1.18	B	25	5	56
$C_{38}H_{28}O$	2,3,4,5,6-Pentaphenylacetophenone	1.86	B[j]	20	0	254

hh) For criticism see Nature 171, 1075 (1953). ii) Plus water. j) Saturated with water.

FORMULA	COMPOUND NAME	μ, D	State, or Solv.	t, °C	Method or P_a, % of P_e	Reference
$C_{38}H_{30}$	Hexaphenylethane	<0.7	B	8-35	-	883
$C_{38}H_{30}As_2NiO_2$	*Bis*(triphenylarsine)nickel dicarbonyl	3.34±0.01	B	n.s.	-	291
$C_{38}H_{30}NiO_2P_2$	*Bis*(triphenylphosphine)nickel dicarbonyl	3.82±0.01	B	n.s.	-	290-1
$C_{38}H_{33}NP_2PtS$	*trans*-Methylthiocyanato*bis*(triphenylphosphine)platinum	6.5±0.1	B	25	15	293
$C_{38}H_{35}IP_2Pt$	*trans*-*Bis*(triphenylphosphine)iodo(ethyl)-platinum	4.25	B	25	15	294
$C_{38}H_{36}P_2Pt$	*cis*-Dimethyl*bis*(triphenylphosphine)-platinum	5.45±0.1	B	25	15	293
$C_{38}H_{52}NiP_2$	*trans*-*Bis*(diethylphenylphosphine)di-(mesityl)nickel	0	B	25	-	295
$C_{38}H_{74}O_4$	Ethylene glycol distearate	2.28	B	30	0	590
$C_{39}H_{36}Si_2$	1,1,1-Triphenyl-2,2,2-tri-*p*-tolyldisilane	0.77±0.02	B	30	0	1488
$C_{39}H_{74}O_6$	Glycerol trilaurate	2.61	B	30	15	1468
$C_{40}H_{56}BN$	Tetra-*n*-butylammoniumtetraphenylboride	17.3±1	ClB	25	-	582
$C_{40}H_{84}O_4Si$	Tetradecyl orthosilicate	1.71	B	20	5	53
$C_{40}H_{100}O_8Si_9$	Eicosaethylnonasiloxane	1.78	Hx	25	0	627
$C_{41}H_{84}O_4$	Tetradecyl orthocarbonate	1.07	B	25	5	56
$C_{42}H_{42}Br_2NiO_6P_2$	*Bis*(tri-*p*-methoxyphenylphosphine)nickel dibromide	9.0±0.3	B	(25)	0	231
$C_{42}H_{42}Br_2NiP_2$	*Bis*(tri-*p*-tolylphosphine)nickel dibromide	8.6±0.2	B	(25)	0	231

FORMULA	COMPOUND NAME	μ, D	State, or Solv.	t, °C	Method or P_a, % of P_e	Reference
$C_{42}H_{42}Cl_2NiO_6P_2$	*Bis*(tri-*p*-methoxyphenylphosphine)nickel dichloride	7.4±0.2	B	(25)	0	231
$C_{42}H_{42}I_2NiO_6P_2$	*Bis*(tri-*p*-methoxyphenylphosphine)nickel diiodide	9.2±0.1	B	(25)	0	231
$C_{42}H_{42}I_2NiP_2$	*Bis*(tri-*p*-tolylphosphine)nickel diiodide	9.8±0.3	B	(25)	0	231
$C_{42}H_{42}O_6W$	Hexa-*p*-tolyltungstate	2.12	B	25	5	60
$C_{42}H_{81}AlO_6$	Aluminum(III) myristate	4.45	B	25	-	1455
$C_{43}H_{30}O$	2,3,4,5,6-Pentaphenylbenzophenone	1.93	B[k]	20	0	255
$C_{44}H_{30}Mn_2O_{14}P_2$	Triphenylphosphite - tetracarbonylmanganese dimer	2.67±0.11	B	25	25	712
$C_{44}H_{30}N_4$	α,β,γ,δ-Tetraphenylporphine	~0	B	25	-	996
$C_{44}H_{42}N_2NiP_2S_2$	*Bis*(tri-*p*-tolylphosphine)nickel dithiocyanate	2.3±0.3	B	(25)	0	231
$C_{45}H_{34}O$	2',3,3',4',5',6'-Hexaphenylpropiophenone	3.00	B[k]	25	0	255
$C_{48}H_{40}P_2Pt$	*cis*-*Bis*(triphenylphosphine)diphenylplatinum	7.0	B	25	15	294
$C_{48}H_{72}Si_2$	1,1,1-Triphenyl-2,2,2-trimenthyldisilane	0.64±0.02	B	30	0	1488
$C_{48}H_{93}AlO_6$	Aluminum(III) palmitate	5.26	B	25	-	1455
$C_{48}H_{108}Cl_6Cr_2P_4$	*Bis*(trichloro*bis*(tri-*n*-butylphosphine)-chromium)	8.00	B	20	15	810
$C_{51}H_{98}O_6$	Glycerol tripalmitate	2.80	B	23	-	1869
		2.93	D	23	-	1869

k) Saturated with water.

FORMULA	COMPOUND NAME	μ, D	State, or Solv.	t, °C	Method or P_a, % of P_e	Reference
$C_{52}H_{40}NiP_2$	*trans-Bis*(triphenylphosphine)di(phenylethynyl)nickel	1.4	B	25	-	295
$C_{54}H_{45}Br_3Cu_3O_9P_3$	Diphenyl phenylphosphate - copper bromide complex trimer	3.39	B	n.s.	-	57
$C_{54}H_{45}Cl_3Cu_3O_9P_3$	Diphenyl phenylphosphate - copper chloride complex trimer	3.21	B	n.s.	-	57
$C_{54}H_{99}AlO_6$	Aluminum(III) oleate	3.81	B	n.s.	-	90
		4.32	B	25	-	1455
$C_{54}H_{105}AlO_6$	Aluminum(III) stearate	5.53	B	25	-	1455
$C_{57}H_{104}O_6$	Glycerol trioleate	3.11	B	23	-	1869, 1871,2046
		3.154	B	2	0	2044-5
		3.5188	B	30	-	1465
		3.09	D	19,23	0	1869, 1871,2046
$C_{57}H_{104}O_9$	Glycerol triricinoleate	4.147	B	29	-	1465
$C_{57}H_{110}O_6$	Glycerol tristearate	2.7	B	35	5	1874
		2.72	B	30	0	590,1468
		2.86	B	23	-	1869,1871
		2.874	B	29	-	1465
		2.98	D	23	-	1869,1871
$C_{64}H_{40}ClFeN_8$	Octaphenylporphyrazine - iron chloride complex	5.80	B	20-60	-	1518
$C_{64}H_{41}ClN_8$	Chlorophenylheptaphenylporphyrazine	2.73	B	60	-	1518
		2.78	B	40	-	1518
		2.92	B	20	-	1518

FORMULA	COMPOUND NAME	μ, D	State, or Solv.	t, °C	Method or P_a, % of P_e	Reference
$C_{72}H_{44}BMnN_6$	Tetraphenylboride of hexa(*p*-tolylisonitrile)manganese	22.29	B	23.9	-	1670
$C_{72}H_{44}BMnN_6O_6$	Tetraphenylboride of hexa(*p*-methoxyisocyanobenzene)manganese	22.37	B	23.9	-	1670
$C_{102}H_{74}BMnN_6$	Tetraphenylboride of hexa(4-isocyanobiphenyl)manganese	23.57	B	23.9	-	1670

TABLE III

COMPOUNDS OF UNSPECIFIED FORMULA

FORMULA	COMPOUND NAME	μ, D	State, or Solv.	t, °C	Method or P_a, % of P_e	Reference
Oils						
	Castor	3.7	B	30	5	1874
		3.71	B	26	-	1467
		3.83	B	n.s.	0	177
		3.65	kk	25	0	1124
	Coconut	2.82	B	n.s.	0	177
		2.85	B	26	-	1467
	Linseed	3.0	B	27	5	1874
		3.02	B	n.s.	0	177
	Lubricating, mineral	0.22-0.41	liq	20	-	695
		0.24-0.49	liq	50	-	695
	naphthenic distillate	1.23-1.64	liq	20-80	-	806
		0.22-0.45	B	20-50	-	695
	Olive	2.25-2.30	liq	20-50	-	695
		2.26-2.35	B	20-50	-	695
		3.04	B	n.s.	5	177
		3.06	B	26	-	1467
	Peanut	2.30-2.35	liq	20-50	-	695
		2.31-2.41	B	20-50	-	695
	Poppy	3.09	B	n.s.	0	177
	Rape	2.59	liq	20	-	695
		2.72	liq	50	-	695
		2.61-2.85	B	20-50	-	695
		2.77	B	n.s.	0	177
	Sesame	2.93	B	26	-	1467
		2.96	B	n.s.	0	177

kk) Gasoline.

FORMULA	COMPOUND NAME	μ, D	State, or Solv.	t, °C	Method or P_a, % of P_t	Reference
Oils (contd.)	Tung	2.1-2.2	B	20-100	-	767
		2.30	B	(20)	-	182
		2.8	B	25	5	1874
		2.95	m	25	0	1124
Polymers						
$(C_2ClF_3)_n$	Poly(chlorotrifluoroethylene)	0.54-0.60[mm]	n.s.	n.s.	-	1928
$(C_2H_3Cl)_n$	Poly(vinyl chloride)	1.67-1.75[p]	D	25-65	0	803
		2.0	n.s.	n.s.	-	2149
		52±2	q	-34 to 16	-	392
		59±1	r	-34 to 16	-	393
$(C_2H_4)_n$	Poly(ethylene)	0	n.s.	n.s.	-	2149
$(C_2H_4O)_n$	Poly(ethylene glycol) MW = 62-400	2.0-3.7	B	n.s.	-	1172
	200	3.64	B	25	-	73
	280	4.10	B	25	-	73
	800	4.9	B	n.s.	-	1172
	800	5.46	B	25	-	73
	1000	5.3	B	n.s.	-	1172
	1500	6.4	B	n.s.	-	1172
	1500	6.97	B	25	-	73
	3440	10.1	B	25	-	73
	("Carbowax 4000") 3750	9.9	B	44	5	1913
	6750	14.0	B	25	-	73
$(C_3H_3N)_n$	Polyacrylonitrile	3.4	n.s.	n.s.	-	2149
$(C_3H_6O)_n$	Poly(vinyl methyl ether)	1.2	n.s.	n.s.	-	2149
$(C_4H_5Cl)_n$	Poly(2-chloro-1,3-butadiene) "Polychloroprene" (MW = 23,800)	24.3	n.s.	n.s.	-	1062
	"Neoprene GN"	1.99	Solid	20-60	-	1715

m) Kerosene. mm) Depending on molecular weight. p) Per monomer unit. q) Tetrahydrofuran.
r) Heptane-tetrahydrofuran.

FORMULA	COMPOUND NAME	μ, D	State, or Solv.	t, °C	Method or P_a, % of P_e	Reference
$(C_4H_6O_2)_n$	Poly(vinyl acetate)	1.7^s	n.s.	n.s.	-	2149
	MW = 10,000	18.4	B	n.s.	-	1076
		18.3	CCl_4	n.s.	-	1076
	18,000	24.0	B	n.s.	-	1076
		24.5	CCl_4	n.s.	-	1076
	24,200	28.9	n.s.	n.s.	-	1062
	45,000	36.9	B	n.s.	-	1076
	60,000	44.7	n.s.	n.s.	-	1062
$(C_4H_8)_n$	Poly(isobutylene)	0	n.s.	n.s.	-	2149
$(C_5H_8)_n$	Rubber	2.47	B	25	-	877
	gel fraction	2.94	B	25	-	877
	sol fraction	2.82	B	25	-	877
		0.73	Ether	25	-	877
	crepe - lightly milled	27	B	n.s.	-	1454
	- strongly milled	33	B	n.s.	-	1454
	- first milling (MW = 20,000-200,000)	$14.5\text{-}45.7^t$	B	25	-	1456
	- second milling (MW = 20,000-200,000)	$18.4\text{-}58.1^t$	B	25	-	1456
	- second milling + reprecipitation	$17.6\text{-}55.8^t$	B	25	-	1456
	smoked sheet - first milling	$14.6\text{-}46.0^t$	B	25	-	1456
	extracted in acetone	0.71	B	25	-	878
	extracted after ultraviolet irradiation	1.10	B	25	-	878
	natural rubber hydrochloride	1.68	B	50	-	805
	squalene hydrochloride	1.76	B	50	-	805
$(C_5H_8O_2)_n$	Poly(methyl methacrylate)	1.3^s	B	n.s.	-	934
		1.7	n.s.	n.s.	-	2149
		1.95	B	18	-	1310

s) Per monomer unit. t) Not corrected for new values of N and k.

FORMULA	COMPOUND NAME	μ, D	State, or Solv.	t, °C	Method or P_a, % of P_e	Reference
$(C_5H_8O_2)_m$-$(C_8H_8)_n$	Copolymer of styrene and methyl methacrylate	1.65[tt]	B	20	-	953
$(C_6H_7N_3O_{11})_n$	Nitrocellulose[u]	2-7.2	n.s.	n.s.	-	833
$(C_6H_{10}O_2)_n$	Poly(ethyl methacrylate)	2.13	B	14	-	1310
$(C_6H_{12}O)_n$	Poly(vinyl isobutyl ether) atactic	1.07±0.03[tt]	B	25	-	804,1925
		1.11±0.03[tt]	B	50	-	804,1925
	isotactic	1.16±0.03[tt]	B	25	-	804,1925
		1.21±0.03[tt]	B	50	-	804,1925
$(C_7H_{12}O_2)_n$	Poly(propyl methacrylate)	2.13	B	14	-	1310
$(C_8H_7Br)_n$	Poly(*p*-bromostyrene)	1.30[tt]	B	20	-	956
		1.34[tt]	B	35	-	956
		1.37[tt]	B	50	-	956
$(C_8H_7Cl)_n$	Poly(*p*-chlorostyrene) atactic	1.66	Solid	130	-	2147
		1.33[tt]	B	5	-	956
		1.37[tt]	B	20	-	956
		1.38	B	20	-	954
		1.40[tt]	B	35	-	956
		1.42[tt]	B	50	-	956
	atactic	1.45±0.02	B	30,50	-	397
$(C_8H_7Cl)_m$-$(C_8H_8)_n$	Copolymers of styrene and *p*-chlorostyrene atactic					
	mole fraction styrene = 0.988	0.38	Solid	130	-	2147
	= 0.965	0.53	Solid	130	-	2147
	= 0.874	0.90	Solid	130	-	2147
	= 0.682	1.29	Solid	130	-	2147
	= 0.477	1.48	Solid	130	-	2147
	= 0.258	1.60	Solid	130	-	2147
	80% styrene	1.68±0.02	B	20	-	954

tt) Per monomer unit. u) Degree of nitration varies.

FORMULA	COMPOUND NAME	μ, D	State, or Solv.	t, °C	Method or P_a, % of P_e	Reference
$(C_8H_8)_n$	Polystyrene atactic	0.26_3	Solid	130	-	2147
	atactic	0.36	Tol	38.4	-	974
	isotactic	0.43_5	Tol	38.4	-	974
	MW = 2,820	6.37	B	25	0	561
	7,750	8.04	B	25	0	561
	12,900	2.9	n.s.	n.s.	-	1062
	23,000	8.67	B	25	0	561
	42,900	5.7	n.s.	n.s.	-	1062
$(C_8H_{14}O_2)_n$	Poly(butyl methacrylate)	2.16	B	14	-	1310
$(C_9H_8)_n$	Polyindene MW = 116	0.40	B	25	0	561
	603	1.38	n.s.	n.s.	-	1062
	855	1.69	B	25	0	561
	855	1.76	n.s.	n.s.	-	1062
	1039	1.76	n.s.	n.s.	-	1062
	1135	1.93	n.s.	n.s.	-	1062
	1490	2.13	B	25	0	561
	1490	2.20	n.s.	n.s.	-	1062
$(C_9H_{16}O_2)_n$	Poly(amyl methacrylate)	2.13	B	18	-	1310
$(C_{10}H_8O_4)_n$	Poly(ethylene *o*-phthalate) MW = 1022	6.2	B	25	5	71
$(C_{10}H_{18}O_2)_n$	Poly(hexyl methacrylate)	2.15	B	18	-	1310
	Poly(10-hydroxy-*n*-decanoic acid)					
	MW = 850	5.0	B	25	0	207
	2,040	6.7	B	25	0	207
	4,090	10.2	B	25	0	207
	7,830	12.4	B	25	0	207
	9,000	15.7	B	25	0	207
	13,900	19.0	B	25	0	207
$(C_{11}H_{20}O_2)_n$	Poly(heptyl methacrylate)	2.15	B	18	-	1310

FORMULA	COMPOUND NAME	μ, D	State, or Solv.	t, °C	Method or P_a, % of P_e	Reference
$(C_{12}H_{15}NO_4)_n$	Poly(γ-benzyl-L-glutamate) MW = 154,000	1650±50	D	30	-	2053
	= 70,000	1040±50	W	n.s.	-	2054
	= 86,000	1404	W	25	-	2055
	= 102,000	1570±50	W	n.s.	-	2054
	= 154,000	2460±50	W	n.s.	-	2054
	= 180,000	2910	W	25	-	2055
		3.4[x]	W	n.s.	-	2054
$(C_{12}H_{18}O_8)_n$	Triacetylcellulose	1.97[y]	$CHCl_3$	20	0	1676
$(C_{12}H_{22}O_5)_n$	Triethylcellulose	3.26[y]	B	20	0	1063,1677
		3.69[y]	B	35	0	1063
		3.92[y]	B	50	0	1063,1677
	MW = 5,000-100,000	9.82-43.9[a]	B	25	-	1456
$(C_{27}H_{28}O_5)_n$	Tribenzylcellulose	2.56[y]	B	20	0	1063,1677
		2.64[y]	B	35	0	1063
		2.69[y]	B	50	0	1063
		3.11	B	20	0	1678
		3.17	B	35	0	1678
		3.32	B	50	0	1678
	MW = 5,000-100,000	9.65-42.8[a]	B	25	-	1456
$[(CH_3)_3Si(OSi(CH_3)_2)_nCH_3]_x$	linear Poly(methylpolysiloxane)					
	n = 1	0.43	liq	25	-	83
		0.74	liq	20	-	1698
	n = 2	0.80	liq	25	-	83
		0.99	liq	20	-	1698
	n = 3	0.43	liq	25	-	83
		1.22	liq	20	-	1698
	n = 4	0.93	liq	25	-	83
		1.40	liq	20	-	1698
	n = 5	1.167	liq	25	-	83
		1.58	liq	20	-	1698

w) Ethylene dichloride. x) Per monomer unit - in direction of helical axis. y) Per monomer unit.
a) Not corrected for new values of N and k.

FORMULA	COMPOUND NAME	μ, D	State, or Solv.	t, °C	Method or P_a, % of P_t	Reference
	n = 6	1.143	liq	25	-	83
	n = 7	1.107	liq	25	-	83
$[(CH_3)_2SiO]_n$	cyclic Poly(methylpolysiloxane)					
	n = 4	1.09	liq	20	-	1698
	n = 5	1.35	liq	20	-	1698
	n = 6	1.56	liq	20	-	1698
	n = 7	1.78	liq	20	-	1698
	n = 8	1.96	liq	20	-	1698
$(C_{54}H_{73}O_5)_n$	Tripalmitylcellulose	1.93	B	20	0	1678
		2.03	B	35	0	1678
		2.38	B	50	0	1678
Formula Uncertain	Cellulose tosylate	30.6	D	n.s.	10	aa
	Poly(ethyl to heptyl esters of methacrylic acid)	2.13-2.16	B	18	-	1310
Proteins						
	Albumin, egg	250	H_2O	25	-	1433
	Carboxyhemoglobin	470-500	H_2O	25	-	1431
	horse	480	H_2O	25	-	1433
	pig	410	H_2O	20	-	1433
	Edestin	1400	H_2O	n.s.	-	1432-3
	Glutin	46	H_2O	22	-	460
	Gliadin	13.5	b	20	RT	75
		190	c	25	-	1433
	Hemoglobin	47	H_2O	11	-	460

aa) Helv. Chim. Acta 23, 627-49 (1940). b) Ethanol. c) Ethanol plus water.

FORMULA	COMPOUND NAME	μ, D	State, or Solv.	t, °C	Method or P_a, % of P_e	Reference
Proteins (contd.)	Insulin	300	d	25	-	1433
		310	e	25	-	1433
		360	f	25	-	1433
	β-Lactoglobulin	730	g	25	-	505,1433
		770	g	0	-	505,1433
		790±26	h	20	RT	1731
	Myoglobulin	170	H_2O	25	-	1433
	Ovalbumin	44	H_2O	18	-	460
	Secalin	440	ii	25	-	1433
	Serum albumin	40	H_2O	18	-	460
		380	H_2O	n.s.	-	1432
		280	H_2O	25	-	504
		510	H_2O	25	-	504
	γ-Serum pseudoglobulin	1100	H_2O	25	-	1433
		1200	H_2O	n.s.	-	504,1432
		1300	H_2O	0	-	1433
	Zein	~60	i	0-80	-	2151
		380	ii	25	-	1433
Miscellaneous	Gasoline	0-0.2	liq	25	-	1129
	Methylbenzylpenicillinate	4.73	D	25	0	1005
	Musk ambrette	3.4_8	B	25	0	1071
	Rosin	8.8	B	25	0	1126

d) Propanol. e) 10% H_2O-90% propanol. f) 20% H_2O-80% propanol. g) Glycine plus water.
h) Glycine. i) Propanol plus water. ii) Ethanol plus water.

FORMULA	COMPOUND NAME	μ, D	State, or Solv.	t, °C	Method or P_a, % of P_e	Reference
	Sesamin	1.78	B	20	-	1655
	Starch	~56.5	H_2O	n.s.	-	425
	Wax					
	carnauba (North Country No. 3, m.p. = 79°C)	4.8	B	24	0	253
	montan (American No. 16)	4.5	B	24	0	253
	ouricuri (m.p. = 69°C)	7.5	B	24	0	253

APPENDIX A

APPROXIMATE DIPOLE MOMENTS OF SOME COMMON CHEMICALS

These values represent an average of experimental measurements; see the main tables for actual results. Symbols: g = gas, l = liquid, B = benzene solution.

acetaldehyde	2.5(B)	methyl benzoate	1.85(B)
acetamide	3.7(B)	methyl ether	1.25(B)
acetic acid	1-1.5(B)	morpholine	1.5(B)
acetone	3.0(l), 2.7(B)	nitric oxide	0.15(g)
acetonitrile	3.5(B)	nitrobenzene	4.0(B)
acetophenone	2.9(B)	nitrogen dioxide	0.4(g)
acetylacetone	2.8(B)	nitromethane	3.1(B)
acetyl chloride	2.4(B)	phenol	1.6(B)
acrolein	2.9(B)	propylene oxide	1.95(B)
acrylonitrile	3.5(B)	pyridine	2.3(l), 2.2(B)
ammonia	0.9(l)	pyrrole	1.8(B)
aniline	1.55(B)	quinoline	2.1(l)
anisole	1.2(l,B)	styrene	0.1(l)
benzamide	3.65(B)	tetrahydrofuran	1.7(B)
benzene	0(g,l)	thiophene	0.5(l,B)
benzonitrile	3.2(l), 3.9(B)	toluene	0.4(l,B)
benzophenone	2.7(B)	triethylamine	0.85(l,B)
bromobenzene	1.5(l,B)	1,3,5-trinitrobenzene	0.4-0.8(B)
bromoethane	1.8(B)	s-trioxane	2.2(B)
bromoform	1.0(B)	water	2-3(l), 1.8(B)
carbon dioxide	0(g)		
carbon disulfide	0(l,B)		
carbon tetrahalides	0(l,B)		
carbonyl sulfide	0.71(g)		
chloroacetic acid	2.3(B)		
chlorobenzene	1.55(l,B)		
chloroethane	2.0(B)		
chloroform	1.1(B)		
cyclohexanol	1.8(B)		
1,2-dibromoethane	1.2(B)		
dibromomethane	1.85(l)		
1,2-dichloroethane	1.75(B)		
dichloromethane	1.8(l)		
diethylamine	1.1(l,B)		
dimethylaniline	1.55(B)		
dimethylformamide	3.8(l,B)		
dimethylsulfoxide	3.9(B)		
dioxane	0.4(B)		
ethanol	1.7(B)		
ethyl acetate	1.85(B)		
ethylamine	1.3(B)		
ethylbenzene	0.35(l)		
ethylene carbonate	4.8(B)		
ethylene glycol	2(B)		
ethylene oxide	1.9(l)		
ethyl ether	1.25(B)		
formamide	3.4(B)		
formic acid	1.5(B)		
furan	0.65(B)		
hexane	0(g,l)		
hydrazine	3.0(l)		
hydrogen peroxide	2.1(dioxane)		
hydrogen sulfide	0.95(g)		
iodoethane	1.75(B)		
methanol	1.65(B)		

APPENDIX B

APPROXIMATE DIPOLE MOMENTS FOR SOME HOMOLOGOUS SERIES

These values represent an average of experimental measurements; see the main tables for actual results. In benzene solution unless marked; g = gas, l = liquid.

	Acid	Alcohol	Aldehyde	Amide	Amine	Bromo-	Chloro-	Fluoro-	Iodo-	Methyl Ester	Nitrile	Nitro-
Total Carbons in Chain												
1	1.5	1.65	2.2 (g)	3.4	1.3 (g)	1.6	1.7	1.8 (g)	1.4	1.8 (g)	2.6	3.1 (g)
2	1-1.5	1.7	2.5	3.6	1.3	1.9	2.0	1.9 (g)	1.75	1.75	3.5	3.2
3	1-1.5	1.65	2.5	3.4	1.35	1.95	1.95	-	1.85	1.7	3.6	3.7 (g)
4	1-1.8	1.7	2.5	3.5	1.4	2.0	1.95	-	1.9	1.7	3.5	3.3
5	-	1.65	2.6	-	1.55	2.0	1.95	1.85	1.9	1.6	3.6	-
6	1.2 (l)	1.65	-	-	1.3-1.6	2.0	1.95	-	1.9	-	3.5[b]	-
8	1.2 (l)	1.7	-	-	1.4	2.0	2.1	-	1.9[a]	-	-	-
10	-	1.65	-	-	1.3	1.9	-	-	-	-	-	-
12	0.8	1.5	-	-	-	1.9 (l)	1.95	-	1.9	-	-	-
Other Hydrocarbon Groups												
Benzene Ring	1-2	1.5	2.85	3.65	1.55	1.55	1.55	1.45	1.3	1.8	3.95	4.0
Cyclohexyl Ring	0.9	1.8	3.0[c]	-	1.3	2.25	2.1	-	2.0	-	-	-

a) In CCl_4. b) Isohexanenitrile. c) A ketone.

APPENDIX C

SELECTION OF RECOMMENDED VALUES

GENERAL PROCEDURE

Many users of dipole moments require a single value for a formula or for a comparison table. For such situations, finding a list of 6 to 30 values in the tables is not helpful. Thus, it seemed desirable to select, for those compounds which have been measured many times, a single value as "recommended." The first criterion for such treatment was that six separate workers had reported a value for the same state or solvent and the same (± 5°C) temperature. There were 106 such compounds. Next, the original data (dielectric constant, density and concentration for solutions, dielectric constant or polarization, and temperature for pure substances) were recorded for each compound. At this stage, a second requirement was that at least four workers' data be available and that there be a total of 15 or more individual sets of data. Approximately 35 compounds were dropped because they did not meet this requirement.

The data were then punched on cards, and the dipole moment was calculated. A detailed description is given below. Actually, two calculations were made. The data of each worker were computed, and the entire set of all data was also treated. This allowed comparison of the machine value for each worker, as well as a first trial value of selected μ value. The values of each worker were required to be within 3% of the reported values. When this occurred for each worker *and* when the accompanying selected value was in reasonable agreement with most of the reported values, the selection was complete.

More frequently there was too large a discrepancy for one or more authors. A reconsideration was made, usually by plotting ε versus concentration. This process located errors in transcribing data, typographic mistakes in the journal entries, and, most important, curvature of the ε versus concentration line. Errors of the first two types were corrected, and those due to curvature were eliminated by discarding the points for the more concentrated solutions until only those defining a straight line remained. Sometimes these measures reduced the number of data sets below 15, and the compound was deleted. Occasionally, a second consideration was required. The final result was that 52 compounds survived and have recommended values given in the Tables.

DETAILED PROCEDURE

For solutions, the Halverstadt-Kumler equation[1] was used in the following form:

$$P_{2\infty} = \frac{3\alpha' v_1}{(\epsilon_1 + 2)^2} M_1 + (M_2 v_1 + M_1 \beta') \frac{(\epsilon_1 - 1)}{(\epsilon_1 + 2)}.$$

where

ε = dielectric constant,
d = density, g/ml,
v = specific volume, ml/g = 1/d,
M = molecular weight,
w = weight fraction.
n = mole fraction,

---

[1] I. F. Halverstadt and W. D. Kumler, J. Am. Chem. Soc. 64, 2988 (1942).

N = Avogadro's number,
k = Boltzmann's constant,
α' = $d\epsilon_{12}/dn_2$ from $\epsilon_{12} = \epsilon_1 + \alpha' n_2$,
β' = dv_{12}/dn_2 from $v_{12} = v_1 + \beta' n_2$,
μ = dipole moment, in Debyes,
T = absolute temperature,
P_e = electronic polarization,
P_a = atomic polarization,

and where subscripts 1, 2, and 12 designate solvent, solute, and solution.

Many workers report solution concentration in weight fraction and a few use molarity. These were converted to mole fraction (by the computer) using the following relations.

Weight fraction, (w):

$$\frac{1}{n_2} = \left(\frac{1}{w_2} - 1\right) \frac{M_2}{M_1} + 1.$$

Molarity, ($\underline{M}$):

$$\frac{1}{n_2} = \frac{\underline{M} M_2}{\underline{M}(M_2 - M_1) + 1000 d_1}.$$

For gases, the Debye equation sufficed:

$$P = A + \frac{B}{T}.$$

The constants were taken from DuMond and Cohen[2]:

$$k = 1.38032 \times 10^{-16},$$

$$N = 6.0251 \times 10^{23},$$

$$0°C = 273.15°K,$$

$$\pi = 3.141592654.$$

The treatment of atomic polarization, P_a, is always troublesome. For these summary calculations the bond electronic polarization values recently suggested by Le Fèvre[3] were used. In every case they led to P_e values near those derived by individual workers. Atomic polarization was then set equal to 10% P_e, or, for a few cases, at zero.

Where different temperatures were used during the various measurements, the most frequent was chosen. The effect of temperature is not large for a 5°C change, amounting perhaps to 0.01 D, an amount smaller than the error of the values which results from these computations. Molecular weights were based on the $^{12}C = 12.0000$ atomic weight scale adopted in 1961.

The program was a least squares linear correlation treatment worked out by G. G. Bejarano, who supervised the calculations on a Datatron 205 computer. Two straight lines, ϵ versus concentration and $v(= 1/d)$ versus concentration, were found along with

---

[2]J. W. DuMond and E. R. Cohen, Rev. Mod. Phys. 20, 82 (1948).

[3]R. J. W. Le Fèvre and K. D. Steel, Chem. and Ind. 1961, 670.

the slope and intercept. The computer also printed the deviations, standard error, and correlation coefficient. The latter three were used to monitor the quality of fit between measured and calculated values. These slopes and intercepts were used to derive P and were then combined with P_e, P_a, and the constants to give dipole moment via the equation

$$\mu = \sqrt{\frac{9kT}{4\pi N}\,(P_{2\infty}-P_e-P_a)} = 0.012812 \times 10^{-18}\ \sqrt{P_oT}.$$

The procedure for gases is identical in theory but simpler in practice because only one straight line is needed and P is determined directly.

DELETED DATA

Even for the compounds that survived all the tests and have recommended values given in the main tables, there are some references that were not included. The following list indicates the *omitted* papers and those for which some part of the data was discarded. The notations have these meanings:

No Data	Dielectric constant and/or density not given in the original work.
Wrong Kind	The data are not suitable for the equation used. For example, all microwave data are omitted for this reason.
No Fit	The data seemed out of line.
Points Omitted	Some data points, usually at high concentration, were omitted. The usual reason for this was that the ϵ versus concentration line curved.

Substance	No Data	Wrong Kind	No Fit	Points Omitted
ClH (g) Hydrogen chloride	189, 326, 1556	466, 552, 1662		
H_2O (g) Water	450	352, 591		
H_3N (g) Ammonia	2001	324, 531, 910, 1959		
N_2O (g) Nitrous oxide	323, 1768, 1963		2071	573
O_2S (g) Sulfur dioxide	189, 925	351, 551, 1302	1804	
CHN (g) Hydrogen cyanide	176, 576, 1769, 1771, 1784, 2069		537	
CH_3Br (g) Bromomethane		1768		

Substance	No Data	Wrong Kind	No Fit	Points Omitted
CH_3Cl (g) Chloromethane	1556, 1689	884, 1768, 1963		1235
CH_3I (g) Iodomethane	1768, 1963			
CH_4O (g) Methanol	242	1878		
CH_4O (B) Methanol	977, 1387 1665, 2051			1108, 1241a, 1875, 2039, 2140
CH_5N (g) Methylamine	1131, 1135, 1867		572	
CO (g) Carbon monoxide	2075	189, 243, 1962	2071	
CO_2 (g) Carbon dioxide	795, 1602, 2070	2002	2071	
C_2H_5Cl (g) Chloroethane	837, 1556 1689	93-4	1564	
C_2H_6O (B) Ethanol	977	1951	2140	1108, 1875
C_3H_6O (g) Acetone		1916, 1962		
C_3H_9N (g) Trimethylamine	139	179, 1144	572	
C_4H_4S (B) Thiophene	1183, 1626			666, 716, 902
$C_4H_{10}O$ (g) Ethyl ether	745			1687
$C_4H_{10}O$ (B) Ethyl ether	422, 671, 683, 1304 1747, 1841		816, 2133	1039, 1653, 1926, 2121
$C_4H_{10}O$ (B) 1-Butanol	977, 1295, 1401	194, 1239		1039, 1830
$C_4H_{10}O$ (B) *t*-Butanol	401, 697			413, 1807
$C_4H_{11}N$ (B) *n*-Butylamine	336, 1579, 1729			
$C_6H_4BrNO_2$ (B) *p*-Bromo-nitrobenzene	481, 570-1			
$C_6H_4Br_2$ (B) *o*-Dibromo-benzene	454-5			

Substance	No Data	Wrong Kind	No Fit	Points Omitted
$C_6H_4ClNO_2$ (B) *m*-Chloronitrobenzene	750-1, 1841			2057
$C_6H_4ClNO_2$ (B) *p*-Chloronitrobenzene	750-1			1949
$C_6H_4Cl_2$ (B) *o*-Dichlorobenzene	454-5, 976, 1262, 1824			
$C_6H_4N_2O_4$ (B) *m*-Dinitrobenzene	750-1, 963, 977, 1673			
C_6H_5Cl (g) Chlorobenzene	1097			
C_6H_5Cl (B) Chlorobenzene	226, 548, 581, 589, 683, 750-1, 904, 976-7, 1082, 1304, 1366, 1384, 1580, 1710, 1824, 1969, 2116	921	1376, 1542	1861
C_6H_5ClN (B) *p*-Chloroaniline	671	472		1864*
C_6H_5F (B) Fluorobenzene	548, 1378, 2017-8			
$C_6H_5NO_2$ (B) Nitrobenzene	589, 683, 838, 976-7, 979, 1080, 1082, 1365, 1377, 1384, 1580, 2020	1513 1951	1375	281 1039
$C_6H_6N_2O_2$ (B) *m*-Nitroaniline	751	1673		1950
$C_6H_6N_2O_2$ (B) *p*-Nitroaniline	751, 1351	375, 471	1950	
$C_6H_6N_2O_2$ (D) *p*-Nitroaniline	471, 2026			
C_6H_6O (CCl_4) Phenol	1221, 1295-6 1718, 1987			
C_6H_7N (B) Aniline	589, 683, 751, 797, 1373, 1384, 1387, 1710, 2051			722, 1095
$C_6H_8N_2$ (B) *p*-Phenylenediamine	2088, 2112			
$C_6H_{10}O$ (B) Cyclohexanone	302, 1113, 1513, 2136			
C_7H_5N (B) Benzonitrile	671, 674, 1856			
$C_7H_7NO_2$ (B) *o*-Nitrotoluene	750-1			1949
$C_7H_7NO_2$ (B) *p*-Nitrotoluene	750-1, 1466			

*Journal not available.

Substance	No Data	Wrong Kind	No Fit	Points Omitted
C_7H_8 (B) Toluene	379, 1080, 1824			359
C_7H_8O (B) Anisole	81, 463, 751, 922, 1241b			
C_7H_9N (B) *N*-Methylaniline	336			
C_7H_9N (B) *o*-Toluidine	530			
$C_8H_{10}O_2$ (B) *p*-Dimethoxybenzene	673, 922, 1531			
$C_8H_{11}N$ (B) *N,N*-Dimethylaniline	336, 508, 750			
$C_{12}H_{11}N$ (B) Diphenylamine	336, 1219, 1412	1710		

BIBLIOGRAPHY

1. Abadie, P., and G. Champetier.
Compt. rend. 200, 1590-3 (1935).

2. Acheson, R. M., B. Adcock, G. M. Glover, and L. E. Sutton.
J. Chem. Soc. 1960, 3367-71.

3. Ackman, R. G., W. H. Brown, and G. F. Wright.
J. Org. Chem. 20, 1147-58 (1955).

4. Adams, D. M., J. Chatt, R. G. Guy, and N. Sheppard.
Proc. Chem. Soc. 1960, 179-80.

5. Allen, J. S., and H. Hibbert.
J. Am. Chem. Soc. 56, 1398-1403 (1934).

6. Allenstein, E.
Z. anorg. allgem. Chem. 308, 3-12 (1961).

7. Allinger, J., and N. L. Allinger.
J. Am. Chem. Soc. 81, 5736-40 (1959).

8. ———.
J. Org. Chem. 25, 262-3 (1960).

9. ———, L. E. Geller, and C. Djerassi.
J. Org. Chem. 26, 3521-3 (1961).

10. Allinger, N. L.
J. Am. Chem. Soc. 79, 3443-6 (1957).

11. ———.
Ibid., 80, 1953-5 (1958).

12. ——— and J. Allinger.
J. Org. Chem. 24, 1613-4 (1959).

13. ———, L. A. Freiberg, R. F. Czaja, and N. A. LeBel.
J. Am. Chem. Soc. 82, 5876-82 (1960).

14. Allinger, N. L., J. Allinger, L. E. Geller, and C. Djerassi.
J. Org. Chem. 25, 6-12 (1960).

15. Allinger, N. L., J. Allinger, and N. A. LeBel.
J. Am. Chem. Soc. 82, 2926-7 (1960).

16. Allinger, N. L., H. M. Blatter, M. A. DaRooge, and L. A. Freiberg.
J. Org. Chem. 26, 2550-2 (1961).

17. Allinger, N. L., and M. A. DaRooge.
Tetrahedron Letters 1961, 676-81.

18. Allinger, N. L., and L. A. Freiberg.
J. Am. Chem. Soc. 83, 5028-9 (1961).

19. Allinger, N. L., and S. Greenberg.
J. Am. Chem. Soc. 81, 5733-6 (1959).

20. Alpen, E. L., and W. D. Kumler.
J. Am. Chem. Soc. 72, 5745-7 (1950).

21. Altona, C., C. Romers, and E. Havinga.
Tetrahedron Letters 1959, 16-20.

22. Altshuller, A. P.
J. Phys. Chem. 58, 392-5 (1954).

23. ———.
Ibid., 59, 32-4 (1955).

24. ——— and L. Rosenblum.
J. Am. Chem. Soc. 77, 272-4 (1955).

25. Amble, E.
Phys. Rev. 83, 210 (1951).

26. ——— and B. P. Dailey.
J. Chem. Phys. 18, 1422 (1950).

27. Amble, E., and O. Hassel.
Research (London) 3, Suppl., 52 (1950).

28. American Cyanamid Co.
Brit. 612,385, Nov. 11, 1948.

29. Amphlett, C. B., L. W. Mullinger, and L. F. Thomas.
Trans. Faraday Soc. 44, 927-38 (1948).

30. Anantakrishnan, S. V., and D. S. Rao.
Proc. Indian Acad. Sci. Sect. A 43, 99-105 (1956).

31. ———.
Current Sci. (India) 29, 51-2 (1960).

32. Andersen, F. A., B. Bak, and J. Rastrup-Andersen.
Acta Chem. Scand. 7, 643-51 (1953).

33. Andersen, P., and O. Hassel.
Acta Chem. Scand. 3, 1180-1 (1949).

34. Anderson, A. G., and B. M. Steckler.
J. Am. Chem. Soc. 81, 4941-6 (1959).

35. Anderson, A. I.
Proc. Phys. Soc. (London) 40, 62-70 (1928).

36. Anderson, J. D. C., R. J. W. Le Fèvre, and I. R. Wilson.
J. Chem. Soc. 1949, 2082-8.

37. Andrianov, K. A., and G. E. Golubkov.
Vysokomolekul. Soedin. 1, 1801-4 (1959).

38. Angoletta, M.
Ann. chim. (Rome) 45, 970-6 (1955).

39. Angus, W. R., A. H. Leckie, C. G. Le Fèvre, R. J. W. Le Fèvre,
and A. Wasserman.
J. Chem. Soc. 1935, 1751-5.

40. Angyal, C. L., and S. J. Angyal.
J. Chem. Soc. 1952, 695-7.

41. Angyal, C. L., G. A. Barclay, A. A. Hukins, and R. J. W. Le Fèvre.
J. Chem. Soc. 1951, 2583-8.

42. Angyal, C. L., G. A. Barclay, and R. J. W. Le Fèvre.
J. Chem. Soc. 1950, 3370-3.

43. Angyal, C. L., and R. J. W. Le Fèvre.
J. Chem. Soc. 1950, 562-4.

44. ———.
Ibid., 1952, 1651-3.

45. Angyal, S. J., and C. L. Angyal.
J. Chem. Soc. 1952, 1461-6.

46. Anzilotti, W. F., and B. C. Curran.
J. Am. Chem. Soc. 65, 607-11 (1943).

47. Aparajithan, K., and V. Baliah.
J. Indian Chem. Soc. 36, 159-64 (1959).

48. Applequist, D. E., T. L. Brown, J. P. Kleiman, and S. T. Young.
Chem. and Ind. (London) 1959, 850-1.

49. Applequist, D. E., E. C. Friedrich, and M. T. Rogers.
J. Am. Chem. Soc. 81, 457-8 (1959).

50. Arbuzov, A. E., and P. I. Rakov.
Izv. Akad. Nauk SSSR, Otd. Khim. Nauk 1950, 237-46.

51. Arbuzov, B. A.
J. chim. phys. 50, 647-51 (1953).

52. ——, A. V. Fuzhenkova, V. S. Vinogradova, and T. G. Tolkacheva.
Khim. i Primenenie Fosfororgan. Soedin. Akad. Nauk SSSR Kazansk. Filial, Tr. 1 Konf. 1955, 62-70 (Pub. 1957).

53. Arbuzov, B. A., and T. G. Shavsha.
Dokl. Akad. Nauk SSSR 68, 859-60 (1949).

54. ——.
Ibid., 68, 1045-8 (1949).

55. ——.
Ibid., 69, 41-3 (1949).

56. ——.
Ibid., 68, 515-17 (1949).

57. ——.
Ibid., 84, 507-8 (1952).

58. ——.
Izv. Akad. Nauk SSSR, Otd. Khim. Nauk 1952, 875-81.

59. ——.
Ibid., 1951, 795-8.

60. ——.
Ibid., 1954, 614-21

61. ——.
Ibid., 812-22.

62. Arcoria, A., H. Lumbroso, and R. Passerini.
Bull. Soc. Chim. France 1959, 754-60.

63. Armstrong, R. S., C. G. Le Fèvre, and R. J. W. Le Fèvre.
J. Chem. Soc. 1957, 371-6.

64. Arndt, F., G. T. O. Martin, and J. R. Partington.
J. Chem. Soc. 1935, 602-4.

65. Arndt, R., Hs. H. Günthard, and T. Gäumann.
Helv. Chim. Acta 41, 2213-16 (1958).

66. Aroney, M., L. H. L. Chia, and R. J. W. Le Fèvre.
J. Chem. Soc. 1961, 4144-7.

67. Aroney, M., D. Izsak, and R. J. W. Le Fèvre.
J. Chem. Soc. 1961, 4148-53.

68. Aroney, M., and R. J. W. Le Fèvre.
J. Chem. Soc. 1958, 3002.

69. ———.
Ibid., 1960, 2161-8.

70. ———.
Ibid., 1960, 3600-8.

71. ——— and Shu-Sing Chang.
J. Chem. Soc. 1960, 3173-81.

72. Aroney, M., R. J. W. Le Fèvre, and P. M. Lenthen.
J. Chem. Soc. 1961, 4140-3.

73. Aroney, M., R. J. W. Le Fèvre, and G. M. Parkins.
J. Chem. Soc. 1960, 2890-5.

74. Aroney, M., R. J. W. Le Fèvre, and K. M. Somasundaram.
J. Chem. Soc. 1960, 1812-4.

75. Arrhenius, S.
J. Chem. Phys. 5, 63-6 (1937).

76. Audrith, L. F., W. Nespital, and H. Ulich.
J. Am. Chem. Soc. 55, 673-8 (1933).

77. Audsley, A., and F. R. Goss.
J. Chem. Soc. 1941, 864-73.

78. ———.
Ibid., 1942, 358-66.

79. ———.
Ibid., 1942, 497-500.

80. ———.
Ibid., 1949, Suppl. Issue, No. 1, S228-9.

81. Baba, H., and S. Nagakura.
J. Chem. Soc. Japan, Pure Chem. Sect., 72, 3-6 (1951).

82. Backer, H. J., and W. G. Perdok.
Rec. trav. chim. 62, 533-49 (1943).

83. Baker, E. B., A. J. Barry, and M. J. Hunter.
Ind. Eng. Chem. 38, 1117-20 (1946).

84. Baker, J. W., and L. G. Groves.
J. Chem. Soc. 1939, 1144-50.

85. Baker, W., A. S. El-Nawawy, and W. D. Ollis.
J. Chem. Soc. 1952, 3163-7.

86. Baker, W., B. Gilbert, and W. D. Ollis.
J. Chem. Soc. 1952, 1443-6.

87. Baker, W., W. D. Ollis, V. D. Poole, J. A. Barltrop,
R. A. W. Hill, and L. E. Sutton.
Nature 160, 366-7 (1947).

88. Baliah, V., and Sp. Shanmuganathan.
Trans. Faraday Soc. 55, 232-4 (1959).

89. Banerjee, B. C., and S. R. Palit.
J. Indian Chem. Soc. 27, 385-94 (1950).

90. ———.
Ibid., 29, 175-82 (1952).

91. Barassin, J., and H. Lumbroso.
Bull. soc. chim. France 1959, 1947-52.

92. ———.
Ibid., 1961, 492-500.

93. Barchukov, A. I., T. M. Minaeva, and A. M. Prokhorov.
Zhur. Eksperim. i Teor. Fiz. 29, 892 (1955).

94. Barchukov, A. I., T. M. Murina, and A. M. Prokhorov.
Opt. i Spektroskopiya 4, 521-3 (1958).

95. Barchukov, A. I., and U. N. Petrov.
Opt. i Spektroskopiya 11, 129 (1961).

96. Barchukov, A. I., and A. M. Prokhorov.
Opt. i Spektroskopiya 4, 799 (1958).

97. ———.
Ibid., 5, 530-4 (1958).

98. Barclay, G. A., and R. J. W. Le Fèvre.
J. Chem. Soc. 1950, 556-62.

99. ———.
Ibid., 1952, 1643-8.

100. Barclay, G. A., R. J. W. Le Fèvre, and B. M. Smythe.
Trans. Faraday Soc. 46, 812-20 (1950).

101. ———.
Ibid., 47, 357-65 (1951).

102. Barrett, A. H., and M. Mandel.
Phys. Rev. 109, 1572-89 (1958).

103. Barrow, F., and F. J. Thorneycroft.
J. Chem. Soc. 1934, 722-6.

104. ———.
Ibid., 1939, 773-7.

105. Bastiansen, O., and O. Hassel.
Acta Chem. Scand. 1, 683 (1947).

106. ——— and E. H. Vihovde.
Acta Chem. Scand. 8, 1951-2 (1954).

107. Bastiansen, O., and J. Markali.
Acta Chem. Scand. 6, 442-4 (1952).

108. Bates, W. W., and M. E. Hobbs.
J. Am. Chem. Soc. 73, 2151-6 (1951).

109. Batt, L., and B. G. Gowenlock.
J. Chem. Soc. 1960, 376-80.

110. Bax, C. M., A. R. Katritzky, and L. E. Sutton.
J. Chem. Soc. 1958, 1254-7.

111. ———.
Ibid., 1258-63.

112. Beard, C. I., and D. R. Bianco.
J. Chem. Phys. 20, 1488-9 (1952).

113. Beard, C. I., and B. P. Dailey.
Phys. Rev. 75, 1318 (1949).

114. ———.
J. Chem. Phys. 18, 1437-41 (1950).

115. Becher, H. J.
Z. anorg. allgem. Chem. 270, 273-86 (1952).

116. Beck, W., W. Hieber, and H. Tengler.
Chem. Ber. 94, 862-72 (1961).

117. Beech, W. F., and H. A. Piggott.
J. Chem. Soc. 1955, 423-9.

118. Beg, M. A. A., and H. C. Clark.
Can. J. Chem. 38, 119-24 (1960).

119. Bèguin, Cl., and T. Gäumann.
Helv. Chim. Acta 41, 1376-86 (1958).

120. ———.
Ibid., 1971-83 (1958).

121. Bell, R. P., and I. E. Coop.
Trans. Faraday Soc. 34, 1209-14 (1938).

122. Bender, P., D. L. Flowers, and H. L. Goering.
J. Am. Chem. Soc. 77, 3463-5 (1955).

123. Bennett, G. M., D. P. Earp, and S. Glasstone.
J. Chem. Soc. 1934, 1179-80.

124. Bennett, G. M., and S. Glasstone.
J. Chem. Soc. 1934, 128-9.

125. Bennett, R. L., and J. M. Scott.
J. Phys. Chem. 60, 1585-6 (1956).

126. Benoit, J., and G. Ney.
Compt. rend. 208, 1888-90 (1939).

127. Bentley, J. B., K. B. Everard, R. J. B. Marsden, and L. E. Sutton.
J. Chem. Soc. 1949, 2957-70.

128. Bergmann, E.
J, Chem. Soc. 1935, 987-9.

129. ———.
Ibid., 1936, 402-11.

130. Bergmann, E. D., and R. Barshai.
J. Am. Chem. Soc. 81, 5641-6 (1959).

131. Bergmann, E. D., G. Berthier, E. Fischer, Y. Hirshberg, D. Lavie, E. Lowenthal, and B. Pullman.
Bull. soc. chim. France 1952, 78-83.

132. Bergmann, E. D., G. Berthier, E. Fischer, Y. Hirshberg, D. Lavie, A. Pullman, and B. Pullman.
Bull. soc. chim. France 1951, 693-7.

133. Bergmann, E., and L. Engel.
Physik. Z. 32, 507-9 (1931).

134. ———.
Z. Elektrochem. 37, 563-9 (1931).

135. ———.
Z. physik. Chem. B13, 232-46 (1931).

136. ———.
Ibid, B15, 85-96 (1931).

137. ——— and H. Hoffman.
Z. physik. Chem. B17, 92-9 (1932).

138. Bergmann, E., L. Engel, and H. Meyer.
Chem. Ber. B65, 446-57 (1932).

139. Bergmann, E., L. Engel, and S. Sándor.
Chem. Ber. B63, 2572-5 (1930).

140. ———.
Z. physik. Chem. B10, 106-20 (1930).

141. ———.
Ibid., 397-413 (1930).

142. Bergmann, E., L. Engel, and H. A. Wolff.
Z. physik. Chem. B17, 81-91 (1932).

143. Bergmann, E. D., and E. Fischer.
Bull. soc. chim. France 1950, 1084-91.

144. ———.
Ibid., 1952, 712.

145. ———.
J. chim. phys. 49, 140 (1952).

146. ———, D. Ginsberg, Y. Hirshberg, D. Lavie, M. Mayot, A. Pullman, and B. Pullman.
Bull. soc. chim. France 1951, 684-92.

147. Bergmann, E. D., E. Fischer, Y. Hirshberg, and D. Lavie.
Bull. soc. chim. France 1952, 709-11.

148. ———, B. Pullman, and D. Shapiro.
Bull. soc. chim. France 1952, 262-7.

149. Bergmann, E. D., E. Fischer, and J. H. Jaffe.
J. Am. Chem. Soc. 75, 3230-3 (1953).

150. Bergmann, E. D., E. Fischer, and B. Pullman.
J. chim. phys. 48, 356-8 (1951).

151. Bergmann, E. D., E. Fischer, E. Zimkin, and S. Pinchas.
Rec. trav. chim. 71, 213-28 (1952).

152. Bergmann, E., and G. C. Hampson.
J. Chem. Soc. 1935, 989-93.

153. Bergmann, E., and J. Hirshberg.
J. Chem. Soc. 1936, 331-6.

154. Bergmann, E., M. Magat, and D. Wagenberg.
Chem. Ber. B63, 2576-84 (1930).

155. Bergmann, E., and W. Schütz.
Nature 128, 1077-8 (1931).

156. ———.
Z. physik. Chem. B19, 389-94 (1932).

157. ———.
Ibid., 395-400 (1932).

158. ———.
Ibid., 401-4 (1932).

159. Bergmann, E., and M. Tschudnowsky.
Z. physik. Chem. B17, 100-6 (1932).

160. ———.
Ibid., 107-15 (1932).

161. ———.
Ibid., 116-19 (1932).

162. ———.
Chem. Ber. B65, 457-63 (1932).

163. Bergmann, E., and A. Weizmann.
J. Am. Chem. Soc. 57, 1755 (1935).

164. ———.
Trans. Faraday Soc. 32, 1318-26 (1936).

165. ———.
Ibid., 1327-31 (1936).

166. ———.
Ibid., 34, 783-6 (1938).

167. ———.
Chem. and Ind. (London) 1938, 364.

168. ———.
J. Am. Chem. Soc. 60, 1801-4 (1938).

169. ———.
Ibid., 61, 3583 (1939).

170. Bergmann, E. D., A. Weizmann, and E. Fischer.
J. Am. Chem. Soc. 72, 5009-12 (1950).

171. Bernoulli, A. L., and H. Stauffer.
Helv. Chim. Acta 23, 615-26 (1940).

172. ———.
Ibid., 627-49 (1940).

173. Berthier, G., H. Lumbroso, and B. Pullman.
Bull. soc. chim. France 1950, 174-6.

174. Beyaert, M.
Natuurw. Tijdschr. (Ghent) 19, 197-213 (1937).

175. ——— and F. Goraert.
Natuurw. Tijdschr. (Ghent) 20, 119-24 (1938).

176. Bhattacharya, B. N., and W. Gordy.
Phys. Rev. 119, 144-9 (1960).

177. Bhattacharyya, G. N.
Indian J. Physics 10, 281-94 (1936).

178. ———.
Ibid., 16, 369-76 (1942).

179. Birnbaum, G.
J. Chem. Phys. 27, 360-8 (1957).

180. ——— and S. K. Chatterjee.
J. Applied Phys. 23, 220-3 (1952).

181. Birtles, R. H., and G. C. Hampson.
J. Chem. Soc. 1937, 10-15.

182. Bless, A. A.
Phys. Rev. 37, 1149-54 (1931).

183. Blinc, R., D. Hadži, and E. Pirkmajer.
In "Hydrogen Bonding," ed. D. Hadži.
London: Pergamon Press (1959) p. 333-8.

184. Bloom, G. I. M., and L. E. Sutton.
J. Chem. Soc. 1941, 727-42.

185. Bodenheimer, W., and K. Wehage.
Z. physik. Chem. B18, 343-6 (1932).

186. Böeseken, J., P. C. Henriquez, and J. J. van der Spek.
Rec. trav. chim. 55, 145-52 (1936).

187. Böeseken, J., and H. V. Takes.
Rec. trav. chim. 56, 858-62 (1937).

188. Böeseken, J., F. Tellegen, and P. C. Henriquez.
Rec. trav. chim. 54, 733-9 (1935).

189. Boggs, J. E., C. M. Crain, and J. E. Whiteford.
J. Phys. Chem. 61, 482-4 (1957).

190. Boggs, J. E., and A. P. Deam.
J. Chem. Phys. 32, 315-6 (1960).

191. Bosworth, R. C. L.
Proc. Cambridge Phil. Soc. 33, 394 (1937).

192. ——— and E. K. Rideal.
Proc. Roy. Soc. (London) A162, 1-31 (1937).

193. Böttcher, C. J. F., C. Altona, and H. F. Van Woerden.
Rec. trav. chim. 80, 5-10 (1961).

194. Boud, A. H., D. Cleverdon, G. B. Collins, and J. W. Smith.
J. Chem. Soc. 1955, 3793-8.

195. Boud, A. H., and J. W. Smith.
J. Chem. Soc. 1956, 4507-13.

196. Bragg, J. K., and A. H. Sharbaugh.
Phys. Rev. 75, 1774-5 (1949).

197. Bramley, R., C. G. Le Fèvre, R. J. W. Le Fèvre,
and B. Purnachandra Rao.
J. Chem. Soc. 1959, 1183-8.

198. Bransford, J. W., A. C. Kunkle, and A. W. Jache.
J. Inorg. Nucl. Chem. 14, 159-60 (1960).

199. Braune, H., and T. Asche.
Z. physik. Chem. B14, 18-26 (1931).

200. Braune, H., and R. Linke.
Z. physik. Chem. B31, 12-17 (1935).

201. Braunstein, R., and J. W. Trischka.
Phys. Rev. 98, 1092-8 (1955).

202. Breckenridge, R. G., and J. G. Jelatis.
Natl. Research Council, Div. Eng. and Ind. Research, Ann. Rept. Conf. on Elec. Insulation 1948, 99-101 (1949).

203. Bretscher, E.
Helv. Phys. Acta 1, 355-61 (1928).

204. ———.
Ibid., 2, 257-70 (1929).

205. ———.
Physik. Z. 32, 765-73 (1931).

206. ——— and T. Wagner-Jauregg.
Helv. Phys. Acta 2, 522-3 (1929).

207. Bridgeman, W. B., and J. W. Williams.
J. Am. Chem. Soc. 59, 1579-80 (1937).

208. Briegleb, G.
Z. physik. Chem. B10, 205-37 (1930).

209. ———.
Ibid., B16, 276-83 (1932).

210. ——— and J. Czekalla.
Naturwissenschaften 41, 448-9 (1954).

211. ———.
Z. Elektrochem. 58, 249-63 (1954).

212. ———.
Ibid., 59, 184-202 (1955).

213. Briegleb, G., and J. Kambeitz.
Z. physik. Chem. B25, 251-6 (1934).

214. ———.
Ibid., B27, 11-14 (1934).

215. ———.
Naturwissenschaften 22, 105-6 (1934).

216. Briner, E., D. Frank, and E. Perrottet.
Helv. Chim. Acta 21, 1312-17 (1938).

217. Briner, E., A. Gelbert, and E. Perrottet.
Helv. Chim. Acta 22, 1491-6 (1939).

218. Briner, E., E. Perrottet, H. Paillard, and B. Susz.
Helv. Chim. Acta 20, 762-7 (1937).

219. Briner, E., K. Ryffel, and E. Perrottet.
Helv. Chim. Acta 22, 927-34 (1939).

220. Brockway, L. O., and I. E. Coop.
Trans. Faraday Soc. 34, 1429-39 (1938).

221. Brook, A. G., R. Donovan, and G. F. Wright.
Can. J. Chem. 31, 536-42 (1953).

222. Brooker, L. G. S., F. L. White, G. H. Keyes, C. P. Smyth, and P. F. Oesper.
J. Am. Chem. Soc. 63, 3192-203 (1941).

223. Brooks, C. S., and M. E. Hobbs.
J. Am. Chem. Soc. 62, 2851-4 (1940).

224. Brown, C. A., and R. C. Osthoff.
J. Am. Chem. Soc. 74, 2340-5 (1952).

225. Brown, F., J. M. A. de Bruyne, and P. M. Gross.
J. Am. Chem. Soc. 56, 1291-3 (1934).

226. Brown, P. E., and T. De Vries.
J. Am. Chem. Soc. 73, 1811-13 (1951).

227. Brown, R. D., A. A. Hukins, R. J. W. Le Fèvre, J. Northcott, and I. R. Wilson.
J. Chem. Soc. 1949, 2812-16.

228. Brown, T. L.
J. Am. Chem. Soc. 81, 3232-5 (1959).

229. ———, D. Y. Curtin, and R. R. Fraser.
J. Am. Chem. Soc. 80, 4339-41 (1958).

230. Brown, T. L., J. G. Verkade, and T. S. Piper.
J. Phys. Chem. 65, 2051-3 (1961).

231. Browning, M. C., R. F. B. Davies, D. J. Morgan, L. E. Sutton, and L. M. Venanzi.
J. Chem. Soc. 1961, 4816-23.

232. Burawoy, A., and C. S. Gibson.
J. Chem. Soc. 1935, 219-23.

233. ———, G. C. Hampson, and H. M. Powell.
J. Chem. Soc. 1937, 1690-3.

234. Buckingham, A. D., J. Y. H. Chau, H. C. Freeman, R. J. W. Le Fèvre, D. A. A. S. Narayana Rao, and J. Tardif.
J. Chem. Soc. 1956, 1405-11.

235. Buckingham, A. D., B. Harris, and R. J. W. Le Fèvre.
J. Chem. Soc. 1953, 1626-7.

236. Buckingham, A. D., H. G. Holland, and R. J. W. Le Fèvre.
J. Chem. Soc. 1954, 1646-8.

237. Buckingham, A. D., and R. J. W. Le Fèvre.
J. Chem. Soc. 1953, 3432-5.

238. ———.
Ibid., 4169-70.

239. Buckingham, A. D., and R. E. Raab.
Trans. Faraday Soc. 55, 377-87 (1959).

240. ———.
J. Chem. Soc. 1961, 5511-23.

241. Burdun, G. D., and P. B. Kantor.
Dokl. Akad. Nauk SSSR 67, 985-8 (1949).

242. Burkhard, D. G., and D. M. Dennison.
Phys. Rev. 84, 408 (1951).

243. Burrus, C. A.
J. Chem. Phys. 28, 427-9 (1958).

244. ———.
Ibid., 31, 1270-2 (1959).

245. ——— and J. D. Graybeal.
Phys. Rev. 109, 1553-6 (1958).

246. Byers, W. H.
J. Chem. Phys. 7, 175-7 (1939).

247. Calderbank, K. E., and R. J. W. Le Fèvre.
J. Chem. Soc. 1948, 1949-52.

248. ———.
Ibid., 1949, 199-202.

249. ———.
Ibid., 1949, 1462-8.

250. ———.
Ibid., 1951, 649-51.

251. Caldwell, C. C., and R. J. W. Le Fèvre.
J. Chem. Soc. 1939, 1614-22.

252. Calingaert, G., A. J. Kolka, and H. D. Orloff.
J. Am. Chem. Soc. 73, 2359 (1951).

253. Callinan, T. D., and A. M. Parks.
J. Electrochem. Soc. 107, 799-803 (1960).

254. Całus, H.
Roczniki Chem. 25, 235-41 (1951).

255. ———.
Ibid., 28, 85-9 (1954).

256. ———, Z. Eckstein, W. Sobótka, and T. Urbański.
Bull. Acad. Polon. Sci., Ser. Sci. Chim. 9, 725-9 (1961).

257. Całus, H., H. Jankowska, H. Piotrowska, and T. Urbański.
Chem. and Ind. (London) 37, 1286 (1959).

258. Całus, H., and B. Życzyńska.
Zeszyty Nauk Politech. Warszaw-Chem. 2, 61-70 (1957).

259. Calas, R., R. Lalande, and P. Mauret.
Bull. soc. chim. France 1960, 148-9.

260. ——— and J. Boissieras.
Compt. rend. 249, 1901-3 (1960).

261. Calas, R., P. Mauret, and R. Lalande.
Compt. rend. 247, 2146-8 (1958).

262. Calvin, G., G. E. Coates, and P. S. Dixon.
Chem. and Ind. (London) 37, 1628 (1959).

263. Cambi, L.
Gazz. chim. ital. 63, 579-84 (1933).

264. Campbell, D. N., and J. L. Hall.
Proc. West Virginia Acad. Sci. 23, 64-8 (1951).

265. Campbell, I. G. M., C. G. Le Fèvre, R. J. W. Le Fèvre, and E. E. Turner.
J. Chem. Soc. 1938, 404-9.

266. Campbell, N., A. W. Henderson, and D. Taylor.
J. Chem. Soc. 1953, 1281-5.

267. Campbell, T. W., W. A. McAllister, and M. T. Rogers.
J. Am. Chem. Soc. 75, 864-5 (1953).

268. Campbell, T. W., and M. T. Rogers.
J. Am. Chem. Soc. 70, 1029-31 (1948).

269. Campbell, T. W., D. A. Young, and M. T. Rogers.
J. Am. Chem. Soc. 73, 5789-91 (1951).

270. Carlson, R. O., C. A. Lee, and B. P. Fabricand.
Phys. Rev. 85, 784-7 (1952).

271. Cartan, F., and C. N. Caughlan.
J. Phys. Chem. 64, 1756-8 (1960).

272. Cass, R. C., H. Spedding, and H. D. Springall.
J. Chem. Soc. 1957, 3451-6.

273. Caughlan, C. N., and F. Cartan.
J. Am. Chem. Soc. 81, 3840-1 (1959).

274. Caughlan, C. N., W. Katz, and W. Hodgson.
J. Am. Chem. Soc. 73, 5654-5 (1951).

275. Cavallaro, L., and L. Felloni.
Atti Accad. Sci. Ferrara 35, 95-104 (1957-8).

276. ——.
Ann. Chim. 49, 579-91 (1959).

277. Cavell, H. J., and S. Sugden.
J. Chem. Soc. 1935, 621-4.

278. Chalandon, P., and B. P. Susz.
Helv. Chim. Acta 41, 697-704 (1958).

279. Challenger, F., and J. B. Harrison.
J. Inst. Petroleum Tech. 21, 135-54, 169-71 (1935).

280. Chan, S. I., J. Zinn, J. Fernandez, and W. D. Gwinn.
J. Chem. Phys. 33, 1643-55 (1960).

281. Chang, K. S., and Y-T. Cha.
J. Chinese Chem. Soc. 1, 107-15 (1933).

282. Charles, R. G., and H. Freiser.
J. Am. Chem. Soc. 72, 2233-5 (1950).

283. Charney, E.
J. Am. Chem. Soc. 83, 578-80 (1961).

284. —— and E. D. Becker.
J. Am. Chem. Soc. 83, 4468-9 (1961).

285. Chatt, J., and L. A. Duncanson.
J. Chem. Soc. 1953, 2939-47.

286. —— and B. L. Shaw.
Chem. and Ind. (London) 1958, 859-60.

287. Chatt, J., R. G. Guy, and L. A. Duncanson.
J. Chem. Soc. 1961, 827-34.

288. Chatt, J., and F. A. Hart.
Nature 169, 673-4 (1952).

289. ______.
J. Chem. Soc. 1953, 2363-71.

290. ______.
Chem. and Ind. (London) 1958, 1474-5.

291. ______.
J. Chem. Soc. 1960, 1378-89.

292. ______.
Ibid., 2807-14.

293. Chatt, J., and B. L. Shaw.
J. Chem. Soc. 1959, 705-16.

294. ______.
Ibid., 4020-33.

295. ______.
Ibid., 1960, 1718-29.

296. Chatt, J., and R. G. Wilkins.
J. Chem. Soc. 1952, 4300-6.

297. Chatt, J., and A. A. Williams.
J. Chem. Soc. 1951, 3061-7.

298. Chau, J. Y. H., C. G. Le Fèvre, and R. J. W. Le Fèvre.
J. Chem. Soc. 1959, 2666-9.

299. Chau, J. Y. H., and R. J. W. Le Fèvre.
Australian J. Chem. 8, 562-4 (1955).

300. ______.
J. Chem. Soc. 1957, 2300-2.

301. Cherrier, C.
Compt. rend. 225, 1306-7 (1947).

302. ______.
Bull. soc. chim. France 1948, 1076.

303. Chiba, T.
Bull. Chem. Soc. Japan 28, 19-34 (1955).

304. ______.
Ibid., 295-9 (1955).

305. ______.
Ibid., 505-9 (1955).

306. ______, T. Shimozawa, I. Miyagawa, and Y. Morino.
Bull. Chem. Soc. Japan 30, 223-7 (1957).

307. Ch'ien, J-Y., I-H. P'ang, and J-C. Wu.
Hua Hsüeh Hsüeh Pao 25, 110-15 (1959).

308. Chien, S. L., and T. C. Lay.
J. Chinese Chem. Soc. 5, 204-13 (1937).

309. Chierici, L., H. Lumbroso, and R. Passerini.
Compt. rend. 237, 611-13 (1953).

310. ———.
Bull. soc. chim. France 1955, 686-94.

311. Chu, T. L., N. C. Li, and C. T. Fujii.
J. Am. Chem. Soc. 77, 2085-7 (1955).

312. Cilento, G., E. de Carvalho Filho, and A. C. G. Albanese.
J. Am. Chem. Soc. 80, 4472-4 (1958).

313. Claeys, E. G., G. P. van der Kelen, and Z. Eeckhaut.
Bull. Soc. Chim. Belg. 70, 462-7 (1961).

314. Clark, D. E., and S. N. Kumar.
Brit. J. Appl. Phys. 7, 282-4 (1956).

315. Clark, R. J.
Proc. Roy. Soc. (London) A124, 689-98 (1929).

316. Clayton, L., Q. Williams, and T. L. Weatherly.
J. Chem. Phys. 30, 1328-34 (1959).

317. Clemo, G. R., and K. H. Jack.
Chem. and Ind. (London) 1953, 195.

318. Cleverdon, D., G. B. Collins, and J. W. Smith.
J. Chem. Soc. 1956, 4499-507.

319. Cleverdon, D., and J. W. Smith.
J. Chem. Soc. 1951, 2321-3.

320. Coates, G. E.
J. Chem. Soc. 1946, 838-9.

321. ——— and L. E. Sutton.
J. Chem. Soc. 1942, 567-70.

322. Coleman, A. M., and H. Freiser.
J. Am. Chem. Soc. 83, 4127-31 (1961).

323. Coles, D. K., and R. H. Hughes.
Phys. Rev. 76, 178 (1949).

324. Coles, D. K., W. E. Good, J. K. Bragg, and A. H. Sharbaugh.
Phys. Rev. 82, 877-8 (1951).

325. Coles, D. K., W. E. Good, and R. H. Hughes.
Phys. Rev. 79, 224 (1950).

326. Compton, K. T., and C. T. Zahn.
Phys. Rev. 23, 781-2 (1924).

327. Conradi, J. J., and N. C. Li.
J. Am. Chem. Soc. 75, 1785-8 (1953).

328. Cook, R. L., and A. P. Mills.
J. Phys. Chem. 65, 252-4 (1961).

329. Coomber, D. I., and J. R. Partington.
J. Chem. Soc. 1938, 1444-52.

330. Coop, I. E., N. R. Davidson, and L. E. Sutton.
J. Chem. Phys. 6, 905 (1938).

331. Coop, I. E., and L. E. Sutton.
J. Chem. Soc. 1938, 1269-86.

332. ______.
Ibid., 1869-72.

333. ______.
Trans. Faraday Soc. 35, 505-11 (1939).

334. Copeland, C. S., and M. W. Rigg.
J. Am. Chem. Soc. 73, 3584-8 (1951).

334a Cotton, F. A., and R. Francis.
J. Am. Chem. Soc. 82, 2986-91 (1960).

335. Coussmaker, C. R. C., M. H. Hutchinson, J. R. Mellor, L. E. Sutton, and L. M. Venanzi.
J. Chem. Soc. 1961, 2705-13.

336. Cowley, E. G.
Nature 168, 705-6 (1951).

337. ______.
J. Chem. Soc. 1952, 3557-70.

338. ______ and J. R. Partington.
J. Chem. Soc. 1933, 1252-4.

339. ______.
Ibid., 1255-7.

340. ______.
Ibid., 1257-9.

341. ______.
Ibid., 1259.

342. ______.
Ibid., 1935, 604-9.

343. ______.
Nature 136, 643 (1935).

344. ______.
J. Chem. Soc. 1936, 45-7.

345. ______.
Ibid., 47-50.

346. ______.
Ibid., 1184-94.

347. ______.
Ibid., 1937, 130-8.

348. ______.
Ibid., 1938, 977-83.

349. ______.
Ibid., 1598-601.

350. Cox, A. P., L. F. Thomas, and J. Sheridan.
Nature 181, 1000-1 (1958).

351. Crable, G. F., and W. V. Smith.
J. Chem. Phys. 19, 502 (1951).

352. Crain, C. M.
Phys. Rev. 74, 691-3 (1948).

353. Criegee, R., and K. Noll.
Ann. 627, 1-14 (1959).

354. Criegee, R., G. Schröder, G. Maier, and H-G. Fischer.
Chem. Ber. 93, 1553-9 (1960).

355. Crowe, R. W., and C. N. Caughlan.
J. Am. Chem. Soc. 72, 1694-7 (1950).

356. Cumper, C. W. N., G. H. Jeffery, J. Leicester, A. I. Vogel, and S. Walker.
Research Correspondence, Suppl. to Research (London) 8, No. 12, S65-6 (1955).

357. Cumper, C. W. N., and A. I. Vogel.
J. Chem. Soc. 1959, 3521-6.

358. ———.
Ibid., 1960, 4723-8.

359. ——— and S. Walker.
J. Chem. Soc. 1957, 3640-3.

360. ———.
Ibid., 1956, 3621-8.

361. Cumper, C. W. N., and S. Walker.
Trans. Faraday Soc. 52, 193-9 (1956).

362. Cunningham, G. L., Jr., A. W. Boyd, R. J. Myers, W. D. Gwinn, and W. I. Le Van.
J. Chem. Phys. 19, 676-85 (1951).

363. Cureton, P. H., C. G. Le Fèvre, and R. J. W. Le Fèvre.
J. Chem. Soc. 1961, 4447-51.

364. Curl, R. F., Jr.
J. Chem. Phys. 30, 1529-36 (1959).

365. Curran, B. C.
J. Am. Chem. Soc. 63, 1470-1 (1941).

366. ———.
Ibid., 64, 830-3 (1942).

367. ———.
Ibid., 67, 1835-7 (1945).

368. ——— and E. P. Chaput.
J. Am. Chem. Soc. 69, 1134-7 (1947).

369. Curran, B. C., and G. K. Estok.
J. Am. Chem. Soc. 72, 4575-8 (1950).

370. Curran, B. C., P. A. McCusker, and H. S. Makowski.
J. Am. Chem. Soc. 79, 5188-9 (1957).

371. Curran, B. C., and H. H. Wenzke.
J. Am. Chem. Soc. 59, 943-4 (1937).

372. Curran, W. J., and H. H. Wenzke.
J. Am. Chem. Soc. 57, 2162-3 (1935).

373. Curran, B. C., R. M. Witucki, and P. A. McCusker.
J. Am. Chem. Soc. 72, 4471-4 (1950).

374. Curtis, H. J.
J. Chem. Phys. 1, 160-5 (1933).

375. Czekalla, J., and G. Wick.
Z. Elektrochem. 65, 727-34 (1961).

376. Czerlinsky, E.
Z. Physik 88, 515-21 (1934).

377. Dailey, B. P., J. M. Mays, and C. H. Townes.
Phys. Rev. 76, 136-7 (1949).

378. ———.
Ibid., 472 (1949).

379. Dailey, C. R.
Phys. Rev. 34, 548 (1929).

380. Dakin, T. W., W. E. Good, and D. K. Coles.
Phys. Rev. 71, 640-1 (1947).

381. Dalbert, R.
J. chim. phys. 50, 329-43 (1953).

382. Danö, M., S. Furberg, and O. Hassel.
Acta Chem. Scand. 4, 965-6 (1950).

383. Darmois, E., and L. Mouradoff.
Bull. soc. chim. France 1949, D446-7.

384. Das, L. M., and S. C. Roy.
Indian J. Physics 5, 441-62 (1930).

385. Dave, L. D., D. F. Evans, and G. Wilkinson.
J. Chem. Soc. 1959, 3684-8.

386. Davidson, D. W.
Can. J. Chem. 39, 2139-54 (1961).

387. Davidson, N. R., and L. E. Sutton.
J. Chem. Soc. 1942, 565-7.

388. Davies, M., and D. G. Jenkin.
J. Chem. Soc. 1954, 2374-7.

389. Davies, M., and G. Williams.
Trans. Faraday Soc. 56, 1619-28 (1960).

390. Davis, R., H. S. Bridge, and W. J. Svirbely.
J. Am. Chem. Soc. 65, 857-62 (1943).

391. de Bièvre, P. J., G. P. van der Kelen, G. Cornille, and Z. Eeckhaut.
Bull. soc. chim. Belges 68, 550-7 (1959).

392. de Brouckere, L., and R. van Nechel.
Bull. soc. chim. Belges 61, 261 (1952).

393. ———.
Ibid., 452-61 (1952).

394. de Bruyne, J. M. A., R. M. Davis, and P. M. Gross.
Physik. Z. 33, 719-24 (1932).

395. ———.
J. Am. Chem. Soc. 55, 3936-44 (1933).

396. de Bruyne, J. M. A., and C. P. Smyth.
J. Am. Chem. Soc. 57, 1203-5 (1935).

397. Debye, P., and F. Bueche.
J. Chem. Phys. 19, 589-94 (1951).

398. de Gaouck, V., and R. J. W. Le Fèvre.
J. Chem. Soc. 1938, 741-5.

399. ———.
Ibid., 1939, 1392-4.

400. ———.
Ibid., 1457-65.

401. de Kowalewski, D. G., P. Kökeritz, and H. Selén.
J. Chem. Phys. 31, 1438 (1959).

402. DeMore, B. B., W. S. Wilcox, and J. H. Goldstein.
J. Chem. Phys. 22, 876-7 (1954).

403. Devoto, G.
Gazz. chim. ital. 63, 495-9 (1933).

404. De Vries, J., and W. H. Rodebush.
J. Am. Chem. Soc. 53, 2888-93 (1931).

405. Di Giacomo, A., and C. P. Smyth.
J. Am. Chem. Soc. 77, 774-7 (1955).

406. ———.
Ibid., 1361-4 (1955).

407. ———.
Ibid., 74, 4411-13 (1952).

408. Dixon, W. B., and E. B. Wilson.
J. Chem. Phys. 35, 191-8 (1961).

409. Dobiński, S.
Bull. intern. acad. Polonaise A, 239-47 (1932).

410. Doborzynski, D.
Z. Physik 66, 657-68 (1930).
Bull. intern. acad. Polonaise A, 97 (1930).

411. Dodd, R. E., and R. Little.
Nature 188, 737 (1960).

412. Dolter, R. J., and C. Curran.
J. Am. Chem. Soc. 82, 4153-5 (1960).

413. Donle, H. L.
Z. physik. Chem. B14, 326-38 (1931).

414. ———.
Ibid., B18, 146-52 (1932).

415. ——— and K. A. Gehrckens.
Z. physik. Chem. B18, 316-26 (1932).

416. Donle, H. L., and G. Volkert.
Z. physik. Chem. B8, 60-71 (1930).

417. Donle, H. L., and K. L. Wolf.
Z. physik. Chem. B8, 55-9 (1930).

418. Dornte, R. W., and C. P. Smyth.
J. Am. Chem. Soc. 52, 3546-52 (1930).

419. Dostrovsky, I., and R. J. W. Le Fèvre.
J. Chem. Soc. 1939, 535-7.

420. Draney, J. J., Jr., and M. Cefola.
J. Am. Chem. Soc. 76, 1975-7 (1954).

421. Drefahl, G., and G. Heublein.
Chem. Ber. 93, 497-500 (1960).

422. Drenth, W., G. L. Hekkert, and B. G. Zwanenburg.
Rec. trav. chim. 79, 1056-65 (1960).

423. Dryden, J. S., and R. J. Meakins.
Proc. Phys. Soc. (London) 69B, 252-4 (1956).

424. Dubenko, R. G., P. S. Pel'kis, and I. A. Sheka.
Ukr. Khim. Zh. 26, 48-52 (1960).

425. Dumanskiĭ, A. V., and O. D. Kurilenko.
Dokl. Akad. Nauk SSSR 60, 1197-9 (1948).

426. Eagle, D. F., T. L. Weatherly, and Q. Williams.
J. Chem. Phys. 30, 603-4 (1959).

427. Earl, J. C., E. W. Leake, and R. J. W. Le Fèvre.
Nature 160, 366 (1947).

428. ———.
J. Chem. Soc. 1948, 2269-75.

429. Ebert, L., R. Eisenschitz, and H. von Hartel.
Z. physik. Chem. B1, 94-114 (1928).

430. Ebert, L., and K. Højendahl.
Z. physik. Chem. B15, 74-8 (1931).

431. Ebert, L., and H. von Hartel.
Naturwissenschaften 15, 669-70 (1927).

432. Eckert, J. M., and R. J. W. Le Fèvre.
J. Chem. Soc. 1961, 2356-9.

433. Eda, B., and K. Ito.
Bull. Chem. Soc. Japan 30, 164-7 (1957).

434. Eda, B., K. Tsuda, and M. Kubo.
J. Am. Chem. Soc. 80, 2426-8 (1958).

435. Edgerley, P. G., and L. E. Sutton.
Nature 164, 1050 (1949).

436. ———.
J. Chem. Soc. 1950, 3394-6.

437. ———.
Ibid., 1951, 1069-74.

438. Edsall, J. T., and J. Wyman.
J. Am. Chem. Soc. 57, 1964-75 (1935).

439. Eide, A. E., and O. Hassel.
Tidsskr. Kjemi Bergvesen 10, 93-5 (1930).

440. Eisenlohr, F., and W. Hass.
Z. physik. Chem. A173, 249-64 (1935).

441. Eisenlohr, F., and L. Hill.
Z. physik. Chem. B36, 30-44 (1937).

442. Eisenlohr, F., and A. Metzner.
Z. physik. Chem. A178, 350-4 (1937).

443. Eisinger, J.
J. Chem. Phys. 27, 1206-7 (1957).

444. ———.
Ibid., 29, 1154-60 (1958).

445. ———.
Ibid., 28, 165-6 (1958).

446. ———.
Ibid., 30, 927-30 (1959).

447. Ekelund, B.
Acta Acad. Aboensis, Math. et Phys. 19, No. 11, 1-93 (1954).

448. ———.
Ibid., 20, No. 7, 10 pp (1955).

449. Emblem, H. G., and C. A. McDowell.
J. Chem. Soc. 1946, 641-2.

450. Epprecht, G. W.
Z. angew. Math. Physik 1, 138-47 (1950).

451. Eric, B., E. V. Goode, and D. A. Ibbitson.
J. Chem. Soc. 1960, 55-61.

452. Erlandsson, G., and H. Selén.
Arkiv. Fysik. 14, 61-4 (1958).

453. Errera, J.
J. phys. radium (6) 6, 390-6 (1925).

454. ———.
Compt. rend. 182, 1623-5 (1926).

455. ———.
Physik. Z. 27, 764-9 (1926).

456. ———.
Ibid., 29, 426-9 (1928).

457. ———.
Ibid., 689-90 (1928).

458. ———.
J. phys. radium (6) 9, 307-9 (1928).

459. ———.
Kolloid-Z. 51, 104-5 (1930).

460. ———.
J. chim. phys. 29, 577-85 (1932).

461. ——— and M. L. Sherrill.
Leipziger Vorträge 1929, 41.

462. ———.
J. Am. Chem. Soc 52, 1993-7 (1930).

463. Estermann, I.
Z. physik. Chem. B1, 134-60 (1928).

464. ———.
Ibid., 161-9 (1928).

465. ———.
Ibid., B2, 287-8 (1929).

466. ——— and R. G. J. Fraser.
J. Chem. Phys. 1, 390-9 (1933).

467. Estermann, I., and M. Wohlwill.
Z. physik. Chem. B20, 195-208 (1933).

468. Estok, G. K., and J. S. Dehn.
J. Am. Chem. Soc. 77, 4769-70 (1955).

469. Estok, G. K., and J. H. Sikes.
J. Am. Chem. Soc. 75, 2745-7 (1953).

470. Estok, G. K., and S. P. Sood.
J. Phys. Chem. 61, 1445-7 (1957).

471. ——— and C. H. Stembridge.
J. Phys. Chem. 62, 1464-6 (1958).

472. Estok, G. K., and C. H. Stembridge.
J. Am. Chem. Soc. 76, 4316-7 (1954).

473. Estok, G. K., and W. W. Wendlandt.
J. Am. Chem. Soc. 77, 4767-9 (1955).

474. Evans, D. E. M., J. A. Godsell, R. Stephens, J. C. Tatlow, and E. H. Wiseman.
Tetrahedron 2, 183-92 (1958).

475. Everard, K. B., L. Kumar, and L. E. Sutton.
J. Chem. Soc. 1951, 2807-15.

476. Everard, K. B., and L. E. Sutton.
Nature 162, 104 (1948).

477. ———.
J. Chem. Soc. 1949, 2312-8.

478. ———.
Ibid., 2318-23.

479. ———.
Ibid., 1951, 16-21.

480. ———.
Ibid., 2816-17.

481. ———.
Ibid., 2826-8.

482. Exner, O., V. Jehlička, and A. Reiser.
Collection Czech. Chem. Commun. 24, 3207-21 (1959).

483. Exner, O., and O. Wichterle.
Chem. Listy 50, 922-30 (1956).

484. Fabricand, B. P., R. O. Carlson, C. A. Lee, and I. I. Rabi.
Phys. Rev. 86, 607 (1952).

485. ———.
Ibid., 91, 1403-8 (1953).

486. Fairbrother, F.
J. Chem. Soc. 1932, 43-55.

487. ———.
Ibid., 1933, 1541-3.

488. ———.
Proc. Roy. Soc. (London) A142, 173-97 (1933).

489. ———.
Nature 134, 458-9 (1934).

490. ———.
Trans. Faraday Soc. 30, 862-70 (1934).

491. ———.
J. Chem. Soc. 1934, 1846-49.

492. ———.
Ibid., 1936, 847-53.

493. ———.
Ibid., 1945, 503-9.

494. ———.
Ibid., 1948, 1051-6.

495. Farmer, E. H., and N. J. H. Wallis.
J. Chem. Soc. 1933, 1304-9.

496. Farmer, E. H., and F. L. Warren.
J. Chem. Soc. 1933, 1297-301.

497. ———.
Ibid., 1302-4.

498. Feichtmayr, F., and F. Würstlin.
Festschrift Carl Wurster zum 60 Geburtstag 1960, 177-87.

499. Felloni, L.
Ann. chim. (Rome) 47, 1163-76 (1957).

500. ——— and F. Pulidori.
Ann. chim. (Rome) 85, 1027-47 (1961).

501. Ferguson, R. C.
J. Am. Chem. Soc. 76, 850-3 (1954).

502. Fernandez, J., R. J. Myers, and W. D. Gwinn.
J. Chem. Phys. 23, 758-9 (1955).

503. Ferraro, C. F., J. J. Draney, and M. Cefola.
J. Am. Chem. Soc. 75, 1206-8 (1953).

504. Ferry, J. D., and J. L. Oncley.
J. Am. Chem. Soc. 60, 1123-32 (1938).

505. ———.
Ibid., 63, 272-8 (1941).

506. Few, A. V., and J. W. Smith.
J. Chem. Soc. 1949, 3057-60.

507. ———.
Ibid., 753-60.

508. ———.
Ibid., 2663-8.

509. Few, A. V., and J. W. Smith.
J. Chem. Soc. 1949, 2781-4.

510. Fielding, P. E., and R. J. W. Le Fèvre.
J. Chem. Soc. 1951, 1811-14.

511. Finn, A. E., G. C. Hampson, and L. E. Sutton.
J. Chem. Soc. 1938, 1254-63.

512. Fischer, E.
J. Chem. Phys. 19, 395-6 (1951).

513. ———.
J. Chem. Soc. 1955, 1382-3.

514. ——— and F. Rogowski.
Physik. Z. 40, 331-7 (1939).

515. Fischer, E. O.
J. Inorg. Nucl. Chem. 8, 268-72 (1958).

516. ——— and G. Bürger.
Chem. Ber. 94, 2409-12 (1961).

517. Fischer, E. O., and W. Fröhlich.
Chem. Ber. 92, 2995-8 (1959).

518. Fischer, E. O., and W. Hafner.
Z. Naturforsch. 10b, 665-8 (1955).

519. Fischer, E. O., and C. Palm.
Z. Naturforsch. 14b, 598-9 (1959).

520. Fischer, E. O., and S. Schreiner.
Chem. Ber. 92, 938-48 (1959).

521. Fischer, E. O., and K. Ulm.
Chem. Ber. 94, 2413-6 (1961).

522. ——— and H. P. Fritz.
Chem. Ber. 93, 2167-73 (1960).

523. Fischer, E. O., and A. Wirzmüller.
Z. Naturforsch. 12b, 737-8 (1957).

524. Fischer, I.
Acta Chem. Scand. 4, 1197-205 (1950).

525. ———.
Nature 165, 239 (1950).

526. ——— and N. Löfgren.
Acta Chem. Scand. 4, 1408-12 (1950).

527. Fitzgerald, E. R., and R. F. Miller.
J. Colloid Sci. 8, 148-69 (1953).

528. Florentine, R. A.
Dissertation Abstr. 13, 671 (1953).

529. ——— and J. G. Miller.
J. Am. Chem. Soc. 81, 5103-6 (1959).

530. Fogelberg, J. M., and J. W. Williams.
Physik. Z. 32, 27-31 (1931).

531. Foley, H. M., and H. M. Randall.
Phys. Rev. 59, 171-3 (1941).

532. Fonteyne, R.
Natuurw. Tijdschr. (Ghent) 20, 275-8 (1938).

533. ——— and M. Ticket.
Natuurw. Tijdschr. (Ghent) 25, 49-66 (1943).

534. Forró, M.
Z. Physik 47, 430-45 (1928).

535. Frank, F. C.
J. Chem. Soc. 1936, 1324-7.

536. Frankel, M., R. Wolovsky, and E. Fischer.
J. Chem. Soc. 1955, 3441-5.

537. Fredenhagen, K., and F. Maske.
Z. physik. Chem. B10, 142-8 (1930).

538. Freeman, H. C., and R. J. W. Le Fèvre.
J. Chem. Soc. 1950, 3128-31.

539. ——— and F. Maramba.
J. Chem. Soc. 1952, 1649-51.

540. Freeman, H. C., R. J. W. Le Fèvre, J. Northcott, and C. V. Worth.
J. Chem. Soc. 1952, 3384-9.

541. Freeman, H. C., R. J. W. Le Fèvre, J. Northcott, and I. Youhotsky.
J. Chem. Soc. 1952, 3381-4.

542. Freeman, H. C., R. J. W. Le Fèvre, and I. R. Wilson.
J. Chem. Soc. 1951, 1977-80.

543. Freiser, H., R. G. Charles, J. Speier, and M. Eagle.
J. Am. Chem. Soc. 73, 5229-30 (1951).

544. Freiser, H., M. V. Eagle, and J. L. Speier, Jr.
J. Am. Chem. Soc. 75, 2821-4 (1953).

545. ———.
Ibid., 2824-7 (1953).

546. Freiser, H., and W. L. Glowacki.
J. Am. Chem. Soc. 71, 514-6 (1949).

547. ———.
Ibid., 70, 2575-8 (1948).

548. Freiser, H., M. E. Hobbs, and P. M. Gross.
J. Am. Chem. Soc. 71, 111-15 (1949).

549. Frey, P. R., and E. C. Gilbert.
J. Am. Chem. Soc. 59, 1344-7 (1937).

550. Fritzky, H. G.
Z. Physik 151, 351 (1958).

551. Frivold, O. E.
Physik. Z. 22, 603-10 (1921).

552. ——— and O. Hassel.
Physik. Z. 24, 82-6 (1923).

553. Fu, H-C., T. Psarras, H. Weidmann, and H. K. Zimmerman.
Ann. Chem. 641, 116-20 (1961).

554. Fuchs, O.
Z. Physik <u>63</u>, 824-48 (1930).

555. ——— and H. L. Donle.
Z. physik. Chem. <u>B22</u>, 1-20 (1933).

556. Fujita, T.
J. Am. Chem. Soc. <u>79</u>, 2471-5 (1957).

557. ——— and M. Hamada.
Botyu Kagaku <u>19</u>, 80-3 (1954).

558. Fukushima, K.
Nippon Kagaku Zasshi <u>79</u>, 1096-103 (1958).

559. Fuoss, R. M.
J. Am. Chem. Soc. <u>60</u>, 1633-7 (1938).

560. Gagnaux, P., D. Janjic, and B. P. Susz.
Helv. Chim. Acta <u>41</u>, 1023-9 (1958).

561. Gallay, W.
Kolloid-Z. <u>57</u>, 1-7 (1931).

562. Gäumann, T.
Helv. Chim. Acta <u>41</u>, 1956-70 (1958).

563. ———, R. W. Schmid, and E. Heilbronner.
Helv. Chim. Acta <u>39</u>, 1985-93 (1956).

564. Geddes, J. A., and C. A. Kraus.
Trans. Faraday Soc. <u>32</u>, 585-93 (1936).

565. Gehrckens, K. A., and E. Müller.
Ann. <u>500</u>, 296-306 (1933).

566. Gent, W. L. G.
J. Chem. Soc. <u>1957</u>, 58-62.

567. George, M. V., R. W. Kierstead, and G. F. Wright.
Can. J. Chem. <u>37</u>, 679-99 (1959).

568. George, M. V., and G. F. Wright.
Can. J. Chem. <u>36</u>, 189-98 (1958).

569. ———.
J. Am. Chem. Soc. <u>80</u>, 1200-4 (1958).

570. Georgian, V.
Chem. and Ind. (London) <u>37</u>, 1626 (1959).

571. Gerson, F., T. Gäumann, and E. Heilbronner.
Helv. Chim. Acta <u>41</u>, 1481-91 (1958).

572. Ghosh, P. N., and T. P. Chatterjee.
Phys. Rev. <u>37</u>, 427-9 (1931).

573. Ghosh, P. N., P. C. Mahanti, and B. C. Mukhergee.
Z. Physik <u>58</u>, 200-4 (1929).

574. Ghosh, P. N., P. C. Mahanti, and D. N. Sen-Gupta.
Z. Physik <u>54</u>, 711-4 (1929).

575. Ghosh, S. N., R. Trambarulo, and W. Gordy.
Phys. Rev. <u>87</u>, 172 (1952).

576. ———.
J. Chem. Phys. <u>21</u>, 308-10 (1953).

577. Gibbs, J. H., and C. P. Smyth.
J. Am. Chem. Soc. 73, 5115-8 (1951).

578. Gibson, C. S.
Sci. J. Roy. Coll. Sci. 5, 54-62 (1935).

579. Giddings, S. A., and R. J. Best.
J. Am. Chem. Soc. 83, 2393-4 (1961).

580. Gilbert, D. A., A. Roberts, and P. A. Griswold.
Phys. Rev. 76, 1723 (1949).

581. Gilkerson, W. R., and K. K. Srivastava.
J. Phys. Chem. 64, 1485-7 (1960).

582. ———.
Ibid., 65, 272-4 (1961).

583. Gilliam, O. R., H. D. Edwards, and W. Gordy.
Phys. Rev. 75, 1014-6 (1949).

584. Glemser, O., H. Schröder, and H. Haeseler.
Naturwissenschaften 42, 44 (1955).

585. Glemser, O., and V. Häusser.
Z. Naturforsch. 3b, 159-63 (1948).

586. Glemser, O., and T. Risler.
Z. Naturforsch. 3b, 1-6 (1948).

587. Goebel, H. L., and H. H. Wenzke.
J. Am. Chem. Soc. 59, 2301-2 (1937).

588. ———.
Ibid., 60, 697-9 (1938).

589. Goethals, C. A.
Rec. trav. chim. 54, 299-306 (1935).

590. Gokhale, S. D., N. L. Phalnikar, and S. D. Bhawe.
J. Univ. Bombay A11, Pt. 5, 56-62 (1943).

591. Golden, S., T. Wentink, R. E. Hillger, and M. W. P. Strandberg.
Phys. Rev. 73, 92-3 (1948).

592. Gol'dshtein, I. P., E. N. Gur'yanova, E. D. Delinskaya, and K. A. Kocheshkov.
Dokl. Akad. Nauk SSSR 136, 1079-81 (1961).

593. Goldsmith, M., and G. W. Wheland.
J. Am. Chem. Soc. 70, 2632-4 (1948).

594. Gol'tsman, F. M., and Sh. Sh. Raskin.
Dokl. Akad. Nauk SSSR 80, 817-9 (1953).

595. Golubkov, G. E., and K. A. Andrianov.
Vysokomolekul. Soedin. 1, 1273-8 (1959).

596. Gomel, M., and H. Lumbroso.
Compt. rend. 252, 3039-41 (1961).

597. ———, N. Marziano, and R. Passerini.
Bull. soc. chim. France 1959, 1908-12.

598. Goode, E. V., and D. A. Ibbitson.
J. Chem. Soc. 1960, 4265-70.

599. Gorman, M., R. M. Davis, and P. M. Gross.
Physik. Z. 39, 181-5 (1938).

600. Goss, F. R.
J. Chem. Soc. 1933, 1341-5.

601. ——.
Ibid., 1934, 1467.

602. ——.
Ibid., 1937, 1915-20.

603. Goubeau, J., and J. Jiménez-Barberá.
Z. anorg. allgem. Chem. 303, 217-26 (1960).

604. Goubeau, J., and H. Sommer.
Z. anorg. allgem. Chem. 289, 1-4 (1957).

605. Grabner, L., and V. Hughes.
Phys. Rev. 79, 819-28 (1950).

606. Gräff, G.
Z. Physik 155, 433-40 (1959).

607. ——, W. Paul, and C. Schlier.
Z. Physik 153, 38-63 (1958).

608. Graffunder, W., and E. Heymann.
Z. physik. Chem. B15, 377-82 (1932).

609. Granier, J.
Compt. rend. 223, 893-4 (1946).

610. Granitova, O. I., and A. P. Toropov.
Dokl. Akad. Nauk Uzbek. SSR 1955, No. 2, 27-8.

611. Grant, R. F., D. W. Davidson, and P. Gray.
J. Chem. Phys. 33, 1713-8 (1960).

612. Grassi, U.
Nuovo cimento 14, 461-73 (1937).

613. ——.
Ibid., 10, 3-20 (1933).

614. Green, G. W., and H. Lew.
Can. J. Physics 38, 482-94 (1960).

615. Greene, E. W., and J. W. Williams.
Phys. Rev. 42, 119-40 (1932).

616. Grosham, [illegible], [illegible] Jansen, and T. W. [illegible].
J. Am. Chem. Soc. 70, 998-9 (1948).

617. Griffith, W. P., J. Lewis, and G. Wilkinson.
J. Chem. Soc. 1961, 2259-60.

618. Grinberg, A. A., and A. D. Troitskaya.
Tr. Radievogo Inst. Akad. Nauk SSSR 7, 5-13 (1956).

619. Grishko, N. I., and E. N. Gur'yanova.
Zhur. Fiz. Khim. 32, 2725-30 (1958).

620. Gross, P.
Physik. Z. 32, 587-92 (1931).

621. Groves, L. G., and S. Sugden.
J. Chem. Soc. 1934, 1094-8.

622. ———.
Ibid., 1935, 971-4.

623. ———.
Ibid., 1937, 158-62.

624. ———.
Ibid., 1779-82.

625. ———.
Ibid., 1782-4.

626. Grubb, W. T., and R. C. Osthoff.
J. Am. Chem. Soc. 75, 2230-2 (1953).

627. Gundyrev, A. A., N. S. Nametkin, G. M. Panchenkov, and A. V. Topchiev.
Dokl. Akad. Nauk SSSR 129, 1325-7 (1959).

628. Gundyrev, A. A., N. S. Nametkin, and A. V. Topchiev.
Dokl. Akad. Nauk SSSR 121, 1031-3 (1958).

629. Günthard, Hs. H., and T. Gäumann.
Helv. Chim. Acta 33, 1985-7 (1950).

630. ———.
Ibid., 34, 39-46 (1951).

631. Gürne, D., and T. Urbanski.
J. Chem. Soc. 1959, 1912-3.

632. Gur'yanova, E. N.
Acta Physicochem. USSR 14, 154-5 (1941).

633. ———.
J. Phys. Chem. USSR 15, 142-3 (1941).

634. ———.
Ibid., 21, 411-21 (1947).

635. ———.
Ibid., 633-42 (1947).

636. ———.
Zhur. Fiz. Khim. 24, 479-86 (1950).

637. ———, I. I. Eĭtingon, M. C. Fel'shteĭn, I. G. Chernomorskaya, and B. A. Dogadkin.
Zhur. Obshcheĭ Khim. 31, 3709-12 (1961).

638. Gur'yanova, E. N., and L. S. Kuzina.
Zhur. Fiz. Khim. 28, 2116-28 (1954).

639. ———.
Ibid., 30, 616-28 (1956).

640. Gur'yanova, E. N., and Ya. K. Syrkin.
Acta Physicochem. USSR 11, 657-8 (1939).

641. ———.
Zhur. Fiz. Khim. 23, 105-14 (1949).

642. Gur'yanova, E. N., L. A. Yanovskaya, and A. P. Terent'ev.
Zhur. Fiz. Khim. 25, 897-902 (1951).

643. Habgood, H. W.
Dissertation Abstr. 13, 296 (1953).

644. Hacket, N., and R. J. W. Le Fèvre.
J. Chem. Soc. 1961, 2612-15.

645. Halmöy, E., and O. Hassel.
Z. physik. Chem. B15, 472-3 (1932).

646. ———.
J. Am. Chem. Soc. 61, 1601-2 (1939).

647. Halverstadt, I. F., and W. D. Kumler.
J. Am. Chem. Soc. 64, 1982 (1942).

648. Hammick, D. L., G. C. Hampson, and G. I. Jenkins.
Nature 136, 990-1 (1935).

649. Hammick, D. L., R. G. A. New, N. V. Sidgwick, and L. E. Sutton.
J. Chem. Soc. 1930, 1876-87.

650. Hammick, D. L., R. G. A. New, and L. E. Sutton.
J. Chem. Soc. 1932, 742-8.

651. Hammick, D. L., R. G. A. New, and R. B. Williams.
J. Chem. Soc. 1934, 29-32.

652. Hammick, D. L., and R. B. Williams.
J. Chem. Soc. 1938, 211-5.

653. Hampson, G. C.
Trans. Faraday Soc. 30, 877-83 (1934).

654. ———, R. H. Farmer, and L. E. Sutton.
Proc. Roy. Soc. (London) A143, 147-68 (1933).

655. Hampson, G. C., and A. Weissberger.
J. Chem. Soc. 1936, 393-8.

656. ———.
J. Am. Chem. Soc. 58, 2111-7 (1936).

657. Hanai, T., N. Koizumi, and R. Gotoh.
Bull. Inst. Chem. Res., Kyoto Univ. 39, 195-201 (1961).

658. Hannay, N. B., and C. P. Smyth.
J. Am. Chem. Soc. 65, 1931-4 (1943).

659. ———.
Ibid., 68, 171-3 (1946).

660. ———.
Ibid., 244-7 (1946).

661. ———.
Ibid., 1005-8 (1946).

662. ———.
Ibid., 1357-60 (1946).

663. Hardung, V.
Helv. phys. Acta 20, 459-61 (1947).

664. Hardy, W. A., and G. Silvey.
Phys. Rev. 95, 385-8 (1954).

665. Harms, H.
Z. physik. Chem. B30, 440-2 (1935).

666. Harris, B., R. J. W. Le Fèvre, and E. P. A. Sullivan.
J. Chem. Soc. <u>1953</u>, 1622-6.

667. Hartley, G. S.
J. Chem. Soc. <u>1938</u>, 633-42.

668. ——— and R. J. W. Le Fèvre.
J. Chem. Soc. <u>1939</u>, 531-5.

669. Hartmann, H., and K. H. Flenner.
Z. physik. Chem. <u>194</u>, 278-83 (1950).

670. Hartmann, H., and H. Gossel.
Z. Elektrochem. <u>61</u>, 337-40 (1957).

671. Hassel, O.
Z. Elektrochem. <u>36</u>, 735-6 (1930).

672. ——— and E. Naeshagen.
Z. physik. Chem. <u>B4</u>, 217-22 (1929).

673. ———.
Ibid., <u>B6</u>, 152-8 (1929).

674. ———.
Ibid., <u>B8</u>, 357-64 (1930).

675. ———.
Ibid., <u>B6</u>, 441-5 (1930).

676. ———.
Tidsskr. Kjemi Bergvesen <u>10</u>, 126-7 (1930).

677. ———.
Ibid., 81-4 (1930).

678. ———.
Z. physik. Chem. <u>B12</u>, 79-88 (1931).

679. ———.
Ibid., <u>B14</u>, 232-6 (1931).

680. ———.
Ibid., <u>B15</u>, 373-6 (1932).

681. ———.
Ibid., 417-20 (1932).

681a ———.
Ibid., <u>B19</u>, 434-42 (1932).

682. Hassel, O., and S. Ore.
Tidsskr. Kjemi Bergvesen Met. <u>6</u>, 72-3 (1946).

683. Hassel, O., and A. H. Uhl.
Z. physik. Chem. <u>B8</u>, 187-206 (1930).

684. ———.
Naturwissenschaften <u>18</u>, 247 (1930).

685. Hata, Y., S. Senoh, and M. Murakami.
Nippon Kagaku Zasshi <u>79</u>, 1531-7 (1958).

686. Havriliak, S., R. W. Swenson, and R. H. Cole.
J. Chem. Phys. <u>23</u>, 134-5 (1955).

687. Hawkins, N. J., V. W. Cohen, and W. S. Koski.
J. Chem. Phys. <u>20</u>, 528 (1952).

688. Hayasaki, K.
J. Chem. Soc. Japan, Pure Chem. Sect. 74, 386-7 (1953).

689. ———.
Ibid., 76, 284-6 (1955).

690. Hayashi, M.
Nippon Kagaku Zasshi 79, 775-80 (1958).

691. ———.
J. Chem. Soc. Japan, Pure Chem. Sect. 80, 1073-8 (1959).

692. Hayman, H. J. G., and I. Eliezer.
J. Chem. Phys. 35, 644-8 (1961).

693. Heil, L.
Phys. Rev. 39, 666-74 (1932).

694. Hein, Fr., A. Schleede, and H. Kallmeyer.
Z. anorg. allgem. Chem. 311, 260-9 (1961).

695. Heinze, R., M. Marder, K. H. Döring, and K. Blechstein.
Oel u. Kohle 37, 8-23 (1941).

696. Henbest, H. B., and B. Nicholls.
Proc. Chem. Soc. 1958, 225.

697. Hennings, C.
Z. physik. Chem. B28, 267-89 (1935).

698. Henrion, J.
Bull. soc. roy. sci. Liége 8, 36-8 (1939).

699. ———.
Ibid., 10, 414-23 (1941).

700. Henriquez, P. C.
Physica 1, 41-52 (1933).

701. ———.
Rec. trav. chim. 53, 1139-40 (1934).

702. Hersh, C. K., G. M. Platz, and R. J. Swehla.
J. Phys. Chem. 63, 1968-9 (1959).

703. Hertel, E., and E. Dumont.
Z. physik. Chem. B30, 139-48 (1935).

704. Hertel, E., and F. Lebok.
Z. physik. Chem. B47, 315-42 (1940).

705. Hertel, E., and M. Schinzel.
Z. physik. Chem. B48, 289-308 (1941).

706. Herweg, J., and W. Pötzsch.
Z. Physik 8, 1-12 (1922).

707. Heston, W. M., Jr., E. J. Hennelly, and C. P. Smyth.
J. Am. Chem. Soc. 72, 2071-5 (1950).

708. Hetland, E.
Acta Chem. Scand. 2, 678-82 (1948).

709. Hibbert, H., and J. S. Allen.
J. Am. Chem. Soc. 54, 4115-6 (1932).

710. Hieber, W., and W. Beck.
Z. Naturforsch. 13b, 194-5 (1958).

711. ———.
Z. anorg. allgem. Chem. 305, 274-8 (1960).

712. Hieber, W., and W. Freyer.
Chem. Ber. 92, 1765-71 (1959).

713. Hieber, W., and G. Wagner.
Z. Naturforsch. 13b, 339-47 (1958).

714. ———.
Ann. 618, 24-30 (1958).

715. Hieber, W., and E. Weiss.
Z. anorg. allgem. Chem. 287, 223-35 (1956).

716. Higashi, K.
Bull. Inst. Phys.-Chem. Res. (Tokyo) 11, 729-40 (1932).

717. ———.
Ibid., 12, 22-34 (1933).

718. ———.
Ibid., 771-9 (1933).

719. ———.
Ibid., 780-9 (1933).

720. ———.
Ibid., 13, 186-94 (1934).

721. ———.
Ibid., 703-15 (1934).

722. ———.
Ibid., 1167-75 (1934).

723. ———.
Sci. Papers Inst. Phys.-Chem. Res. (Tokyo) 24, 57-78 (1934).

724. ———.
Bull. Inst. Phys.-Chem. Res. (Tokyo) 15, 776-86 (1936).

725. ———.
Sci. Papers Inst. Phys.-Chem. Res. (Tokyo) 31, 311-6 (1937).

726. ———.
Ibid., 317-20 (1937).

727. ———.
Bull. Chem. Soc. Japan 13, 158-66 (1938).

728. ———.
Bull. Inst. Phys.-Chem. Res. (Tokyo) 20, 218-24 (1941).

729. ———.
Sci. Papers Inst. Phys.-Chem. Res. (Tokyo) 38, 331-40 (1941).

730. ———.
Bull. Chem. Soc. Japan 25, 159-60 (1952).

731. ———.
Ibid., 26, 248-50 (1953).

732. Higashi, K., and S. Uyeo.
Bull. Chem. Soc. Japan 14, 87-101 (1939).

733. ———.
J. Chem. Soc. Japan 62, 396-9 (1941).

734. ———.
Ibid., 400 (1941).

735. Hill, R. A. W., and L. E. Sutton.
J. Chem. Soc. 1949, 746-53.

736. ———.
J. chim. phys. 46, 244-8 (1949).

737. ———.
J. Chem. Soc. 1953, 1482-90.

738. Hillger, R. E., and M. W. P. Strandberg.
Phys. Rev. 83, 575-81 (1951).

739. ———, T. Wentink, and R. L. Kyhl.
Phys. Rev. 72, 157 (1947).

740. Hirota, E., and Y. Morino.
Bull. Chem. Soc. Japan 33, 158-62 (1960).

741. ———.
Ibid., 34, 341-8 (1961).

742. Hirota, E., T. Oka, and Y. Morino.
J. Chem. Phys. 29, 444-5 (1958).

743. Hitchcock, C. H. S., F. G. Mann, and A. Vanterpool.
J. Chem. Soc. 1957, 4537-46.

744. Hobbs, M. E., and W. W. Bates.
J. Am. Chem. Soc. 74, 746-9 (1952).

745. Hobbs, M. E., J. W. Jacokes, and P. M. Gross.
Rev. Sci. Instr. 11, 126-33 (1940).

746. Hobbs, M. E., and A. J. Weith.
J. Am. Chem. Soc. 65, 967-71 (1943).

747. Hoecker, F. E.
J. Chem. Phys. 4, 431-4 (1936).

748. ———.
Ibid., 5, 372 (1937).

749. Häfelin, J.
Arch. sci. phys. et nat. 28, 19-59 (1946).

750. Højendahl, K.
Nature 117, 892 (1926).

751. ———.
Physik. Z. 30, 391-7 (1929).

752. ———.
Kgl. Danske Videnskab. Selskab, Mat.-Fys. Medd. 24, No. 2 (1946).

753. ——— and K. K. Møller.
Dansk Tids. Farm. 20, 197-200 (1946).

754. Holland, H. G., and R. J. W. Le Fèvre.
J. Chem. Soc. 1950, 2166-9.

755. Holland, R. S., and C. P. Smyth.
J. Am. Chem. Soc. 77, 268-71 (1955).

756. Hollis, N. R. S., and R. L. McIntosh.
Can. J. Chem. 29, 494-7 (1951).

757. Holm, R. H., and F. A. Cotton.
J. Inorg. Nucl. Chem. 15, 63-6 (1960).

758. Honig, A., M. Mandel, M. L. Stitch, and C. H. Townes.
Phys. Rev. 96, 629-42 (1954).

759. Hooper, G. S., and C. A. Kraus.
J. Am. Chem. Soc. 56, 2265-8 (1934).

760. Hovey, R. J.
Dissertation Abstr. 19, 1578 (1959).

761. ——— and A. E. Martell.
J. Am. Chem. Soc. 82, 364-6 (1960).

762. ———.
Ibid., 2697-700 (1960).

763. Howe, J. A., and J. H. Goldstein.
J. Chem. Phys. 23, 1223-5 (1955).

764. Hrostowski, H. J., and R. J. Myers.
J. Chem. Phys. 22, 262-5 (1954).

765. ——— and G. C. Pimentel.
J. Chem. Phys. 20, 518 (1952).

766. Hrynakowski, C., and C. Kalinowski.
Compt. rend. 197, 483-4 (1933).

767. Hsü, T-Y., and C-T. Kwei.
J. Chinese Chem. Soc. 4, 105-16 (1936).

768. Hübel, W., and E. H. Braye.
J. Inorg. Nucl. Chem. 10, 250-68 (1959).

769. Hückel, W., J. Datow, and E. Simmersbach.
Z. physik. Chem. A186, 129-79 (1940).

770. Hückel, W., and W. Jahnentz.
Chem. Ber. B74, 652-6 (1941).

771. Hückel, W., and W. Rothkegel.
Chem. Ber. 81, 71-9 (1948).

772. Hückel, W., and C. M. Salinger.
Chem. Ber. 77, 810-6 (1944).

773. Hückel, W., and W. Wenzke.
Z. physik. Chem. A193, 132-62 (1944).

774. Hughes, E. D., C. G. Le Fèvre, and R. J. W. Le Fèvre.
J. Chem. Soc. 1937, 202-7.

775. Hughes, H. K.
Phys. Rev. 72, 614-23 (1947).

776. ———.
Ibid., 76, 1675-7 (1949).

777. Hughes, R. H.
Phys. Rev. 85, 717 (1952).

778. ——————.
J. Chem. Phys. 21, 959-60 (1953).

779. ——————.
Ibid., 24, 131-8 (1956).

780. Hugill, J. A. C., I. E. Coop, and L. E. Sutton.
Trans. Faraday Soc. 34, 1518-34 (1938).

781. Huisgen, R., and H. Ott.
Angew. Chem. 70, 312 (1958).

782. Huisgen, R., and J. Reinertshofer.
Ann. 575, 197-216 (1952).

783. Huisgen, R., and H. Walz.
Chem. Ber. 89, 2616-29 (1956).

784. Huitric, A. C., and W. D. Kumler.
J. Am. Chem. Soc. 78, 614-22 (1956).

785. ——————.
Ibid., 1147-51 (1956).

786. Hukins, A. A., and R. J. W. Le Fèvre.
J. Chem. Soc. 1949, 898-901.

787. ——————.
Ibid., 2088-91.

788. ——————.
Nature 164, 1050-1 (1949).

789. Hulett, J. R., J. A. Pegg, and L. E. Sutton.
J. Chem. Soc. 1955, 3901-8.

790. Hunter, E. C. E., and J. R. Partington.
J. Chem. Soc. 1931, 2062-70.

791. ——————.
Ibid., 1932, 2812-29.

792. ——————.
Ibid., 1933, 87-90.

793. ——————.
Ibid., 309-13.

794. Hurdis, E. C., and C. P. Smyth.
J. Am. Chem. Soc. 64, 2212-6 (1942).

795. ——————.
Ibid., 2829-34 (1942).

796. ——————.
Ibid., 65, 89-96 (1943).

797. Hurwic, J.
Roczniki Chem. 25, 114-31 (1951).

798. Hutchinson, M. H., and L. E. Sutton.
J. Chem. Soc. 1958, 4382-6.

799. Hüttel, R., and J. Kratzer.
Angew. Chem. 71, 456 (1959).

800. Ibbitson, D. A., T. Jackson, A. McCarthy, and C. W. Stone.
J. Chem. Soc. 1960, 5127-32.

801. Ibbitson, D. A., P. C. James, and R. A. Knaust.
J. Chem. Soc. 1961, 5163-7.

802. Ilias, D., and G. Boudouris.
Compt. rend. 250, 1833-5 (1960).

803. Imamura, Y.
J. Chem. Soc. Japan, Pure Chem. Sect. 76, 217-19 (1955).

804. ———.
Nippon Kagaku Zasshi 81, 1509-12 (1960).

805. ———.
Ibid., 1662-5 (1960).

805a Ingham, C. E., and G. C. Hampson.
J. Chem. Soc. 1939, 981-6.

806. Irving, R., and C. N. Thompson.
Chem. and Ind. (London) 1952, 975-80.

807. Ishiguro, T., T. Chiba, and N. Gotoh.
Bull. Chem. Soc. Japan 30, 25-7 (1957).

808. Ishiguro, T., M. Hamada, and I. Miyagawa.
Botyu Kagaku 16, 220-6 (1951).

809. Isnardi, H.
Z. Physik 9, 153-79 (1922).

810. Issleib, K., and H. O. Fröhlich.
Z. anorg. allgem. Chem. 298, 84-99 (1958).

811. Issleib, K., and B. Mitscherling.
Z. anorg. allgem. Chem. 304, 73-88 (1960).

812. Issleib, K., A. Tzschach, and H. O. Fröhlich.
Z. anorg. allgem. Chem. 298, 164-75 (1958).

813. Jache, A. W., P. W. Moser, and W. Gordy.
J. Chem. Phys. 25, 209-10 (1956).

814. Jacobs, T. L., J. D. Roberts, and W. G. MacMillan.
J. Am. Chem. Soc. 66, 656-7 (1944).

815. Jagielski, A.
Bull. intern. acad. polon. sci., Classe sci. math. nat. A, 312-9 (1937).

816. ——— and J. Weslowski.
Bull. intern. acad. Polonaise A, 260-9 (1935).

817. Janetzky, E. F. J., and M. C. Lebret.
Rec. trav. chim. 63, 123-6 (1944).

818. Jannelli, L., and R. A. Nicolaus.
Gazz. chim. ital. 89, 1457-66 (1959).

819. Jannelli, L., and P. G. Orsini.
Gazz. chim. ital. 86, 1104-10 (1956).

820. ———.
Ibid., 88, 331-41 (1958).

821. Jannelli, L., and P. G. Orsini.
Gazz. chim. ital. 89, 1467-72 (1959).

822. ——— and G. Daniele.
Rend. accad. sci. fis. e mat. (Soc. nazl. sci. Napoli) 20, 153-60 (1953).

823. Jatkar, S. K. K., and C. M. Deshpande.
J. Univ. Poona, Sci. and Technol. No. 6, 48-58 (1954).

824. ———.
J. Indian Chem. Soc. 37, 1-10 (1960).

825. ———.
Ibid., 11-4 (1960).

826. ———.
Ibid., 15-8 (1960).

827. ———.
Ibid., 19-24 (1960).

828. Jatkar, S. K. K., and B. R. Y. Iyengar.
Indian J. Phys. 23, 145-52 (1949).

829. Jatkar, S. K. K., and S. B. Kulkarni.
Science and Culture 14, 482 (1949).

830. ———.
J. Indian Chem. Soc. 27, 273-80 (1950).

831. Jatkar, S. K. K., and V. K. Phansalkar.
J. Univ. Poona, Sci. and Technol. No. 4, 45-54 (1953).

832. ———.
Ibid., No. 6, 59-66 (1954).

833. Jatkar, S. K. K., and D. S. Sastry.
J. Univ. Poona, Sci. and Technol. No. 4, 55-63 (1953).

834. Javan, A., and A. Engelbrecht.
Phys. Rev. 96, 649-58 (1954).

835. Jehlička, V., and A. Reiser.
Chem. listy 52, 1877-81 (1958);
Collection Czechoslov. Chem. Commun. 24, 1250-5 (1959).

836. Jelatis, J. G.
Mass. Inst. Technol., Lab. Insulation Res., Tech. Rept. No. 7, (1947).

837. ———.
J. Applied Phys. 19, 419-25 (1948).

838. Jenkins, H. O.
Nature 133, 106 (1934).

839. ———.
Trans. Faraday Soc. 30, 739-52 (1934).

840. ———.
J. Chem. Soc. 1934, 480-5.

841. ———.
Ibid., 1936, 862-7.

842. ———.
Ibid., 1049-50.

843. Jensen, F. R., and W. E. Coleman.
J. Org. Chem. 23, 869-73 (1958).

844. Jensen, K. A.
Z. anorg. allgem. Chem. 225, 97-114 (1935).

845. ———.
Ibid., 229, 225-51 (1936).

846. ———.
Ibid., 265-81 (1936).

847. ———.
Ibid., 282-4 (1936).

848. ———.
Ibid., 231, 365-71 (1937).

849. ———.
Ibid., 250, 245-56 (1943).

850. ———.
Ibid., 257-67 (1943).

851. ———.
Ibid., 268-76 (1943).

852. ———.
Acta Chem. Scand. 3, 1447-8 (1949).

853. ———.
Ibid., 7, 868-70 (1953).

854. ——— and B. Bak.
J. prakt. Chem. 151, 167-76 (1938).

855. Jensen, K. A., and N. H. Bang.
Ann. 548, 95-105 (1941).

856. ———.
Ibid., 106-110 (1941).

857. Jensen, K. A., and A. Berg.
Ann. 548, 110-7 (1941).

858. Jensen, K. A., and N. Clauson-Kaas.
Z. anorg. allgem. Chem. 250, 277-86 (1943).

859. Jensen, K. A., and A. Friediger.
Kgl. Danske Videnskab. Selskab, Mat.-Fys. Medd. 20, No. 20, 1-54 (1943).

860. ———.
Dansk. Tids. Farm. 16, 280-4 (1942).

861. Jensen, K. A., and C. H. Holten.
Acta Chem. Scand. 3, 1446-7 (1949).

862. Jensen, K. A., and B. Nygaard.
Acta Chem. Scand. 3, 474-80 (1949).

863. Joerges, M., and A. Nikuradse.
Z. Naturforsch. 5a, 259-69 (1950).

864. Johnson, A. W.
J. Org. Chem. 24, 282-4 (1959).

865. Johnson, H. R.
Mass. Inst. Technol. Research Lab., Electronics Quart. Progr. Rept. No. 18, 44-46 (1950).

866. _____ and M. W. P. Strandberg.
Mass. Inst. Technol. Research Lab., Electronics Tech. Rept. No. 192, 20 pp (1951).

867. _____.
J. Chem. Phys. 20, 687-95 (1952).

868. Johnson, R. C., T. L. Weatherly, and Q. Williams.
J. Chem. Phys. 35, 2261-2 (1961).

869. Johnson, R. D., R. J. Myers, and W. D. Gwinn.
J. Chem. Phys. 21, 1425 (1953).

870. Jolliffe, B. R., and C. P. Smyth.
J. Am. Chem. Soc. 80, 1064-6 (1958).

871. Jona, M.
Physik. Z. 20, 14-21 (1919).

872. Kadomtzeff, I.
Bull. soc. chim. France 1949, D394-6.

873. _____.
Compt. rend. 228, 681-3 (1949).

874. _____.
Ibid., 230, 443-5 (1950).

875. _____.
Ibid., 536-7 (1950).

876. Kalinowski, K.
Roczniki Chem. 13, 384-98 (1933).

877. Kambara, S.
J. Soc. Chem. Ind. Japan 38, Suppl. 10, 506-10 (1935).

878. _____.
Ibid., 39, Suppl. 4, 138-40 (1936).

879. _____.
Ibid., 42, Suppl. Binding, 314-5 (1939).

880. _____.
Ibid., 47, 518-9 (1944).

881. Kanda, Y., and R. Shimada.
Mem. Fac. Sci., Kyushu Univ., Ser. C, 3, 127-77 (1960).

882. Kapustin, A. P.
J. Exptl. Theoret. Phys. USSR 17, 30-40 (1947).

883. Karagunis, G., and T. Jannakopoulos.
Z. physik. Chem. B47, 343-56 (1940).

884. Karplus, R., and A. H. Sharbaugh.
Phys. Rev. 75, 1449 (1949).

885. Kartsev, G. N., and Ya. K. Syrkin.
Izv. Akad. Nauk SSSR, Otd. Khim. Nauk 1960, 374-5.

886. _____ and V. F. Mironov.
Izv. Akad. Nauk SSSR, Otd. Khim. Nauk 1960, 948-9.

887. ——— and E. A. Chernyshev.
Dokl. Akad. Nauk SSSR 122, 99-102 (1958).

888. Kasai, P. H., R. J. Myers, D. F. Eggers, Jr., and K. B. Wiberg.
J. Chem. Phys. 30, 512-6 (1959).

889. Kasuya, T., and T. Oka.
J. Phys. Soc. Japan 15, 296-303 (1960).

890. Katagiri, S.
Sci. Repts. Tohoku Univ., First Ser. 44, 165-76 (1960).

891. Katritzky, A. R., E. W. Randall, and L. E. Sutton.
J. Chem. Soc. 1957, 1769-75.

892. Kaufman, M. H., F. M. Ernsberger, and W. S. McEwan.
J. Am. Chem. Soc. 78, 4197-201 (1956).

893. Kaufman, M. H., and A. L. Woodman.
J. Phys. Chem. 62, 508-9 (1958).

894. Kazakova, V. M., and L. S. Fel'dshtein.
Russian J. Phys. Chem. 35, 239-41 (1961).

895. Kazakova, V. M., and Ya. K. Syrkin.
Izv. Akad. Nauk SSSR, Otd. Khim. Nauk 1958, 673-8.

896. ——— and A. N. Shidlovskaya.
Izv. Akad. Nauk SSSR, Otd. Khim. Nauk 1954, 562-3.

897. Kempa, R., and W. H. Lee.
J. Chem. Soc. 1958, 1936-8.

898. Kende, A. S.
Tetrahedron Letters 1959, No. 14, 13-6.

899. Kennard, Sister M. S., and P. A. McCusker.
J. Am. Chem. Soc. 70, 1039-43 (1948).

900. Kenney, M. E., and A. W. Laubengayer.
J. Am. Chem. Soc. 76, 4839-41 (1954).

901. Kesler, M.
Arhiv kem. 27, 67-72 (1955).

902. Keswani, R., and H. Freiser.
J. Am. Chem. Soc. 71, 218-20 (1949).

903. ———.
Ibid., 1789-91 (1949).

904. Ketelaar, J. A. A.
Rec. trav. chim. 59, 757-60 (1940).

905. ———.
Ibid., 62, 289-92 (1943).

906. ———, H. R. Gersmann, and F. Hartog.
Rec. trav. chim. 77, 982-9 (1958).

907. Ketelaar, J. A. A., and N. van Meurs.
Rec. trav. chim. 76, 437-79 (1957).

909. ———.
Ibid., 495-505 (1957).

910. Keyes, F. G., and J. G. Kirkwood.
Phys. Rev. 36, 1570-5 (1930).

911. Kilb, R. W., C. C. Lin, and E. B. Wilson.
J. Chem. Phys. 26, 1695-703 (1957).

912. Kilb, R. W., and L. Pierce.
J. Chem. Phys. 27, 108-12 (1957).

913. Kimura, K., and R. Fujishiro.
Bull. Chem. Soc. Japan 32, 433-9 (1959).

914. Kimura, M., M. Aoki, and Y. Kurita.
Bull. Chem. Soc. Japan 27, 360-4 (1954).

915. King, F. E., and D. G. I. Felton.
J. Chem. Soc. 1949, 274-7.

916. King, W. H., and H. A. Smith.
J. Am. Chem. Soc. 72, 3459-62 (1950).

917. Kisliuk, P.
J. Chem. Phys. 22, 86-92 (1954).

918. Kivelson, D., E. B. Wilson, and D. Lide.
J. Chem. Phys. 32, 205-9 (1960).

919. Klages, G., and E. Klöpping.
Z. Elektrochem. 57, 369-73 (1953).

920. Klages, G., and R. Langpape.
Z. Elektrochem. 63, 533-6 (1959).

921. ———.
Z. Naturforschung 15a, 964-74 (1960).

922. Klages, G., and A. Zentek.
Z. Naturforschung 16a, 1016-21 (1961).

923. Kletenik, Yu. B., and O. A. Osipov.
Zhur. Obshcheĭ Khim. 29, 1423-9 (1959).

924. ——— and E. E. Kravtsov.
Zhur. Obshcheĭ Khim. 29, 11-6 (1959).

925. Kliefoth, W.
Z. Physik 39, 402-14 (1926).

926. Knowles, H. L.
J. Phys. Chem. 36, 2554-66 (1932).

927. Koehl, S. M., and H. H. Wenzke.
J. Am. Chem. Soc. 59, 1418-20 (1937).

928. Kofod, H.
Acta Chem. Scand. 7, 928-32 (1953).

929. ———, L. Kumar, and L. E. Sutton.
J. Chem. Soc. 1951, 1790-4.

930. Kofod, H., L. E. Sutton, W. A. de Jong, P. E. Verkade, and B. M. Wepster.
Rec. trav. chim. 71, 521-4 (1952).

931. Kofod, H., L. E. Sutton, and J. Jackson.
J. Chem. Soc. 1952, 1467-76.

932. Kofod, H., L. E. Sutton, P. E. Verkade, and B. M. Wepster.
Rec. trav. chim. 78, 790-3 (1959).

933. Koizumi, N., and T. Hanai.
J. Phys. Chem. 60, 1496-500 (1956).

934. Kojima, K.
J. Chem. Soc. Japan 62, 903-6 (1941).

935. ——— and S. Mizushima.
Sci. Papers Inst. Phys.-Chem. Res. (Tokyo) 31, No. 697, 296-310 (1937).

936. Kojima, K., K. Sakashita, and S. Maeda.
J. Am. Chem. Soc. 76, 1965-9 (1954).

937. Kojima, K., K. Sakashita, and Y. Takeishi.
Bull. Chem. Soc. Japan 27, 287-90 (1954).

938. Kojima, K., and T. Yoshino.
J. Chem. Soc. Japan, Pure Chem. Sect., 72, 20-2 (1951).

939. Kojima, T., E. L. Breig, and C. C. Lin.
J. Chem. Phys. 35, 2139-44 (1961).

940. Kökeritz, P. G., and H. Selén.
Arkiv Fysik 16, 197-8 (1959).

941. Kokoreva, I. U., and V. M. Kazakova.
Zhur. Obshcheĭ Khim. 31, 371-2 (1961).

942. Kokoshko, Z. Yu., and Z. V. Pushkareva.
Zhur. Obshcheĭ Khim. 24, 877-81 (1954).

943. Koliński, R., H. Piotrowska, and T. Urbański.
Roczniki Chem. 32, 1289-300 (1958).

944. Komandin, A. V., and A. K. Bonetskaya.
Zhur. Fiz. Khim. 28, 1113-9 (1954).

945. ———.
Ibid., 1789-94 (1954).

946. Komandin, A. V., and V. Ya. Rosolovskiĭ.
Zhur. Fiz. Khim. 28, 2215-21 (1954).

947. Kortüm, G.
J. chim. phys. 49, No. 7/8, C127-9 (1952).

948. ——— and M. Buck.
Z. Elektrochem. 60, 53-8 (1956).

949. Kortüm, G., and H. Walz.
Z. Elektrochem. 57, 73-81 (1953).

950. Kosolapoff, G. M.
J. Chem. Soc. 1954, 3222-5.

951. ———.
J. Am. Chem. Soc. 76, 615-7 (1954).

952. ———.
J. Chem. Soc. 1955, 3092-4.

953. Kotera, A.
J. Chem. Soc. Japan 63, 364-7 (1942).

954. ———.
Chem. High Polymers (Japan) 1, 1-7 (1944).

955. Kotera, A.
J. Chem. Soc. Japan, Pure Chem. Sect., 70, 213-6 (1949).

956. ———.
Ibid., 118-21 (1949).

957. ——— and Y. Go.
J. Chem. Soc. Japan 61, 455-62 (1940).

958. Kotera, A., S. Nishimura, and Y. Oto.
J. Chem. Soc. Japan 65, 527-32 (1944).

959. Kotera, A., S. Shibata, and K. Sone.
J. Am. Chem. Soc. 77, 6183-6 (1955).

960. Kozima, K., and T. Yoshino.
J. Am. Chem. Soc. 75, 166-71 (1953).

961. Kraitchman, J., and B. P. Dailey.
J. Chem. Phys. 23, 184-90 (1955).

962. Krasil'nikov, V. Y.
J. Phys. Chem. (USSR) 18, 174-82 (1944).

963. Kraus, C. A., and G. S. Hooper.
Proc. Nat. Acad. Sci. 19, 939-43 (1933).

964. Kraus, G., and A. B. Conciatori.
J. Am. Chem. Soc. 72, 2283-4 (1950).

965. Krause, H. J.
Z. Elektrochem. 59, 1004-8 (1955).

966. Krchma, I. J., and J. W. Williams.
J. Am. Chem. Soc. 49, 2408-16 (1927).

967. Kresze, G., and H. Goetz.
Chem. Ber. 90, 2161-76 (1957).

968. Kresze, G., and H. Smalla.
Chem. Ber. 92, 1042-8 (1959).

969. ———.
Ibid., 1048-55 (1959).

970. Krieger, F. J., and H. H. Wenzke.
J. Am. Chem. Soc. 60, 2115-9 (1938).

971. Krieger, H.
Suomen Kemistilehti 31B, 348-53 (1958).

972. ———.
Ibid., 32B, 109-14 (1959).

973. ——— and J. J. Lindberg.
Suomen Kemistilehti 33B, 117-8 (1960).

974. Krigbaum, W. R., and A. Roig.
J. Chem. Phys. 31, 544-5 (1959).

975. Krisher, L. C., and E. B. Wilson.
J. Chem. Phys. 31, 882-9 (1959).

976. Krishna, B., and K. K. Srivastava.
J. Chem. Phys. 27, 835-7 (1957).

977. ———.
Ibid., 32, 663-4 (1960).

978. Krishna, K. V. G.
Indian J. Phys. 31, 283 (1957).

979. ———.
Trans. Faraday Soc. 53, 767-70 (1957).

980. Krishnamurthy, S.
J. Indian Chem. Soc. 28, 193-7 (1951).

981. Kruh, R., and K. H. Stern.
J. Am. Chem. Soc. 78, 278-81 (1956).

982. Kubo, H., T. Kobayashi, and T. Wada.
Med. J. Osaka Univ. 6, 575-604 (1955).

983. Kubo, M.
Sci. Papers Inst. Phys.-Chem. Res. (Tokyo) 26, 242-57 (1935).

984. ———.
Ibid., 27, 65-73 (1935).

985. ———.
Ibid., 29, 122-8 (1936).

986. ———.
Ibid., 179-87 (1936).

987. ———.
Ibid., 30, 169-79 (1936).

988. ———.
Ibid., 238-43 (1936).

989. ———.
Ibid., 32, 26-32 (1937).

990. ———, Y. Morino, and S. Mizushima.
Sci. Papers Inst. Phys.-Chem. Res. (Tokyo) 32, 129-37 (1937).

991. Kubo, M., T. Nozoe, and Y. Kurita.
Nature 167, 688-9 (1951).

992. Kulkarni, S. B.
J. Indian Chem. Soc. 26, 207-10 (1949).

993. ———.
Ibid., 215-8 (1949).

993a Kumler, W. D.
J. Am. Chem. Soc. 58, 1049-50 (1936).

994. ———.
Ibid., 62, 3292-5 (1940).

995. ———.
Ibid., 64, 1948-50 (1942).

996. ———.
Ibid., 2993-4 (1942).

997. ———.
Ibid., 67, 1901-6 (1945).

998. ———.
Ibid., 74, 261-2 (1952).

999. Kumler, W. D.
J. Am. Chem. Soc. 75, 3092-3 (1953).

1000. ______.
J. Org. Chem. 18, 676-9 (1953).

1001. ______ and G. M. Fohlen.
J. Am. Chem. Soc. 64, 1944-8 (1942).

1002. ______.
Ibid., 67, 437-41 (1945).

1003. Kumler, W. D., and I. F. Halverstadt.
J. Am. Chem. Soc. 64, 1941-3 (1942).

1004. ______.
Ibid., 63, 2182-7 (1941).

1005. ______ and E. L. Alpen.
J. Am. Chem. Soc. 71, 3382-3 (1949).

1006. Kumler, W. D., and A. C. Huitric.
J. Am. Chem. Soc. 78, 3369-74 (1956).

1007. ______ and H. K. Hall, Jr.
J. Am. Chem. Soc. 78, 4345-7 (1956).

1008. Kumler, W. D., and C. M. Lee.
J. Am. Chem. Soc. 82, 6305-6 (1960).

1009. Kumler, W. D., A. Lewis, and J. Meinwald.
J. Am. Chem. Soc. 83, 4591-3 (1961).

1010. Kumler, W. D., N. Pearson, and F. V. Brutcher.
J. Am. Chem. Soc. 83, 2711-4 (1961).

1011. Kumler, W. D., and C. W. Porter.
J. Am. Chem. Soc. 56, 2549-54 (1934).

1012. Kumler, W. D., and P. P. T. Sah.
J. Org. Chem. 18, 669-75 (1953).

1013. Kurita, Y.
Science Repts. Tohoku Univ., First Ser. 38, 85-96 (1954).

1014. ______ and M. Kondo.
Bull. Chem. Soc. Japan 27, 160-3 (1954).

1015. Kurita, Y., and M. Kubo.
Bull. Chem. Soc. Japan 27, 364-7 (1954).

1016. ______.
J. Am. Chem. Soc. 79, 5460-3 (1957).

1017. Kurita, Y., T. Mizuno, T. Mukai, and M. Kubo.
Bull. Chem. Soc. Japan 26, 192-4 (1953).

1018. Kurita, Y., T. Nozoe, and M. Kubo.
J. Chem. Soc. Japan, Pure Chem. Sect. 71, 543-5 (1950).

1019. ______.
Bull. Chem. Soc. Japan 24, 10-3 (1951).

1020. ______.
Ibid., 99-100 (1951).

1021. ______.
Ibid., 26, 242-4 (1953).

1022. Kurita, Y., S. Seto, T. Nozoe, and M. Kubo.
Bull. Chem. Soc. Japan 26, 272-5 (1953).

1023. Kurland, R. J.
Bull. Am. Phys. Soc.[2], 1, 12 (1956).

1024. ______ and E. B. Wilson.
J. Chem. Phys. 27, 585-90 (1957).

1025. Kurosaki, K.
Nippon Kagaku Zasshi 79, 1339-43 (1958).

1026. Kushner, L. M., G. Gorin, and C. P. Smyth.
J. Am. Chem. Soc. 72, 477-9 (1950).

1027. Kushner, L. M., and C. P. Smyth.
J. Am. Chem. Soc. 71, 1401-6 (1949).

1028. Kuz'min, K. I., and G. Kamaĭ.
Dokl. Akad. Nauk SSSR 73, 709-10 (1950).

1029. Kwak, N., J. H. Goldstein, and J. W. Simmons.
J. Chem. Phys. 25, 1203-5 (1956).

1030. Kwart, H.
J. Am. Chem. Soc. 75, 5942-4 (1953).

1031. ______, R. K. Miller, and J. L. Nyce.
J. Am. Chem. Soc. 80, 887-93 (1958).

1032. Kwei, G. H., and D. R. Hershbach.
J. Chem. Phys. 32, 1270-1 (1960).

1033. Kwestroo, W., F. A. Meijer, and E. Havinga.
Rec. trav. chim. 73, 717-36 (1954).

1034. Lal, K. C.
J. Sci. Ind. Res. (India) 20B, 181-2 (1961).

1035. Lander, J. J., and W. J. Svirbely.
J. Am. Chem. Soc. 66, 235-9 (1944).

1036. ______.
Ibid., 67, 322-4 (1945).

1037. Landt, E.
Naturwissenschaften 22, 809 (1934).

1038. Lane, T. J., P. A. McCusker, and B. C. Curran.
J. Am. Chem. Soc. 64, 2076-8 (1942).

1039. Lange, L.
Z. Physik 33, 169-82 (1925).

1040. La Rochelle, J. H.
Dissertation Abstr. 17, 46 (1957).

1041. Larsson, E.
Trans. Chalmers Univ. Technol, Gothenburg, No. 115, 15-9 (1951).

1042. Latosh, N. I., and Z. V. Pushkareva.
Dokl. Akad. Nauk SSSR 124, 98-101 (1958).

1043. Laubengayer, A. W., and R. Bottei.
J. Am. Chem. Soc. 74, 1618-9 (1952).

1044. Laubengayer, A. W., and G. R. Finlay.
J. Am. Chem. Soc. 65, 884-9 (1943).

1045. Lauer, K.
Chem. Ber. B70, 1127-33 (1937).

1046. Laurie, V. W.
J. Chem. Phys. 24, 635-6 (1956).

1047. ______.
Ibid., 26, 1359-62 (1957).

1048. ______.
Ibid., 30, 1210-4 (1959).

1049. ______.
Ibid., 31, 1500-5 (1959).

1050. ______.
Ibid., 34, 291-4 (1961).

1051. ______.
Ibid., 1516-9 (1961).

1052. Lautsch, W.
Z. physik. Chem. B1, 115-33 (1928).

1053. Lawrence, A. R., and A. J. Matuszko.
J. Phys. Chem. 65, 1903-4 (1961).

1054. Lawrance, R. B., and M. W. P. Strandberg.
Mass. Inst. Technol. Research Lab. Electronics,
Tech. Rept. No. 177, 16 pp (1950).

1055. ______.
Phys. Rev. 83, 363-9 (1951).

1056. LeBel, N. A.
J. Am. Chem. Soc. 82, 623-7 (1960).

1057. LeBlanc, O. H., V. W. Laurie, and W. D. Gwinn.
J. Chem. Phys. 33, 598-600 (1960).

1058. Lee, C. A., B. P. Fabricand, R. O. Carlson, and I. I. Rabi.
Phys. Rev. 91, 1395-403 (1953).

1059. Lee, C. M., and W. D. Kumler.
J. Am. Chem. Soc. 83, 4586-90 (1961).

1060. ______.
Ibid., 4593-6 (1961).

1061. ______.
Ibid., 4596-600 (1961).

1062. Lee, S.
J. Soc. Chem. Ind., Japan 43, Suppl., 190-1 (1940).

1063. ______ and I. Sakurada.
Kolloid-Z. 82, 72-5 (1938).

1064. Le Fèvre, C. G., and R. J. W. Le Fèvre.
J. Chem. Soc. 1935, 957-65.

1065. _____.
Ibid., 1696-701.

1066. _____.
Ibid., 1936, 398-9.

1067. _____.
Ibid., 487-91.

1068. _____.
Ibid., 1130-7.

1069. _____.
Ibid., 1937, 196-202.

1070. _____.
Ibid., 1088-90.

1071. _____.
Ibid., 1950, 1829-33.

1072. _____.
Ibid., 3373-5.

1073. _____.
Australian J. Chem. 7, 33-9 (1954).

1074. _____.
J. Chem. Soc. 1956, 3549-63.

1075. _____ and W. T. Oh.
Australian J. Chem. 10, 218-26 (1957).

1076. Le Fèvre, C. G, R. J. W. Le Fèvre, and G. M. Parkins.
J. Chem. Soc. 1960, 1814-9.

1077. Le Fèvre, C. G., R. J. W. Le Fèvre, and B. P. Rao.
J. Chem. Soc. 1959, 2340-4.

1078. _____ and A. J. Williams.
J. Chem. Soc. 1960, 123-8.

1079. Le Fèvre, C. G., R. J. W. Le Fèvre, B. P. Rao, and M. R. Smith.
J. Chem. Soc. 1959, 1188-92.

1080. Le Fèvre, C. G., R. J. W. Le Fèvre, and K. W. Robertson.
J. Chem. Soc. 1935, 480-8.

1081. Le Fèvre, C. G., R. J. W. Le Fèvre, and M. R. Smith.
J. Chem. Soc. 1958, 16-23.

1082. Le Fèvre, R. J. W.
Trans. Faraday Soc. 46, 1-5 (1950).

1083. _____ and T. H. Liddicoet.
J. Chem. Soc. 1951, 2743-8.

1084. Le Fèvre, R. J. W., and F. Maramba.
J. Chem. Soc. 1952, 235-40.

1085. Le Fèvre, R. J. W., F. Maramba, and R. L. Werner.
J. Chem. Soc. 1953, 2496-8.

1086. Le Fèvre, R. J. W., J. W. Mulley, and B. M. Smythe.
J. Chem. Soc. 1950, 290-5.

1087. Le Fèvre, R. J. W., and J. Northcott.
J. Chem. Soc. 1949, 333-7.

1088. ______.
Ibid., 2235-9.

1089. ______.
Ibid., 2374-5.

1090. Le Fèvre, R. J. W., and C. A. Parker.
J. Chem. Soc. 1939, 677.

1091. Le Fèvre, R. J. W., and D. A. A. S. N. Rao.
Australian J. Chem. 7, 135-45 (1954).

1092. ______.
Ibid., 8, 140-2 (1955).

1093. Le Fèvre, R. J. W., and G. J. Rayner.
J. Chem. Soc. 1938, 1921-5.

1094. Le Fèvre, R. J. W., W. P. H. Roberts, and B. M. Smythe.
J. Chem. Soc. 1949, 902-4.

1095. Le Fèvre, R. J. W., and I. G. Ross.
J. Chem. Soc. 1950, 283-90.

1096. ______ and B. M. Smythe.
J. Chem. Soc. 1950, 276-83.

1097. Le Fèvre, R. J. W., and P. Russell.
J. Chem. Soc. 1936, 491-5.

1098. ______.
Ibid., 496-7.

1099. ______.
Trans. Faraday Soc. 43, 374-93 (1947).

1100. Le Fèvre, R. J. W., and J. W. Smith.
J. Chem. Soc. 1932, 2239-45.

1101. ______.
Ibid., 2810-1.

1102. Le Fèvre, R. J. W., and H. Vine.
J. Chem. Soc. 1937, 1805-9.

1103. ______.
Ibid., 1938, 431-8.

1104. ______.
Ibid., 967-72.

1105. ______.
Ibid., 1790-5.

1106. ______.
Ibid., 1795-801.

1107. ______.
Ibid., 1878-82.

1108. Le Fèvre, R. J. W., and A. J. Williams.
J. Chem. Soc. 1960, 108-15.

1109. ______.
Ibid., 1825-9.

1110. Le Fèvre, R. J. W., and C. V. Worth.
J. Chem. Soc. 1951, 1814-7.

1111. Lehn, J., J. Levisalles, and G. Ourisson.
Tetrahedron Letters 1961, 682-6.

1112. Leis, D. G., and B. C. Curran.
J. Am. Chem. Soc. 67, 79-81 (1945).

1113. Leonard, N. J., D. F. Morrow, and M. T. Rogers.
J. Am. Chem. Soc. 79, 5476-9 (1957).

1114. Leonard, N. J., and L. E. Sutton.
J. Am. Chem. Soc. 70, 1564-71 (1948).

1115. Lertes, P.
Z. Physik 6, 56-68 (1921).

1116. Leto, J. R., and F. A. Cotton.
J. Am. Chem. Soc. 81, 2970-3 (1959).

1117. Lew, H., D. Morris, F. E. Geiger, and J. T. Eisinger.
Can. J. Phys. 36, 171-83 (1958).

1118. Lewis, G. L., P. F. Oesper, and C. P. Smyth.
J. Am. Chem. Soc. 62, 3243-6 (1940).

1119. Lewis, G. L., and C. P. Smyth.
J. Am. Chem. Soc. 61, 3063-6 (1939).

1120. ______.
Ibid., 3067-70 (1939).

1121. ______.
J. Chem. Phys. 7, 1085-93 (1939).

1122. ______.
J. Am. Chem. Soc. 62, 1529-33 (1940).

1123. Li, N. C. C.
J. Chem. Phys. 7, 1068 (1939).

1124. ______.
J. Chinese Chem. Soc. 13, 8-10 (1946).

1125. ______, C. V. An, and W. H. Wu.
J. Am. Chem. Soc. 69, 2558-9 (1947).

1126. Li, N. C. C., and F-K. Chen.
J. Chinese Chem. Soc. 13, 1-7 (1946).

1127. Li, N. C. C., and T-L. Chu.
J. Am. Chem. Soc. 69, 558-9 (1947).

1128. Li, N. C. C., and P-C. Hsu.
J. Chinese Chem. Soc. 13, 11-3 (1946).

1129. Li, N. C. C., and T. D. Terry.
J. Am. Chem. Soc. 70, 344-5 (1948).

1130. Lide, D. R.
J. Chem. Phys. 19, 1605-6 (1951).

1131. ______.
Ibid., 20, 1812-3 (1952).

1133. Lide, D. R.
J. Am. Chem. Soc. 74, 3548-52 (1952).

1134. ———.
J. Chem. Phys. 22, 1577-8 (1954).

1135. ———.
Ibid., 27, 343-52 (1957).

1136. ———.
Spectrochim. Acta 1959, 473-6.

1137. ———.
J. Chem. Phys. 33, 1514-8 (1960).

1138. ———.
Ibid., 1519-22 (1960).

1139. ——— and D. K. Coles.
Phys. Rev. 80, 911 (1950).

1140. Lide, D. R., and D. E. Mann.
J. Chem. Phys. 25, 595 (1956).

1141. ———.
Ibid., 26, 734-9 (1957).

1142. ———.
Ibid., 27, 874-7 (1957).

1143. ———.
Ibid., 868-73 (1957).

1144. ———.
Ibid., 28, 572-6 (1958).

1145. ———.
Ibid., 29, 914-20 (1958).

1146. ———.
Ibid., 31, 1129-30 (1959).

1147. Lieber, E., J. Ramachandran, C. N. R. Rao, and C. N. Pillai.
Can. J. Chem. 37, 563-74 (1959).

1148. Liese, E.
Quoted in Z. Elektrochem. 47, 813-9 (1941).

1149. Lifschitz, I.
Rec. trav. chim. 69, 1495-503 (1950).

1150. Lin, W-C., and F-T. Tuan.
J. Chinese Chem. Soc. (Taiwan) 5, 33-45 (1958).

1151. Lind, E. L., M. E. Hobbs, and P. M. Gross.
J. Am. Chem. Soc. 72, 4474-7 (1950).

1152. Lindberg, J. J.
Acta Chem. Scand. 14, 379-84 (1960).

1153. Lindquist, F. E., and C. L. A. Schmidt.
Compt. rend. trav. Lab. Carlsberg, Sér. chim. 22, 307-16 (1938).

1154. Linke, R.
Z. physik. Chem. B46, 251-60 (1940).

1155. ______.
Ibid., 261-9 (1940).

1156. ______.
Ibid., B48, 193-6 (1940).

1157. ______ and W. Rohrmann.
Z. physik. Chem. B35, 256-60 (1937).

1158. Linton, E. P.
J. Am. Chem. Soc. 62, 1945-8 (1940).

1159. ______ and O. Maass.
Can. J. Res. 7, 81-5 (1932).

1160. Littlejohn, A. C., and J. W. Smith.
J. Chem. Soc. 1953, 2456-63.

1161. ______.
Ibid., 1954, 2552-6.

1162. ______.
Ibid., 1957, 2476-82.

1163. Lobunez, W.
Dissertation Abstr. 14, 1939 (1954).

1164. ______, J. R. Rittenhouse, and J. G. Miller.
J. Am. Chem. Soc. 80, 3505-9 (1958).

1165. Longster, G. F., and E. E. Walker.
Trans. Faraday Soc. 49, 228-33 (1953).

1166. Loomis, C. C.
Mass. Inst. Technol. Research Lab., Electronics Progr. Rept. July 15, 1950, p 43-4.

1167. ______ and M. W. P. Strandberg.
Phys. Rev. 81, 798-807 (1951).

1168. Lotspeich, J. F.
Dissertation Abstr. 19, 340-1 (1958).

1169. ______, A. Javan, and A. Engelbrecht.
J. Chem. Phys. 31, 633-43 (1959).

1170. Lowry, T. M., and J. Hofton.
J. Chem. Soc. 1932, 207-11.

1171. Loveluck, G. D.
J. Phys. Chem. 64, 385-7 (1960).

1172. ______.
J. Chem. Soc. 1961, 4729-32.

1173. Lovering, W. F., and L. Wiltshire.
Proc. Inst. Elec. Engrs. 1951, 557-63.

1174. Luce, R. G., and J. W. Trischka.
Phys. Rev. 83, 851-2 (1951).

1175. ______.
J. Chem. Phys. 21, 105-9 (1953).

1176. Luferova, M. A., and Ya. K. Syrkin.
Dokl. Akad. Nauk SSSR 59, 79-82 (1948).

1177. Luft, K. F.
Z. Physik 84, 767-82 (1933).

1178. Lumbroso, H.
Compt. rend. 225, 1003-5 (1947).

1179. ______.
Ibid., 226, 1365-6 (1948).

1180. ______.
Ibid., 228, 77-9 (1949).

1181. ______.
Ibid., 1425-7 (1949).

1182. ______.
Bull. soc. chim. France 1949, D387-93.

1183. ______.
Ann. Fac. Sci. Univ. Toulouse Sci. Math. Sci. Phys. 14, 31-4 (1950) (Pub. 1951).

1184. ______.
Ibid., 35-47 (1950) (Pub. 1951).

1185. ______.
Ibid., 48-55 (1950) (Pub. 1951).

1186. ______.
Ibid., 56-92 (1950) (Pub. 1951).

1187. ______.
Ibid., 93-4 (1950) (Pub. 1951).

1188. ______.
Ibid., 95-107 (1950) (Pub. 1951).

1189. ______.
Ibid., 108-29 (1950) (Pub. 1951).

1190. ______.
Cahiers phys. No. 42, 25-41 (1953).

1191. ______.
Bull. soc. chim. France 1959, 887-90.

1192. ______ and R. Dabard.
Bull. soc. chim. France 1959, 749-53.

1193. Lumbroso, H., and G. Dumas.
Bull. soc. chim. France 1955, 851-9.

1194. Lumbroso, H., L. Gasco, and C. Malén.
Bull. soc. chim. France 1951, 823-8.

1195. Lumbroso, H., R. Golse, and A. Liermain.
Bull. soc. chim. France 1956, 1608-11.

1196. Lumbroso, H., and D. Lauransan.
Bull. soc. chim. France 1959, 513-21.

1197. Lumbroso, H., and C. Marschalk.
J. chim. phys. 48, 123-34 (1951).

1198. ______.
Ibid., 49, 385-93 (1952).

1199. Lumbroso, H., A. Pacault, and B. Pullman.
Bull. soc. chim. France 1950, 34-6.

1200. Lumbroso, H., and G. Pappalardo.
Bull. soc. chim. France 1961, 1131-5.

1201. Lumbroso, H., and R. Passerini.
Bull. soc. chim. France 1955, 1179-85.

1202. ———.
Ibid., 1957, 311-9.

1203. Lumbroso, H., and P. Rumpf.
Bull. soc. chim. France 1950, 371-5.

1204. ———.
Ibid., 1951, 628-32.

1205. ———.
Ibid., 1953, 827-36.

1206. Lütgert, H.
Z. physik. Chem. B14, 27-30 (1931).

1207. ———.
Ibid., 31-5 (1931).

1208. ———.
Ibid., 350-8 (1931).

1209. ———.
Ibid., B17, 460-2 (1932).

1210. Luther, H., and J. Operskalski.
Naturwissenschaften 37, 376-7 (1950).

1211. Lutskiĭ, A. E.
Zhur. Fiz. Khim. 23, 361-7 (1949).

1212. ———, V. T. Alekseeva, and B. P. Kondratenko.
Zhur. Fiz. Khim. 35, 1706-9 (1961).

1213. Lutskiĭ, A. E., and V. V. Dorofeev.
Zhur. Fiz. Khim. 33, 331-4 (1959).

1214. Lutskiĭ, A. E., and L. A. Kochergina.
Zhur. Fiz. Khim. 33, 174-9 (1959).

1215. ———.
Ibid., 2135-9 (1959).

1216. Lutskiĭ, A. E., and B. P. Kondratenko.
Zhur. Fiz. Khim. 33, 2017-23 (1959).

1217. ———.
Zhur. Obshcheĭ Khim. 29, 2073-6 (1959).

1218. ———.
Ibid., 2077-9 (1959).

1219. Lutskiĭ, A. E., and V. N. Konel'skaya.
Zhur. Fiz. Khim. 35, 1938-43 (1961).

1220. Lutskiĭ, A. E., L. M. Volova, and P. A. Chernyavskiĭ.
Zhur. Obshcheĭ Khim. 30, 4085-8 (1960).

1221. Lüttke, W., and R. Mecke.
Z. Elektrochem. 53, 241-9 (1949).

1222. Lutton, J. M., and R. W. Parry.
J. Am. Chem. Soc. 76, 4271-4 (1954).

1223. Lüttringhaus, A., and J. Grohmann.
Z. Naturforsch. 10b, 365-7 (1955).

1224. Madden, R. P., and W. S. Benedict.
J. Chem. Phys. 23, 408-9 (1955).

1225. Magnuson, D. W.
J. Chem. Phys. 19, 1071 (1951).

1226. ——.
Ibid., 1614 (1951).

1227. ——.
Ibid., 20, 229-32 (1952).

1228. ——.
Ibid., 24, 344-7 (1956).

1229. ——.
Ibid., 27, 223-6 (1957).

1234. Mahanti, P. C.
J. Indian Chem. Soc. 6, 743-7 (1929).

1235. ——.
Physik. Z. 31, 546-55 (1930).

1236. ——.
Phil. Mag. (7) 20, 274-87 (1935).

1237. ——.
Z. Physik 94, 220-3 (1935).

1238. —— and R. N. Das-Gupta.
J. Indian Chem. Soc. 6, 411-7 (1929).

1239. ——.
Indian J. Physics 3, 467-75 (1929).

1240. Mahanti, P. C., and D. N. Sen-Gupta.
Indian J. Physics 3, 181-96 (1928).

1241. ——.
J. Indian Chem. Soc. 5, 673-81 (1928).

1241a Maĭbaum, B. K.
J. Exptl. Theoret. Phys. (USSR) 9, 1383-7 (1939).

1241b Maier, W.
Arch. Sci. (Geneva) fasc. spec. 12, 20-3 (1959).

1241c —— and G. Baumgartner.
Z. Naturforsch. 7a, 172-5 (1952).

1241d Maier, W., and G. Meier.
Z. physik. Chem. (Frankfurt) 13, 251-4 (1957).

1241e ——.
Z. Naturforsch. 16a, 470-7 (1961).

1242. Malatesta, L.
Gazz. chim. ital. 69, 629-39 (1939).

1243. ——.
Ibid., 70, 541-53 (1940).

1244. ——.
Ibid., 734-7 (1940).

1245. ——.
Rend. ist. lombardo sci. 78, 103-10 (1944-5).

1246. ——.
Gazz. chim. ital. 76, 182-6 (1946).

1247. —— and R. Pizzotti.
Gazz. chim. ital. 72, 491-6 (1942).

1248. ——.
Ibid., 73, 143-8 (1943).

1249. ——.
Ibid., 344-9 (1943).

1250. ——.
Ibid., 349-55 (1943).

1251. Malone, J. G.
J. Chem. Phys. 1, 197-9 (1933).

1252. Malone, M. G., and A. L. Ferguson.
J. Chem. Phys. 2, 99-104 (1934).

1253. Mann, F. G., and A. Senior.
J. Chem. Soc. 1954, 4476-80.

1254. Marchal, J., and C. Lapp.
J. Polymer Sci. 27, 571-3 (1958).

1255. Marinangeli, A.
Ann. chim. (Rome) 44, 219-31 (1954).

1256. ——.
Ibid., 880-3 (1954).

1257. Markby, R., I. Wender, R. A. Friedel, F. A. Cotton, and H. W. Sternberg.
J. Am. Chem. Soc. 80, 6529-33 (1958).

1258. Marsden, R. J. B., and L. E. Sutton.
J. Chem. Soc. 1936, 599-606.

1259. ——.
Ibid., 1383-90.

1260. Marshall, S. A., and J. Weber.
Phys. Rev. 105, 1502-6 (1957).

1261. Martin, A. R.
Nature 135, 909 (1935).

1262. Martin, G.
Physik. Z. 37, 665-77 (1936).

1263. Martin, G. T. O., and J. R. Partington.
J. Chem. Soc. 1936, 158-63.

1264. Martin, G. T. O., and J. R. Partington.
J. Chem. Soc. 1936, 1175-8.

1265. ———.
Ibid., 1178-82.

1266. ———.
Ibid., 1182-4.

1267. Maryott, A. A.
J. Am. Chem. Soc. 63, 3079-82 (1941).

1268. ———.
J. Research Natl. Bur. Standards 41, 1-6 (1948).

1269. ———.
Ibid., 7-9 (1948).

1270. ——— and S. F. Acree.
J. Res. U.S. Natl. Bur. Standards 33, 71-4 (1944).

1271. ———.
Ibid., 38, 505 (1947).

1272. Maryott, A. A., and G. Birnbaum.
J. Chem. Phys. 24, 1022-6 (1956).

1273. Maryott, A. A., and F. Buckley.
Natl. Bur. Standards (U.S.) Circ. No. 537 (1953).

1274. Maryott, A. A., M. E. Hobbs, and P. M. Gross.
J. Am. Chem. Soc. 62, 2320-4 (1940).

1275. ———.
Ibid., 63, 659-63 (1941).

1276. ———.
Ibid., 71, 1671-4 (1949).

1277. Maryott, A. A., and S. J. Kryder.
J. Chem. Phys. 27, 1221-2 (1957).

1278. ———.
Ibid., 31, 617-21 (1959).

1279. Massey, J. T., and D. R. Bianco.
J. Chem. Phys. 22, 442-8 (1954).

1280. Mathias, S., and E. de Carvalho Filho.
J. Phys. Chem. 62, 1427-30 (1958).

1281. ——— and R. G. Cecchini.
J. Phys. Chem. 65, 425-7 (1961).

1282. Mautner, H. G., and W. D. Kumler.
J. Am. Chem. Soc. 78, 97-101 (1956).

1283. Mays, J. M., and B. P. Dailey.
J. Chem. Phys. 20, 1695-702 (1952).

1284. McAlpine, K. B., and C. P. Smyth.
J. Am. Chem. Soc. 55, 453-62 (1933).

1285. ———.
J. Chem. Phys. 3, 55-7 (1935).

1286. McCarthy, P. J.
Dissertation Abstr. 15, 1506 (1955).

1287. ——— and A. E. Martell.
J. Am. Chem. Soc. 78, 264-6 (1956).

1288. ———.
Ibid., 2106-8 (1956).

1289. McCusker, P. A., and B. C. Curran.
J. Am. Chem. Soc. 64, 614-7 (1942).

1290. McCusker, P. A., T. J. Lane, and M. S. Kennard.
J. Am. Chem. Soc. 81, 2974-6 (1959).

1291. McDowell, C. A., H. G. Emblem, and E. A. Moelwyn-Hughes.
J. Chem. Soc. 1948, 1206-8.

1292. McLaughlin, E. P., and R. L. Scott.
J. Phys. Chem. 60, 674-6 (1956).

1293. Mead, T. E., and L. B. Clapp.
J. Org. Chem. 23, 921-2 (1958).

1294. Meakins, R. J.
Trans. Faraday Soc. 52, 320-7 (1956).

1295. Mecke, R., and A. Reuter.
Z. Naturforsch. 4a, 368-78 (1949).

1296. ———.
Naturwissenschaften 36, 251-2 (1949).

1297. Mecke, R., and H. Specht.
Z. Elektrochem. 62, 500-5 (1958).

1299. Melander, B.
Farm. Revy 47, 503-4 (1948).

1300. Meredith, C. C., and G. F. Wright.
Can. J. Chem. 38, 1177-90 (1960).

1301. Meredith, C. C., L. Westland, and G. F. Wright.
J. Am. Chem. Soc. 79, 2385-90 (1957).

1302. Meschi, D. J., and R. J. Myers.
J. Mol. Spectr. 3, 405-16 (1959).

1303. Meuche, D., T. Gäumann, and E. Heilbronner.
Helv. Chim. Acta 41, 2230 (1958).

1304. Meyer, L.
Z. physik. Chem. B8, 27-54 (1930).

1305. ——— and A. Büchner.
Physik. Z. 33, 390-1 (1932).

1306. Meyer, R. T., and R. J. Myers.
J. Chem. Phys. 34, 1074-5 (1961).

1307. Middleton, B. A., and J. R. Partington.
Nature 141, 516-7 (1938).

1308. Mikhaĭlov, G. P., and D. Tishchenko.
Acta Physicochem. URSS 12, 129 (1940).

1309. Mikhaĭlov, G. P., and L. L. Burshtein.
Soviet Phys., Tech. Phys. 4, 165-70 (1959).

1310. Mikhaĭlov, G. P., and L. V. Krasner.
Zhur. Tekh. Fiz. 23, 1931-5 (1953).

1311. Miles, J. B.
Phys. Rev. 34, 964 (1929).

1312. Millar, I. T., C. T. Mortimer, and H. D. Springall.
J. Chem. Soc. 1957, 3456-7.

1313. Millen, D. J., and J. R. Morton.
Chem. and Ind. (London) 1956, 954.

1314. ———.
J. Chem. Soc. 1960, 1523-8.

1315. Millen, D. J., and K. M. Sinnott.
J. Chem. Soc. 1958, 350-5.

1316. Miller, J. G., and H. S. Angel.
J. Am. Chem. Soc. 68, 2358-9 (1946).

1317. Miller, R. C., and C. P. Smyth.
J. Chem. Phys. 24, 814-7 (1956).

1318. Miller, R. F.
Phys. Rev. 89, 341 (1953).

1319. Milone, M.
Gazz. chim. ital. 65, 94-102 (1935).

1320. ———.
Ibid., 152-8 (1935).

1321. ——— and E. Borello.
Ann. chim. (Rome) 41, 320-2 (1951).

1322. Milone, M., and G. Tappi.
Atti Congr. intern. chim., 10th, Rome 2, 352-7 (1938).

1323. ———.
Atti accad. sci. Torino, I. Classe Sci.Fis., Mat. Nat. 75, 454-60 (1940).

1324. ——— and E. Borello.
Ann. chim. (Rome) 41, 333-9 (1951).

1325. Mirri, A. M., A. Guarnieri, and P. Favero.
Nuovo cimento 19, 1189-94 (1961).

1326. Miyagawa, I.
J. Chem. Soc. Japan, Pure Chem. Sect. 75, 1162-5 (1954).

1327. ———.
Ibid., 1165-9 (1954).

1328. ———, Y. Morino, and R. Riemschneider.
Bull. Chem. Soc. Japan 27, 177-81 (1954).

1329. Miyagawa, I., T. Chiba, S. Ikeda, and Y. Morino.
Bull. Chem. Soc. Japan 30, 218-22 (1957).

1330. Miyazaki, H.
Osaka Daigaku Igaku Zasshi 11, 4306-9 (1959).

1331. Mizushima, S., and K. Higashi.
Proc. Imp. Acad. (Tokyo) 8, 482-5 (1932).

1332. Mizushima, S., and M. Kubo.
Bull. Chem. Soc. Japan 13, 174-81 (1938).

1333. Mizushima, S., Y. Morino, and K. Kojima.
Sci. Papers Inst. Phys.-Chem. Res. (Tokyo) 29, 111-21 (1936).

1334. Mizushima, S., Y. Morino, and M. Kubo.
Physik. Z. 38, 459-62 (1937).

1335. Mizushima, S., Y. Morino, and S. Noziri.
Sci. Papers Inst. Phys.-Chem. Res. (Tokyo) 25, 159-221 (1934).

1336. Mizushima, S., Y. Morino, and H. Okazaki.
Sci. Papers Inst. Phys.-Chem. Res. (Tokyo) 34, 1147-63 (1938).

1337. Mizushima, S., T. Shimanouchi, I. Ichishima, T. Miyazawa, I. Nakagawa, and T. Araki.
J. Am. Chem. Soc. 78, 2038-41 (1956).

1338. Mizushima, S., T. Shimanouchi, K. Kuratani, M. Tsuboi, T. Sugita, I. Nakagawa, and K. Kurosaki.
Nature 169, 1058-9 (1952).

1339. Mizushima, S., T. Shimanouchi, S. Nagakura, K. Kuratani, M. Tsuboi, H. Baba, and O. Fujioka.
J. Am. Chem. Soc. 72, 3490-4 (1950).

1340. ———.
J. Chem. Soc. Japan, Pure Chem. Sect. 71, 35-7 (1950).

1341. Mizushima, S., K. Suenaga, and K. Kojima.
Bull. Chem. Soc. Japan 10, 167-8 (1935).

1342. Mizutani, M., J. Shiraishi, and Y. Koyama.
Osaka Daigaku Igaku Zasshi 10, 1189-90 (1958).

1343. Mockler, R. C., and G. R. Bird.
Phys. Rev. 98, 1837-9 (1955).

1344. Moede, J. A., and C. Curran.
J. Am. Chem. Soc. 71, 852-8 (1949).

1345. Mohler, H.
Helv. Chim. Acta 21, 67-72 (1938).

1346. ———.
Ibid., 784-6 (1938).

1347. ———.
Ibid., 787-8 (1938).

1348. ———.
Ibid., 789-92 (1938).

1349. ——— and J. Sorge.
Helv. Chim. Acta 20, 1447-57 (1937).

1350. ———.
Ibid., 23, 1200-11 (1940).

1351. Moll, F., and E. Lippert.
Z. Elektrochem. 58, 853-9 (1954).

1352. Moore, E. M., and M. E. Hobbs.
J. Am. Chem. Soc. 71, 411-13 (1949).

1353. Morgan, S. O., and H. H. Lowry.
J. Phys. Chem. 34, 2385-432 (1930).

1354. Morino, Y., I. Miyagawa, T. Chiba, and T. Shimozawa.
Bull. Chem. Soc. Japan 30, 222-3 (1957).

1355. Morino, Y., I. Miyagawa, T. Haga, and S. Mizushima.
Bull. Chem. Soc. Japan 28, 165-71 (1955).

1356. Morino, Y., I. Miyagawa, and T. Oiwa.
Botyu Kagaku 15, 181-9 (1950).

1357. Morino, Y., I. Miyagawa, and A. Wada.
J. Chem. Phys. 20, 1976-7 (1952).

1358. Morino, Y., H. Shiio, and I. Miyagawa.
Repts. Radiation Chem. Research Inst., Tokyo Univ. 5, 6-7 (1950).

1359. Mortimer, C. T., H. Spedding, and H. D. Springall.
J. Chem. Soc. 1957, 188-91.

1360. Moureu, H.
Compt. rend. 202, 314-6 (1936).

1361. Mousseron, M., R. Granger, H. Bourrel, E. Canals, and J. Cabanes.
Bull. soc. chim. France 1947, 605-15.

1362. Mousseron, M., R. Richaud, R. Granger, F. Winternitz, G. Combes, E. Canals, L. Souche, J. Cabanes, and P. Froger.
Bull. soc. chim. France 1946, 629-39

1363. Müller, A.
Proc. Roy. Soc. (London) A158, 403-14 (1937).

1364. Müller, F. H.
Physik. Z. 35, 1009-11 (1934).

1365. ———.
Ibid., 38, 283-92 (1937).

1366. Müller, H., and H. Sack.
Physik. Z. 31, 815-22 (1930).

1367. Murty, C. R.
J. Sci. Ind. Research (India) 15B, 260 (1956).

1368. ———.
Ibid., 16B, 334-6 (1957).

1369. ———.
Indian J. Phys. 31, 256-60 (1957).

1370. ———.
Ibid., 32, 365-8 (1958).

1371. ———.
Ibid., 492-6 (1958).

1372. ———.
Ibid., 516-20 (1958).

1373. ———.
J. Sci. Ind. Research (India) 17B, 441-4 (1958).

1374. ———.
Ibid., 18B, 268-70 (1959).

1375. ——— and D. V. G. L. Narasimha Rao.
J. Sci. Ind. Research (India) 15B, 346-9 (1956).

1376. ———.
Ibid., 350-2 (1956).

1377. ———.
Current Sci. 25, 49 (1956).

1378. Myers, A. L., and T. De Vries.
J. Am. Chem. Soc. 73, 1813-5 (1951).

1379. Myers, R. J., and W. D. Gwinn.
J. Chem. Phys. 20, 1420-7 (1952).

1380. Nachod, F. C., A. R. Surrey, G. Y. Lesher, C. M. Martini,
J. R. Mayer, M. Priznar, and W. G. Webb.
J. Am. Chem. Soc. 81, 2897-8 (1959).

1381. Nace, H. R., and R. B. Turner.
J. Am. Chem. Soc. 75, 4063-6 (1953).

1382. ———.
J. Org. Chem. 25, 1403-5 (1960).

1383. Naeshagen, E. Quoted by A. Langseth and B. Quiller.
Z. physik. Chem. B27, 79-99 (1934).

1384. Naeshagen, E.
Z. physik. Chem. B25, 157-60 (1934).

1385. Nagakura, S., and H. Baba.
Repts. Radiation Chem. Research Inst., Tokyo Univ. 5, 19-22 (1950).

1386. ———.
J. Chem. Soc. Japan, Pure Chem. Sect. 72, 217-20 (1951).

1387. ———.
J. Am. Chem. Soc. 74, 5693-8 (1952).

1388. Nagakura, S., and A. Kuboyama.
Rept. Inst. Sci. Technol., Univ. Tokyo 5, 27-31 (1951).

1389. ———.
J. Chem. Soc. Japan, Pure Chem. Sect. 74, 499-503 (1953).

1390. ———.
J. Am. Chem. Soc. 76, 1003-5 (1954).

1391. Nakagawa, I.
Nippon Kagaku Zasshi 79, 1353-7 (1958).

1392. ———.
Ibid., 1358-61 (1958).

1393. Nakashima, R., S. Watarai, and T. Kinugasa.
Bull. Chem. Soc. Japan 34, 1740-1 (1961).

1394. Nakata, N.
Chem. Ber. B64, 2059-69 (1931).

1395. ———.
Bull. Chem. Soc. Japan 10, 318-23 (1935).

1396. Narasimhan, P. T.
Proc. Indian Acad. Sci. 37A, 551-6 (1953).

1397. ———.
J. Indian Inst. Sci. 37A, 30-4 (1955).

1398. ———.
Ibid., 35-8 (1955).

1399. ———.
Proc. Natl. Inst. Sci. India 24, Pt A, 121-9 (1958).

1400. ——— and S. V. Anantakrishnan.
Proc. Indian Acad. Sci. 37A, 747-51 (1953).

1401. Narayana, B. L.
J. Sci. Ind. Research (India) 18B, 304 (1959).

1402. Naves, Y. R., and E. Perrottet.
Helv. Chim. Acta 23, 912-25 (1940).

1403. ———.
Ibid., 24, 3-29 (1941).

1404. Naylor, N. E., and E. B. Wilson.
J. Chem. Phys. 26, 1057-60 (1957).

1405. Nazarova, L. M.
Zhur. Fiz. Khim. 28, 36-41 (1954).

1406. ——— and Ya. K. Syrkin.
Izv. Akad. Nauk SSSR, Otd. Khim. Nauk 1949, 35-43.

1407. ———.
Zhur. Obshcheĭ Khim. 19, 777-80 (1949).

1408. ———.
Ibid., 23, 478-81 (1953).

1409. Nechaeva, O. N., and Z. V. Pushkareva.
Zhur. Obshcheĭ Khim. 28, 2702-5 (1958).

1410. Neckel, A., and H. Volk.
Z. Elektrochem. 62, 1104-15 (1958).

1411. Nederbragt, G. W., and J. Pelle.
Mol. Phys. 1, 97-8 (1958).

1412. Neil, D. E., and C. R. Estee.
Proc. S. Dakota Acad. Sci. 33, 182-7 (1954).

1413. Nerdel, F., and I. Huldschinsky.
Chem. Ber. 86, 1005-10 (1953).

1414. Nesmeyanov, A. N., and A. E. Borisov.
Dokl. Akad. Nauk SSSR 60, 67-72 (1948).

1415. Nespital, W.
Z. physik. Chem. B16, 153-79 (1932).

1416. New, R. G. A., and L. E. Sutton.
J. Chem. Soc. 1932, 1415-22.

1417. Nickerson, J. D., and R. McIntosh.
Can. J. Chem. 35, 1325-31 (1957).

1418. Niini, A.
Ann. Acad. Sci. Fennicae Ser. I, A46, 108 pp (1936).

1419. Nikuradse, A., and A. Berger.
Physik. Z. 45, 71-81 (1944).

1420. Nöth, H., and H. Bayer.
Chem. Ber. 93, 939-44 (1960).

1421. Novak, A., and E. Whalley.
Can. J. Chem. 36, 1116-20 (1958).

1422. Nozoe, T., T. Mukai, and K. Matsui.
Proc. Japan Acad. 27, 646-8 (1951).

1423. Nozoe, T., T. Mukai, and I. Murata.
J. Am. Chem. Soc. 76, 3352-3 (1954).

1424. Nukada, N.
Nia Kemio 5, 41-6 (1932).

1425. Nukada, K., and Y. Bansho.
Bull. Chem. Soc. Japan 26, 454-7 (1953).

1426. Oehme, F.
Farbe Lack 64, 183-5 (1958).

1427. Oesper, P. F., and C. P. Smyth.
J. Am. Chem. Soc. 64, 173-5 (1942).

1428. ———.
Ibid., 768-71 (1942).

1429. ——— and M. S. Kharasch.
J. Am. Chem. Soc. 64, 937-40 (1942).

1430. Oesper, P. F., G. L. Lewis, and C. P. Smyth.
J. Am. Chem. Soc. 64, 1130-3 (1942).

1431. Oncley, J. L.
J. Am. Chem. Soc. 60, 1115-23 (1938).

1432. ———.
J. Phys. Chem. 44, 1103-13 (1940).

1433. ———.
Chap. 22 of "Proteins and Amino Acids," Reinhold, New York, N.Y., Edited by Cohn and Edsall.

1434. O'Reilly, J. M., and L. Pierce.
J. Chem. Phys. 34, 1176-81 (1961).

1435. Oriani, R. A., and C. P. Smyth.
J. Am. Chem. Soc. 70, 125-30 (1948).

1436. ———.
J. Chem. Phys. 17, 1174-8 (1949).

1437. Orthner, L., and G. Freyss.
Ann. 484, 131-54 (1930).

1438. Osipov, O. A.
Zhur. Obshcheĭ Khim. 26, 322-8 (1956).

1439. ———, V. M. Artemova, and N. G. Bedarev.
Zhur. Obshcheĭ Khim. 29, 975-9 (1959).

1440. Osipov, O. A., and V. M. Artemova.
Dokl. Akad. Nauk SSSR 133, 166-9 (1960).

1441. Osipov, O. A., and O. E. Kashireninov.
Zhur. Obshcheĭ Khim. 31, 1755-9 (1961).

1442. Osipov, O. A., and Yu. B. Kletenik.
Zhur. Obshcheĭ Khim. 27, 2921-7 (1957).

1443. ———.
Zhur. Neorg. Khim. 2, 2406-9 (1957).

1444. ———.
Zhur. Obshcheĭ Khim. 29, 2119-24 (1959).

1445. ———.
Zhur. Neorg. Khim. 4, 1494-7 (1959).

1446. ———.
J. Gen. Chem. USSR 29, 2085-9 (1960).

1447. Osipov, O. A., and Yu. A. Lysenko.
Zhur. Obshcheĭ Khim. 30, 3866-9 (1960).

1448. Osipov, O. A., and M. A. Panina.
Zhur. Fiz. Khim. 32, 2287-93 (1958).

1449. Osipov, O. A., G. S. Samofalova, and E. I. Glushko.
Zhur. Obshcheĭ Khim. 27, 1428-33 (1957).

1450. Osipov, O. A., and I. K. Shelomov.
Nauchn. Dokl. Vyssheĭ Shkoly, Khim. i Khim. Tekhnol. 1959, No. 2, 253-5.

1451. Osipov, O. A., A. M. Simonov, V. I. Minkin, and A. D. Garnovskiĭ.
Dokl. Akad. Nauk SSSR 137, 1374-6 (1961).

1452. Osthoff, R. C., C. A. Brown, and J. A. Hawkins.
J. Am. Chem. Soc. 73, 5480 (1951).

1453. Osthoff, R. C., and E. G. Rochow.
J. Am. Chem. Soc. 74, 845 (1952).

1454. Ostwald, W., and R. Riedel.
Kolloid-Z. 59, 150 (1932).

1455. ———.
Ibid., 69, 185-99 (1934).

1456. ———.
Ibid., 70, 75-9 (1935).

1457. Otto, M. M.
J. Am. Chem. Soc. 57, 693-5 (1935).

1458. ———.
Ibid., 1476-8 (1935).

1459. ———.
Ibid., 59, 1590-2 (1937).

1460. ——— and H. H. Wenzke.
Ind. Eng. Chem. Anal. Ed. 6, 187-8 (1934).

1461. ———.
J. Am. Chem. Soc. 56, 1314-5 (1934).

1462. ———.
Ibid., 57, 294-5 (1935).

1463. Pal, N. N.
Phil. Mag. (7) 10, 265-80 (1930).

1464. Palm, K., and H. Dunken.
Z. physik. Chem. (Leipzig) 217, 248-62 (1961).

1465. Paranjpe, G. R., and D. J. Davar.
Indian J. Physics 12, 283-8 (1938).

1466. ———.
Ibid., 15, 173-83 (1941).

1467. Paranjpe, G. R., and P. Y. Deshpande.
Proc. Indian Acad. Sci. A1, 880-6 (1935).

1468. ———.
J. Univ. Bombay 9, Pt. 3, 24-37 (1940).

1469. Paranjpe, G. R., and M. B. Vijifdar.
Indian J. Phys. 20, 197-204 (1946).

1470. Parsons, G. S., and C. W. Porter.
J. Am. Chem. Soc. 55, 4745-6 (1933).

1471. Partington, J. R.
Trans. Faraday Soc. 30, 822-3 (1934).

1472. ——— and D. I. Coomber.
Nature 139, 510 (1937).

1473. ———.
Ibid., 141, 918 (1938).

1474. Partington, J. R., and E. G. Cowley.
Nature 135, 474 (1935).

1475. ———.
Ibid., 1038 (1935).

1476. Parts, A.
Z. physik. Chem. B4, 227-33 (1929).

1477. ———.
Ibid., B7, 327-38 (1930).

1478. ———.
Ibid., B10, 264-72 (1930).

1479. ———.
Ibid., B12, 312-22 (1931).

1480. ———.
Ibid., 323-6 (1931).

1483. Pearce, J. N., and L. F. Berhenke.
Proc. Iowa Acad. Sci. 41, 141-2 (1934).

1484. ———.
Ibid., 40, 93 (1933).

1485. ———.
J. Phys. Chem. 39, 1005-10 (1935).

1486. Petro, A. J., C. P. Smyth, and L. G. S. Brooker.
J. Am. Chem. Soc. 78, 3040-3 (1956).

1487. Petro, A. J., and C. P. Smyth.
J. Am. Chem. Soc. 79, 6142-7 (1957).

1488. Petro, A. J., and C. P. Smyth.
J. Am. Chem. Soc. 79, 6147-9 (1957).

1489. ———.
J. Am. Chem. Soc. 80, 73-6 (1958).

1490. Petrov, A. A., B. S. Kupin, T. V. Yakovleva, and K. S. Mingaleva.
J. Gen. Chem. USSR 29, 3689-94 (1959).

1491. Petrov, A. A., and K. S. Mingaleva.
J. Gen. Chem. USSR 29, 2785-8 (1959).

1492. ——— and B. S. Kupin.
Dokl. Akad. Nauk SSSR 123, 298-300 (1958).

1493. Petrov, A. A., K. S. Mingaleva, M. D. Stadnichuk, and I. A. Maretina.
Zhur. Obshcheĭ Khim. 31, 3521-4 (1961).

1494. Pflaum, D. J., and H. H. Wenzke.
J. Am. Chem. Soc. 56, 1106-7 (1934).

1495. Phadke, R. P.
J. Indian Inst. Sci. 34, 189-207 (1952).

1496. Phadke, R. S.
J. Indian Inst. Sci. 35A, 123-9 (1953).

1497. Phadke, S. R., S. D. Gokhale, N. L. Phalnikar, and B. V. Bhide.
J. Indian Chem. Soc. 22, 235-8 (1945).

1498. Phadke, S. R., N. L. Phalnikar, and B. V. Bhide.
J. Indian Chem. Soc. 22, 239-42 (1945).

1499. Phalnikar, N. L.
J. Univ. Bombay 11, Pt. 3, 87-91 (1942).

1500. ——— and B. V. Bhide.
Current Sci. 8, 473-4 (1939).

1501. ——— and K. S. Nargund.
J. Univ. Bombay 10, Pt. 3, 48-52 (1941).

1502. Phillips, G. M., J. S. Hunter, and L. E. Sutton.
J. Chem. Soc. 1945, 146-62.

1503. Piekara, A.
Phys. Rev. 42, 449-50 (1932).

1504. ———.
Bull. intern. acad. polonaise A, 33 (1933).

1505. ——— and B. Piekara.
Compt. rend. 198, 1018-20 (1934).

1506. Pierce, L.
J. Chem. Phys. 29, 383-8 (1958).

1507. ———, R. Jackson, and N. Di Cianni.
J. Chem. Phys. 35, 2240-1 (1961).

1508. Pierce, L., and M. Hayashi.
J. Chem. Phys. 35, 479-85 (1961).

1509. Pierce, L., and L. C. Krisher.
J. Chem. Phys. 31, 875-82 (1959).

1510. Pierce, L., and J. M. O'Reilly.
J. Mol. Spectr. 3, 536-47 (1959).

1511. Pierce, L., and D. H. Petersen.
J. Chem. Phys. 33, 907-13 (1960).

1512. Pilpel, N.
J. Am. Chem. Soc. 77, 2949-53 (1955).

1513. ———.
Trans. Faraday Soc. 56, 893-902 (1960).

1514. Pinchas, S., E. Zimkin, and E. D. Bergmann.
Bull. Res. Council Israel 1, No. 1-2, 143-5 (1951).

1515. Pino, P., and G. Speroni.
Rend. ist. lombardo sci., Pt. I, Classe sci. mat. e nat. 88, 331-46 (1955).

1516. Pitt, D. A., A. J. Petro, and C. P. Smyth.
J. Am. Chem. Soc. 79, 5633-4 (1957).

1517. Pitt, D. A., and C. P. Smyth.
J. Am. Chem. Soc. 80, 1061-4 (1958).

1518. ———.
J. Phys. Chem. 63, 582-7 (1959).

1519. Pliev, T. N.
Uchenye Zapiski Yakutsk. Univ. 1958, No. 4, 23-38.

1520. ———.
Ibid., 39-43.

1521. ———.
Dokl. Akad. Nauk SSSR 125, 1044-7 (1959).

1522. ———.
Zhur. Fiz. Khim. 35, 2144-5 (1961).

1523. Plotnikov, V. A., I. A. Sheka, and Z. A. Yankelevich.
Mem. Inst. Chem., Acad. Sci. Ukrain. SSR 4, 363-83 (1938).

1524. ———.
J. Gen. Chem. (USSR) 9, 868-79 (1939).

1525. Podleschka, P., L. Westland, and G. F. Wright.
Can. J. Chem. 36, 574-80 (1958).

1526. Pohl, H. A., M. E. Hobbs, and P. M. Gross.
Ann. N.Y. Acad. Sci. 40, 389-428 (1940).

1527. Poltz, H.
Z. physik. Chem. B20, 351-6 (1933).

1528. ———, O. Steil, and O. Strasser.
Z. physik. Chem. B17, 155-60 (1932).

1529. Popov, A. I., and R. D. Holm.
J. Phys. Chem. 65, 774-6 (1961).

1530. Potapenko, G., and D. Wheeler.
Rev. Mod. Phys. 20, 143 (1948).

1531. Price, A. H.
Arch. Sci. (Geneva) fasc. spec. 12, 31-5 (1959).

1532. Price, A. H.
J. Phys. Chem. 64, 1442-4 (1960).

1533. Price, C. C., and R. H. Michel.
J. Am. Chem. Soc. 74, 3652-7 (1952).

1534. Prober, M.
J. Am. Chem. Soc. 77, 3224-8 (1955).

1535. Prokhorov, A. M., and G. P. Shipulo.
Optics and Spectroscopy 8, 218-9 (1960).

1536. Prosser, R. A.
Dissertation Abstr. 21, 778 (1960).

1537. Puchalik, M.
Physik. Z. 33, 341-5 (1932).

1538. ———.
Acta Phys. Polon. 2, 305-10 (1933).

1539. ———.
Ibid., 3, 179-85 (1934).

1540. ———.
Ibid., 4, 145-50 (1935).

1541. ———.
Ibid., 10, 89-92 (1950).

1542. ———.
Ibid., 93-101 (1950).

1543. Pullman, A., B. Pullman, E. D. Bergmann, G. Berthier, E. Fischer, D. Ginsburg, and Y. Hirshberg.
Bull. soc. chim. France 1951, 707-13.

1544. Pullman, B., A. Pullman, E. D. Bergmann, H. Berthod, E. Fischer, Y. Hirshberg, D. Lavie, and M. Mayot.
Bull. soc. chim. France 1952, 73-8.

1545. Pullman, A., B. Pullman, E. D. Bergmann, D. Ginsburg, and D. Lavie.
Bull. Res. Council Israel 1, No. 4, 85-7 (1952).

1546. Purcell, W. P., K. Fish, and C. P. Smyth.
J. Am. Chem. Soc. 82, 6299-301 (1960).

1547. Purcell, W. P., and C. P. Smyth.
J. Am. Chem. Soc. 83, 1060-3 (1961).

1548. ———.
Ibid., 1063-6 (1961).

1549. Pushkareva, Z. V., and Z. Yu. Kokoshko.
Dokl. Akad. Nauk SSSR 93, 77-80 (1953).

1550. ———.
Zhur. Obshcheĭ Khim. 24, 870-6 (1954).

1551. Pushkareva, Z. V., and L. B. Radina.
Dokl. Akad. Nauk SSSR 123, 301-4 (1958).

1552. Pushkareva, Z. V., L. V. Varyukhina, and Z. Yu. Kokoshko.
Dokl. Akad. Nauk SSSR 108, 1098-101 (1956).

1553. Pyle, W. R.
Phys. Rev. 38, 1057-70 (1931).

1554. Quan, D. Q.
Compt. rend. 252, 2247-9 (1961).

1555. Radina, L. B., Z. V. Pushkareva, and Z. Yu. Kokoshko.
Dokl. Akad. Nauk SSSR 123, 483-6 (1958).

1556. Raman, C. V., and K. S. Krishnan.
Phil. Mag. (7) 3, 713-23 (1927).

1557. Raman, R., and S. Soundararajan.
Proc. Indian Acad. Sci. 47A, 357-64 (1958).

1558. ———.
Ibid., 54A, 41-50 (1961).

1559. ———.
Can. J. Chem. 39, 1247-52 (1961).

1560. Ramakrishna, V.
Kolloid-Z. 154, 152-3 (1957).

1561. ———.
Ibid., 132, 30-4 (1953).

1562. ———.
Ibid., 133, 4-7 (1953).

1563. Ramaswamy, K. L.
Proc. Indian Acad. Sci. 2A, 364-77 (1935).

1564. ———.
Ibid., 4A, 108-33 (1936).

1565. Ramirez, F., and S. Levy.
J. Am. Chem. Soc. 79, 6167-72 (1957).

1566. Rampolla, R. W., and C. P. Smyth.
J. Am. Chem. Soc. 80, 1057-61 (1958).

1567. Randall, E. W., and L. E. Sutton.
J. Chem. Soc. 1958, 1266-9.

1568. ———.
Proc. Chem. Soc. 1959, 93-4.

1569. Rao, B. S. S., and S. Soundararajan.
Proc. Indian Acad. Sci. 50A, 149-56 (1959).

1570. Rao, D. A. A. S. N., and M. V. S. S. Murty.
J. Sci. Ind. Research (India) 16B, 183-5 (1957).

1571. Rao, D. V. G. L. N.
Current Sci. (India) 24, 407 (1955).

1572. ———.
Indian J. Phys. 29, 398-402 (1955).

1573. ———.
Current Sci. (India) 25, 217 (1956).

1574. ———.
Indian J. Phys. 30, 91-4 (1956).

1575. ———.
Ibid., 582-3 (1956).

1576. ———.
Ibid., 31, 60 (1957).

1577. Rao, D. V. G. L. N.
Indian J. Phys. 31, 334 (1957).

1578. ——.
Ibid., 33, 51-61 (1959).

1579. ——.
Trans. Faraday Soc. 55, 1324-8 (1959).

1580. ——.
J. Sci. Ind. Research (India) 18B, 133-7 (1959).

1581. Rao, R. J. R. M., and S. R. Palit.
Indian J. Phys. 34, 55-6 (1960).

1582. Ratuský, J., A. Reiser, and F. Šorm.
Chem. Listy 48, 1794-9 (1954).

1583. Rau, M. A. G.
Current Sci. 5, 132 (1936).

1584. ——.
Proc. Indian Acad. Sci. A4, 687-97 (1936).

1585. —— and N. Anantanarayanan.
Proc. Indian Acad. Sci. A5, 185-92 (1937).

1586. Rau, M. A. G., and B. N. Narayanaswamy.
Proc. Indian Acad. Sci. A1, 14-27 (1934).

1587. ——.
Ibid., 217-23 (1934).

1588. ——.
Ibid., 489-97 (1934).

1589. ——.
Z. physik. Chem. B26, 23-44 (1934).

1590. Rau, M. A. G., and S. S. Rao.
Proc. Indian Acad. Sci. A2, 232-5 (1935).

1591. Razumov, A. I., and N. N. Bankovskaya.
Dokl. Akad. Nauk SSSR 116, 241-3 (1957).

1592. Reilly, E. L., C. Curran, and P. A. McCusker.
J. Am. Chem. Soc. 76, 3311-2 (1954).

1593. Reiser, A., V. Jehlička, and K. Dvořák.
Collection Czech. Chem. Communs. 16, 13-22 (1951).

1594. Reppe, W., O. Schichting, K. Klager, and T. Toepel.
Ann. 560, 1-92 (1948).

1595. Reutler, H., and F. Semmler.
Monatsh. 85, 460-1 (1954).

1596. Richards, J. H., and S. Walker.
Trans. Faraday Soc. 57, 399-405 (1961).

1597. ——.
Ibid., 406-11 (1961).

1598. ——.
Ibid., 412-7 (1961).

1599. ——.
Ibid., 418-24 (1961).

1600. Richmond, H. H., and H. Freiser.
J. Am. Chem. Soc. 77, 2022-3 (1955).

1601. Riedinger, A.
Physik. Z. 39, 380-4 (1938).

1602. Riegger, H.
Ann. Physik. (4) 59, 753-60 (1919).

1603. Riemschneider, R.
Z. Naturforsch. 8b, 701-5 (1953).

1604. ———.
Monatsh. 85, 417-23 (1954).

1605. ———.
Ibid., 86, 101-16 (1955).

1606. ———.
Chem. Ber. 88, 1694-6 (1955).

1607. ———.
Ibid., 92, 894-9 (1959).

1608. ———, I. Ahrlé, W. Cohnen, and E. Heilman.
Chem. Ber. 92, 900-9 (1959).

1609. Riemschneider, R., and S. Bäker.
Z. Naturforsch. 9b, 751-3 (1954).

1610. ———, J. T. Shimozawa, and D. Lamparsy.
Z. Naturforsch. 9b, 799 (1954).

1611. Riemschneider, R., and A. Brand.
Monatsh. 84, 1240-2 (1953).

1612. Riemschneider, R., and W. Cohnen.
Chem. Ber. 89, 2702-12 (1956).

1613. Riemschneider, R., and P. Geschke.
Monatsh. 83, 1281-4 (1952).

1614. Riemschneider, R., and R. Oswald.
Monatsh. 85, 972-5 (1954).

1615. Riemschneider, R., and E. Scheppler.
Monatsh. 86, 548-50 (1955).

1616. Rittenhouse, J. R., W. Lobunez, D. Swern, and J. G. Miller.
J. Am. Chem. Soc. 80, 4850-2 (1958).

1617. Roberti, D. M., O. F. Kalman, and C. P. Smyth.
J. Am. Chem. Soc. 82, 3523-6 (1960).

1618. Roberti, D. M., and C. P. Smyth.
J. Am. Chem. Soc. 82, 2106-10 (1960).

1619. Roberts, A., and W. F. Edgell.
J. Chem. Phys. 17, 742-3 (1949).

1620. Roberts, J. D.
J. Am. Chem. Soc. 72, 3300-2 (1950).

1621. ———, R. Armstrong, R. F. Trimble, Jr., and M. Burg.
J. Am. Chem. Soc. 71, 843-7 (1949).

1622. Roberts, J. D., and V. C. Chambers.
J. Am. Chem. Soc. 73, 5030-4 (1951).

1623. Roberts, J. D., F. O. Johnson, and R. A. Carboni.
J. Am. Chem. Soc. 76, 5692-9 (1954).

1624. Roberts, J. D., E. A. McElhill, and R. Armstrong.
J. Am. Chem. Soc. 71, 2923-6 (1949).

1625. Roberts, J. D., R. L. Webb, and E. A. McElhill.
J. Am. Chem. Soc. 72, 408-11 (1950).

1626. Robles, H. de V.
Rec. trav. chim. 58, 111-24 (1939).

1627. Rodebush, W. H., L. A. Murray, and M. E. Bixler.
J. Chem. Phys. 4, 372-6 (1936).

1628. Rodinov, A. N., V. N. Vasil'eva, T. V. Talalaeva, D. N. Shigorin, E. N. Gur'yanova, and K. A. Kocheshkov.
Dokl. Akad. Nauk SSSR 125, 562-4 (1959).

1629. Rogers, J. D., W. J. Pietenpol, and D. Williams.
Phys. Rev. 82, 431-4 (1951).

1630. Rogers, M. T.
J. Am. Chem. Soc. 69, 457-9 (1947).

1631. ———.
Ibid., 1243-6 (1947).

1632. ———.
Ibid., 2544-8 (1947).

1633. ———.
Ibid., 77, 3681-4 (1955).

1634. ———.
J. Phys. Chem. 60, 125-6 (1956).

1635. ———.
Ibid., 61, 1442-3 (1957).

1636. ———, G. M. Barrow, and F. G. Bordwell.
J. Am. Chem. Soc. 78, 1790-2 (1956).

1637. Rogers, M. T., and T. L. Brown.
J. Phys. Chem. 61, 366-7 (1957).

1638. Rogers, M. T., and T. W. Campbell.
J. Am. Chem. Soc. 69, 2039-41 (1947).

1639. ———.
Ibid., 74, 4742-3 (1952).

1640. ———.
Ibid., 75, 1209-10 (1953).

1641. ———.
Ibid., 77, 4527-8 (1955).

1642. Rogers, M. T., and J. M. Canon.
J. Phys. Chem. 65, 1417-9 (1961).

1643. Rogers, M. T., and S. J. Cristol.
J. Am. Chem. Soc. 77, 764-5 (1955).

1644. Rogers, M. T., and K. J. Gross.
J. Am. Chem. Soc. 74, 5294-6 (1952).

1645. Rogers, M. T., and M. B. Panish.
J. Am. Chem. Soc. 77, 3684-6 (1955).

1646. ———.
Ibid., 4230-2 (1955).

1647. Rogers, M. T., and R. D. Pruett.
J. Am. Chem. Soc. 77, 3686-8 (1955).

1648. ——— and J. L. Speirs.
J. Am. Chem. Soc. 77, 5280-2 (1955).

1649. Rogers, M. T., R. D. Pruett, H. B. Thompson, and J. L. Speirs.
J. Am. Chem. Soc. 78, 44-5 (1956).

1650. Rogers, M. T., and J. D. Roberts.
J. Am. Chem. Soc. 68, 843-6 (1946).

1651. Rogers, M. T., H. B. Thompson, and J. L. Speirs.
J. Am. Chem. Soc. 76, 4841-3 (1954).

1652. Rogers, M. T., and A. Young.
J. Am. Chem. Soc. 68, 2748 (1946).

1653. Rolinski, J.
Physik. Z. 29, 658-67 (1928).

1654. Rolla, M., P. Fontana, and A. M. Marinangeli.
Gazz. chim. ital. 79, 491-502 (1949).

1655. Rolla, M., and A. M. Marinangeli.
Boll. sci. facoltà chim. ind. Bologna 7, 48-9 (1949).

1656. Rolla, M., and M. Sanesi.
Ricerca Sci. Rend. A. [2] 1, 43-55 (1961).

1657. ———.
Ibid., 289-97 (1961).

1658. ——— and G. Leandri.
Ann. chim. (Rome) 42, 664-72 (1952).

1659. ———.
Ibid., 44, 424-9 (1954).

1660. Rolla, M., M. Sanesi, and G. Traverso.
Ann. chim. (Rome) 42, 673-80 (1952).

1661. ———.
Ibid., 44, 430-6 (1954).

1662. Rollefson, R., and A. H. Rollefson.
Phys. Rev. 48, 779 (1935).

1663. Rollier, M. A.
Gazz. chim. ital. 77, 366-72 (1947).

1664. ———.
Ibid., 372-4 (1947).

1665. Romanov, V. I., and I. A. Eltzin.
Physik. Z. Sowjetunion 11, 526-38 (1937).

1666. Romans, J. B., and T. D. Callinan.
J. Electrochem. Soc. 104, 359-65 (1957).

1667. Rooney, C. S., and A. N. Bourns.
Can. J. Chem. 33, 1633-7 (1955).

1668. Rumpf, P., and H. Lumbroso.
Bull. soc. chim. France 1950, 283-5.

1669. Rutner, E., and S. H. Bauer.
J. Am. Chem. Soc. 82, 298-304 (1960).

1670. Sacco, A., and L. Naldini.
Rend. ist. lombardo sci. Pt. I, 91, 286-90 (1957).

1671. Sacconi, L., M. Ciampolini, F. Maggio, and G. Del Re.
J. Am. Chem. Soc. 82, 815-8 (1960).

1672. Sahney, R. C., R. Barucha, and H. R. Sarna.
J. Indian Chem. Soc. 25, 285-87 (1948).

1673. Sahney, R. C., R. M. Beri, H. R. Sarna, and M. Singh.
J. Indian Chem. Soc. 26, 329-34 (1949).

1674. Sakashita, K.
J. Chem. Soc. Japan, Pure Chem. Sect. 74, 315-8 (1953).

1675. ———.
Nippon Kagaku Zasshi 80, 13-7 (1959).

1676. Sakurada, I., and S. Lee.
J. Soc. Chem. Ind. Japan 37, Suppl. 331 (1934).

1677. ———.
Kolloid-Z. 72, 320-5 (1935).

1678. ———.
Ibid., 82, 67-72 (1938).

1679. Sakurada, I., and M. Taniguchi.
Bull. Inst. Phys.-Chem. Res. (Tokyo) 12, 224-34 (1933).

1680. Salovey, R.
J. Polymer Sci. 50, S7-9 (1961).

1681. Sanesi, M.
Gazz. chim. ital. 86, 1246-56 (1956).

1682. ——— and G. Leandri.
Ann. chim. (Rome) 45, 1106-17 (1955).

1683. ———.
Ibid., 46, 1127-32 (1956).

1684. Sanesi, M., and G. Traverso.
Chem. Ber. 93, 1566-72 (1960).

1685. Sänger, R.
Physik. Z. 27, 165-74 (1926).

1686. ———.
Ibid., 556-63 (1926).

1687. ———.
Ibid., 28, 455-7 (1927).

1688. ———.
Leipziger Vorträge 1929, p 1.

1689. ———.
Helv. Phys. Acta 3, 162 (1930).

1690. ———.
Physik. Z. 31, 306-15 (1930).

1691. ———.
Ibid., 32, 21-6 (1931).

1692. ——— and O. Steiger.
Helv. Phys. Acta 1, 369-84 (1928).

1693. ———.
Ibid., 2, 136-44 (1929).

1694. ———.
Ibid., 411-8 (1929).

1695. ——— and K. Gächter.
Helv. Phys. Acta 5, 200-10 (1932).

1696. Sängewald, R., and A. Weissberger.
Physik. Z. 30, 268 (1929).

1697. Sato, T.
Bull. Chem. Soc. Japan 33, 501-3 (1960).

1698. Sauer, R. O., and D. J. Mead.
J. Am. Chem. Soc. 68, 1794-6 (1946).

1699. Sawatzky, H., G. K. White, and G. F. Wright.
Can. J. Chem. 37, 1132-45 (1959).

1700. Sawatzky, H., and G. F. Wright.
Can. J. Chem. 36, 1555-69 (1958).

1701. Schaaffs, W.
Z. physik. Chem. 196, 413-26 (1951).

1702. Schaarschmidt, K.
Z. anorg. allgem. Chem. 310, 78-85 (1961).

1703. Scheele, W., and W. Meine.
Kolloid-Z. 129, 39-49 (1952).

1704. Scheffers, H.
Physik. Z. 35, 425-33 (1934).

1705. Scheibe, G., and O. Stoll.
Chem. Ber. B71, 1571-5 (1938).

1706. Schenck, G. O.
Chem. Ber. 77B, 741-7 (1944).

1707. Schmelz, M. J., M. A. G. Hill, and C. Curran.
J. Phys. Chem. 65, 1273-4 (1961).

1708. Schmidt, C. H.
Chem. Ber. 90, 1352-6 (1957).

1709. Schmitz, E.
Chem. Ber. 91, 1133-41 (1958).

1710. Schneider, J.
J. Chem. Phys. 32, 665-8 (1960).

1711. Schneider, R. G., and H. Heuer.
Z. Naturforsch. 8b, 695-6 (1953).

1712. Schneider, W. C.
J. Am. Chem. Soc. 70, 627-30 (1948).

1713. ———.
Ibid., 72, 761-3 (1950).

1714. Schneider, W. C.
J. Am. Chem. Soc. 77, 2796-7 (1955).

1715. ———, W. C. Carter, M. Magat, and C. P. Smyth.
J. Am. Chem. Soc. 67, 959-63 (1945).

1716. Schneider, W. C., and I. F. Halverstadt.
J. Am. Chem. Soc. 70, 2626-31 (1948).

1717. Schulz, R. W.
Z. Physik 109, 517-37 (1938).

1718. Schupp, L., and R. Mecke.
Z. Elektrochem. 52, 54-60 (1948).

1719. Schurz, J., H. Koren, and E. Treiber.
Monatsh. 86, 986-94 (1955).

1720. Schwabe, K.
Z. phys. Chem. (NF) 20, 68-82 (1959).

1721. Schwingel, C. H., and E. W. Greene.
J. Am. Chem. Soc. 56, 653-4 (1934).

1722. Schwingel, C. H., and J. W. Williams.
Phys. Rev. 35, 855-62 (1930).

1723. Semenow, D. A., and J. D. Roberts.
J. Am. Chem. Soc. 79, 2741-2 (1957).

1724. Senatore, S. J.
Phys. Rev. 78, 293-4 (1950).

1725. Sen-Gupta, D. N.
Nature 125, 600 (1930).

1726. Seward, R. P., and E. C. Vieira.
J. Phys. Chem. 62, 127-8 (1958).

1727. Seyer, W. F., and G. M. Barrow.
J. Am. Chem. Soc. 70, 802-5 (1948).

1728. Sharbaugh, A. H., V. G. Thomas, and B. S. Pritchard.
Phys. Rev. 78, 64-5 (1950).

1729. Sharpe, A. N., and S. Walker.
J. Chem. Soc. 1961, 2974-81.

1730. ———.
Ibid., 4522-30.

1731. Shaw, T. M., E. F. Jansen, and H. Lineweaver.
J. Chem. Phys. 12, 439-48 (1944).

1732. Shaw, T. M., and J. J. Windle.
J. Chem. Phys. 19, 1063-4 (1951).

1733. Sheka, I. A.
Zapiski Inst. Khim., Akad. Nauk Ukr. RSR 7, No. 1, 57-67 (1940).

1734. ———.
J. Phys. Chem. (USSR) 16, 99-105 (1942).

1735. ———.
Zhur. Fiz. Khim. 23, 885-8 (1949).

1736. ———.
Ibid., 1180-6 (1949).

1737. ———.
Ibid., 24, 519-23 (1950).

1738. ———.
Zhur. Obshcheĭ Khim. 25, 2401-5 (1955).

1739. ———.
Ibid., 26, 26-30 (1956).

1740. ——— and K. F. Karlysheva.
Zhur. Obshcheĭ Khim. 21, 833-9 (1951).

1741. ———.
Zhur. Fiz. Khim. 30, 1316-8 (1956).

1742. Sheka, I. A., and Z. A. Sheka.
Dokl. Akad. Nauk SSSR 69, 197-200 (1949).

1743. Sheridan, J., J. K. Tyler, E. E. Aynsley, R. E. Dodd, and R. Little.
Nature 185, 96 (1960).

1744. Sherrill, M. L., M. E. Smith, and D. D. Thompson.
J. Am. Chem. Soc. 56, 611-4 (1934).

1745. Shidlovskaya, A. N., M. I. Gostev, and Ya. K. Syrkin.
Dokl. Akad. Nauk SSSR 87, 101-3 (1952).

1746. Shidlovskaya, A. N., and Ya. K. Syrkin.
Compt. rend. acad. sci. URSS 55, 231-2 (1947).

1747. ———.
Zhur. Fiz. Khim. 22, 913-9 (1948).

1748. ———.
Dokl. Akad. Nauk SSSR 139, 418-20 (1961).

1749. ——— and N. K. Kochetkov.
Izv. Akad. Nauk SSSR, Otd. Khim. Nauk 1956, 254-6.

1750. Shidlovskaya, A. N., Ya. K. Syrkin, and I. N. Nazarov.
Dokl. Akad. Nauk SSSR 94, 905-7 (1954).

1751. ——— and V. F. Kucherov.
Dokl. Akad. Nauk SSSR 118, 967-9 (1958).

1752. Shidlovskaya, A. N., Ya. K. Syrkin, I. N. Nazarov, and D. V. Sokolov.
Izv. Akad. Nauk SSSR, Otd. Khim. Nauk 1958, 241.

1753. Shidlovskaya, A. N., Ya. K. Syrkin, S. S. Novikov, A. A. Fainzil'berg, V. V. Sevost'yanova, and V. I. Gulevskaya.
Dokl. Akad. Nauk SSSR 132, 1376-7 (1960).

1754. Shimozawa, T.
Bull. Chem. Soc. Japan 28, 389-92 (1955).

1755. ———.and Y. Morino.
J. Chem. Soc. Japan 81, 20-1 (1960).

1756. Shimozawa, T., and Y. Nagai.
Nippon Kagaku Zasshi 81, 22-3 (1960).

1757. Shirai, M.
Sci. Papers Coll. Gen. Educ., Univ. Tokyo 6, 147-51 (1956).

1758. Shoolery, J. N., and A. H. Sharbaugh.
Phys. Rev. 82, 95 (1951).

1759. Shoolery, J. N., R. G. Shulman, W. F. Sheehan, Jr., V. Schomaker, and D. M. Yost.
J. Chem. Phys. 19, 1364-9 (1951).

1760. Shoolery, J. N., R. G. Shulman, and D. M. Yost.
J. Chem. Phys. 19, 250-1 (1951).

1761. Shott-L'vova, E. A., and Ya. K. Syrkin.
Acta Physicochem. URSS 11, 659-60 (1939).

1762. ———.
Dokl. Akad. Nauk SSSR 87, 639-41 (1952).

1763. ———.
Izv. Akad. Nauk SSSR, Otd. Khim. Nauk 1954, 381-2.

1764. ———.
Ibid., 1956, 127-8.

1765. ———.
Ibid., 1960, 139-40.

1766. ———, I. I. Levkoev, and M. V. Deĭchmeĭster.
Dokl. Akad. Nauk USSR 121, 1048-51 (1958).

1767 Shott-L'vova, E. A., Ya. K. Syrkin, I. I. Levkoev, and Z. P. Sytnik.
Dokl. Akad. Nauk SSSR 116, 804-7 (1957).

1768. Shulman, R. G., B. P. Dailey, and C. H. Townes.
Phys. Rev. 78, 145-8 (1950).

1769. Shulman, R. G., and C. H. Townes.
Phys. Rev. 77, 421-2 (1950).

1770. ———.
Ibid., 500-6 (1950).

1771. ———.
Ibid., 78, 347 (1950).

1772. ———.
Quoted by P. Kisliuk and C. H. Townes in J. Res. Natl. Bur. Std. 44, 611-41 (1950).

1773. Sidgwick, N. V., and H. D. Springall.
J. Chem. Soc. 1936, 1532-7.

1774. Sidgwick, N. V., L. E. Sutton, and W. Thomas.
J. Chem. Soc. 1933, 406-12.

1775. Siegel, S.
J. Chem. Phys. 27, 989-90 (1957).

1776. Simons, J. H., and G. Jessop.
J. Am. Chem. Soc. 53, 1263-6 (1931).

1777. Sipos, J. C., H. Sawatzky, and G. F. Wright.
J. Am. Chem. Soc. 77, 2759-62 (1955).

1778. Sircar, S. C.
Indian J. Phys. 3, 197-208 (1928).

1779. Sirvetz, M. H.
J. Chem. Phys. 19, 1609-10 (1951).

1780. ——— and R. E. Weston, Jr.
J. Chem. Phys. 21, 898-902 (1953).

1781. Skita, A., and W. Faust.
Chem. Ber. B72, 1127-38 (1939).

1782. Skita, A., and R. Rössler.
Chem. Ber. B72, 265-72 (1939).

1783. Slayton, G. R., J. W. Simmons, and J. H. Goldstein.
J. Chem. Phys. 22, 1678-9 (1954).

1784. Smith, A. G., H. Ring, W. V. Smith, and W. Gordy.
Phys. Rev. 74, 370-2 (1948).

1785. Smith, D. F.
Proc. U. N. Intern. Conf. Peaceful Uses At. Energy, 2nd, Geneva 28, 130-8 (1958).

1786. ——— and D. W. Magnuson.
Phys. Rev. 87, 226 (1952).

1787. Smith, D. F., M. Tidwell, and D. V. P. Williams.
Phys. Rev. 77, 420-1 (1950).

1788. ———.
Ibid., 79, 1007-8 (1950).

1789. Smith, H. A., and W. H. King.
J. Am. Chem. Soc. 70, 3528 (1948).

1790. Smith, H. A., and L. E. Line, Jr.
J. Am. Chem. Soc. 72, 5434-6 (1950).

1791. Smith, J. W.
Proc. Roy. Soc. (London) A136, 256-63 (1932).

1792. ———.
Ibid., A138, 154-61 (1932).

1793. ———.
J. Chem. Soc. 1933, 1567-70.

1794. ———.
Ibid., 1950, 3532-5.

1795. ———.
Ibid., 1953, 109-13.

1796. ———.
Ibid., 1957, 4050-3.

1797. ———.
Ibid., 1961, 4700-4.

1798. ——— and W. R. Angus.
Proc. Roy. Soc. (London) A137, 372-9 (1932).

1799. Smith, J. W., and S. M. Walshaw.
J. Chem. Soc. 1957, 3217-22.

1800. Smith, J. W., and S. M. Walshaw.
J. Chem. Soc. 1957, 4527-31.

1801. ———.
Ibid., 1959, 3784-8.

1802. Smith, J. W., and L. B. Witten.
Trans. Faraday Soc. 47, 1304-18 (1951).

1803. Smith, R. P., and J. C. Tatlow.
J. Chem. Soc. 1957, 2505-11.

1804. Smits, A., N. F. Moerman, and J. C. Pathius.
Z. physik. Chem. B35, 60-8 (1937).

1805. Smyth, C. P.
J. Am. Chem. Soc. 63, 57-66 (1941).

1806. ———.
Ibid., 51, 2380-8 (1929).

1807. ——— and R. W. Dornte.
J. Am. Chem. Soc. 53, 545-55 (1931).

1808. ———.
Ibid., 1296-304 (1931).

1809. ———.
Ibid., 2005-6 (1931).

1810. ——— and E. B. Wilson.
J. Am. Chem. Soc. 53, 4242-60 (1931).

1811. Smyth, C. P., A. J. Grossman, and S. R. Ginsberg.
J. Am. Chem. Soc. 62, 192-5 (1940).

1812. Smyth, C. P., and N. B. Hannay.
U.S. At. Energy Comm. TID-5290, Book 2, 437-40 (1958).

1813. Smyth, C. P., and S. E. Kamerling.
J. Am. Chem. Soc. 53, 2988-98 (1931).

1814. Smyth, C. P., and G. L. Lewis.
J. Am. Chem. Soc. 62, 721-7 (1940).

1815. ———, A. J. Grossman, and F. B. Jennings.
J. Am. Chem. Soc. 62, 1219-23 (1940).

1816. Smyth, C. P., and K. B. McAlpine.
J. Chem. Phys. 1, 60-1 (1933).

1817. ———.
Ibid., 190-6 (1933).

1818. ———.
Ibid., 2, 499-502 (1934).

1819. ———.
Ibid., 571-3 (1934).

1820. ———.
J. Am. Chem. Soc. 56, 1697-700 (1934).

1821. ———.
Ibid., 57, 979-83 (1935).

1822. ———.
J. Chem. Phys. 3, 347-50 (1935).

1823. Smyth, C. P., and S. A. McNeight.
J. Am. Chem. Soc. 58, 1723-8 (1936).

1824. Smyth, C. P., and S. O. Morgan.
J. Am. Chem. Soc. 49, 1030-8 (1927).

1825. ———.
Ibid., 50, 1547-60 (1928).

1826. ——— and J. C. Boyce.
J. Am. Chem. Soc. 50, 1536-47 (1928).

1827. Smyth, C. P., and H. E. Rogers.
J. Am. Chem. Soc. 52, 1824-30 (1930).

1828. ———.
Ibid., 2227-40 (1930).

1829. Smyth, C. P., and W. N. Stoops.
J. Am. Chem. Soc. 50, 1883-90 (1928).

1830. ———.
Ibid., 51, 3312-29 (1929).

1831. ———.
Ibid., 3330-41 (1929).

1832. Smyth, C. P., and W. S. Walls.
J. Am. Chem. Soc. 53, 527-39 (1931).

1833. ———.
Ibid., 2115-22 (1931).

1834. ———.
Ibid., 54, 1854-62 (1932).

1835. ———.
Ibid., 2261-70 (1932).

1836. ———.
Ibid., 3230-40 (1932).

1837. ———.
J. Chem. Phys. 1, 200-4 (1933).

1838. ———.
Ibid., 3, 557-9 (1935).

1839. Smyth, C. P., and C. T. Zahn.
J. Am. Chem. Soc. 47, 2501-6 (1925).

1840. Snoek, J. L.
Physik. Z. 35, 196-203 (1934).

1841. Sobczyk, L.
Roczniki Chem. 33, 743-54 (1959).

1842. ———.
Ibid., 34, 567-72 (1960).

1843. ———.
Trans. Faraday Soc. 57, 1041-3 (1961).

1844. Sobhanadri, J.
J. Sci. Ind. Res. 18B, 508-11 (1959).

1845. Soffer, H., and T. De Vries.
J. Am. Chem. Soc. 73, 5817-9 (1951).

1846. Soundararajan, S.
Trans. Faraday Soc. 53, 159-66 (1957).

1847. ———.
Ibid., 54, 1147-50 (1958).

1848. ——— and S. V. Anantakrishnan.
Proc. Indian Acad. Sci. 37A, 578-83 (1953).

1849. ———.
Ibid., 38A, 176-83 (1953).

1850. Soundararajan, S., and M. J. Vold.
Trans. Faraday Soc. 54, 1151-4 (1958).

1851. ———.
Ibid., 1155-9 (1958).

1852. Spaght, M. E., F. Hein, and H. Pauling.
Physik. Z. 34, 212-4 (1933).

1853. Spauschus, H. O., A. P. Mills, J. M. Scott, and C. A. MacKenzie.
J. Am. Chem. Soc. 72, 1377-9 (1950).

1854. Spauschus, H. O., and J. M. Scott.
J. Am. Chem. Soc. 73, 210-2 (1951).

1855. Speroni, G.
Chimica e industria (Milan) 33, 543-6 (1951).

1856. ———.
Ricerca sci. 27, 1199-203 (1957).

1857. ——— and L. Mori.
Atti accad. nazl. Lincei, Rend., Classe sci. fis., mat. e nat. 12, 704-12 (1952).

1858. Speroni, G., and P. Pino.
Gazz. chim. ital. 80, 549-71 (1950).

1859. ———.
Atti accad. nazl. Lincei, Rend., Classe sci. fis., mat. e nat. 13, 39-45 (1952).

1860. ——— and L. Mori.
Gazz. chim. ital. 82, 269-77 (1952).

1861. Spinrad, B. I.
J. Am. Chem. Soc. 68, 617-20 (1946).

1862. Springall, H. D., G. C. Hampson, G. G. May, and H. Spedding.
J. Chem. Soc. 1949, 1524-32.

1863. Spurr, R. A., and H. Zeitlin.
J. Am. Chem. Soc. 72, 4832 (1950).

1864. Srivastava, H. N.
J. Sci. Ind. Res. (India) 19B, 149-51 (1960).

1865. Starobinets, G. L., and K. S. Starobinets.
Zhur. Fiz. Khim. 25, 759-67 (1951).

1866. Stearn, A. E., and C. P. Smyth.
J. Am. Chem. Soc. 56, 1667-70 (1934).

1867. Steiger, O.
Helv. Phys. Acta 3, 161 (1930).

1868. Stepanenko, N. N.
J. Exptl. Theoret. Phys. (USSR) 14, 163-70 (1944).

1869. ———, B. A. Agranat, and T. Novikova.
Acta Physicochem. URSS 20, 923-32 (1945).

1870. Stepanenko, N. N., B. A. Agranat, and V. F. Yakovlev.
J. Phys. Chem. USSR 21, 893-7 (1947).

1871. Stepanenko, N. N., and V. Agranat.
J. Exptl. Theoret. Phys. (USSR) 14, 226-31 (1944).

1872. Stepanenko, N. N., and L. I. Bogdanov.
Zhur. Fiz. Khim. 26, 1472-6 (1952).

1873. Sterzer, F.
J. Chem. Phys. 22, 2094 (1954).

1874. Stoops, W. N.
J. Phys. Chem. 35, 1704-11 (1931).

1875. Stranathan, J. D.
Phys. Rev. 31, 653-71 (1928).

1876. ———.
Ibid., 48, 538-44 (1935).

1877. ———.
J. Chem. Phys. 5, 828-30 (1937).

1878. ———.
Ibid., 6, 395-8 (1938).

1879. Strandberg, M. W. P.
Phys. Rev. 74, 1245 (1948).

1880. ———.
J. Chem. Phys. 17, 901-4 (1949).

1881. ——— et al.
MITRLE Quarterly Progress Report (Oct. 15, 1948).
Quoted in J. Res. Natl. Bur. Std. 44, 611-41 (1950).

1882. Strandberg, M. W. P., C. S. Pearsall, and M. T. Weiss.
J. Chem. Phys. 17, 429 (1949).

1883. Strandberg, M. W. P., T. Wentink, Jr., and A. G. Hill.
Phys. Rev. 75, 827-32 (1949).

1884. Strandberg, M. W. P., T. Wentink, Jr., R. E. Hillger, G. H. Wannier, and M. L. Deutsch.
Phys. Rev. 73, 188 (1948).

1885. Strandberg, M. W. P., T. Wentink, Jr., and R. L. Kyhl.
Phys. Rev. 75, 270-8 (1949).

1886. Strohmeier, W.
Z. Elektrochem. 60, 58-61 (1956).

1887. ——— and K. Hümpfner.
Z. Elektrochem. 60, 1111-4 (1956).

1888. ———.
Ibid., 61, 1010-4 (1957).

1889. ———, K. Miltenberger, and F. Seifert.
Z. Elektrochem. 63, 537-9 (1959).

1890. Strohmeier, W., and H. Länghauser.
Z. physik. Chem. (Frankfort) 25, 427-9 (1960).

1891. ———.
Z. Phys. Chem. 28, 268-71 (1961).

1892. Strohmeier, W., and K. Miltenberger.
Z. Phys. Chem. 17, 274-8 (1958).

1893. Strohmeier, W., and K. Nützel.
Z. Elektrochem. 59, 538-42 (1955).

1894. Strohmeier, W., and D. von Hobe.
Z. Elektrochem. 64, 945-51 (1960).

1895. Stuart, H. A.
Z. Physik 47, 457-78 (1928).

1896. ———.
Ibid., 51, 490-510 (1928).

1897. Suenga, K., and A. Kotera.
J. Chem. Soc. Japan, Pure Chem. Sect. 70, 116-8 (1949).

1898. Sugden, S.
Nature 133, 415-6 (1934).

1899. Sukigara, K., Y. Hata, Y. Kurita, and M. Kubo.
Tetrahedron 4, 337-41 (1958).

1900. Sun, C. E., and C. Liu.
J. Chinese Chem. Soc. 5, 39-40 (1937).

1901. Sundhoff, D., and H-J. Schumacher.
Z. physik. Chem. B28, 17-30 (1935).

1902. Sutton, L. E.
Nature 128, 639 (1931).

1903. ———.
Proc. Roy. Soc. (London) A133, 668-95 (1931).

1904. ———.
Bull. soc. chim. France 1949, D448-56.

1905. ——— and J. B. Bentley.
Nature 130, 314-5 (1932).

1906. Sutton, L. E., and G. C. Hampson.
Trans. Faraday Soc. 31, 945-57 (1935).

1907. Sutton, L. E., R. G. A. New, and J. B. Bentley.
J. Chem. Soc. 1933, 652-8.

1908. Sutton, L. E., and T. W. J. Taylor.
J. Chem. Soc. 1931, 2190-5.

1909. Suzuki, I.
Nippon Kagaku Zasshi 80, 353-6 (1959).

1910. ———.
Ibid., 697-700 (1959).

1911. Suzuki, S.
Chem. High Polymers (Japan) 11, 41-6 (1954).

1912. Svirbely, W. J., J. E. Ablard, and J. C. Warner.
J. Am. Chem. Soc. 57, 652-5 (1935).

1913. Svirbely, W. J., and J. J. Lander.
J. Am. Chem. Soc. 67, 2189-90 (1945).

1914. ———.
Ibid., 70, 4121-3 (1948).

1915. ———.
Ibid., 72, 3756-61 (1950).

1916. Swalen, J. D., and C. C. Costain.
J. Chem. Phys. 31, 1562-74 (1959).

1917. Swalen, J. D., and D. R. Herschbach.
J. Chem. Phys. 27, 100-8 (1957).

1918. Swalen, J. D., and B. P. Stoicheff.
J. Chem. Phys. 28, 671-4 (1958).

1919. Sweeting, O. J., and J. R. Johnson.
J. Am. Chem. Soc. 68, 1057-61 (1946).

1920. Syrkin, Ya. K., and K. M. Anisimova.
Dokl. Akad. Nauk SSSR 59, 1457-9 (1948).

1921. Syrkin, Ya. K., and E. A. Shott-L'vova.
Acta Physicochem. URSS 19, 379-84 (1944).

1922. ———.
Ibid., 20, 397-406 (1945).

1923. Szczucki, E.
Biul. Wojskowej Akad. Tech. im J. Dąbrowskiego 8, No. 2, 45-63 (1959).

1924. Takeda, K., S. Nagakura, and K. Kitahonoki.
Pharm. Bull. (Japan) 1, 135-8 (1953).

1925. Takeda, M., Y. Imamura, S. Okamura, and T. Higashimura.
J. Chem. Phys. 33, 631-2 (1960).

1926. Tamamushi, B., H. Akiyama, and S. Umezawa.
Bull. Chem. Soc. Japan 14, 310-7 (1939).

1927. ———.
Ibid., 318-22 (1939).

1928. Taniguchi, I.
Doshisha Kogaku Kaishi 10, 116-28 (1959).

1929. Tannenbaum, E., R. J. Myers, and W. D. Gwinn.
J. Chem. Phys. 25, 42-7 (1956).

1930. Tappi, G.
Gazz. chim. ital. 71, 111-7 (1941).

1931. ——— and U. di Vajo.
Gazz. chim. ital. 69, 615-20 (1939).

1932. Tappi, G., and C. Springer.
Gazz. chim. ital. 70, 190-6 (1940).

1933. Tate, P. A., and M. W. P. Strandberg.
J. Chem. Phys. 22, 1380-3 (1954).

1934. Tatsuno, T.
Pharm. Bull. (Japan) 2, 140-6 (1954).

1935. Taylor, A. M., and E. K. Rideal.
Proc. Roy. Soc. (London) A115, 589-609 (1927).

1936. Taylor, T. G., and A. W. Baker.
Tetrahedron Letters 1959, No. 19, 14-8.

1937. Taylor, T. I., E. R. Smith, J. L. Torgesen, H. Mathesen, and J. K. Taylor.
U.S. At. Energy Comm. TID-5290, Book 1, 369-428 (1958).

1938. Taylor, T. W. J., and L. E. Sutton.
J. Chem. Soc. 1933, 63-5.

1939. Thiec, J., and J. Wiemann.
Bull. soc. chim. France 1956, 177-80.

1940. ———.
Ibid., 1958, 207-11.

1941. Theilacker, W., and K. Fauser.
Ann. 539, 103-15 (1939).

1942. Thomas, J. R., and W. D. Gwinn.
J. Am. Chem. Soc. 71, 2785-90 (1949).

1943. Thompson, H. B., and S. L. Hanson.
J. Phys. Chem. 65, 1005-9 (1961).

1944. Thompson, H. B., and C. W. Lawson.
J. Phys. Chem. 64, 1788-9 (1960).

1945. Thompson, H. B., and C. C. Sweeney.
J. Phys. Chem. 64, 221-4 (1960).

1946. Thomson, G.
J. Chem. Soc. 1939, 1118-23.

1947. ———.
Ibid., 1944, 404-8.

1948. Thomson, R. H.
J. Org. Chem. 13, 371-6 (1948).

1949. Tiganik, L.
Z. physik. Chem. B13, 425-61 (1931).

1950. ———.
Ibid., B14, 135-48 (1931).

1951. Tomassi, W., and J. Młodzki.
Roczniki Chem. 25, 505-8 (1951).

1952. Tomisawa, H.
Yakugaku Zasshi 79, 1170-3 (1959).

1953. Topchiev, K. S., M. M. Yakshin, and R. E. Shindel.
Compt. rend. acad. sci. URSS 30, 502-4 (1941).

1954. Tourky, A. R., and H. A. Rizk.
J. Phys. Chem. 61, 231-3 (1957).

1955. ———.
Ibid., 1255-6 (1957).

1956. ———.
Can. J. Chem. 35, 630-3 (1957).

1957. ——— and Y. M. Girgis.
J. Phys. Chem. 64, 565-7 (1960).

1958. ———.
Ibid., 65, 40-2 (1961).

1959. Townes, C. H.
Phys. Rev. 70, 665-71 (1946).

1960. ———, A. N. Holden, and F. R. Merritt.
Phys. Rev. 73, 1334-7 (1948).

1961. Townes, C. H., F. R. Merritt, and B. D. Wright.
Phys. Rev. 73, 1334-7 (1948).

1962. Townes, C. H., and A. L. Schawlow.
Microwave Spectroscopy, McGraw-Hill Book Company, Inc., New York, 1955.

1963. Townes, C. H., R. G. Shulman, and B. P. Dailey.
Phys. Rev. 76, 472 (1949).

1964. Toyoda, M., and I. Taniguchi.
Bull. Inst. Chem. Res., Kyoto Univ. 35, 16-20 (1957).

1965. Trambarulo, R., S. N. Ghosh, C. A. Burrus, Jr., and W. Gordy.
J. Chem. Phys. 21, 851-5 (1953).

1966. Trautteur, P.
Nuovo cimento 14, 265-71 (1937).

1967. Trieschmann, H. G.
Z. physik. Chem. B33, 283-9 (1936).

1968. Trinh, N. Q.
Compt. rend. 227, 393-5 (1948).

1969. ———.
Bull. soc. chim. France 1949, D397-9.

1970. Trischka, J. W.
Phys. Rev. 76, 1365-8 (1949).

1971. ———.
J. Chem. Phys. 25, 784-5 (1956).

1972. Truce, W. E., and D. L. Goldhamer.
J. Am. Chem. Soc. 81, 5795-8 (1959).

1973. ——— and R. B. Kruse.
J. Am. Chem. Soc. 81, 4931-5 (1959).

1974. Trunel, P.
Compt. rend. 200, 557-9 (1935).

1975. ———.
Ibid., 2186-7 (1935).

1976. ———.
Ibid., 202, 37-9 (1936).

1977. ———.
Ibid., 203, 563-5 (1936).

1978. ———.
Ibid., 205, 236-8 (1937).

1979. ———.
Ann. chim. (11) 12, 93-168 (1939).

1980. Tschamler, H., and R. Reiberger.
Monatsh. 79, 394-409 (1948).

1981. Tschamler, H., F. Wettig, and E. Richter.
Monatsh. 80, 572-82 (1949).

1982. Tseng, C-L., C. Liu, and C. E. Sun.
J. Chinese Chem. Soc. 4, 473-6 (1936).

1983. Tseng, C-L., C. E. Sun, and C-H. Yao.
J. Chinese Chem. Soc. 5, 236-8 (1937).

1984. Tsintsevich, V. M., G. P. Khomchenko, and G. D. Vovchenko.
Vestn. Mosk. Univ., Ser. Mat., Mekhan., Astron., Fiz. i Khim. 1959, No. 6, 205-9.

1985. Tsoucaris, G.
J. phys. chim. 58, 619-24 (1961).

1986. Tsubomura, H.
Bull. Chem. Soc. Japan 27, 1-4 (1954).

1987. ———.
Ibid., 31, 435-40 (1958).

1988. ——— and S. Nagakura.
J. Chem. Phys. 27, 819-20 (1957).

1989. Tsuda, K., B. Eda, and M. Kubo.
Pharm. Bull. (Tokyo) 5, 624-5 (1957).

1990. Tsudzuki, Y., and K. Higashi.
Sci. Papers Inst. Phys. Chem. Res. (Tokyo) 39, 185 (1941).

1991. Tsvetkov, V. N., and V. Marinin.
Zhur. Eksptl. Teoret. Fiz. 18, 641-50 (1948).

1992. Tulinskie, A., A. Di Giacomo, and C. P. Smyth.
J. Am. Chem. Soc. 75, 3552-4 (1953).

1993. Tuomikoski, P.
J. phys. radium 16, 347-8 (1955).

1994. ——— and A. Niini.
Ann. Acad. Sci. Fennicae A48, No. 11 (1937).

1995. Turkevich, J., P. F. Oesper, and C. P. Smyth.
J. Am. Chem. Soc. 64, 1179-80 (1942).

1996. Tyler, J. K., and J. Sheridan.
Proc. Chem. Soc. 1960, 119-20.

1997. Tyler, J. K., L. F. Thomas, and J. Sheridan.
Proc. Chem. Soc. 1959, 155-6.

1998. Tyrell, H. J. V., and Mr. Mills.
Chem. and Ind. (London) 1951, 30-1.

1999. Uchida, T., Y. Kurita, N. Koizumi, and M. Kubo.
J. Polymer Sci. 21, 313-22 (1956).

2000. Uchida, T., Y. Kurita, and M. Kubo.
J. Polymer Sci. 19, 365-72 (1956).

2001. Uenishi, R. K.
Dissertation Abstr. 16, 1327-8 (1956).

2002. Uhlig, H. H., J. G. Kirkwood, and F. G. Keyes.
J. Chem. Phys. 1, 155-9 (1933).

2003. Ulich, H., E. Hertel, and W. Nespital.
Z. physik. Chem. B17, 21-45 (1932).

2004. ———.
Ibid., 369-79 (1932).

2005. Ulich, H., and G. Heyne.
Z. physik. Chem. B49, 284-92 (1941).

2006. Ulich, H., and N. Nespital.
Z. Elektrochem. 37, 559-63 (1931).

2007. ———.
Z. angew. Chem. 44, 750-3 (1931).

2008. Ulich, H., H. Peisker, and L. F. Audrieth.
Chem. Ber. B68, 1677-82 (1935).

2009. Uyeo, S.
Bull. Chem. Soc. Japan 16, 177-9 (1941).

2010. ——— and K. Higashi.
J. Chem. Soc. Japan 60, 199-203 (1939).

2011. ———.
Ibid., 204-11 (1939).

2012. van Arkel, A. E.
Rec. trav. chim. 52, 733-41 (1933).

2013. ———, P. Meerburg, and C. R. van der Handel.
Rec. trav. chim. 61, 767-70 (1942).

2014. van Arkel, A. E., and J. L. Snoek.
Z. physik. Chem. B18, 159-66 (1932).

2015. ———.
Rec. trav. chim. 52, 719-32 (1933).

2016. ———.
Ibid., 53, 91-4 (1934).

2017. ———.
Physik. Z. 35, 187-96 (1934).

2018. ———.
Trans. Faraday Soc. 30, 707-23 (1934).

2019. Van Blaricom, L., and E. C. Gilbert.
J. Am. Chem. Soc. 61, 3238-9 (1939).

2020. van den Berg, J. A.
J. S. African Chem. Inst. 9, 46-50 (1956).

2021. van Itterbeek, A., and K. de Clippeleir.
Physica 14, 349-56 (1948).

2022. van Tiggelen, P., and P. W. Lobo.
Bull. Soc. Chim. Belg. 70, 285-90 (1961).

2023. Vasil'ev, V. G., and Ya. K. Syrkin.
Acta Physicochem. URSS 6, 639-60 (1937).

2024. ———.
Ibid., 9, 203-4 (1938).

2025. Vasil'ev, V. G., and Ya. K. Syrkin.
J. Phys. Chem. (USSR) 12, 153-4 (1938).

2026. ———.
Acta Physicochem. URSS 14, 414-6 (1941).

2027. ——— and I. Kenez.
Nature 135, 71 (1935).

2028. Vasil'eva, V. N., V. P. Bazov, and M. A. Geiderikh.
Zhur. Fiz. Khim. 33, 1516-20 (1959).

2029. Vasil'eva, V. N., and E. N. Gur'yanova.
Zhur. Fiz. Khim. 28, 1319-26 (1954).

2030. ———.
Ibid., 33, 1976-81 (1959).

2031. Vasil'eva, V. N., V. V. Perekalin, and V. G. Vasil'ev.
Akad. Nauk SSSR 141, 620-3 (1961).

2032. ———.
Zhur. Obshcheĭ Khim. 31, 2171-5 (1961).

2033. ———.
Ibid., 2175-8 (1961).

2034. Vasil'eva, V. N., T. V. Talalaeva, and E. N. Gur'yanova.
Izv. Akad. Nauk SSSR, Otd. Khim. Nauk 1960, 1549-52.

2035. Vaugh, W. C.
Phil. Mag. (7) 27, 669-76 (1939).

2036. Vaughn, W. E., W. P. Purcell, and C. P. Smyth.
J. Am. Chem. Soc. 83, 571-4 (1961).

2037. Veibel, S., K. Eggersen, and S. C. Linholt.
Acta Chem. Scand. 8, 768-76 (1954).

2038. Veibel, S., I. Refn, and A. Friediger.
Acta Chem. Scand. 2, 927-32 (1948).

2039. Velasco, M.
Anales soc. españ. fís. quím. 28, 1228-38 (1930).

2040. Venanzi, L. M.
J. Chem. Soc. 1958, 719-24.

2041. Venturello, G.
Atti. accad. sci. Torino, classe sci. fis. math. nat. 77, I, 57-63 (1941).

2042. Veselago, V. G.
Zhur. Eksptl. i Teoret. Fiz. 32, 620 (1957).

2043. ———.
Opt. i Spektroskopiya 6, 450-7 (1959).

2044. Volarovich, M. P., and N. N. Stepanenko.
Acta Physicochem. URSS 13, 647-58 (1940).

2045. ———.
J. Exptl. Theoret. Phys. (USSR) 10, 817-22 (1940).

2046. ———.
Ibid., 14, 313-7 (1944).

2047. Vol'pin, M. E., U. D. Koreshkov, and D. N. Kursanov.
Izv. Akad. Nauk USSR, Otd. Khim. Nauk 1959, 560.

2048. Vol'pin, M. E., and A. F. Plate.
Dokl. Akad. Nauk SSSR 70, 843-6 (1950).

2049. Voltz, S. E.
J. Am. Chem. Soc. 76, 1025 (1954).

2050. von Braunmühl, H-J.
Physik. Z. 28, 141-9 (1927).

2051. Vyas, A., and H. N. Srivastava.
J. Sci. Ind. Research (India) 17B, 377-9 (1958).

2052. ——.
Ibid., 18B, 399-401 (1959).

2053. Wada, A.
J. Chem. Phys. 29, 674-5 (1958).

2054. ——.
Ibid., 30, 328-9 (1959).

2055. ——.
Bull. Chem. Soc. Japan 33, 822-30 (1960).

2056. Wagner, R., J. Fine, J. W. Simmons, and J. H. Goldstein.
J. Chem. Phys. 26, 634-7 (1957).

2057. Walden, P., and O. Werner.
Z. physik. Chem. B2, 10-26 (1929).

2058. Walker, R., and D. W. Davidson.
Can. J. Chem. 37, 492-5 (1959).

2059. Walls, W. S., and C. P. Smyth.
J. Chem. Phys. 1, 337-40 (1933).

2060. Wang, Y. L.
Z. physik. Chem. B45, 323-8 (1940).

2061. Wanzlick, H. W., G. Gollmer, and H. Milz.
Chem. Ber. 88, 69-73 (1955).

2062. Warren, F. L.
J. Chem. Soc. 1937, 1858.

2063. ——.
Ibid., 1938, 1100.

2064. Washino, M., K. Yamada, and Y. Kurita.
Bull. Chem. Soc. Japan 31, 552-5 (1958).

2065. Watanabe, H., and M. Kubo.
J. Am. Chem. Soc. 82, 2428-30 (1960).

2066. Watanabe, I., S. Mizushima, and Y. Morino.
Sci. Papers Inst. Phys. and Chem. Res. (Tokyo) 39, 401-9 (1942).

2067. Watson, H. E.
Proc. Roy. Soc. (London) A117, 43-62 (1927).

2068. ——, G. P. Kane, and K. L. Ramaswamy.
Proc. Roy. Soc. (London) A156, 137-43 (1936).

2069. Watson, H. E., and K. L. Ramaswamy.
Proc. Roy. Soc. (London) A156, 130-7 (1936).

2070. Watson, H. E., G. G. Rao, and K. L. Ramaswamy.
Proc. Roy. Soc. (London) A132, 569-85 (1931).

2071. ———.
Ibid., A143, 558-88 (1934).

2072. Weaver, J. R., C. W. Heitsch, and R. W. Parry.
J. Chem. Phys. 30, 1075-6 (1959).

2073. Weaver, J. R., S. G. Shore, and R. W. Parry.
J. Chem. Phys. 29, 1-2 (1958).

2074. Webb, R. L., S. Frank, and W. C. Schneider.
J. Am. Chem. Soc. 77, 3491-3 (1955).

2075. Weigt, H.
Physik. Z. 22, 643 (1921).

2076. Weiss, E.
Z. anorg. allgem. Chem. 287, 236-41 (1956).

2077. Weissberger, A.
J. Am. Chem. Soc. 67, 778-9 (1945).

2078. ——— and H. Bach.
Chem. Ber. B65, 24-32 (1932).

2079. Weissberger, A., and R. Sängewald.
Z. physik. Chem. B5, 237-40 (1929).

2080. ———.
Physik. Z. 30, 792-801 (1929).

2081. ———.
Z. physik. Chem. B9, 133-40 (1930).

2082. ———.
Ibid., B12, 399-407 (1931).

2083. ———.
Ibid., B13, 383-6 (1931).

2084. ———.
Chem. Ber. B65, 701-4 (1932).

2085. ———.
Z. physik. Chem. B20, 145-57 (1933).

2086. ———.
J. Chem. Soc. 1935, 855.

2087. ——— and G. C. Hampson.
Trans. Faraday Soc. 30, 884-93 (1934).

2088. Weissberger, A., and J. W. Williams.
Z. physik. Chem. B3, 367-76 (1929).

2089. Weith, A. J., M. E. Hobbs, and P. M. Gross.
J. Am. Chem. Soc. 70, 805-11 (1948).

2090. Weizmann, A.
Trans. Faraday Soc. 36, 329-33 (1940).

2091. ———.
Ibid., 978-82 (1940).

2092. ———.
J. Am. Chem. Soc. 70, 2342-3 (1948).

2093. Wentink, T., M. W. P. Strandberg, and R. E. Hillger.
Bull. Am. Phys. Soc. 23, No. 2, 18 (1948).

2094. Wenzke, H. H., and R. P. Allard.
J. Am. Chem. Soc. 56, 858-60 (1934).

2096. Werner, O.
Z. physik. Chem. B4, 371-92 (1929).

2097. ———.
Ibid., 393-400 (1929).

2098. ———.
Z. anorg. allgem. Chem. 181, 154-8 (1929).

2099. West, W., and R. B. Killingsworth.
J. Chem. Phys. 6, 1-8 (1938).

2100. Westenberg, A. A., J. H. Goldstein, and E. B. Wilson.
J. Chem. Phys. 17, 1319-21 (1949).

2101. Westenberg, A. A., and E. B. Wilson.
J. Am. Chem. Soc. 72, 199-200 (1950).

2102. Westlake, H. E., Jr., H. L. Laquer, and C. P. Smyth.
J. Am. Chem. Soc. 72, 436-8 (1950).

2103. Wharton, L., L. P. Gold, and W. Klemperer.
J. Chem. Phys. 33, 1255 (1960).

2104. Wheland, G. W., and D. E. Mann.
J. Chem. Phys. 17, 264-8 (1949).

2105. White, A. H., and S. O. Morgan.
Physics 2, 313-21(1932).

2106. Wilcox, W. S., J. H. Goldstein, and J. W. Simmons.
J. Chem. Phys. 22, 516-8 (1954).

2107. Wild, H.
Helv. Chim. Acta 29, 497-6 (1946).

2108. Wildschut, A. J.
Physica 12, 194-210 (1932).

2109. Williams, A. J., and R. J. W. Le Fèvre.
J. Chem. Soc. 1957, 2425-6.

2110. Williams, J. W.
Physik. Z. 29, 204-5 (1928).

2111. ———.
Ibid., 271-2 (1928).

2112. ———.
Ibid., 683-8 (1928).

2113. ———.
Z. physik. Chem. A138, 75-84 (1928).

2114. ———.
J. Am. Chem. Soc. 50, 2350-7 (1928).

2115. ———.
Ibid., 52, 1831-7 (1930).

2116. Williams, J. W.
J. Am. Chem. Soc. 52, 1838-41 (1930).

2117. ——— and R. J. Allgeier.
J. Am. Chem. Soc. 49, 2416-22 (1927).

2118. Williams, J. W., and J. M. Fogelberg.
Physik. Z. 31, 363-5 (1930).

2119. ———.
J. Am. Chem. Soc. 52, 1356-63 (1930).

2120. ———.
Ibid., 53, 2096-104 (1931).

2121. Williams, J. W., and I. J. Krchma.
J. Am. Chem. Soc. 49, 1676-86 (1927).

2122. Williams, J. W., and E. F. Ogg.
J. Am. Chem. Soc. 50, 94-101 (1928).

2123. Williams, J. W., and C. H. Schwingel.
J. Am. Chem. Soc. 50, 362-8 (1928).

2124. ——— and C. H. Winning.
J. Am. Chem. Soc. 58, 197-203 (1936).

2125. Williams, J. W., and A. Weissberger.
J. Am. Chem. Soc. 50, 2332-6 (1928).

2126. Wilson, C. J., and H. H. Wenzke.
J. Chem. Phys. 2, 546-7 (1934).

2127. ———.
J. Am. Chem. Soc. 56, 2025-7 (1934).

2128. ———.
Ibid., 57, 1265-7 (1935).

2129. Wilson, W., and R. Woodger.
J. Chem. Soc. 1955, 2943-8.

2130. Winstein, S., and R. E. Wood.
J. Am. Chem. Soc. 62, 548-51 (1940).

2131. Wirth, H. E., and E. D. Palmer.
J. Phys. Chem. 60, 914-6 (1956).

2132. Wiswall, R. H., and C. P. Smyth.
J. Chem. Phys. 9, 356-61 (1941).

2133. Witt, D. J., and C. R. Estee.
Proc. S. Dakota Acad. Sci. 31, 229-36 (1952).

2134. Wohlwill, M.
Z. Physik 80, 67-79 (1933).

2135. Wolf, E.
Z. physik. Chem. B17, 46-67 (1932).

2136. Wolf, K. L.
Z. physik. Chem. B3, 128-38 (1929).

2137. ———.
Physik. Z. 31, 227-9 (1930).

2138. ———.
Trans. Faraday Soc. 26, 315-20 (1930).

2139. ——— and W. Bodenheimer.
Z. physik. Chem. Bodenstein Festband, 620-6 (1931).

2140. Wolf, K. L., and W. J. Gross.
Z. physik. Chem. B14, 305-25 (1931).

2141. Wolf, K. L., and E. Lederle.
Physik. Z. 29, 948 (1928).
[Wolf, K. L.; Z. physik. Chem. B2, 39-76 (1929).]

2142. Wolf, K. L., and O. Strasser.
Z. physik. Chem. B21, 389-409 (1933).

2143. Wolf, K. L., and H. G. Trieschmann.
Z. physik. Chem. B14, 346-9 (1931).

2144. Wolf, K. L., and H. Volkmann.
Z. physik. Chem. B3, 139-48 (1929).

2145. Woodrow, C. C., M. Carmack, and J. G. Miller.
J. Chem. Phys. 19, 951-4 (1951).

2146. Woods, G. F., and L. H. Schwartzman.
J. Am. Chem. Soc. 70, 3394-6 (1948).

2147. Work, R. N., and Y. M. Tréhu.
J. Appl. Phys. 27, 1003-11 (1956).

2148. Worsham, J. E., and M. E. Hobbs.
J. Am. Chem. Soc. 76, 206-8 (1954).

2148a Wrede, E.
Z. Physik 44, 261-8 (1927).

2149. Würstlin, F.
Kunststoffe 40, 158-60 (1950).

2150. Wüsthoff, P.
Ann. Physik (5) 27, 312-28 (1936).

2151. Wyman, J.
J. Biol. Chem. 90, 443-76 (1931).

2152. ——— and T. L. McMeekin.
J. Am. Chem. Soc. 55, 915-22 (1933).

2153. Yamada, S., and T. Ikegawa.
Tôkai Denkyoku Gihô 18, No. 2, 26-30 (1957).

2154. Yamakawa, M., H. Watanabe, T. Mukai, T. Nozoe, and M. Kubo.
J. Am. Chem. Soc. 82, 5665-7 (1960).

2155. Yao, C-H., and C. E. Sun.
J. Chinese Chem. Soc. 5, 22-4 (1937).

2156. Yaroslavsky, S., and E. D. Bergmann.
J. Chem. Phys. 33, 635 (1960).

2157. Yasumi, M.
J. Chem. Soc. Japan 60, 1208-14 (1939).

2158. Zaeschmar, G.
Kolloid-Z. 160, 107-14 (1958).

2159. Zahn, C. T.
Phys. Rev. 24, 400-17 (1924).

2160. Zahn, C. T.
Phys. Rev. 27, 455-9 (1926).

2161. ——.
Ibid., 35, 848-54 (1930).

2162. ——.
Ibid., 1047-55 (1930).

2163. ——.
Ibid., 37, 1516-26 (1931).

2164. ——.
Ibid., 38, 521-7 (1931).

2165. ——.
Ibid., 40, 291-8 (1932).

2166. ——.
Physik. Z. 33, 525-30 (1932).

2167. ——.
Ibid., 686-7 (1932).

2168. ——.
Ibid., 730-1 (1932).

2169. ——.
Ibid., 34, 461-2 (1933).

2170. ——.
Ibid., 570-4 (1933).

2171. —— and J. B. Miles.
Phys. Rev. 32, 497-504 (1928).

2172. Zakrzewski, C., and D. Doborzyński.
Bull. intern. acad. Polonaise A, 300-8 (1930).

2173. Zenitz, B. L., C. M. Martini, M. Priznar, and F. C. Nachod.
J. Am. Chem. Soc. 74, 5564-6 (1952).

2174. Zhdanov, U. A., O. A. Osipov, O. E. Shelepin, and V. A. Kogan.
Dokl. Akad. Nauk SSSR 128, 719-21 (1959).

2175. Zijp, D. H., and H. Gerding.
Rec. trav. chim. 77, 682-91 (1958).

2176. Zimmermann, H.
Z. Elektrochem. 63, 601-8 (1959).

2177. Zwartsenberg, J. W.
Rec. trav. chim. 62, 148-50 (1943).

2178. ——.and J. A. A. Ketelaar.
Rec. trav. chim. 61, 877-80 (1942).

AUTHOR INDEX

The entries in this index refer to the bibliography. The names of the first author of each paper are not included, but an asterisk before a name here indicates authors who have been first on some papers. A complete collection of articles by a given author can be obtained from this index and the bibliography.

To place names in order diacritical marks were ignored, ø was considered as o, names such as MacNab and McClellan were alphabetized as spelled, and names containing von and van are placed under V.

INDEX OF CARBON COMPOUNDS
CONTAINING METALLIC ELEMENTS

For this index, metallic elements are defined to be B, Si, As, Sb, and Bi and any element to the left of these in a row of the expanded form of the periodic table. Compounds containing metals but no carbon are in Table I (pp. 9-31) and are not included here.

GOLD (Contd.)

INDIUM

IRON

LANTHANUM

MANGANESE (Contd.)

MERCURY

MOLYBDENUM

NICKEL

Carbonyl complexes

Halogen complexes (all *bis*), R_mNiX_n

Other nickel salt complexes (all *bis*)

SILICON (Contd.)

SILVER

SODIUM

THALLIUM

THORIUM

TIN

TITANIUM

TITANIUM (Contd.)

TUNGSTEN

URANIUM

VANADIUM

ZINC